FRET and FLIM Techniques

LABORATORY TECHNIQUES IN BIOCHEMISTRY AND MOLECULAR BIOLOGY

Series Editors

P.C. van der Vliet—*Department for Physiological Chemistry, University of Utrecht, Utrecht, The Netherlands*

and

S. Pillai—*MGH Cancer Center, Boston, Massachusetts, USA*

Volume 33

ELSEVIER

AMSTERDAM • BOSTON • HEIDELBERG • LONDON • NEW YORK • OXFORD
PARIS • SAN DIEGO • SAN FRANCISCO • SINGAPORE • SYDNEY • TOKYO

FRET AND FLIM TECHNIQUES

Edited by

T. W. J. Gadella
Section of Molecular Cytology and
Centre for Advanced Microscopy
Swammerdam Institute for Life Sciences
University of Amsterdam
Amsterdam, The Netherlands

ELSEVIER

AMSTERDAM • BOSTON • HEIDELBERG • LONDON • NEW YORK • OXFORD
PARIS • SAN DIEGO • SAN FRANCISCO • SINGAPORE • SYDNEY • TOKYO

Cover photo credit: Sensitized emission calculated from confocal images. Cells expressing CFP- and YFP-tagged Pleckstrin homology domains were seeded together with control cells expressing either CFP or YFP; image shows all input file. Image courtesy of Dr. Kees Jalink and Jacco van Rheenen. Laboratory Techniques in Biochemistry and Molecular Biology. (2009) **33,** pp 289–350.

Elsevier
Radarweg 29, PO Box 211, 1000 AE Amsterdam, The Netherlands
Linacre House, Jordan Hill, Oxford OX2 8DP, UK

First edition 2009

Library of Congress Cataloging-in-Publication Data
A catalog record for this book is available from the Library of Congress

British Library Cataloguing in Publication Data
A catalogue record for this book is available from the British Library

ISBN: 978-0-08-054958-3
ISSN: 0075-7535

Printed and bound by CPI Group (UK) Ltd, Croydon, CR0 4YY

Transferred to Digital Print 2012

Contents

CONTRIBUTORS

A. V. Agronskaia
Molecular Biophysics Group, Debye Institute, Utrecht University, NL 3508 TA, Utrecht, The Netherlands

Egidijus Auksorius
Photonics Group, Department of Physics, Imperial College London, South Kensington Campus, London SW7 2AZ, UK

Brian J. Bacskai
Harvard Medical School, Massachusetts General Hospital, Mass General Institute for Neurodegenerative Disorders, Charlestown, Massachusetts 02129

A. N. Bader
Molecular Biophysics Group, Debye Institute, Utrecht University, NL 3508 TA, Utrecht, The Netherlands

Richard K. P. Benninger
Photonics Group, Department of Physics, Imperial College London, South Kensington Campus, London SW7 2AZ, UK

Riyaz A. Bhat
Department of Biology, Indiana University, Bloomington, Indiana 47405

Paul S. Blank
National Institute of Child Health and Human Development, National Institutes of Health, 10 Center Drive Bldg. 10 Bethesda, Maryland 20892

Robert M. Clegg
Department of Physics, University of Illinois at Urbana-Champaign Loomis, Loomis Laboratory of Physics, Urbana, IL 61801-3080, USA

Amanda Cobos Correa
European Molecular Biology Laboratory, Meyerhofstr. 1, 69117 Heidelberg, Germany

Pieter de Beule
Photonics Group, Department of Physics, Imperial College London, South Kensington Campus, London SW7 2AZ, UK

Andrew J. de Mello
Photonics Group, Department of Physics, Imperial College London, South Kensington Campus, London SW7 2AZ, UK

Christopher Dunsby
Photonics Group, Department of Physics, Imperial College London, South Kensington Campus, London SW7 2AZ, UK

Daniel S. Elson
Photonics Group, Department of Physics, Imperial College London, South Kensington Campus, London SW7 2AZ, UK

A. Esposito
Laser Analytics Group, Department of Chemical Engineering and Biotechnology, University of Cambridge, Cambridge, UK; Molecular Biophysics Group, Debye Institute, Utrecht University, NL 3508 TA, Utrecht, The Netherlands

Paul M. W. French
Photonics Group, Department of Physics, Imperial College London, South Kensington Campus, London SW7 2AZ, UK

T. W. J. Gadella Jr.
Section of Molecular Cytology and Centre for Advanced Microscopy, Swammerdam Institute for Life Sciences, University of Amsterdam, Kruislaan 316, 1098 SM Amsterdam, The Netherlands

Neil Galletly
Photonics Group, Department of Physics, Imperial College London, South Kensington Campus, London SW7 2AZ, UK

H. C. Gerritsen
Molecular Biophysics Group, Debye Institute, Utrecht University, NL 3508 TA, Utrecht, The Netherlands

Joachim Goedhart
Section Molecular Cytology and Centre for Advanced Microscopy, Swammerdam Institute for Life Sciences, University of Amsterdam, Kruislaan 316, NL-1098 SM Amsterdam, The Netherlands

David Grant
Photonics Group, Department of Physics, Imperial College London, South Kensington Campus, London SW7 2AZ, UK

Quentin S. Hanley
School of Science and Technology, Nottingham Trent University, Nottingham NG11 8NS, UK

Oliver Hofmann
Photonics Group, Department of Physics, Imperial College London, South Kensington Campus, London SW7 2AZ, UK

Bradley T. Hyman
Harvard Medical School, Massachusetts General Hospital, Mass General Institute for Neurodegenerative Disorders, Charlestown, Massachusetts 02129

Kees Jalink
Department of Cell Biology, The Netherlands Cancer Institute, Plesmanlaan 121, 1066 CX Amsterdam, The Netherlands

Elizabeth A. Jares-Erijman
Departamento de Química Orgánica, Facultad de Ciencias Exactas y Naturales (FCEyN), Universidad de Buenos Aires (UBA), Ciudad Universitaria, Pabellón II/Piso 3, 1428 Buenos Aires, Argentina

Phill B. Jones
Harvard Medical School, Massachusetts General Hospital, Mass General Institute for Neurodegenerative Disorders, Charlestown, Massachusetts 02129

Thomas M. Jovin
Laboratory of Cellular Dynamics, Max Planck Institute for Biophysical Chemistry, am Fassberg 11, 37077 Göttingen, Germany

Gordon Kennedy
Photonics Group, Department of Physics, Imperial College London, South Kensington Campus, London SW7 2AZ, UK

Srinagesh V. Koushik
National Institute on Alcohol Abuse and Alcoholism, National Institutes of Health, 5625 Fishers Lane, Rockville, Maryland 20892

Gert-Jan Kremers
Department of Molecular Physiology and Biophysics, Vanderbilt University Medical Center, 702 Light Hall, Nashville, Tennessee 37232, USA

Sunil Kumar
Photonics Group, Department of Physics, Imperial College London, South Kensington Campus, London SW7 2AZ, UK

Peter M. P. Lanigan
Photonics Group, Department of Physics, Imperial College London, South Kensington Campus, London SW7 2AZ, UK

M. John Lever
Photonics Group, Department of Physics, Imperial College London, South Kensington Campus, London SW7 2AZ, UK

Hugh Manning
Photonics Group, Department of Physics, Imperial College London, South Kensington Campus, London SW7 2AZ, UK

James McGinty
Photonics Group, Department of Physics, Imperial College London, South Kensington Campus, London SW7 2AZ, UK

Ian Munro
Photonics Group, Department of Physics, Imperial College London, South Kensington Campus, London SW7 2AZ, UK

Mark A. A. Neil
Photonics Group, Department of Physics, Imperial College London, South Kensington Campus, London SW7 2AZ, UK

Björn Önfelt
Photonics Group, Department of Physics, Imperial College London, South Kensington Campus, London SW7 2AZ, UK

Dylan Owen
Photonics Group, Department of Physics, Imperial College London, South Kensington Campus, London SW7 2AZ, UK

Jose Requejo-Isidro
Photonics Group, Department of Physics, Imperial College London, South Kensington Campus, London SW7 2AZ, UK

Carsten Schultz
European Molecular Biology Laboratory, Meyerhofstr. 1, 69117 Heidelberg, Germany

P. Soutter
Photonics Group, Department of Physics, Imperial College London, South Kensington Campus, London SW7 2AZ, UK

Gordon S. Stamp
Photonics Group, Department of Physics, Imperial College London, South Kensington Campus, London SW7 2AZ, UK

Klaus Suhling
Photonics Group, Department of Physics, Imperial College London, South Kensington Campus, London SW7 2AZ, UK

Clifford B. Talbot
Photonics Group, Department of Physics, Imperial College London, South Kensington Campus, London SW7 2AZ, UK

Christopher Thaler
National Institute on Alcohol Abuse and Alcoholism, National Institutes of Health, 5625 Fishers Lane, Rockville, Maryland 20892

Jacco van Rheenen
Hubrecht Institute-KNAW and University Medical Center Utrecht, Uppsalalaan 8, 3584CT, Utrecht, The Netherlands

Peter J. Verveer
Department of Systemic Cell Biology, Max Planck Institute of Molecular Physiology, Otto Hahn Straße 11, D-44227 Dortmund, Germany

Steven S. Vogel
National Institute on Alcohol Abuse and Alcoholism, National Institutes of Health, 5625 Fishers Lane, Rockville, Maryland 20892

Preface

On September 13 2005 I received the invitation from professor P.C. van der Vliet (the editor of the Laboratory Techniques in Biochemistry and Molecular Biology Series) to become the editor of a new volume in the series on "FRET and FLIM". In the letter it was mentioned that "in view of the rapid developments in single cell technology, we feel that a book on imaging techniques in living cells, such as FRET and FLIM, is appropriate and timely".

Indeed FLIM and FRET (fluorescence lifetime imaging microscopy and Förster resonance energy transfer) have experienced a strong and (still exponentially) growing interest during the past years (for a quantitative assessment see Chapter 10, Fig. 10.1). The three major driving forces for this uplift are (i) the ease of in situ fluorescent labeling using the visible fluorescent proteins (since ~1996); (ii) the commercial availability of advanced fluorescence microscopes with FRET acquisition software and with special detectors capable of acquiring complete spectra or fluorescence lifetimes (since ~2000), and most importantly (iii) the unique information on in situ molecular conformation and-proximity that FLIM and FRET can extract from single living cells. In the more early days, factor (iii) was the only driving force available: for performing in situ FRET measurements it was required to first go through the burden of chemical (fluorescent) labeling of molecules and/or building an imaging microscope capable of acquiring digital (lifetime) images. My personal first FRET experiment dates back to ~1990 in the laboratory of Karel Wirtz in Utrecht, when I measured FRET between tryptophan residues of a membrane protein and (self synthesized) pyrene-labeled polyphosphoinositides (see the Verbist et al. reference in Chapter 6). At that time almost the only reference to FRET work in biochemistry was the famous spectroscopic ruler review of Lubert Stryer of 1978 (see Chapter 1 for reference). During my postdoctorate time in the Laboratory of Tom Jovin (1992–1994) I worked on (donor) photobleaching FRET (see Section 8.1 of Chapter 1) and I was introduced to FLIM by Bob Clegg (together being the α and ω of this book). To get it to work, software for analysis had to be written and a dedicated (non-commercial) FLIM setup was required, but in the end we could measure the oligomerization of EGF receptors in situ with both FRET techniques using fluorescein- and rhodamine-labeled EGF (see the reference by Gadella and Jovin in Chapter 5). To perform FLIM–FRET in the pre-GFP era, it was required that a laboratory (or scientist) combined the skills of biochemical labeling, microscope equipment construction and analysis software programming with relevant knowledge of (molecular) biology/biochemistry. Albeit this was a serious limitation for the technique to become more widespread, FRET-laboratories at that time typically could identify the sources of error, pitfalls and workarounds because they usually covered every aspect of the FRET experiment. In that respect I feel very privileged to have worked at that time with some of the pioneers of FRET–FLIM microscopy and fluorescence spectroscopy, most notably Tom Jovin, Bob Clegg and Ton Visser (incidentally, all of them worked in the lab of the ancestor of all fluorescence spectroscopy in biology: Professor Gregorio Weber).

Nowadays, with some background in molecular biology, almost any scientist can perform a FRET experiment using commercial microscopes. Some microscopes even are equipped with a "FRET-button" so that all image acquisition and data processing is automated but "hidden" for the experimenter. The ease of performing automated FRET experiments by non-FRET experts also encompasses a danger: the underlying principles and pitfalls are often not well understood, leading to all sorts of misinterpretations, errors and frustration. Whether or not the correct filters, probes, laser sources, acquisition strategies, or image processing routines were used is often not completely (to completely not) known by the much larger community of FRET-scientists nowadays. Some of these frustrations culminate into statements that FRET technology is "unreliable" or produces "false negatives".

Hence, for modern FRET and FLIM techniques in Molecular Biology and Biochemistry it is important to keep the enthusiasm for the in situ technique, yielding unprecedented rich information on molecular states in live cells, and to keep the advantages of easy labeling techniques, modern microscopes and automated data processing. However, we need to "educate" the new generations of FRET scientists in the theoretical background of the technique, how it should be done correctly, and what the sources of errors are. Only then it will be clear that FRET–(FLIM) is a very direct, robust, extremely sensitive, and reliable technique.

This thought convinced me that I should accept Professor van Vliet's invitation to become editor of a FRET–FLIM volume. What I intended with this FLIM–FRET volume is to make a compilation of chapters that would be useful for the new generation of FRET scientists, but also interesting enough for the experts. So while nowadays, a variety of "exotic" FRET applications, theory, and instrumentation is around, I aimed to highlight the most straightforward and mainstream FRET work in this volume with sometimes also giving a peek into more advanced and future directions. Hence, this volume will not cover every aspect of FLIM–FRET. For instance, it does not cover single molecule, low temperature, detailed spectroscopic FRET work or very detailed hardware issues. In addition, triplet state conversion, or saturation effects (ground state depletion) giving rise to nonlinear excitation power-fluorescence intensity relationships are generally ignored (the reader is referred to Chapter 12, the FRET calculator for further information). Finally treatment of very complex FRET situations with (i) movement between donor and acceptor within the donor lifetime, (ii) multiple (n) acceptors for one donor (multiplying Eq. (1) of Chapter 1 with n), (iii) situations of a coexistence of many energy transfer states due to different conformations, (iv) effects of geometry leading for instance to a fourth power distance dependency for energy transfer to a planar surface of acceptors, and (v) changes in local refractive index in the cell leading to different local values of R_0 are not considered. It is of note however that in reality, in cells, (a multitude of) these situations will be applicable. So the basic FRET concepts and measurement strategies are illustrated for a situation with uniform FRET efficiency E in the cell in the absence of the above "problems" but with (local) variation in fractions of molecules (f_D and f_A) experiencing FRET, since these parameters are most interesting for biologists. In the more complex example cases (i)–(v) listed above, the equations listed in this volume will still be correct for defining an "average" or "apparent" energy transfer value in a microscopy image; this value will still represent an interaction or conformational state(s); but it will not be possible to make a statement on the fraction of molecules involved in such interactions and/or states with these formulas: simulation or more exotic equations are required.

The first chapter written by Bob Clegg introduces the FRET theory and basic equations. Also the original work and historic background of the FRET theory is presented in this chapter. Because sometimes it is difficult to picture a situation from an equation, Bob Clegg describes an analogy between FRET and monkeys escaping through doors of a dark room. Another highlight of this chapter is a description of a measurement of FRET without measuring fluorescence (arguing strongly in favor of FRET being "Förster resonance energy transfer" instead of "fluorescence resonance energy transfer").

The second chapter by Peter Verveer and Quentin Hanley describes frequency domain FLIM and global analysis. While the frequency domain technique for fluorescence lifetime measurement is sometimes counterintuitive, "the majority of the 10 most cited papers using FLIM have taken advantage of the frequency domain method" as stated by these authors. The global analysis of lifetime data in the frequency domain, resolving both E and f_D has contributed significantly to this advantage.

The third chapter by Alessandro Esposito et al., describes the time domain counterpart of FLIM. When photon economy and fast decaying components are considered, the time-domain implementation of FLIM is the method of choice. Most commercial (multiphoton) confocal FLIM systems implement this technology.

The fourth chapter by James McGuinty et al. describes the more advanced forms of time-domain FLIM. While not immediately available on commercial instruments this chapter should give the reader an idea what the current state-of-the-art is in terms of FLIM instrumentation, and perhaps what to expect on future commercial instruments. Real-time FLIM, combined FLIM-spectral imaging, hyperspectral FLIM-imaging, combined lifetime-anisotropy imaging and some of their applications are covered here.

Besides FRET theory and instrumentation, also probes are a key issue for performing a valid FRET experiment in cells. Chapter 5 by Gert-Jan Kremers and Joachim Goedhart highlights the various visible fluorescent proteins (VFPs) for the use of FRET. These authors argue that "with the large number of spectral classes and several variants within each spectral class, choosing the right VFP FRET pair for FRET can be a daunting task". To assist in this choice, a unique table with all Förster radii (R_0) between the major monomeric VFPs (Chapter 5, Table 5.1) and spectra highlighting five different VFP combinations with theoretical FRET spectra, are included.

While VFPs have boosted the applications of FRET–FLIM, chemical FRET probes should not be dismissed. The advantage of chemical probes is that they are much smaller in size and that they often have much better spectral readout than VFP probes. In Chapter 6, Amanda Cobos Correa and Carsten Schultz highlight the various small molecule-based FRET probes and their use in bioimaging.

For many scientists dedicated FLIM instruments are too expensive and/or too complicated to work with. Therefore, Chapter 7 by Jacco van Rheenen and Kees Jalink is included dealing with "low budget" but "high quality" Filter FRET. Filter FRET has the advantage that it is fast, sensitive, direct and inexpensive. However, if you want to do it quantitatively and without errors, you need to go through a lot of formulas and correction factors. In this chapter, the reader is guided through these issues and a full comprehensive description is given to perform correct calibration of a filterFRET microscope (both wide-field and confocal).

Chapter 8 written by Steve Vogel et al. also deals with sensitized emission based FRET methodology, but now using a spectral imaging detector device. Because a spectral detector and spectral unmixing software nowadays are standard options on the major commercial confocal microscopes, here a complete description is given how to quantify FRET from unmixed spectral components.

The smallest Chapter 9 (written by undersigned) deals with total internal reflection FRET–(FLIM) imaging. This technique enables the measurement of FRET with high contrast in a layer of only 80 nm above the cover glass, which is very useful for cellular membrane-related events. It is explained how an existing FLIM-system can be "upgraded" to incorporate the TIRF contrast (with thanks to Carsten Schultz for proofreading/editing).

In Chapter 10, Riyaz Bhat highlights FLIM–FRET applications in plant systems. Particularly plant sciences suffering from notoriously difficult biochemistry can profit most from the detailed in situ molecular imaging and contrast provided by FRET–(FLIM) imaging. With the help of genetically encoded probes and the ease of plant transformation and (back)crossing, plant scientists increasingly see the benefit of FRET–FLIM. For non-plant scientists it may be interesting to read how shooting gold bullets into plant material can be used for performing FRET microscopy.

In Chapter 11, by Phill Jones et al., biomedical FRET–FLIM applications are reviewed and illustrated. The molecular background of a variety of diseases (e.g., Alzheimer's disease) can be uncovered by using FRET–FLIM. In this major funding area, the "killer"-applications of the technology are and will be found, leading to a further boost of the implementation and commercial availability of high-end microscopes with automated acquisition and standardized analysis features.

Chapter 12, by Eli Erijman and Tom Jovin concludes the volume. This special chapter introduces a new and quantitative definition of FRET-measurements without requiring knowledge of the donor quantum yield Q_D or the energy transfer efficiency E. Furthermore, it highlights the recent explosion in labeling strategies, ranging from genetic encoded FlAsH labeling, through AGT, NTA technology, photochromic labels to quantum dots and further. Hence, the potential of getting the most sensitive probe attached to a biomolecule of interest nowadays is phenomenal (and still increasing). They conclude their chapter with "Quo vadis"...

So approximately 3 years after receiving the invitation letter from professor van Vliet, the FLIM and FRET volume is ready and I believe the combined chapters make an excellent statement for FRET and FLIM technology. Hereby, I want to thank all chapter authors for their efforts and the fine work they have delivered. It was a pleasure from my side to work on the diverse chapters and for me it was also a good learning experience especially to go through the equation rich Chapters 1, 7, 8, and 12. For that I am especially grateful for the list of common symbols that we could agree on (see Table). The equations in the volume may scare off scientists with training mainly in (molecular) biology. Although they may appear as difficult, the vast majority of the equations represent "simple mathematics" being not more difficult than subtracting, adding, multiplying or dividing (elementary school stuff). Let the reward of understanding the equation (quantitative information on molecular interactions and conformation in situ) be a motivation to go through the math.

Having said this, I hope the above thoughts and chapters highlights make you curious and eager for reading the volume, contributing to more good, reliable and enthusiastic future FRET–FLIM work in Biochemistry and Molecular Biology.

Dorus Gadella, 5 August 2008

Laboratory Techniques in Biochemistry and Molecular Biology, Volume 33
FRET and FLIM Techniques
T. W. J. Gadella (Editor)

CHAPTER 1

Förster resonance energy transfer—FRET what is it, why do it, and how it's done

Robert M. Clegg

Department of Physics, University of Illinois at Urbana-Champaign Loomis, Loomis Laboratory of Physics, Urbana, IL 61801-3080, USA

The applications of Förster resonance energy transfer (FRET) have expanded tremendously in the last 25 years, and the technique has become a staple technique in many biological and biophysical fields. Many publications appear weekly using FRET and most of the applications use FRET as a spectroscopic research tool. In this chapter, we have examined some general salient features of resonance energy transfer by stressing the kinetic competition of the FRET pathway with all other pathways of de-excitation. This approach emphasizes many of the biotechnological and biophysical uses of FRET, as well as emphasizing the important competing processes and biological functions of FRET in photosynthesis.

1.1. Introduction

There are numerous excellent reviews and original literature about Förster resonance energy transfer (FRET) where one can read detailed descriptions and get lists of earlier references [1–11]. This chapter is neither a review of the literature, nor a detailed account

DOI: 10.1016/S0075-7535(08)00001-6

of experimental techniques and methods of analysis, nor a how-to-do manual, nor an appraisal of theoretical descriptions of energy transfer for the specialist. The chapter focuses on a few critical essentials concerning the fundamentals of energy transfer and the methods of measurement. These basic aspects of FRET are helpful for understanding the important features and interpretations of energy transfer measurements. It is also useful to see historically how these ideas were developed. If one understands these simple fundamentals, it is usually straightforward to appreciate specific theoretical and experimental details pertaining to methods of acquisition and analysis.

1.1.1. Fluorescence and FRET are popular methods

Fluorescence has exquisite sensitivity for detecting very low concentrations of molecules over broad spatial and temporal dimensions. By choosing luminophores (fluorophores and phosphors) with lifetimes from subnanoseconds to milliseconds, molecular dynamics can be observed over a large time scale; nevertheless, the FRET measurement can be carried out on macrosystems. FRET is frequently applied to determine molecular distances or to show whether or not molecular complexes are present. Lifetime-resolved FRET has been carried out on fluorescence images [12–20]. FRET is increasingly occupying a center stage in biological studies and in biotechnology (especially dealing with DNA chips and other massively parallel assay systems). It has been applied in single molecule studies to provide information on conformational changes [21, 22] and the pharmaceutical industry has developed major microscopic fluorescence assay detection systems with very low sample volumes, even on the single molecule level, using fluorescence correlation spectroscopy [23–29].

Information about molecular interactions, spatial juxtapositions, and distributions of molecular and supramolecular components constituting biological structures are of crucial importance for understanding functions on a molecular scale in biology. This information is especially vital when we consider that a major part of biology takes place at the interface between interacting molecules and supramolecular organizations. Many of these macromolecular systems are ideally suited for FRET applications. For this reason, FRET has received so much interest in biotechnology and medicine as well as in biophysics [30–38]. Applications for FRET extend from more traditional cuvette spectroscopic measurements on larger volumes (from 100 μm to 100 ml) to FRET imaging experiments in the fluorescence microscope [16, 17, 39–44] and single molecule experiments [21, 22]. The recent applications of FRET in the optical microscope have become very popular because of its interpretive power on the molecular scale with regard to statically and dynamically associating molecular systems in cellular biology [45–48]. Using geometrical and stereochemical information that can be attained from FRET measurements, we can more confidently propose models how biological structures carry out their functions, for instance in ribosomes [49, 50]. Knowing the spatial distribution of the parts of a structure makes it possible to ask more specific questions concerning the dynamics of intermolecular interactions.

1.1.2. FRET—A molecular detective, transmitting molecular dimensional information to the experimenter

Fluorescence molecules are analogous to roaming molecular spies with radio transmitters, radiating information to the experimenter about the state of affairs on the molecular scale, and informing us where the spies are located and how many there are. A feature unique to FRET is the capability to inform us whenever two or more molecules (usually biological macromolecules) are close to one another on a molecular scale ($\leq$80 Å), and whether they are moving relative to each other (Section 1.3.1.1). It is even sometimes possible to detect how the *D* and *A* transition moments are oriented relative to each other, because the efficacy of transfer depends on the relative angular dispositions of the two dipoles (Section 1.3.1.2). As with all other fluorescence methods, we can couple FRET with other physical and biological methods, and this greatly extends the usefulness. Such broad application is characteristic to fluorescence. Very importantly, FRET (and fluorescence in general) can be carried out in most laboratories, whether the "samples" are large (such as in cuvettes, or even on whole mammalian bodies) or small (such as in the fluorescence microscope, and on the single molecule level). No matter what scale of the sample, the information on the molecular scale derivable from FRET remains accessible. In this regard, FRET is like a spectroscopic microscope, providing us with distance and orientation information on the molecular scale regardless of the size of the sample (Stryer [9] dubbed FRET a molecular ruler). In addition, by observing FRET over time (such as in stopped-flow), we can follow the dynamics of changes in molecular dimensions and proximities.

As should be apparent from the above discussion, and a perusal of the contents of the recent literature, the range of applications of FRET is extremely broad.

In Section 1.2, we first introduce some historical facts concerning the development of FRET and indicate how FRET is interrelated with several scientific disciplines. If one is not interested in this historical account, then you can skip to Section 1.3 without losing the thread.

1.2. Historical background; setting the groundwork

In a series of remarkable papers, Theodor Förster revealed the correct theoretical explanation for nonradiative energy transfer [7, 51–55]. He was partly motivated by his familiarity with the extreme efficiency of photosynthetic systems in funneling the energy of absorbed photons to a relatively small number of reaction sites [51]. The average number of photons striking the total area of a leaf is much larger than that expected, considering the small area of the leaf containing the reaction centers, where the photosynthetic electron transfer reactions take place. In order for the absorbed energy to be channeled efficiently into the reaction centers, he, and others, reasoned that the excitation energy is rapidly and efficiently transferred throughout an area that is large compared with the reaction center. Eventually, this energy is captured by a reaction center. This process increases the effective capture area of the reaction center. This process was pictured as a random diffusive spreading of the absorbed photon energy that is captured by a sink (the reaction center). Although such a mechanism was suspected at the time, the physical mechanism responsible for this energy transport in photosynthesis was not understood.

1.2.1. Pre-Förster: Dipole–dipole interaction; the Perrins

Early dipole–dipole models of energy transfer were developed by the Perrins (father (J.) and son (F.)) [56, 57]. Dipole interactions had already been used in descriptions of molecular interactions in bulk matter, including dipole-induced–dipole-induced van der Waals forces, dipole–dipole-induced forces, and dipole–dipole interactions [58]. Classically, the electric field emanating from an oscillating dipole, $\vec{E}\tilde{\mu}$, (real dipoles or transition dipoles) decreases as distance between the dipoles, R, is increased. The functional form of the oscillating dipole can be divided into three zones: the near-field zone ($\vec{E}\tilde{\mu}^{nf} \propto R^{-3}$), the far-field zone ($\vec{E}\tilde{\mu}^{ff} \propto R^{-1}$), and in the intermediate transition zone between the near- and far-field. We are concerned here only with the interaction energy between two dipoles in near field, which is very large compared with the other two zones. The near-field dipole–dipole transfer mechanism had been first proposed to explain energy transfer between atoms [59]. A nonradiative dipole–dipole model successfully accounted for energy transfer in gas mixtures arising from near collision processes between the atoms [58, 60–63]. It was found that energy could be transferred over distances beyond the hard core collision distances between molecules, in the near-field zone of dipole–dipole interactions.

The Perrins were the first to attempt a quantitative description of nonradiative (no emission of a photon) energy transfer *in solution* between an excited molecule (originally called the sensitizer; in this chapter called the *donor*) and a neighboring molecule in the ground state (originally called the activator; in this chapter called the *acceptor*). The Perrins' reasoned that the depolarization that occurs in a solution of a fluorophore at higher concentrations resulted from the transfer of excitation energy between molecules with different orientations, before a photon was emitted. The Perrins' model involved a near-field energy of interaction, E_{int}, between the oscillating dipoles of two molecules, D and A ($E_{\text{int}} \propto \vec{E}^{nf}_{\vec{\mu}_D} \cdot \vec{\mu}_A$); $\vec{\mu}_A$ is the dipole of the acceptor. This is simply the general form of the interaction energy of a dipole in an electric field. This interaction is identical to the perturbation employed in a quantum mechanical representation of FRET. An interesting account of the early history of energy transfer is given in a recent review of the Perrins' accomplishments [64], and a recent historical review of FRET [58]. They assumed identical molecules [56, 57, 65]. Initially, they considered a classical model involving oscillating point dipoles; later they presented a quantum mechanical model. In modern quantum mechanical descriptions, the dipoles are the transition dipoles [1, 6, 7]. They were correct that the energy transfer involved dipole perturbations, as had also been realized by earlier researches studying molecular interactions. However, their model did not account quantitatively for the energy transfer between identical molecules in solution. It was known that fluorescence becomes depolarized at concentrations of fluorophores where they are separated by ~2–5 nm; and that depolarization can be detected below concentrations required for quenching effects. However, their models for explaining this depolarization predicted energy transfer (and therefore depolarization) between two molecules separated by a much larger distance, on the order of a fractional wavelength of light (~100–500 nm). Because of this large discrepancy, F. Perrin's theory of energy transfer lay dormant for about 20 years.

1.2.2. Förster: Equilibrium energy distributions of the donor and acceptor to describe resonance

Förster published his first account of nonradiative energy transfer shortly after the Second World War in 1946 [51] (interestingly, the address on this paper is a small village outside Göttingen, Niedernjesa). A later paper [53] is usually referenced as his first publication (this 1948 paper has been translated into English [66]); however, his original publication [51] is a lucid account of the resonance requirement for energy transfer between dipoles of molecules exhibiting broad spectra. Because the frequency dispersion of the fluorescence and absorption spectra of the donor and acceptor is central to FRET, and because this is the reason why Perrins' model predicted the wrong distance of transfer, we shortly review his reasoning in his 1946 paper. Unfortunately, it has been published only in German, and has never appeared in an English translation. A more detailed account of the 1946 paper has been published in English [6]; here we discuss only the fundamentals.

Förster appreciated that the premise of dipole–dipole interactions was correct. However, he also realized that the condition of resonance is not limited to just one frequency (as assumed by the Perrins). Molecules in both ground and excited states have distributions of energy values; in solution, these electronic energy levels are significantly broadened, mainly by vibrational and interactions with the surrounding solvent. Following excitation of a chromophore to its Frank–Condon state, the excited molecule rapidly relaxes to the equilibrium Boltzmann distribution of its vibrational and rotation states in less than a picosecond. These distributions are responsible for the shape and spectral breath of the measured absorption and emission spectra. Even molecules that are isolated from all surrounding influences have a narrow distribution of energy [67, 68], and of course in solution, the spectra are much broader. This energy distribution has a profound effect on the probability per unit time of spectroscopic transitions, as well as on the rate of energy transfer.

Of course, energy must be conserved during the transfer; therefore, energy can only be exchanged between donor and acceptor states with identical energy differences between their instantaneous initial and final states. However, this resonance condition is distributed over the total overlapping frequency spectra of the donor and acceptor, and this drastically lowers the probability that the resonance condition will be met. Because of this spectral breadth, the energy that can be donated by an excited donor molecule only exactly matches the energy that can be gained by a neighboring acceptor molecule in a small fraction of the time (i.e., only a fraction of the molecules are exactly in "resonance" at any time). Förster realized that he could use statistics to calculate the simultaneous probability that both oscillators are in exact resonance, provided that during the time of energy transfer both the donor and acceptor molecules are at equilibrium with the surrounding solvent. In most cases molecules immediately lose their "memory" of how they were excited; that is, they lose their time-coherence with the excitation perturbation, and they become thermally equilibrated with the environment. Then any following process, including the transfer of energy, can be calculated by using a time-independent probability of transfer per unit time for the probabilities of transitions. Another condition for Förster transfer (as we are concerned with here) is that the perturbation between the two molecules is very weak, such that there is no influence on their absorption and emission spectra; that is, the molecules interact, but still act as separate molecular species. Förster discussed the different behavior corresponding to the different levels of interaction; he divided these into very weak, weak, and strong interactions [69]. We are concerned here only with his very weak interaction case with dipole–dipole coupling. There is much recent literature concerning the case of stronger interactions in biology, mostly regarding photosynthesis (see e.g., [2, 70–73]).

The probability that the frequency (energy transition) of the donor exactly matches the frequency (energy transition) of the acceptor is calculated by integrating the multiplication of the donor's emission spectrum and the acceptor's absorption spectrum over all frequencies where there is an overlap, always requiring the exact resonance condition. This is known as the "overlap integral" (Eq. (1.2)). Förster's first paper presented only a simple probabilistic calculation, and he soon published a thorough quantum mechanical account [53]. But his original paper is a clear account of why he emphasized the term "resonance."

Förster's theory agreed with experiment. His final expression predicted a drastic decrease in the distance over which the molecules can effectively exchange energy without emitting and absorbing photons compared to Perrin's results. The integration over the spectral breadths (which is identical to the same procedure in the derivation of what is known as the "Fermi Golden Rule" for calculating rates in quantum mechanics) increases the rate of diminution of energy transfer from an inverse cube of the D (donor)–A (acceptor) separation to the reciprocal distance to the sixth power (Eq. (1.1)). Förster's theory also specifically included the relative orientation of the interacting dipoles (Eqs. (1.1a) and (1.3b)) [7, 53], and this has since been rigorously checked.

If the conditions for Förster transfer are not applicable, then the theory must be extended. There is recently experimental evidence that coherent energy transfer participates in photosynthesis [74, 75]. In this case, the participating molecules are very close together. The excited state of the donor does not completely relax to the Boltzmann distribution before the energy can be shared with the acceptor, and the transfer can no longer be described by a Förster mechanism. We will not discuss this case. There has been active discussion of coherent transfer and very strong interactions in the literature for a longer time [69], and references can be found in some more recent papers [70–72, 76, 77].

1.2.3. Post-Förster: Subsequent derivations of the rate of energy transfer

Following the pioneering studies of the Perrins and Förster there have been many publications and reviews on FRET, both theoretical and experimental. A large literature list is available in several recent reviews; see the references in [6, 10, 11], so it is not necessary to list many of the original references separately here. An interesting, but not commonly recognized, treatment by Oppenheimer [78]) dealt correctly with the transfer process; however, the publication was not noticed by the spectroscopic community and the treatment was very cursory. A later account was published after the major papers by Förster [79]. Unfortunately because of the brief treatment and place of publication, those who were most interested in FRET did not know this early correct treatment. Therefore, Oppenheimer's publication was not influential, at least in the field of biological physics and physical chemistry. Dexter [80] made a very important contribution. He generalized and extended Förster's energy transfer model to include *D* and *A* molecules with overlapping electron orbitals. Transfer on this very small distance scale is appropriately called "Dexter transfer" or transfer by "electron exchange". The distance dependence of Dexter transfer is quite different from Förster transfer, and the rate of the Dexter process is only efficient for very small *D–A* separations (usually <0.5 nm). We will briefly describe the basics of Dexter transfer in Section 1.11.

1.3. FRET basics

Now that we have discussed the basic premises behind FRET, and indicated the broad applicability, we will discuss the quantitative physical descriptions in more detail. The transfer of energy usually takes place within *D–A* separations of 0.5–10 nm. A Coulombic perturbation between the excited *D* molecule on the *A* molecule takes place electrodynamically through space, and this is approximated as an electric dipole–dipole interaction (classically this is an oscillating dipole). If the conditions are right—for instance, if the spectral overlap between the emission and absorption spectra of *D* and *A* is sufficient, if the fluorescence quantum yield of *D* and the absorption coefficient of *A* are great enough, and if the two dipoles are close enough—there is a significant probability that the excitation energy of D will be transferred nonradiatively to A. There is no emission or absorption of a real photon during FRET, so the participation of the optical absorption coefficient of the acceptor and the quantum yield of the donor (see Eq. (1.2)) may seem perplexing; we discuss this below (Section 1.3.3). The efficiency of energy transfer is a strong function of the separation distance between D and A (Eq.(1.1)). If energy transfer can take place between an excited D molecule and a nearby acceptor, the probability of photon emission of D is diminished (i.e., the emission intensity is lowered) and the lifetime of D in the excited state is measurably shortened. It is not required that the acceptor fluoresces; it must only accept the energy from the donor. However, if A fluoresces then this emission, called *sensitized emission*, can be detected following the transfer. The emission from A is characteristic of A's fluorescence spectrum even though only D has been initially excited. This is one of the major ways to detect FRET, and sensitized emission is direct proof of energy transfer.

1.3.1. Dipole–dipole energy transfer between an isolated donor and acceptor pair

1.3.1.1. Distance dependence

Equation (1.1) describes the rate constant of energy transfer between a separate donor–acceptor pair of molecules through a dipole–dipole interaction (Figs. 1.1 and 1.2) [7].

$$k_t = \left(\frac{1}{\tau_D}\right)\left(\frac{R_0}{r_{DA}}\right)^6 \tag{1.1}$$

k_t is the rate constant for transferring energy from an excited donor molecule to an acceptor molecule a distance r_{DA} away. R_0 is a distance that pertains to this individual pair of molecules. R_0 represents the distance r_{DA} where the rate of energy transfer is equal to the rate (probability per unit time) at which the excited

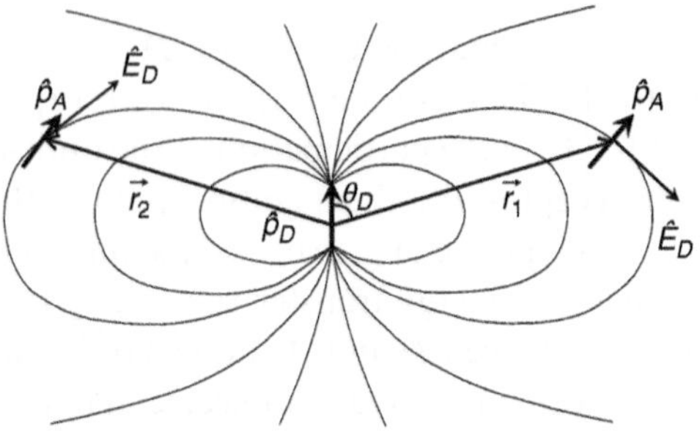

Fig. 1.1. Shown is the spatial variation of the electric near-field of a classical oscillating dipole, a Hertzian dipole. This also represents the correct near-field perturbation between two transition dipoles in a quantum mechanical derivation of FRET. Tangents to the lines show the direction of the field, which at every point oscillates in intensity and changes direction once every half period. The field is symmetrical around $\hat{p}_D$. The vectors $\hat{p}_D$, $\hat{p}_A$, and $\hat{E}_D$ are the unit vectors of the donor and acceptor dipoles and the electric field of the donor dipole. $\vec{r}_1$ and $\vec{r}_2$ are distance vectors from the donor to the acceptor point dipoles. θ_D is the angle between the donor dipole axis and the vector $\vec{r}$ from the donor and acceptor. Two positions of the acceptor are shown to demonstrate that the value of κ^2 will depend on the position in space of the acceptor relative to the donor, even if the relative angle between $\hat{p}_D$ and $\hat{p}_A$ stays the same. This representation makes it easy to visualize the effectiveness of the orientation of the acceptor anywhere in the near-field zone of the donor. The maximum effectiveness is when $\hat{p}_A \cdot \hat{E}_D = 1$. This is, of course, the condition for the interaction energy of the two dipoles to be a maximum, for any particular $\vec{r}$.

donor molecule would decay from the excited state were the acceptor molecule not present. This latter rate is equal to the inverse of the fluorescence lifetime τ_D of the donor molecule in the absence of an acceptor. $1/\tau_D$ is the rate of deactivation from the excited state that includes *all pathways of de-excitation other than the pathway of energy transfer* to the acceptor (including internal conversion rates, intersystem crossing, and dynamic quenching effects). FRET competes kinetically with all the other possible deactivation processes from the donor excited state. τ_D is *not* in general the *intrinsic radiative lifetime* τ_f of the donor. τ_f^{-1} is the rate of deactivation

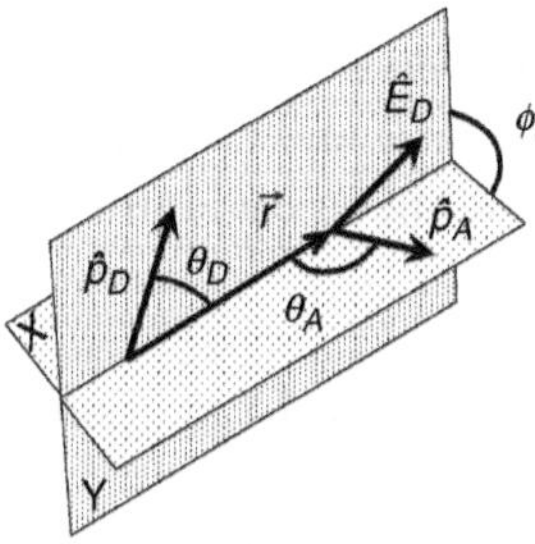

Fig. 1.2. The relative orientation of a donor and acceptor dipole are shown, in order to define the angles θ_D, θ_A, the unit vector $\hat{E}_D$ and the distance $\vec{r}$ vector. Planes ***X*** and ***Y*** are the planes in which $\hat{p}_A$ and $\hat{p}_D$ lie, respectively. $\hat{E}_D$ also lies in the ***Y*** plane. The planes are at an angle of ϕ_{DA} which is defied by the orientations of $\hat{p}_A$ and $\hat{p}_D$.

only through the fluorescence pathway, and can be calculated from the "Einstein *A* coefficient" [67, 81–84]. The value of R_0 for a singular pair of *D* and *A* molecules is:

$$R_0^6 = \frac{9000(\ln 10)\kappa^2 Q_D}{N_A 128\pi^5 n^4}\left[\int_0^\infty \varepsilon_A(\bar{\nu})F_D(\bar{\nu})\bar{\nu}^{-4}d\bar{\nu} \Big/ \int_0^\infty F_D(\bar{\nu})d\bar{\nu}\right] \quad (1.2)$$

where, $\bar{\nu}$ is wave number (in cm^{-1} units), Q_D is the quantum yield of the donor, N_A is Avogadro's number, n is the index of refraction pertaining to the transfer, $\varepsilon_A(\bar{\nu})$ is the molar absorption coefficient of the acceptor (in units of $\mathrm{cm}^{-1}\ \mathrm{mol}^{-1}$), $F_D(\bar{\nu})$ is the fluorescence intensity of the measured fluorescence spectrum of the donor, and κ is the angular function resulting from the inner product between the near unit vector of the electric field of the donor and the unit vector of the absorption dipole of the acceptor. The ratio of integrals in the bracket has units of $\mathrm{cm}^6/\mathrm{mol}$. Using the units given in the paragraph above, we have:

$$R_0^6 = 8.79 \times 10^{-25} n^{-4} Q_D \kappa^2 J(\bar{\nu})\mathrm{cm}^6 \quad (1.3)$$

$J(\bar{\nu})$ is the ratio of integrals given in square brackets in Eq. (1.2).

1.3.1.2. Orientation dependence—κ^2

The parameter κ^2 in Eqs.(1.2) and (1.3) is the orientation factor of FRET. The field surrounding an oscillating classical electric dipole $\vec{p}_D$ is shown in Fig. 1.1,

$$\vec{E}_1 = \frac{|p_D|}{r_{DA}^3}\{2\cos\theta_1 \hat{r}_{DA} + \sin\theta_1 \hat{\theta}_1\} \tag{1.3a}$$

where $|p_D|$ is the time-independent dipole strength. r_{DA} is the distance between the point dipoles, and $\hat{r}_{DA}$ is the unit vector from the dipole to the position $\vec{r}_{DA}$. θ_D is the angle between $\hat{p}_D$ and $\hat{r}_{DA}$. $\hat{\theta}_D$ is the unit vector in the direction of the end of the $\hat{r}_{DA}$ vector as it increases in the direction of θ_D. The caps designate unit vectors. Figure 1.2 shows the juxtaposition of two dipoles, and defines the parameters used in the Eqs. (1.1a), (1.3a), and (1.3b).

k_t in Eq.(1.1) is proportional to the interaction energy squared between the two dipoles $\vec{p}_D$ and $\vec{p}_A$,

$$k_t \propto (U_{\vec{p}_D \to \vec{p}_A})^2 = (\vec{E}_D \cdot \vec{p}_A)^2$$

$$= \frac{|p_D|^2|p_A|^2}{r_{DA}^6}\{2\cos\theta_D \hat{r}_{DA}\cdot\hat{p}_A + \sin\theta_D \hat{\theta}_D\cdot\hat{p}_A\}^2 = \frac{|p_D|^2|p_A|^2}{r_{DA}^6}\kappa^2 \tag{1.1a}$$

$$\kappa^2 = \{2\cos\theta_D \hat{r}_{DA}\cdot\hat{p}_A + \sin\theta_D \hat{\theta}_D\cdot\hat{p}_A\}^2 \tag{1.3b}$$

κ^2 is symmetrical with respect to D and A, and we could have defined it the other way around. κ^2 can have values between 0 and 4. We see that the value of κ^2 depends on the orientation between the acceptor dipole (transition dipole), $\vec{p}_A$, and the field $\vec{E}_D$ of the donor dipole (transition dipole), $\vec{p}_D$ (Eq. (1.1a) and Fig. 1.1). For any selected positions of the donor and acceptor, the value of κ^2 does depend on the angle between the donor and acceptor dipole. However, it is important to realize that for any constant selected angle between the donor and acceptor dipoles, the value of κ^2 will depend on the distance and the position in space where the acceptor dipole is relative to the donor. The field of the donor molecule will change with the distance $|\vec{r}_{DA}|$ for any particular value of θ_D; that is, for any particular direction of $\hat{r}_{DA}$. As illustrated in Fig. 1.1, for a particular angle between the orientations of the donor and acceptor dipole directions, the angle between the acceptor dipole and the electric field of the donor depends on the position in space of the acceptor. This is important to keep in mind. A few examples will clearly make this point. If the two dipoles have orientations in space perpendicular to each other, and if the acceptor dipole is juxtaposed next to the donor, but in the direction perpendicular to the direction of the donor dipole, then $\kappa^2 = 0$ (i.e., because, $\theta_D = \pi/2$, so cos $\theta_D = 0$, and $\hat{\theta}_D \cdot \hat{p}_A = 0$). However, there are other positions in space (almost all other positions) where $\kappa^2 \neq 0$, even though the relative orientations of the donor and acceptor dipole directions are still perpendicular. Another trivial example is when the dipoles are parallel and $\hat{r}_{DA} \cdot \hat{p}_A = 1$. For instance, when $\theta_D = \pi/2$ and $\hat{r}_{DA} \cdot \hat{p}_A = 1$ (parallel dipoles, stacked on each other), $\kappa^2 = 4$. But when $\theta_D = \pi/2$ and $\hat{\theta}_D \cdot \hat{p}_A = 1$ (again parallel dipoles, but now next to each other), $\kappa^2 = 1$. Also, for every position in space of $\vec{p}_A$ relative to $\vec{p}_D$, there are an orientations of $\vec{p}_A$ such that $\kappa^2 = 2/3$. κ^2 varies quite a bit for small movements of the donor and acceptor relative to each other. Since there are always fluctuations in positions and angles of the *D* and *A* molecules, the actual value of κ^2 is an ensemble average, or time average. It has been found that the approximation $\kappa^2 = 2/3$ is usually quite satisfactory [5, 10, 50, 85, 86]. If during the excited state lifetime of the donor the orientations of the donor and acceptor can each individually fully reorient randomly, the κ^2 factor is rigorously 2/3. It is often discussed in the literature as though $\kappa^2 = 2/3$ pertains only to the case of very rapidly rotating *D* and *A* molecules; however, as we see, this is not the only case where this is true. Also, many of the dyes used for FRET have more than a single transition dipole, which can be excited at the same wavelengths, and since different transition dipoles of a fluorophore are usually not parallel to each other (they are often perpendicular to each other), this leads again to an averaging of κ^2. In most cases, $\kappa^2 = 2/3$ is a good approximation, and there are many examples in the literature where the distances derived are quite reasonable using this approximation. κ^2 is also often assumed to be constant when interpreting conformational changes of macromolecules [50]. Nevertheless, one should keep in mind that the assumptions that $\kappa^2 = 2/3$ and that κ^2 remains constant are not rigorous assumptions, by any means. For instance, when fluorescent proteins are used in FRET experiments, κ^2 can become a very important variable, and averages are often not applicable. This is because the chromophores are fairly rigidly held in the fluorescent protein structure, and the fluorescent proteins may have specific interactions either with each other, or with other components of the complex under study [87, 88].

1.3.1.3. Summary of the basic equations of FRET

Equations (1.1)–(1.3) are the quintessential theoretical descriptions of FRET. These equations, together with the well-known theoretical descriptions of spectroscopic transitions (see below) are the basics from which all the experimental acquisition and analysis methods are derived.

As we mentioned in the last paragraph, for most experiments with solution samples there is much less uncertainty in the κ^2 parameter than is often supposed, or suspected. Even for chromophores with orientations solidly fixed, a large fraction of the relative orientation space of the chromophores' transition moments are such that κ^2 is often not too far from 2/3 [6, 10, 89]. It is unlikely that the donor and acceptor molecules will be oriented such that the extreme values of κ^2 apply because the orientation configurational space for values close to these extreme values is relatively small [6]. However, this does not discount, especially for fixed orientations and distances between D and A molecules, that κ^2 can assume a particular value very different from 2/3, and then this must be known to make a reasonable estimate of r_{DA}.

1.3.2. Energy transfer modeled as a classical event; orientation effects

The energy transfer process can be envisioned as absorption by the acceptor of the near-field time-dependent electric field disturbance by a classically oscillating electric dipole of the donor. The classical model does not depict absorption as a quantum change of a photon, but the result describing the energy absorption is the same. This is a very useful way to think about and understand FRET, and agrees exactly quantitatively with the results of a quantum mechanical derivation. In the classical picture, the excited D molecule is regarded as a Hertzian electric dipole oscillator [6, 90, 91]. The near field of the classically oscillating electric dipole of D interacts with the dipole of A (quantum mechanically, these are the transition emission and absorption dipoles, and they interact through a dipole interaction operator). The effect on A is the same as the field of propagating electromagnetic (EM) radiation oscillating at the same frequency as the assumed oscillating classical dipole D, leading to a spectroscopic absorption of energy from the oscillating EM field. This process can be envisioned as a spectroscopic absorption event whereby A interacts with the electrodynamic perturbation of the oscillating electric near-field of an oscillating D-dipole. The shape of the field lines of the near field of a Hertzian dipole is identical to that of the field lines of a static electric dipole. The field of a static dipole and of a Hertzian dipole can be found in any book on electromagnetism [92–94]. The form, amplitude, and distance dependence of the time-dependent electric field of a dynamic Hertzian dipole changes as the distance from the dipole increases (near-, intermediate- and far-field). Propagating waves (emission of radiation) are only formed in the far field at distances of approximately one-half the wavelength from the dipole center. The near-field amplitude is immense compared to the fields farther away. Classically, the energy "leaks" out into the far field as radiation (fluorescence) from the oscillating dipole only very slowly compared to the oscillation frequency. In this classical paradigm, the *D* molecule can be thought of as an antenna that communicates with the *A* molecule, the receiver, except that the oscillating electric field in the near-field zone is not propagating. The communication between *D* and *A* is effectual if the resonance condition is satisfied (conservation of energy). As discussed above, the resonance requirement holds when integrating the energy exchange over the whole spectral distributions of the *D* and *A* molecules. As described following Eq. (1.3b), the relative orientation of the *A* dipole to the E-field lines emanating from the *D*-dipole is critical for efficient transfer [6, 86]. If for symmetry reasons, or because of exact perpendicular orientation, there is no electric dipole–dipole interaction, energy transfer can take place by quadrupole interactions [95], or interactions between magnetic dipole fields and electric dipoles [80, 96, 97]. However, these interactions are orders of magnitude smaller.

1.3.3. No emission/absorption of a photon in FRET; spectroscopic transition dipoles

FRET is a nonradiative process; that is, the transfer takes place without the emission or absorption of a photon. And yet, the transition dipoles, which are central to the mechanism by which the ground and excited states are coupled, are conspicuously present in the expression for the rate of transfer. For instance, the fluorescence quantum yield and fluorescence spectrum of the donor and the absorption spectrum of the acceptor are part of the overlap integral in the Förster rate expression, Eq. (1.2). These spectroscopic transitions are usually associated with the emission and absorption of a photon. These dipole matrix elements in the quantum mechanical expression for the rate of FRET are the same matrix elements as found for the interaction of a propagating EM field with the chromophores. However, the origin of the EM perturbation driving the energy transfer and the spectroscopic transitions are quite different. The source of this interaction term in the case of FRET is the coulomb interaction in the nonpropagating near-field zone between the D and A electron distributions (D in the excited state, and A in the ground state). The Coulomb charge interaction between D and A is usually simplified using a dipole approximation, and this results in an expression consisting of the transition dipole. When the D and A are too close together, this dipole approximation cannot be used [80, 95, 98]. The dipole approximation for spectroscopic transitions involving propagating EM waves is a consequence of the large wavelength of the EM perturbation compared to the extent of the molecular electronic system. In a quantum mechanical description of the interaction matrix the momentum operator of the electrons can be written in terms of the position operator of the electrons [99–101]. Serendipitously—if you will—the interaction term in the case of FRET turns out to be identical to the spectroscopic transition dipoles. This is not surprising when one goes through the derivations; however, it is essential to note the reason. In both cases, a first-order approximation produces mathematical expressions where the dipole transition moment is used. In this sense, it is clear that the term "fluorescence" in FRET is a misnomer. FRET has nothing directly to do with fluorescence. Fluorescence is only a convenient way to measure the energy transfer; and the efficiency of FRET can be determined without ever using fluorescence (see Section 1.9.2).

1.3.4. Noncoherent mechanism and vibrational relaxations; cooling off to equilibrium

It has been assumed so far that the nuclei configurations are in equilibrium in the excited state during the time that the different de-excitation kinetic pathways are active. This means that before de-excitation from the excited to the ground state takes place, the atomic nuclei have found new steady-state equilibrium positions corresponding to the new electronic state. The electronic transition from the ground state to the excited state happens so fast that the positions of the nuclei are left in the equilibrium positions corresponding to the ground state. This "out-of-equilibrium" state is called the Franck–Condon state [84]. The heavier nuclei cannot follow the electron movement immediately. Therefore, the initial atomic structure of the excited state is under stress and the nuclei relax rapidly to new positions commensurate with the potential of the first electronic excited state. The ground state is of course always in an equilibrium configuration. We have tacitly assumed that this relaxation process to an equilibrium configuration of the nuclei always takes place before any of the mechanisms of de-excitation transpire. In general, because biology takes place in a condensed environment, there is always a tight coupling of the thermal motions of the environment with the thermal motions of the chromophore. Therefore, these internal conversion relaxations within an electronic state potential are extremely rapid (on the order of picoseconds, or less). Since the lifetimes of the excited state are of the order of nanoseconds, the very rapid internal conversion from the Franck–Condon state to the thermally relaxed excited state (cooling off) is always rapid compared to the pathways out of the excited state.

1.3.5. Size makes a difference; propagating and nonpropagating energy; near- and far-field zones; radiative transfer

Perrin's expression predicted energy transfer over a distance comparable to the wavelength of light. This is approximately the distance when the EM energy departing from an excited molecule can be considered as a photon [102]. This is a manifestation of the uncertainty principle, which limits the momentum spread of the light quantum depending on the confinement volume. A photon has essentially a single frequency, or momentum $p_v = h/\lambda$. This is not true of the energy packet in the near-field zone. The energy in the near-field zone is composed of a very broad spectrum of frequencies, because of the small confinement. This encompasses the broad frequency bands of the absorption and fluorescence spectra in the overlap integral.

However, radiative energy transfer from *D* to *A* by the emission and absorption of real photons is possible (this "trivial" transfer is called radiative transfer). The rate is very small compared to the rate by the near-field Förster dipole–dipole mechanism. Radiative transfer is not likely at all unless concentrations in an extended sample are extremely high. This was already known from experiments; indeed because the experimental distance between dye molecules calculated for efficient depolarization of fluorescence was 100x shorter than the wavelength of light was the reason why Perrin suggested the near-field dipole–dipole mechanism (although he was not able to calculate the correct distance dependence). Reabsorption of emitted photons does occur experimentally when the concentration of acceptor is so high that before some photons can "get out of" the cuvette/sample, an acceptor (or another donor molecule) absorbs them—this phenomenon is called *internal filtering*. Internal filtering also depolarizes the emitted light. However, transfer by internal filtering is much less efficient by orders of magnitude than true nonradiative energy transfer.

If the energy transferred from the donor to the acceptor in the near zone were due to a photon-emission/reabsorption process, the angular distribution of emitted photons from the donor would depend on the presence or absence of the acceptor molecule. This is not observed in FRET. That is, the angular distribution of the donor emission is not affected by the presence of acceptors. In near-field zone where FRET takes place, only the *probability per unit time* that an excited donor molecule will leave the excited state is affected by the presence of the acceptor molecule. If a photon is emitted and not transferred to the acceptor, then the emission of the photon takes place as though the acceptor were not there. This is because the energy in the near-field zone of an oscillating electric dipole does not propagate; it is, so to speak, volume filling, and the position of the acceptor between the donor molecule and the observer is not important. Only the *number of photons emitted by an excited donor* is decreased when an acceptor is in the near-field zone, depending on its location and the relative angle between the two transition dipoles. The *intrinsic radiative rate* of the donor's fluorescence emission is not affected by the presence of a nearby acceptor in the near-field zone.

If the energy is transferred by trivial emission/reabsorption, it will lengthen the measured lifetime of the donor emission, not shorten it as happens in resonance energy transfer. This comes about because intervening absorption and emission processes take place prior to the final fluorescence emission (the reabsorption cannot take place until the photon has been emitted); the two processes do not compete dynamically, but follow in a serial fashion. In FRET, such an emission/reabsorption process does not occur, and the fluorescence lifetime of the donor decreases. This is an experimental check for reabsorption/reemission.

1.4. What can we learn from energy transfer?

1.4.1. Efficiency of FRET and distance measurements

FRET efficiency can be quantitatively measured over *D–A* separations of 0.5–10 nm. For shorter distances the assumption of point dipoles may not be valid (although this does not seem to be noticeable until very short distances). Very short distances comparable to the spatial extension of electron orbitals can lead to energy transfer by electron transfer (Section 1.11).

The *efficiency of energy transfer* (E) is the ratio of the number of energy transfer occurrences from *D* to *A* divided by the total number of excitations of a donor molecule. This is the same as the ratio of the rate of energy transfer to the total rate of deactivation of the excited donor. The rate of energy transfer between single donor and acceptor molecules is proportional to $1/r_{DA}^6$ (Eq. (1.1)); this is a very sensitive measurement of distance if r_{DA} is of the same order as R_0. Therefore, from the efficiency, we can gain *quantitative* information about the distance between D and A, and how this distance depends on experimental conditions. The distance R_0 is usually changed by choosing different dye pairs, or varying τ_D, for instance by dynamic quenching [50].

1.4.2. FRET in many different environments; following dynamics of macromolecules

FRET can take place whenever D and A molecules are separated by distances comparable to the size of macromolecules. However, the FRET efficiency is usually measured by observing fluorescence in the far field. Therefore, distance information on the scale of molecular dimensions is available from spectroscopic measurements in cuvettes, in optical microscopes, on whole biological organisms, as well as in single-molecule experiments. This gives us a direct handle on molecular structures, conformational changes, and binding interactions in a wide variety of complex biological environments and experiments. FRET can sometimes tell us about underlying ensemble molecular conformational distributions. And because FRET is a kinetic process, we can gain direct knowledge of macromolecular dynamics, molecular rotations, and changes in the relative orientation of D–A dipole axes that take place on a time scale comparable to the transfer. By choosing the chromophores judiciously, this time window can stretch from picoseconds to milliseconds, a remarkably broad temporal range. The dynamics of the macromolecular movements and conformational changes can be observed by FRET if the fluorescence emission of the donor happens on the same time scale of energy transfer. These dynamics can be observed by measuring the changes in the efficiency of FRET with time-resolved fluorescence measurements [103]. FRET measurements have been coupled to a wide variety of biological assays that yield specific information about the environments of the chromophores. These few considerations show why the FRET method has become so widespread and why it has become the choice for many experiments.

1.5. Simple depiction of basic fluorescence; how you determine the FRET efficiency

1.5.1. Rates of the individual pathways of de-excitation of the excited donor

In order to estimate the distance between *D* and *A* (Eqs. (1.1)–(1.3)), it is necessary to determine the *rate* or *efficiency* of the *energy transfer*. The efficiency of energy transfer is the ratio of the number of times that an energy transfer event transpires divided by the number of times that a donor is excited; this is sometimes called the *quantum yield of energy transfer*, in analogy to the quantum yield of fluorescence. Energy transfer competes dynamically with fluorescence emission. Therefore, FRET efficiency is usually determined indirectly by measuring the effect of FRET on the decrease of the fluorescence of the donor, or the increase in the fluorescence of the acceptor. The many different ways to measure FRET can be best presented by considering a very simple model representing the possible kinetic pathways of de-excitation from a molecule in the excited state. Using the kinetic competition between all the different pathways of de-excitation, this will allow us to describe easily and simply almost all the different ways of measuring FRET.

1.5.2. Calculating quantum yields: Fluorescence lifetimes, monkeys, and exit doors

To make the statistical ideas of basic fluorescence events clear, I describe a situation very different from fluorescence, but which has all the important statistical arguments. The toy model is easy to visualize, and some rather subtle effects become evident.[1] A clear appreciation of this simple model makes it straightforward to understand easily almost all different methods of measuring FRET. This is a small detour, which can be skipped if you want to proceed to Section 1.5.2.5. A word at the beginning: the description of the dynamics of standard spectroscopy, such as fluorescence, assumes that all molecules are in at least a steady-state equilibrium. This means we can apply probability methods. This was a very important prerequisite of Förster's original theoretical description of FRET. We are not considering coherent effects of fluorescence, which usually are only observable at subpicosecond times (see Section 1.2.2).

The model is a simple probability calculation of the average time required for a trapped monkey X_D^* (or excited donor molecule) to escape from a room through different doors, which can be opened or closed (see Fig.1.3). We imagine a total of N doors in the room (they do not change their state—open or closed—during the duration of the experiment). When open, each door presents a separate pathway for the monkey to escape. Because the room is dark, the monkey randomly searches the confinement in order to find an open door. The monkey also has no recollection of where the doors could be, and he is so excited that his search is random; that is, he frantically searches the wall locations completely randomly. You can think of this as the "excited state room." As the monkey runs around rapidly, but incoherently, to find an open passage out of the room, there will be a distinct probability per unit time (rate) that the monkey will escape through any particular door. Call this rate k_i for passing through the ith door. This is a rate, a probability per unit time, and depends on his search speed and the size of the doors. If the doors have different sizes, the monkey will find the

[1]This toy model depicts the basic statistical ideas of fluorescence decay that is critical for understanding FRET. I apologize to all who already know all this. You can skip it, or just read it over for fun.

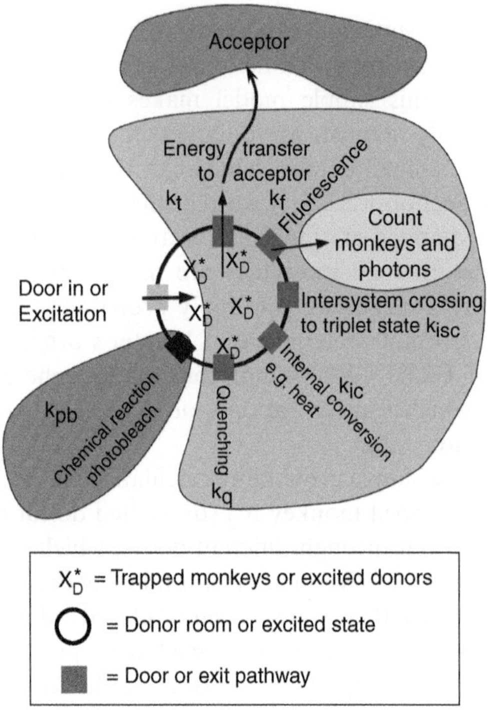

Fig. 1.3. The excited state is depicted as a circle in which the excited molecules are confined. Each monkey or molecule is labeled as X_D^*. The excited monkeys or molecules can leave the excited state through several pathways (or doors). The doors represent (k_i represents the rate of leaving through the ith door): energy transfer (k_t), fluorescence emission (k_f), intersystem crossing (k_{isc}), internal conversion (k_{ic}), quenching (k_{q}), or photolysis (k_{pb}). Every time a molecule leaves the excited state by the fluorescence emission pathway it emits a photon of energy $h\nu_{\mathrm{em}}$. In the toy model, we just count the exiting monkeys. Except for the photobleach pathway (from which the molecule is irreversibly chemically destroyed) and intersystem crossing, every molecule that exits from the excited state returns to the ground state, and can be excited again. If the molecule undergoes intersystem crossing then it enters a triplet state (parallel electron spins; see Fig.1.4) where it again can exit through one of several pathways. From the triplet state the emission is called phosphorescence, and has a much longer lifetime (sometimes up to seconds). For this reason the triplet state is much more likely to undergo and excited state chemical reaction

larger doors more often than the smaller doors, and therefore the rate (probability per unit time) will be faster for the larger doors. These sizes of the individual doors do not change, and for the duration of the experiment each door is either open or closed. Understanding this example is all we need to analyze most fluorescence experiments, including most FRET experiments. The parallel with fluorescence is clear when it is realized that the probability of fluorescence decay from the excited state of a molecule is a random variable; that is, once in the excited state, there is only a probability per unit time that a photon will be emitted. It is this probability we want to calculate.

1.5.2.1. Single monkey and one open door

Say we do the experiment first with just one door open; call this door "f" (which stands for fluorescence). We sit outside of the f-door, and after we start the experiment (at time $T_0 = 0$), we record whether the monkey is still in the room at some time T, and record this. This is the same as a photon counting experiment. We divide up our observational/recording times into Δt equal increments. The probability that the monkey will still be in the room at time $T_0 + \Delta t$, if he had started there at time T_0, is $P(T_0 + \Delta t) = (1 - k_f \Delta t)$, where k_f is the time-independent rate (probability per unit time) for finding the f-door. From now on we define $T_0 = 0$. We repeat this experiment a large number of times, each time we record the time $T + \Delta t$ when the monkey emerges through the f-door (so he was still in there at time T). For each experiment,

---

(often leading to destruction). However, if the transfer to the triplet state from the singlet state takes place, then it must be rapid so that it can compete with the other kinetic pathways in the nanosecond time region. The total rate of de-excitation from the excited state is the sum of all the individual rates. The analogy is made in the text of monkeys being confined in an "excited state room" and the exit pathways are depicted as "exit doors."

the probability that the monkey will still be in the room after a time T (which corresponds to M time intervals of Δt) is $P(T) = P(M\,\Delta t) = (1 - k_f\,\Delta t)^M = (1 - k_f(T/M))^M$. Now, we divide the time intervals into smaller and smaller intervals, so that we can assume $\Delta t \rightarrow 0$ and therefore $M \rightarrow \infty$. We then have $P(T) = \exp(-k_f T)$; this results from the definition of an exponential. So, as we might expect, the probability that the monkey emerges at a time $T + \Delta t$ decreases exponentially with time and with a rate constant k_f. This can also be written as $P(T) = \exp(-k_f T) = \exp(-T/\tau_f)$, where τ_f is the average duration time the monkey is in the room. This is good for all times T. Of course, for the case of fluorescence, τ_f represents the intrinsic radiative lifetime (where every excited molecule only decays by the emission of a photon). k_f is usually a fundamental constant of the molecule, and is independent of the environment of the fluorophore. The important part of this exercise is that the rate of passage through the door is constant, and that the process of finding the door is *random*. This leads to the exponential decay, and is true for fluorescence just as it is for radioactivity decay. What we have described is a single monkey (single molecule) experiment. It is also the way that a single-photon time-correlated fluorescence experiment (also on ensembles of molecules) is carried out, because in this case the arrival times of only a single photon from a single molecule is recorded at any time.

1.5.2.2. An ensemble of monkeys and one open door

There is another way we can do the experiment with many, many monkeys (an ensemble of M_{all} monkeys). We can send them all into the room at time zero, and count how many monkeys emerge from the f-door between the time T and $(T + \Delta t)$. This is proportional to the *rate of emergence*, or the "intensity" of emergence. Now we only have to do the experiment once (we have many monkeys) but we record the number of monkeys emerging per Δt at many times T, and then plot our results. We now get an exponential that decays at the same rate as before (k_f); we call it the intensity of monkeys (the number of monkeys emerging per unit time at some time T). We calculate this intensity function as follows. We know at time T only M_{all} exp($-k_fT$) monkeys are left in the room. So the *intensity* (or rate) at which monkeys are emerging at time T is I_{f} ($T = k_fM_{\text{all}}$ exp($-k_fT$). We can check if this is correct, by integrating the rate at which the monkeys are emerging over all times, which should give the total number of monkeys. This is $\int_0^\infty I_{\text{f}}(T)\text{d}T = k_f M_{\text{all}} \int_0^\infty \exp(-k_f T)\text{d}T = M_{\text{all}}$. Obviously this is the same functional form we would measure if the intensity of fluorescence (number of photons per unit time, which is proportional to the fluorescence intensity) were the only pathway of de-excitation, and if the total number of molecules excited at time zero was M_{all}.

1.5.2.3. Single monkey and two open doors

Now we consider what happens when we open another door—we call this door "t" (for energy transfer). In analogy with k_f, we define k_t as the rate (probability per unit time) of going out the ET door. This is also a constant, but unlike k_f, k_t depends on the experimental design; for example, in fluorescence, how close the molecules are. Now the probability of leaving per unit time is $k_f + k_t$. We still sit outside the f-door, and record the times when the monkey comes out the f-door (we cannot observe the energy transfer door). The probability he is still in the room at time $T = M\,\Delta t$ is now $P(T) = (1 - (k_f + k_t)\Delta t)^M = (1 - k_f')^M$. We again make a histogram of the time the single monkey emerges out the F-door (repeating the experiment many times), and we notice that when we compare to the previous experiment, the histogram is still an exponential, $\exp(-T/\tau_{\text{f}}') = \exp(-k_f' T)$, but now $\tau_{\text{f}}' < \tau_{\text{f}}$, and $k_f' > k_f$. So, on the average, he stays in the room less time. And notice that the monkey emerges from the f-door fewer times than we send him in. Of course this is expected, because the probabilities of independent random events add, and we know with two doors $k_f' = k_f + k_t$. Because the fraction of times the monkey goes through any particular door depends on the relative rate for that door, we can calculate a "quantum yield" for the monkey to emerge through the f-door: $Q_f = k_f/(k_f + k_t)$. This is just as we would expect in fluorescence.

1.5.2.4. An ensemble of monkeys and two open doors

The extension to the second method, where we measure the time dependence of the intensity of monkeys emerging, is straightforward. At time T there are only $M_{all}\exp(-(k_f + k_t)T)$ monkeys left in the room, so the *intensity* of monkeys emerging at time *T from the f-door* is $I_f(T) = k_f M_{all}\exp(-(k_f + k_t)T) = k_f M_{all}\exp(-T(1/\tau_f + 1/\tau_t)) = k_f M_{all}\exp(-T(1/\tau_f'))$, where $1/\tau_f' = 1/\tau_f + 1/\tau_t$. Now from this experiment, let's calculate the quantum yield of monkeys taking the f-door. For this, we simply have to integrate the expression $I_f(T)$ to give the total number of monkeys emerging from the f-door, and then divide by the total number of monkeys that were in the room in the first place. This is

$$Q_f = \frac{1}{M_{all}}\int_0^\infty I_f(T)\mathrm{d}T = \frac{1}{M_{all}}\int_0^\infty k_f M_{all}\exp(-(k_f + k_t)T)\mathrm{d}T = \frac{k_f}{k_f + k_t},$$

where now

$M_{all} = \int_0^\infty (k_f + k_t)M_{all}\exp(-(k_f + k_t)\ T)\mathrm{d}T$. This agrees with the calculation just making the experiment with single monkeys many times. The quantum yield is of course less than when only the f-door was open, because some of the monkeys go through the t-door.

Again, the calculation is identical to what we would make for fluorescence if the total number of molecules excited at time zero was M_{all}, but there were now two pathways of de-excitation. Notice that we can tell whether the t-door (the t-pathway) is open by measuring either the lifetime τ_f' and comparing to τ_f, or measuring the intensity with the door open and again with the door closed, and then taking the ratio $\frac{\left[\int_0^\infty I_f(T)\mathrm{d}T\right]_{\text{door open}}}{\left[\int_0^\infty I_f(T)dT\right]_{\text{door closed}}} = \frac{k_f}{k_f + k_t} = Q_f$. Notice also that the constant k_f is the *same* in the presence or absence of the energy transfer pathway. This trivial observation is important—the measured lifetime of fluorescence decay τ_f' becomes shorter and the integrated intensity $\int_0^\infty I(T)\mathrm{d}T$ becomes less, only because we have opened another pathway of escape, not because the rate constant of fluorescence has decreased. All the other rate constants remain constant in the presence and absence of the acceptor. Therefore, measuring the lifetime of fluorescence, or the quantum yield of fluorescence, and using these simple ideas, we can calculate the rate k_t; then we can determine the distance between the donor and acceptor, using Eqs. (1.1)–(1.3).

1.5.2.5. General application to fluorescence with an arbitrary number of deactivation pathways

We now end the discussion of this toy monkey model, and discuss only fluorescence, keeping in mind the results of this simple model. We will then discuss the different ways of measuring FRET.

In Fig. 1.3, we depict the excited state as a bounded region in a "state space." When a molecule is in the excited state, this is indicated by placing the molecule into the "excited state" region. In Fig. 1.3, we have called the molecule "X_D^*" for "donor" and put a star "*" to indicate that the D-molecule is excited. It is not necessary that the molecule get into the excited state by absorbing a photon. The ensuing process of energy transfer does not depend on how the donor molecule becomes excited. It is possible that the donor molecule could get into the excited state by receiving the energy of excitation from another excited molecule. Or the molecule could arrive in the excited state through means of a chemical or biochemical reaction—as in bio- and chemiluminescence. Bioluminescence has also been used to study energy transfer [104, 105]. Most FRET experiments are done such that the excitation is accomplished by absorbing a photon. Of course experimental details may depend on how the donor is excited, but this does not concern our description of the transfer of energy once the donor is in the excited state.

Once a donor molecule is in the excited state region (Fig. 1.3), it can only exit through one of the allowed exit pathways. All events that take place in the excited state will contribute to the overall rate at which an excited molecule leaves the excited state. The total rate (the probability per unit time) for an excited donor molecule to leave the excited state region is the sum of all possible rates for leaving by any of the possible exit pathways. Some of these individual rates can be affected by the molecular environment of the excited molecule (e.g., solvent effects, proximity of quenchers or energy acceptors). Some of the rates are intrinsic to the molecular structure, such as the intrinsic radiative rate of fluorescence decay, which can be calculated quantum mechanically for an isolated molecule. The rates can also depend on the time. This happens if the molecules move or change significantly during the time the donor molecule is in the excited state. However, if conditions remain constant, then the probability that a particular pathway of de-excitation will be chosen is independent of the other pathways.

Just as above, we can derive expressions for any fluorescence lifetime for any number of pathways. In this chapter we limit our discussion to cases where the excited molecules have relaxed to their lowest excited-state vibrational level by internal conversion (ic) before pursuing any other de-excitation pathway (see the Perrin–Jablonski diagram in Fig. 1.4). This means we do not consider coherent effects whereby the molecule decays, or transfers energy, from a higher excited state, or from a non-Boltzmann distribution of vibrational levels, before coming to steady-state equilibrium in its ground electronic state (see Section 1.2.2). Internal conversion only takes a few picoseconds, or less [82–84, 106]. In the case of incoherent decay, the method of excitation does not play a role in the decay by any of the pathways from the excited state; the excitation scheme is only peculiar to the method we choose to measure the fluorescence (Sections 1.7–1.11).

Now with this simple picture, all we have to do is observe the process of the excited molecule passing through one of the exit pathways (Fig. 1.3). For instance, in order to measure the rate of

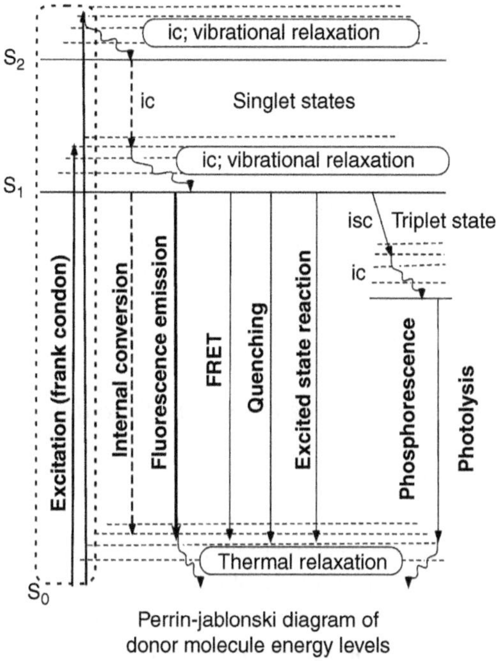

Fig. 1.4. This is a modified Perrin–Jablonski diagram depicting the ground and excited states. The vertical axis is energy. The excited molecules can leave the excited state in several ways. The rates of each pathway are labeled. If the molecules are excited to a higher electronically excited state than S1, or if they are excited to higher vibrational states, the energy is very quickly and efficiently dissipated (thermal relaxation, internal conversion) so that the molecules generally end up in the lowest energy level of the first excited state before any of the pathways of de-excitation (discussed in Fig.1.3) can happen. The different exit pathways from S1 all happen at a particular rate depending on the molecular surroundings. Similar exits from the triplet state can also occur (not shown). The ultimate probability that energy transfer will take place is proportional to the product $\tau_D k_t$. Even though the actual transfer rate, kT, is very small, the lifetime of the triplet, τ_D, can be very long (if it is not quenched). Thus, energy transfer can still be efficient (albeit slow) if there is some spin–orbit coupling such that the transition is not fully nonallowed. The rate is small, but the time is long, and efficient transfer can ensue; the acceptor passes from a singlet or triplet state to a corresponding singlet or triplet state, respectively (i.e., the spin of the acceptor does not change). There is no

energy transfer we measure the rate or extent (efficiency) of passing through the fluorescence exit pathway in the presence and absence of acceptor. This measured rate and the number of events passing through an exit pathway change when the acceptor is present. The rate constant k_t depends on how close the acceptor is. However, we emphasize again, the probability per unit time (the intrinsic radiative rate) of passing through the fluorescence exit pathway remains the same, whether the acceptor is present or not. We will consider a few cases.

1.6. Measuring fluorescence emission of the donor

1.6.1. Donor lifetimes

Every time an excited molecule exits the excited state region by the fluorescence pathway it emits a photon. We can either count the number of photons in a longer time interval (by a steady-state measurement of the fluorescence intensity) or make a time-resolved measurement of the fluorescence decay. These measurements can be done in an ensemble mode or on single molecules—the basic process is the same. The number of photons collected from the donor emission will be depicted by I_{DA} and I_D, where we mean the fluorescence intensity of D in the presence (I_{DA}) and absence (I_D) of acceptor. All other conditions, other than the presence or absence of acceptor, remain the same. During the same time of the experiment where we have measured the photons emitted by D, many of the excited D molecules have exited from the excited state by a pathway other than fluorescence. Obviously, the number of times a pathway has been chosen as an exit pathway is proportional to the

---

corresponding efficient transfer from a singlet donor state that changes the spin of the acceptor, because the lifetime of the excited singlet donor is too short [6]. See the text for more detailed explanations of the different processes.

rate at which the molecules exit by this pathway. We designate the total rate constant of *D* molecules exiting by all pathways to be $\sum_{i,i \neq t} k_i$ in the absence of *A*, and $\sum_i k_i$ in the presence of *A*. Here the notation can refer either to the *number of times* that exits from the excited state have taken place by pathway "*i*" in a set time interval, or the *rate* by which exits happen through pathway *i*. We assume that we know the number of times a *D* molecule has been excited, so we can compare these numbers in the presence and absence of *A*. All the *rates* of the pathways of de-excitation other than by FRET are *independent of the presence of A*. This is important; if this is not true then the experiment cannot be done as we describe. If we subtract the two sums, we get:

$$\sum_i k_i - \sum_{i \neq t} k_i = k_t \tag{1.4}$$

The important conclusion is that we can measure the rate of energy transfer without ever measuring the energy transfer directly, provided we could measure the total rate constant of decay from the excited state in both cases. In the case of fluorescence, this total rate is the reciprocal of the measured fluorescence lifetime $\tau_{D,\text{meas}}$; the rate can be measured indirectly in a steady-state measurement of the fluorescence intensity. It should be understood that we are not limited to measuring fluorescence to get this information. The rate at which the excited *D* molecules leave the excited state is the same no matter which pathway we observe; it is the sum of the individual rates of all pathways of de-excitation. We can determine this rate by observing *any* convenient pathway, it does not have to be fluorescence (see Section 1.9.2); but usually fluorescence is the most convenient. Remember though, the rate constants of each individual pathway are constant. The fraction of excited *D* molecules following any particular pathway is the ratio of the rate constant for following this pathway divided by the sum of all the rate constants of de-excitation. This is the quantum yield of that particular pathway.

Thus, when we determine the rate of *fluorescence decay*, by *measuring* the fluorescence lifetime, we are measuring the total rate.

That is,

$$\frac{1}{\tau_{DA}} = \sum_i k_i \tag{1.5a}$$

and

$$\frac{1}{\tau_D} = \sum_{i \neq t} k_i \tag{1.5b}$$

τ_{DA} is the fluorescence lifetime of the donor in the presence of acceptor, and τ_D; is the lifetime of the donor in the absence of acceptor. Note also that $\tau_{DA} < \tau_D$. The rates are different in the presence and absence of acceptor because the sums of the rates are different. Therefore, we can write:

$$\sum_i k_i - \sum_{i \neq t} k_i = \frac{1}{\tau_{DA}} - \frac{1}{\tau_D} = k_t \tag{1.6}$$

The rate of energy transfer is simply the difference in the reciprocal fluorescence lifetimes of the donor, measured in the presence and absence of acceptor. Of course, if we measure the fluorescence lifetime of the donor, we do not have to keep track of the number of donors that have been excited; this is a great advantage of measuring lifetimes.

The *efficiency of energy transfer* is the same as the quantum yield of energy transfer. It is the number of times that molecules take the energy transfer pathway divided by the number of times that the donor molecules have been excited. This is the same as the ratio of the number of times the excited donor exits by transferring energy to the number of times the excited molecules exit by any process to return to the ground state. In terms of the lifetimes of the donor, this is:

$$E = \frac{\text{Rate of energy transfer}}{\text{Total de-excitation rate}} = \frac{k_t}{(\tau_{DA})^{-1}} = \frac{[1/\tau_{DA} - 1/\tau_D]}{1/\tau_{DA}} = \frac{\tau_D - \tau_{DA}}{\tau_D} \tag{1.7}$$

This is a very simple equation.

Using the Förster's derivation of k_t, see Eq. (1.1), we can write the efficiency in terms of the distances (which is the result that is normally desired):

$$E = \frac{k_t}{{}^1/_{\tau_D} + k_t} = \frac{{}^1/_{\tau_D}\left(\frac{R_0}{r_{DA}}\right)^6}{{}^1/_{\tau_D}\left[1+\left(\frac{R_0}{r_{DA}}\right)^6\right]} = \frac{1}{1+\left(\frac{r_{DA}}{R_0}\right)^6} \tag{1.8}$$

1.6.2. Steady-state fluorescence intensities

We can also count the total number of donor-emitted photons, or measure the corresponding analog intensity, in the presence and absence of energy transfer. From these intensities we can calculate the efficiency of energy transfer. The fluorescence intensity of the donor is proportional to the rate constant through the fluorescence pathway divided by the sum of the rates of leaving the excited state by all pathways. That is,

$$I_{DA} \propto k_f \sum_i k_i = k_f \tau_{DA} \tag{1.9}$$

and

$$I_D \propto k_f \sum_{i \neq t} k_i = k_f \tau_D \tag{1.10}$$

It is convenient, but not necessary, if the intensity of excitation light is the same in both cases. If this is true and if the concentration of donor molecules is also the same for both measurements, then the number of initially excited donor molecules is the same. If these conditions are not met, then the value of detected donor fluorescence can be corrected so the measured values can be compared. Once these corrections have been made (if they are necessary), we can form the following ratio:

$$\frac{I_D - I_{DA}}{I_D} = \frac{\left[\sum\limits_{i \neq t} k_i\right]^{-1} - \left[\sum\limits_{i} k_i\right]^{-1}}{\left[\sum\limits_{i \neq t} k_i\right]^{-1}} = \frac{\sum\limits_{i} k_i - \sum\limits_{i \neq t} k_i}{\sum\limits_{i} k_i} \tag{1.11}$$

$$= \frac{k_t}{\sum_i k_i} \equiv \text{Efficiency of energy transfer}$$

This is the most common method for measuring the energy transfer efficiency.

Of course, this can be derived from the lifetime equations as follows. Because the fluorescence intensity of the donor in the presence of A, I_{DA}, and the absence of A, $I_{\rm D}$, is proportional to τ_{DA} and τ_D we have from Eq. (1.7):

$$E = 1 - \frac{I_{DA}}{I_D} \tag{1.12}$$

1.7. Measuring any de-excitation process in order to determine FRET efficiency; the acronym FRET

1.7.1. FRET measurements with and without fluorescence

We never measure directly the rate of energy transfer. Although we want to determine the efficiency of energy transfer, another process—for example, fluorescence emission—is measured. When the fluorescence intensity of the donor molecules is measured, we are measuring a pathway for molecules that have *not* undergone energy transfer. This method of measuring competitive kinetic processes yields at once the efficiency of energy transfer as we have shown above. Analogous methods can be used to measure the rate or the extent of *any process* happening in the excited state. We do not have to measure fluorescence. For instance, consider the case that the donor undergoes a chemical reaction in the excited state. We can measure either the rate of this excited state reaction, or more easily the extent of this reaction, in the presence and absence of the acceptor. The extent of the photoreaction can be measured for instance by simple chromatography. Then we can use the same method described above to determine the extent of energy transfer without ever measuring fluorescence! We just measure the extent of the excited state reaction in the presence and absence of energy transfer, and carry out similar calculations as just shown. Obviously, not only energy transfer can be measured with this "competitive-rate" method. For instance, similar methods are used to measure dynamic molecular collisions that cause dynamic quenching of fluorescence. We can measure the rate of quenching by observing the photons that are emitted (which are not quenched) from the fluorophore—we just have to do this in the presence and absence of the quencher. Once the rate of dynamic quenching is known, we can estimate diffusion constants and the size of quenching target areas. This is usually done by analyzing a Stern-Volmer plot [82, 106]. Förster used the known properties of dynamic quenching in order to show the dynamic nature of FRET by the competition with quenching [54, 55].

In general, consider two competing processes leading to de-excitation of D^*, process A and process B. Process A can be measured by determining the rate or extent of process B in the presence or absence of process A. Effectively, we measure B in order to determine A. These simple considerations cover essentially all the indirect methods of measuring the energy transfer process. Because the excited donor molecule can be de-excited through many different pathways of *kinetically competing processes*, we can use any of these pathways to determine the rate of exiting from the excited state. Usually fluorescence is used to measure the energy transfer. But as we have shown, fluorescence (the emission of a photon) is not the mechanism of the transfer, and one can measure FRET without ever measuring fluorescence.

1.7.2. The acronym FRET and extensions beyond Förster transfer

FRET is often referred to as the acronym for "*fluorescence* resonance energy transfer". This name is a misnomer, but this use has become common, and is spread throughout the literature (also by the author). However, it is better to reserve the acronym FRET for "Förster resonance energy transfer", because then it is clear when one is referring to the conditions for which the Förster mechanism is valid. This is becoming more important as energy transfer experiments are being pushed beyond the normal Förster conditions [1, 2, 69, 73, 74, 107–111].

1.8. Measuring photobleaching to determine FRET efficiency

1.8.1. The rate of donor photobleaching

Photodestruction of fluorophores often limits the accuracy with which experiments can be carried out, and sometimes makes certain experiments impossible, especially in microscopy. However, photobleaching is a process competing with all the other de-excitation pathways out of the excited state of the donor. Therefore, we can use it to measure FRET efficiency according to the above discussion. The photobleaching technique for measuring FRET efficiency has been well described [40, 40, 41, 112–116] and is especially useful in fluorescence microscopy to image FRET efficiency in biological cells. The time over which photobleaching takes place is usually in the range of seconds or minutes; that is, almost *10 orders of magnitude slower* than the nanosecond emission. This is usually the time it takes for an ensemble of molecules (or, the average time for single molecules) to bleach when illuminated with excitation light; although for higher intensities of light this can take place in microseconds. The reason why we can observe it in the presence of all the other much faster processes is that it is *irreversible*. There is a certain probability every time a molecule enters the excited state that it will be destroyed by photobleaching. Because the pathway of photodestruction competes with all the other pathways of de-excitation (on the nanosecond time scale), we can apply a similar analysis as described above to calculate the effect of an acceptor on the rate of photodestruction of the donor. In general, one follows the slowly decaying steady-state fluorescence intensity of the donor, or we observe the random emission from one molecule if we are making single molecule measurements, and determine the time required for this fluorescence to disappear. Most fluorophores will photodestruct every $10^4 - 10^7$ excitation events; this depends on the molecular structure and the environment. This means that the probability of photodestruction is very small compared to the probability to fluoresce, undergo energy transfer, or be quenched by collision (but not photodestructed). However, over a long time, because photolysis is irreversible, the number of potentially fluorescent molecules will decrease. To measure FRET, the rate of photodestruction is measured in the presence and absence of an acceptor. In analogy with Eq. (1.7), the following ratio is constructed, where we designate photobleaching with the acronym pb; and $\tau_{\mathrm{pb,D}}$ means the measured photobleaching time constant, which is in the second to minute range. The variable k_{pb}, is the intrinsic rate through the photobleaching pathway.

$$\begin{aligned}\frac{\left[\tau_{\mathrm{pb},D}\right]^{-1}-\left[\tau_{\mathrm{pb},DA}\right]^{-1}}{\left[\tau_{\mathrm{pb},D}\right]^{-1}} &= \frac{\dfrac{k_{\mathrm{pb},D}}{\sum\limits_{i\neq t} k_i}-\dfrac{k_{\mathrm{pb},D}}{\sum\limits_{i} k_i}}{\dfrac{k_{\mathrm{pb},D}}{\sum\limits_{i\neq t} k_i}} = \frac{\sum\limits_{i} k_i-\sum\limits_{i\neq t} k_i}{\sum\limits_{i} k_i} \\ &= \frac{k_t}{\sum\limits_{i} k_i} = \text{Efficiency of FRET}\end{aligned} \tag{1.13}$$

The probability per unit time for photodestruction of the donor ($k_{\mathrm{pb},D}$) is always the same, in the presence and absence of the acceptor. However, in the presence of the competing process of energy transfer the overall rate of photobleaching is less. Therefore, we can use the rate of photobleaching to measure the rate of energy transfer. This method uses measurements recorded in the second to minute range in order to measure rates in the nanosecond range.

1.8.2. Photobleaching the acceptor

This method is usually performed in the microscope, on whole images. If we measure the fluorescence intensity or the fluorescence lifetime of the donor in a sample where both D and A are present, and then photobleach the acceptor (not the donor) and repeat the measurement of the donor, we can use these two measurements to measure the efficiency of energy transfer. The equations that were derived above when measuring steady-state donor fluorescence can be used for this, because, essentially by photobleaching the acceptor, we have made the equivalent measurements in the presence and absence of acceptor. The same equations apply. The assumption is made that the photodestructed acceptor does not interfere spectroscopically with the experiment, what is usually the case. This is a very useful method of determining FRET in imaging experiments when one cannot easily determine the spectroscopic properties that are necessary for making corrections in the data for overlapping spectra of D and A. One has to be careful not to photobleach the donor; however, when measuring lifetimes of the donor this is not important. If we assume there has been 100% photobleaching of the acceptor, then the efficiency of transfer can be calculated from,

$$E = 1 - \frac{I_{DA}^{\mathrm{pre}}}{I_{DA}^{\mathrm{post}}} = 1 - \frac{I_{DA}}{I_D} \tag{1.14}$$

where I_{DA}^{pre} is the intensity of the donor in the presence of the acceptor before photobleaching, and I_{DA}^{post} is the intensity of the donor in the presence of the acceptor after photobleaching of the acceptor.

1.9. Measuring the acceptor fluorescence or reaction products to determine the efficiency of energy transfer

1.9.1. Detecting the acceptor fluorescence

Energy transfer can be detected directly by measuring the number or concentration of acceptor molecules that have received their excitation energy from the donor. This is called *sensitized excitation of the acceptor*, or if we can detect the fluorescence emanating from the acceptor molecules that have been excited by energy transfer from the donor, it is called *sensitized emission of the acceptor* fluorescence. In order to accomplish this, account of direct excitation of the acceptor by means other than energy transfer must be made during the analysis, and the fluorescence of the donor contributing to a signal must be known [6]. Determining the contribution of the donor to the fluorescence signal is easy if the fluorescence spectrum is measured [5]. The fluorescence spectrum can be decomposed into the separate contributions from donor and acceptor by linear decomposition if we have spectra of the donor and acceptor alone (in the exact environment that the FRET measurement is made). The degree of direct excitation of the acceptor when exciting with wavelengths for donor excitation can be determined by measuring the acceptor fluorescence using an excitation wavelength high enough not to excite the donor. Depending on the experimental situation, this can be unproblematic. Observing sensitized emission of the acceptor is a reliable and very discerning measurement. Sensitized emission of the acceptor is direct evidence of energy transfer because sensitized emission of the acceptor can *only* arise by energy transfer.

It is very difficult to excite the donor without also exciting some of the acceptor population. This is because the absorption spectra of dyes extend significantly into the blue side of their absorption maxima, so the absorption spectra of the donor and acceptor usually overlap. The donor fluorescence can typically be observed without acceptor fluorescence interference; therefore, when measuring FRET efficiency by observing the donor fluorescence, this overlap is not important. However, when observing the acceptor fluorescence the overlap of the donor and acceptor absorption must be taken into account. The total steady-state fluorescence of the acceptor, assuming that $[A] = [D]$ (i.e., a equal donor and acceptor concentrations, and 100% labeling) is

$$
\begin{aligned}
I_{AD} &= Q_A\{\varepsilon_A[A] + \varepsilon_D E[D]\} = I_A + \varepsilon_D Q_A E[D] \\
&= I_A + (\varepsilon_A Q_A[D])E\frac{\varepsilon_D}{\varepsilon_A} = I_A\left\{1 + E\frac{\varepsilon_D}{\varepsilon_A}\right\}
\end{aligned}
\tag{1.15}
$$

I_{AD} is the intensity of the acceptor in the presence of the donor. From this it is easy to isolate E:

$$
E = \frac{\varepsilon_A}{\varepsilon_D}\left(\frac{I_{AD}}{I_A} - 1\right) \tag{1.16}
$$

Therefore, when measuring the fluorescence of the acceptor, one must know the intensity of the acceptor fluorescence that is excited directly the wavelength of light used to excite the donor, and then the efficiency is calculated from Eq. (1.16). If there is not 100% labeling, then $[A] \neq [D]$, and Eqs. (1.15) and (1.16) must be corrected to take this into account. Such cases have been discussed in the literature [5].

1.9.2. Detecting reaction products of the acceptor

Another method to detect energy transfer directly is to measure the concentration or amount of acceptor that has undergone an excited state reaction by means other than detecting its fluorescence. For instance, by chemical analysis or chromatographic analysis of the product of a reaction involving excited A [117, 118]. An early application of this determined the photolyzed A molecules by absorption spectroscopic analysis. [119–121]. This can be a powerful method, because it does not depend on expensive instrumentation; however, it lacks real-time observation, and requires subsequent manipulation. For this reason, fluorescence is the usual method of detection of the sensitized excitation of the acceptor. If it is possible to excite the donor without exciting the acceptor, then the rate of photolysis of the acceptor (which is an excited state reaction) can be used to calculate the FRET efficiency [122].

1.9.3. Detecting lifetime-resolved fluorescence of the acceptor

Transfer from the donor perturbs the normal emission decay kinetics of the acceptor. The energy transfer from the donor acts as a time-dependent excitation light source for the acceptor. Just as if the acceptor were excited directly by a time-dependent light pulse, the kinetics of the energy transfer from the donor is convoluted with time-dependent fluorescence decay of the acceptor. The rates for leaving the acceptor excited state by any pathway are not affected by the transfer. That is, following the energy transfer, the probability per unit time for any excited acceptor molecule to leave its excited state does not depend on how the acceptor molecule becomes excited. The energy transfer kinetics can be determined by deconvoluting the known acceptor fluorescence decay (determined from an experiment with direct excitation of the acceptor) from the time-dependent acceptor signal. However, unless the time decay kinetics of the acceptor is very well determined, this deconvolution is often difficult to do accurately. There are a few special cases that are of special interest.

If the emission lifetime of the donor is very slow compared to the lifetime of the acceptor, it can be convenient to observe the emission of the acceptor by time-delaying the detection. This has the great advantage that one is observing the emission of the acceptor after all the directly excited acceptor molecules have decayed, and the only acceptor emission arises from excitation by energy transfer. It is still necessary to employ a filter to block the emission of the donor fluorescence. In this case it is clear that the delayed emission from the acceptor fluorescence originates only from the energy transfer. If very long lifetime probes are used, such as lanthanides (with millisecond lifetimes) the time resolution is easy to achieve in most laboratories with simple rotating light choppers [8, 123, 124]. In addition, if lanthanides are used as donors, then it is relatively easy to block the very narrow fluorescence bands of the donor.

In a related method, also very powerful, one can measure the emission of the acceptor following transfer from a donor that has a very short lifetime compared to the acceptor. In this case the emission of the acceptor does not mirror the lifetime of the donor (because the donor fluorescence decay is much faster), and fluorescence from the acceptor can arise from transfer from the donor and from direct excitation. But the advantage is that the interfering fluorescence from the fast decaying donor does not interfere with the detection of the acceptor fluorescence so that the sensitized emission can be detected without donor interference. One must excite the sample with wavelengths that do not excite the acceptor significantly. If this is not practical, then by varying the wavelength of excitation one can determine the fraction of directly excited acceptor, and separate this from the fraction of acceptor that is excited by energy transfer from the donor. Also, if the direct fluorescence of the acceptor can be taken into account, for instance by exciting where only the acceptor absorbs (this is often possible because absorption spectra have a low energy cutoff at the 0–0 position), it is straightforward—if there is no contribution of the donor fluorescence to the signal—to calculate the acceptor fluorescence coming only from energy transfer. In addition, the measurement of the direct excitation of the acceptor will provide normalization for the concentration of the acceptor molecules so that the fluorescence energy transfer efficiency can be calculated.

1.10. Transfer between identical molecules: Fluorescence anisotropy

If the donor and acceptor molecules are chemically identical, then transfer from the excited molecule to the unexcited molecule of the pair can take place (more than once during an single excitation event—back and forth), provided that the spectroscopic requirements of equation 2 are valid. This is called homotransfer [5]. The fluorescence lifetime and the fluorescence quantum yield do not change from that of the singly excited molecule. Because the probability of decay from the excited state does not depend on the amount of time the excited molecule is in the excited state, it does not depend on which molecule the excitation is on, as long as the molecules are identical. However, the energy transfer can still be observed by measuring the anisotropy of the fluorescence emission. As mentioned in the introduction, the experimental parameter that led Perrin to the idea of intermolecular energy transfer was the decrease in polarization of the fluorescence of higher concentrations of dyes caused by homotransfer [57]. Transfer between identical or near-identical molecules plays a pivotal role in the funneling of the energy into the reaction center of photosynthetic systems. The detection of homo-FRET in microscope images using time-resolved anisotropy imaging microscopy (FLIM) offers a powerful tool for cellular imaging [125].

1.11. Energy transfer by electron exchange

If the charge distributions of the D and A overlap than a new class of interactions has to be considered, namely the exchange interaction between the electrons on D and on A. This type of energy transfer is called Dexter transfer [80, 96, 98]. Here we briefly outline the physical principles involved.

The Dexter transfer probability is determined by the exchange integral:

$$\int \left[\Psi_A^{\mathrm{e}}(\vec{r}_1)\Psi_D^{0}(\vec{r}_2) \right]^* \frac{e^2}{|\vec{r}_1 - \vec{r}_2|} \left[\Psi_A^{0}(\vec{r}_2)\Psi_D^{\mathrm{e}}(\vec{r}_1) \right] \mathrm{d}\vec{r}_1 \mathrm{d}\vec{r}_2 \quad (1.17)$$

The first square bracket describes the final state of the system, where the acceptor is excited, and the donor is not. The second bracket represents the initial state with the donor excited and the acceptor not. However, notice that electron 1 starts on the donor, and after transfer ends up on the acceptor. For this reason this is called transfer by an "exchange" mechanism—the electrons of the D and A exchange. The interaction takes place by the usual Coulomb interaction, $e^2/|\vec{r}_1 - \vec{r}_2|$, between a pair of electrons.

One electron has the charge distribution, $[\Psi_D^0(\vec{r}_2)^* \Psi_A^0(\vec{r}_2)]$, and the other with the charge distribution$[\Psi_A^e(\vec{r}_1)^* \Psi_D^e(\vec{r}_1)]$. If the wave functions centered on A do not overlap with the wave functions centered on D, then these charge distributions are everywhere zero, and the integrals vanish, and the transfer cannot take place via this mechanism. Because the wave functions usually decrease exponentially with distance from the nuclei, the transfer probability will vary roughly exponentially with the D–A separation. Thus this type of transfer only can take place over very short distances.

Acknowledgments

It is a pleasure to thank the many people who have worked on FRET projects and the many scientists with whom I have had many enjoyable discussions. I thank especially Thomas Jovin, David Lilley, György Vámosi, Frank Stümeier, Christoph Gohlke, Oliver Holub, Manfredo Seufferfeld, Govindjee, Sophia Breusegem, Elliot Tan, Zigurts Majumdar, Chittanon Buranachai, and Bryan Spring, all of whom have contributed greatly by personal collaborative experimentation and discussions about FRET. I am grateful to Lambert Chao for careful, critical reading of the manuscript. I also thank many colleagues from other labs for fruitful discussions, and of course I am indebted to those scientists, who, either known personally to me or not, have contributed much to the literature of FRET, which continually guides my interests and endeavors. Of course, I take full final responsibility for what is in the text. I would also like to acknowledge the NIH grant PHS P41 RRO3155-16 for partial support.

References

[1] Agranovich, V. M. and Galanin, M. D. (1982). Electronic Excitation Energy Transfer in Condensed Matter. North-Holland Publishing Company, Amsterdam.

[2] Andrews, D. L. and Demidov, A. A. (eds.) (1999). Resonance Energy Transfer. Wiley, Chichester.

[3] Bennett, G. and Kellogg, R. (1967). Mechanisms and rates of radiationless energy transfer. Prog. React. Kinet. *4*, 215–38.

[4] Cheung, H. (1991). Resonance energy transfer. Topics Fluoresc. Spectr. *3*, 127–76.
[5] Clegg, R. M. (1992). Fluorescence resonance energy transfer and nucleic acids. Methods. Enzymol. *211*, 353–358.
[6] Clegg, R. M. (1996). Fluorescence resonance energy transfer. In: Fluorescence Imaging. Spectroscopy and Microscopy. Vol. 137. (Wang, X. F. and Herman, B., eds.). John Wiley & Sons, New York, pp. 179–252.
[7] Förster, T. (1951). Fluoreszenz Organischer Verbindungen. Vandenhoeck & Ruprecht, Göttingen.
[8] Selvin, P. (1995). Fluorescence resonance energy transfer. Academic Press, San Diego.
[9] Stryer, L. (1978). Fluorescence energy transfer as a spectroscopic ruler. Annu. Rev. Biochem. *47*, 819–46.
[10] Van Der Meer, W. B., Coker, G., III and Chen, S.-Y. (1994). Resonance Energy Transfer: Theory and Data. Wiley, New York.
[11] Wu, P. and Brand, L. (1994). Resonance energy transfer: Methods and applications. Anal. Biochem. *218*, 1–13.
[12] Chen, Y., Mills, J. D. and Periasamy, A. (2003). Protein localization in living cells and tissues using FRET and FLIM. Differentiation *71*, 528–41.
[13] Gadella, T., Jovin, T. and Clegg, R. (1994). Fluorescence lifetime imaging microscopy (FLIM): Spatial resolution of microstructures on the nanosecond time scale. Biophys. Chem. *48*, 221–39.
[14] Harpur, A. G., Wouters, F. S. and Bastiaens, P. I. H. (2001). Imaging FRET between spectrally similar GFP molecules in single cells. Nat. Biotechnol. *19*, 167–9.
[15] Holub, O., Seufferheld, M., Gohlke, C., Govindjee, R. M. and Clegg, R. M. (2000). Fluorescence lifetime-resolved imaging (FLI) in real-time – a new technique in photosynthetic research. Photosynthetica *38*, 581–99.
[16] Holub, O., Seufferheld, M. J., Gohlke, C., Govindjee, G. J., Heiss, G. J. and Clegg, R. M. (2007). Fluorescence lifetime imaging microscopy of Chlamydomonas reinhardtii: Non-photochemical quenching mutants and the effect of photosynthetic inhibitors on the slow chlorophyll fluorescence transients. J. Microsc. *226*, 90–120.
[17] Jares-Erijman, E. A. and Jovin, T. M. (2003). FRET imaging. Nat. Biotechnol. *21*, 1387–95.
[18] Redford, G. and Clegg, R. M. (2005). Real time fluorescence lifetime imaging and FRET using fast-gated image intensifiers. In: Methods in

physiology; molecular imaging: FRET microscopy and spectroscopy (Society, T. A. P., ed.). Oxford University Press, Oxford, pp. 193–226.
[19] van Munster, E. B. and Gadella, T. W. J. (2005). Fluorescence lifetime imaging microscopy (FLIM). In: Microscopy Techniques (Rietdorf, J., ed.). Vol. 95. Springer, Berlin/Heidelberg, pp. 143–75.
[20] Wallrabe, H. and Periasamy, A. (2005). Imaging protein molecules using FRET and FLIM microscopy. Curr. Opin. Biotechnol. *16*, 19–27.
[21] Ha, T. (2001). Single-molecule fluorescence methods for the study of nucleic acids. Curr. Opin. Struct. Biol. *11*, 287–92.
[22] Ha, T. (2001). Single-molecule fluorescence resonance energy transfer. Methods *25*, 78–86.
[23] Buehler, C., Stoeckli, K. and Auer, M. (2001). The integration of single molecule detection technologies into miniturized drug screening: Current status and future perspectives. In: New Trends in Fluorescence Spectroscopy (Valeur, B. and Brochon, J. C., eds.). Springer, Berlin, pp. 331–79.
[24] Chen, Y., Müller, J. D., Eid, J. S. and Gratton, E. (2001). Two photon fluorescence fluctuation spectroscopy. In: New Trends in Fluorescence Spectroscopy. Applications to Chemical and Life Sciences. (Valeur, B. and Brochon, J. C., eds.). Springer, Berlin, pp. 277–96.
[25] Eigen, M. and Rigler, R. (1994). Sorting single molecules: Applications to diagnostics and evolutionary biotechnology. Proc. Natl. Acad. Sci. USA *91*, 5740–9.
[26] Elson, E. and Magde, D. (1974). Fluorescence correlation spectroscopy I: Conceptual basis and theory. Biopolymers *13*, 1–28.
[27] Grunwell, J. R., Glass, J. L., Lacoste, T. D., Deniz, A. A., Chemla, D. S. and Schultz, P. G. (2001). Monitoring the conformational fluctuations of DNA hairpins using single-pair fluorescence resonance energy transfer. J. Am. Chem. Soc. *123*, 4295–303.
[28] Rigler, R., Foldes-Papp, Z., Meyer-Almes, F. J., Sammet, C., Volcker, M. and Schnetz, A. (1998). Fluorescence cross-correlation: A new concept for polymerase chain reaction. J. Biotechnol. *63*, 97–109.
[29] Schwille, P., Bieschke, J. and Oehlenschlager, F. (1997). Kinetic investigations by fluorescence correlation spectroscopy: The analytical and diagnostic potential of diffusion studies. Biophys. Chem. *66*, 211–28.
[30] Blomberg, K., Hurskainen, P. and Hemmilä, I. (1999). Terbium and rhodamine as labels in a homogeneous time-resolved fluorometric energy transfer assay of the b-subunit of human chorionic gonadotropin in serum. Clin. Chem. *45*, 855–61.
[31] Chen, X. and Kwok, P. Y. (1999). Homogeneous genotyping assays for single nucleotide polymorphisms with fluorescence resonance energy transfer detection. Genet. Anal. *14*, 157–63.

[32] Mein, C. A., Barratt, B. J., Dunn, M. G., Siegmund, T., Smith, A. N., Esposito, L., Nutland, S., Stevens, H. E., Wilson, A. J., Phillips, M. S., Jarvis, N. Law, S. *et al.* (2000). Evaluation of single nucleotide polymorphism typing with Invader on PCR amplicons and its automation. Genome Res. *10*, 330–43.
[33] Mere, L., Bennett, T., Coassin, P., Hamman, B., Rink, T., Zimmerman, S. and Nelulescu, P. (1999). Miniturized FRET assays and microfluidics: Key components for ultra-high-throughput screening. Drug Discov. Today *4*, 363–9.
[34] Noeh, S. H., Brisco, M. J., Figaira, F. A., Trainor, K. J., Turner, D. R. and Morley, A. A. (1999). Rapid detection of the factor V Leiden (1691 G > A) and haemochromatosis (845 G > A) mutation by fluorescence resonance energy transfer (FRET) and real time PCR. J. Clin. Pathol. *52*, 766–9.
[35] Oswald, B., Lehmann, F., Simon, L., Terpetschnig, E. and Wolfbeis, O. S. (2000). Red laser-induced fluorescence energy transfer in an immunosystem. Anal. Biochem. *280*, 272–7.
[36] Schobel, U., Egelhaaf, H. J., Brecht, A., Oelkrug, D. and Gauglitz, G. (1999). New donor-acceptor pair for fluorescent immunoassays by energy transfer. Bioconjug. Chem. *15*, 1107–14.
[37] Szöllõsi, J., Damjanovich, S. and Matyus, L. (1998). Application of fluorescence resonance energy transfer in the clinical laboratory: Routine and research. Cytometry *34*, 159–79.
[38] Xu, X., Gerard, A. L. V., Huang, B. C. B., Anderson, D. C. and Payan, D. G. (1998). Detection of programmed cell death using fluorescence energy transfer. Nucl. Acid. Res. *26*, 2034–5.
[39] Gordon, G. W., Berry, G., Liang, X. H., Levine, B. and Herman, B. (1998). Quantitative fluorescence resonance energy transfer measurements using fluorescence microscopy. Biophys. J. *74*, 2702–13.
[40] Jovin, T. and Arndt-Jovin, D. (1989). FRET microscopy: Digital imaging of fluorescence resonance energy transfer. Application in cell biology. In: Cell Structure and Function by Microspectrofluorometry (Kohen, E., JG, H. and Ploem, J., eds.). Academic Press, London, pp. 99–117.
[41] Jovin, T. and Arndt-Jovin, D. (1989). Luminescence digital imaging microscopy. Ann. Rev. Biophys. Chem. *18*, 271–308.
[42] Periasamy, A. (2001). Fluorescence resonance energy transfer microscopy: A mini review. J. Biomed. Opt. *6*, 287–91.
[43] Vickery, S. A. and Dunn, R. C. (1999). Scanning near-field fluorescence resonance energy transfer microscopy. Biophys. J. *76*, 1812–8.

[44] Xia, Z. and Liu, Y. (2001). Reliable and global measurement of fluorescence resonance energy transfer using fluorescence microscopes. Biophys. J. *81*, 2395–402.
[45] Cacciatore, T. W., Brodfuehrer, P. D., Gonzalez, J. E., Jiang, J. E., Adams, S. R., Tsien, R. Y., Kristan, W. B. and Kleinfeld, D. (1999). Identification of neural circuits by imaging coherent electrical activity with FRET-based dyes. Neuron *23*, 449–59.
[46] Knowles, R. B., Chin, J., Ruff, C. T. and Hyman, B. T. (1999). Demonstration by fluorescence resonance energy transfer of a close association between activated MAP kinase and neurofibrillary tangles: Implications for MAP kinase activation in Alzheimer disease. J. Neuropathol. Exp. Neurol. *58*, 1090–8.
[47] Miyawaki, A. and Tsien, R. (2000). Monitoring protein conformations and interactions by fluorescence resonance energy transfer between mutants of green fluorescent protein. Methods Enzymol. *2000*, 472–500.
[48] Pollok, B. A. and Heim, R. (1999). Using GFP in FRET-based applications. Trends Cell Biol. *9*, 57–60.
[49] Hickerson, R., Majumdar, Z. K., Baucom, A., Clegg, R. M. and Noller, H. F. (2005). Measurement of internal movements within the 30 S ribosomal subunit using Forster resonance energy transfer. J. Mol. Biol. *354*, 459–72.
[50] Majumdar, Z., Hickerson, R., Noller, H. and Clegg, R. (2005). Measurements of internal distance changes of the 30S ribosome using FRET with multiple donor-acceptor pairs: Quantitative spectroscopic methods. J. Mol. Biol. *351*, 1123–45.
[51] Förster, T. (1946). Energiewanderung und Fluoreszenz. Naturwissenschaften *6*, 166–75.
[52] Förster, T. (1947). Fluoreszenzversuche an Farbstoffmischungen. Angew. Chem. A *59*, 181–7.
[53] Förster, T. (1948). Zwischenmolekulare Energiewanderung und Fluoreszenz. Ann. Phys. *2*, 55–75.
[54] Förster, T. (1949). Expermentelle und theoretische Untersuchung des zwischengmolekularen Übergangs von Elektronenanregungsenergie. A. Naturforsch. *4A*, 321–7.
[55] Förster, T. (1949). Versuche zum zwischenmolekularen Übergang von Elektronenanregungsenergie. Z. Elektrochem. *53*, 93–100.
[56] Perrin, F. (1933). Interaction entre atomes normal et activité. Transferts d'activitation. Formation d'une molécule activitée. Ann. Institut Poincaré *3*, 279–318.
[57] Perrin, J. (1927). Fluorescence et induction moleculaire par resonance. C. R. Hebd. Seances Acad. Sci. *184*, 1097–100.

[58] Clegg, R. M. (2006). The history of FRET. In: Reviews in Fluorescence. (Geddes, C. D. and Lakowicz, J. R., eds.). Vol. 3. Springer, New York, pp. 1–45.
[59] Kallmann, H. and London, F. (1928). Über quantenmechanische Energieübertragungen zwischen atomaren Systemen. Z. Physik. Chem. *B2*, 207–43.
[60] Cairo, G. (1922). Über Entstehung wahrer Lichtabsorption und scheinbare Koppelung von Quantensprüngen. Z. Phys. *10*, 185–99.
[61] Cairo, G. and Frank, J. (1922). Über Zerlegugen von Wasserstoffmolekülen durch angeregte Quecksilberatome. Z. Phys. *11*, 161–6.
[62] Frank, J. (1922). Einige aus der Theorie von Klein und Rosseland zu ziehende Folgerungen über Fluorescence, photochemische Prozesse und die Electronenemission glühender Körper. Z. Physik. *9*, 259–66.
[63] Pringsheim, P. (1928). Luminescence und Phosphorescence im Lichte der neueren Atomtheorie. Interscience, Berlin.
[64] Berberan-Santos, M. N. (2001). Pioneering contributions of Jean and Francis Perrin to molecular luminescence. In: New Trends in Fluorescence Spectroscopy (Valeur, B. and Brochon, J. C., eds.). Springer, Berlin, pp. 7–33.
[65] Perrin, F. (1932). Théorie quantique des transferts d'activation entre molécules de méme espèce. Cas des solutions fluorescentes. Ann. Chim. Phys. (Paris) *17*, 283–314.
[66] Förster, T. (1993). Intermolecular energy migration and fluorescence. In: Biological Physics (Mielczarek, E. V., Greenbaum, E. and Knox, R. S., eds.). American Institute of Physics, New York, pp. 148–60.
[67] Atkins, P. W. and Friedman, R. S. (1997). Molecular Quantum Mechanics. Oxford University Press, Oxford.
[68] Landau, L. D. and Lifshitz, E. M. (1965). Quantum Mechanics. Non-Relativistic Theory. Pergamon Press, Oxford.
[69] Förster, T. (1965). Delocalized excitation and excitation transfer. In: Modern Quantum Chemistry; Part III; Action of Light and Organic Molecules. (Sunanoglu, O., ed.). Vol. 3. Academic, New York, pp. 93–137.
[70] Kenkre, V. M. and Knox, R. S. (1974). Generalized master-equation theory of excitation transfer. Phys. Rev. B *9*, 5279–90.
[71] Kenkre, V. M. and Knox, R. S. (1974). Theory of fast and slow excitation transfer rates. Phys. Rev. Lett. *33*, 803–6.
[72] Leegwater, J. A. (1996). Coherent versus incoherent energy transfer and trapping in photosynthetic antenna complexes. J. Phys. Chem. *100*, 14403–9.

[73] May, V. and Kühn, O. (2004). Charge and Energy Transfer Dynamics in Molecular Systems. Wiley-VCH Verlag GnbH & Co. KGaA, Weinheim.
[74] Engel, G. S., Calhoun, T. R., Read, E. L., Manal, T., Cheng, Y.-C., Blankenship, R. E. and Fleming, G. R. (2007). Evidence for wavelike energy transfer through quantum coherence in photosynthetic systems. Nature *446*, 782–6.
[75] Lee, H., Cheng, Y.-C. and Fleming, G. R. (2007). Coherence dynamics in photosynthesis: Protein protection of excitonic coherence. Science *316*, 1462–5.
[76] Klimov, V., Sekatskii, S. K. and Dietler, G. (2004). Coherent fluorescence resonance energy transfer between two dipoles: Full quantum electrodynamics approach. J. Mod. Opt. *51*, 1919–47.
[77] Renger, T., May, V. and Kühn, O. (2001). Ultrafast excitation energy dynamics in photosynthetic pigment protein complexes. Phys. Rep. *343*, 137–254.
[78] Oppenheimer, J. R. (1941). Internal conversion in photosynthesis. Phys. Rev. *60*, 158.
[79] Arnold, W. and Oppenheimer, J. R. (1950). Internal conversion in the photosynthetric mechanism of blur-green algae. J. Gen. Physiol. *33*, 423–35.
[80] Dexter, D. (1953). A theory of sensitized luminescence in solids. J. Chem. Phys. *21*, 836–50.
[81] Birks, J. B. (1970). Photophysics of Aromatic Molecules. Wiley, London.
[82] Lakowicz, J. R. (1999). Principles of Fluorescence Spectroscopy. Kluwer Academic, New York.
[83] Mataga, N. and Kubota, T. (1970). Molecular Interactions and Electronic Spectra. Marcel Dekker, Inc., New York.
[84] Stepanov, B. I. and Gribkovskii, V. P. (1968). Theory of Luminescence. ILIFFE Books Ltd., Bristol.
[85] dos Remedios, C., Miki, M. and Barden, J. (1987). Fluorescence resonance energy transfer measurements of distances in actin and myosin. A critical evaluation. J. Muscle Res. Cell Motil. *8*, 97–117.
[86] van der Meer, B. W. (2002). Kappa-squared: From nuisance to new sense. J. Biotechnol. *82*, 181–96.
[87] Miyawaki, A. (2003). Visualization of the spatial review and temporal dynamics of intracellular signaling. Dev. Cell *4*, 295–305.
[88] Nagai, T., Yamada, S., Tominaga, T., Ichikawa, M. and Miyawaki, A. (2004). Expanded dynamic range of fluorescent indicators for Ca2+ by circularly permuted yellow fluorescent proteins. Proc. Natl. Acad. Sci. USA *101*, 10554–9.

[89] Dale, R., Eisinger, J. and Blumberg, W. (1979). The orientational freedom of molecular probes. The orientation factor in intramolecular energy transfer. Biophys. J. *26*, 161–94.
[90] Ketskemety, I. (1962). Zwischenmolekulare Energieubertragung in fluoreszierenden Losungen. Z. Naturforsch. *17A*, 666–70.
[91] Kuhn, H. (1970). Classical aspects of energy transfer in molecular systems. J. Chem. Phys. *53*, 101–8.
[92] Rojansky, V. (1979). Electromagnetic Fields and Waves. Dover Publications, Inc., New York.
[93] Scott, W. T. (1966). The Physics of Electricity and Magnetism. Wiley, New York.
[94] Shadowitz, A. (1975). The Electromagnetic Field. Dover Publications, Inc., New York.
[95] Dexter, D. L. and Schulman, J. H. (1954). Theory of concentration quenching in inorganic phosphores. J. Chem. Phys. *22*, 1063–70.
[96] Ganguly, S. and Chaudhury, N. (1959). Energy transport in organic phosphors. Rev. Mod. Phys. *31*, 990–1017.
[97] Inacker, O., Kuhn, H., Bucher, H., Meyer, H. and Tews, K. H. (1970). Monolayer assembling technique used to determine the multipole nature of the phosphorescence of a dye molecule. Chem. Phys. Lett. *7*, 213–8.
[98] Dexter, D. L. and Knox, R. S. (1965). Excitons. Intrerscience Publishers, New York.
[99] Chen, S.-H. and Kotlarchyk, M. (1997). Interactions of Photons and Neutrons with Matter - An Introduction. World Scientific, New Jersey.
[100] Craig, D. P. and Thirunamachandran, T. (1984). Molecular Quantum Electrodynamics. An Introduction to Radiation Molecule Interactions. Dover Publications, Inc., Mineola.
[101] Schiff, L. (1968). Quantum Mechanics. McGraw Hill Book Co., New York.
[102] Heitler, W. (1984). The Quantum Theory of Radiation. Dover, New York.
[103] Haas, E. and Steinberg, I. (1984). Intramolecular dynamics of chain molecules monitored by fluctuations in efficiency of excitation energy transfer. Biophys. J. *46*, 429–37.
[104] Morin, J. G. and Hastings, J. W. (1971). Energy transfer in a bioluminescent system. Cell Phys. *77*, 313–8.
[105] Wilson, T. and Hastings, J. W. (1998). Bioluminescence. Annu. Rev. Cell Dev. Biol. *14*, 197–230.
[106] Valeur, B. (2002). Molecular Fluorescence; Principles and Applications. Wiley-VCH, New York.
[107] Damjanovic, A., Ritz, T. and Schulten, K. (2000). Excitation transfer in the peridinin-chlorophyll-protein of Amphidinium carterae. Biophys. J. *79*, 1695–705.

[108] Förster, T. (1959). Transfer mechanisms of electronic excitation. Discuss. Faraday Soc *27*, 7–17.
[109] Förster, T. (1960). Transfer mechanisms of electronic excitation energy. Radiat. Res. Suppl. *2*, 326–39.
[110] Ray, J. and Makri, N. (1999). Short range coherence in the energy transfer of photosynthetic light harvest-ing systems. J. Phys. Chem. *103*, 9417.
[111] Schulten, K. (ed.) (1999). In From Simplicity to Complexity and Back. Function, Architecture, and Mechanism of Light-Harvesting Systems in Photosynthetic Systems. Simplicity and Complexity in Proteins and Nucleic Acids. Dahlem University Press.
[112] Benson, D., Bryan, J., Plant, A., Gotta, A. J. and Smith, L. (1985). Digital imaging fluorescence microscopy: Spatial heterogeneity of photobleaching rate constants in individual cells. J. Cell Biol. *100*, 1309–23.
[113] Hirschfeld, T. (1976). Quantum efficiency independence of the time integrated emission from a fluorescent molecule. Appl. Opt. *15*, 3135–9.
[114] Kenworthy, A. K., Petranova, N. and Edidin, M. (2000). High resolution FRET microscopy of cholera toxin B-subunit and GPI-anchored proteins in cell plasma membranes. Mol. Biol. Cell *11*, 1645–55.
[115] Szabo, G., Pine, P., Weaver, J., Kasari, M. and Aszalos, A. (1992). Epitope mapping by photobleaching fluorescence resonance energy transfer measurements using a laser scanning microscope system. Biophys. J. *61*, 661–70.
[116] Young, R., Arnette, J., Roess, D. and Barisas, B. (1994). Quantitation of fluorescence energy transfer between cell surface proteins via fluorescence donor photobleaching kinetics. Biophys. J. *67*, 881–8.
[117] Kuhn, H. and Möbius, D. (1993). Monolayer assemblies. In: Investigations of Surfaces and Interfaces. (Rossiter, B. and Baetzold, R., eds.). Part B. Vol. 9B. Wiley, New York, pp. 375–542.
[118] Kuhn, H., Mobius, D. and Bucher, H. (1972). Spectroscopy of monolayer assemblies. In: Phys Methods of Chem. (Weissberger, A. and Rossiter, B., eds.). Vol. 1. Wiley, New York, pp. 577–702.
[119] Barth, P., Beck, K., Drexhage, K., Kuhn, H., Mobius, D., Molzahn, D., Rollig, K., Schafer, F., Sperling, W. and Zwick, M. (1966). Optische und elektrische Phanomen an monomolekularen Farbstoffschichten. Verlag Chemie, Weinheim.
[120] Drexhage, K., Zwick, M. and Kuhn, H. (1963). Sensibilisierte Fluoreszenz nach strahlungslosem Energieubergang durch dunne schiechten. Ber. Bunsenges Phys. Chem. *67*, 62–7.

[121] Kuhn, H. (1965). Versuche zur Herstellung einfacher organisierter Systeme von Molekulen. Pure Appl. Chem. *11*, 345–57.

[122] Mekler, V. (1994). A photochemical technique to enhance sensitivity of detection of fluorescence resonance energy transfer. Photochem. Photobiol. *59*, 615–20.

[123] Marriott, G., Clegg, R. M., Arndt-Jovin, D. J. and Jovin, T. M. (1991). Time resolved imaging microscopy. Phosphorescence and delayed fluorescence imaging. Biophys. J. *60*, 1374–87.

[124] Marriott, G., Heidecker, M., Diamandis, E. P. and Yan-Marriott, Y. (1994). Time-resolved delayed luminescence image microscopy using an europium ion chelate complex. Biophys. J. *67*, 957–65.

[125] Clayton, A. H. A., Hanley, Q. S., Arndt-Jovin, D. J., Subramaniam, V. and Jovin, T. M. (2002). Dynamic fluorescence anisotropy imaging microscopy in the frequency domain (rFLIM). Biophys. J. *83*, 1631–49.

Laboratory Techniques in Biochemistry and Molecular Biology, Volume 33
FRET and FLIM Techniques
T. W. J. Gadella (Editor)

CHAPTER 2

Frequency domain FLIM theory, instrumentation, and data analysis

Peter J. Verveer[1] and Quentin S. Hanley[2]

[1]*Department of Systemic Cell Biology, Max Planck Institute of Molecular Physiology, Otto Hahn Straße 11, D-44227 Dortmund, Germany*
[2]*School of Science and Technology, Nottingham Trent University, Nottingham NG11 8NS, UK*

Practical frequency domain fluorescent lifetime imaging microscopy has enjoyed considerable success in the analysis of biological systems. Appreciation of the strength of the method requires an understanding of kinetics, instrumentation, calibration, data processing, and subsequent analysis. This chapter presents an overview of the governing equations of frequency domain lifetime imaging, specifically: apparent lifetimes, fluorophore mixtures, chi-squared minimization, discrete Fourier processing, and treatment of binary mixtures. Emphasis is placed on instrumentation variations, trends in the field, and finishes with methods of interest for the future.

While publications on fluorescence lifetime imaging microscopy (FLIM) have been relatively evenly divided between time and frequency domain methods, a majority of the 10 most highly cited papers using FLIM have taken advantage of the frequency domain method [1, 2–9]. Both techniques have confronted similar challenges as they have developed and, as such, common themes may be found in both approaches to FLIM. One of the most important criteria is to retrieve the maximum information out of a FLIM

DOI: 10.1016/S0075-7535(08)00002-8

measurement in the shortest period of time. There are many reasons for this interest: minimization of motion artifacts, reduction of sample changes during measurement, generation of time series data sets, and photochemical damage to the specimen [9–15].

In the time domain, this has driven investigations of methods for sampling the fluorescence decay with time in a variety of ways—a topic treated elsewhere in this book. Similarly, in the frequency domain, rapid and minimally intrusive measurement involves a set of tradeoffs between lifetime accuracy and speed. Recently, considerable effort has gone into obtaining the most information about a sample from measurements made at a single frequency [16–20]. To better appreciate these developments and the instrumentation that underpins them, this chapter will review the theory of frequency domain measurements, describe instrumentation for making the measurement, look at some of the emerging areas in the technique, and discuss analysis strategies. The analysis of fluorescence resonance energy transfer using FLIM in single cells is demonstrated with an example.

The chapter will present data primarily from wide field frequency domain measurements; however, a wide range of confocal solutions in the frequency domain have been reported, including: point scanning [21], programmable arrays [22], and spinning disks [23]. The principles of frequency domain lifetime imaging in both wide field and confocal measurements are the same, so the treatment is this chapter is general. The implementation of all types is very similar with the exception of the point scanning reference [21], which uses a lock-in technique. Readers interested in the lock-in approach should consult the original report. The goal of this chapter is to present the collection and processing of frequency domain data and briefly describe subsequent processing of single or multiple lifetime images to provide information to users. This last part is particularly important as there are a growing number of FLIM users who have little or no fundamental interest in fluorescence lifetimes. These users are interested in answers to questions such as: "Is my protein phosphorylated?" "Are my proteins interacting?" "Has my receptor dimerized?" "Where and when is my protein activating its partner?" For these users, the fluorescence lifetime is simply a tool to answer a biological question and the goal is to provide these users with answers as unambiguously as possible.

2.1. Rates, time constants, and lifetimes

When a molecule absorbs a photon of light, this mediates a process in which an electron is promoted from a ground state to an excited state. The energy difference between the two states is equal to the energy of the photon absorbed. The molecule spends some period of time in the excited state before returning to the ground state. The amount of time that the molecule, on an average, spends in the excited state is called the fluorescence lifetime. The excited state may deactivate through a variety of processes (see also Chapters 1 and 12). If the molecule returns to the ground state with the emission of a photon, the process is said to be radiative. If no light is released, it is said to be nonradiative. Lifetime imaging makes it possible to investigate the rates of molecules returning to the ground state by observing the behavior of emitted light. Measured fluorescence lifetimes change in response to a variety of processes. Chemical changes in the structure of a molecule, such as addition of a proton, can result in a fundamental change in the rate constant of emission. More commonly, measured fluorescence lifetimes change in response to nonradiative processes competing with the radiative rate constant for deactivation of the excited state. The two most important nonradiative processes are dynamic quenching of the fluorophore and fluorescence resonance energy transfer (FRET). Quenching effects assess the degree of accessibility of a fluorophore to a quencher. FRET is an indicator of proximity between a fluorophore and the molecule that accepts the excited state energy.

For a simple fluorophore, the release of light proceeds through a first-order process. By this we mean that the rate of return to the ground state depends on the number of molecules in the excited state times a rate constant.

$$\frac{dM^*}{dt} = -kM^* \tag{2.1}$$

In this expression, M^* is the number of molecules in the excited state and k is the first-order rate constant. The differential equation has a solution of the form (see also Chapter 1, Section 1.5.2):

$$M(t) = M_0 e^{-kt} \tag{2.2}$$

The fluorescence lifetime, τ, is the reciprocal of the rate constant k and is the parameter of interest for lifetime imaging:

$$\tau = \frac{1}{k} \tag{2.3}$$

An intuitive way to measure the rate constant, k, is to prepare some number of molecules in the excited state, perhaps with a pulse of light, and observe the amount of light given off per unit of time after the pulse. A logarithmic plot of intensity with time will give a straight-line graph with a slope of $-k$. This approach is the basis of time domain measurements.

If, instead of a pulse of light, a light source modulated at a particular frequency or set of frequencies is used to excite the molecules, the result is a frequency domain measurement. Although frequency domain methods for the analysis of fluorescence decays appeared as early as 1927 with the work of Gaviola [24], the modern development of this approach began with the work of Spencer and Weber [25]. The structure of the modulation function with time can provide information from one or many frequencies. Typically, experiments are done by modulating the excitation light source using a sinusoidal function with a single frequency. This results in a modification of Eq. (2.1) to include a driving function that adds a time-dependent increase in the number of excited state molecules [26, 27].

$$\frac{dM^*}{dt} = -kM^* + f(t) \tag{2.4}$$

The term $f(t)$ is the time-dependent driving function provided by the modulated light source. This differential equation may be solved in terms of the measured signal relative to the driving function. Details of the solutions to these types of equations for a variety of systems may be found in the literature (c.f.: [27]). For lifetime imaging, the goal is a set of measurements allowing the recovery of the rate constant or set of constants. For a single k, the lifetime may be computed from the modulation and phase of the driving function and the equivalent values obtained from the sample. The apparent decay rates are computed from m and ϕ:

$$k = \frac{\omega}{\tan\phi} \text{ and } k = \omega\sqrt{\frac{m^2}{1-m^2}} \tag{2.5}$$

With the rate constant in hand, the lifetime can be reported. Rearrangement of these expressions yields the more familiar expression for phase and modulation lifetimes in terms of τ.

$$\tau_\phi = \frac{1}{\omega} \tan\phi \tag{2.6}$$

$$\tau_m = \frac{1}{\omega}\sqrt{\frac{1}{m^2} - 1} \tag{2.7}$$

where ω is the circular frequency of modulation ($= 2\pi \times$ frequency in Hz). These two estimations yield the same value if only a single lifetime is present, that is, if the decay is truly monoexponential. This will be the case if there is a single fluorescent species of fluorophore with a monoexponential decay. In special cases, including energy transfer or excited state reactions, a complex mixture may also cause the two estimators to yield a single value. In most cases of interest, there will be multiple species that possibly have nonexponential decays. In such a case, these single-frequency lifetime estimations are a function of the lifetimes of the various fluorescent species and their respective amounts. Nevertheless, Eqs. (2.6 and 2.7) are informative, as with an increasing amount of short-lifetime species, these estimations will become lower. For a single species, the lifetimes are reasonably constant with wavelength. However, mixtures of fluorophores will exhibit wavelength-dependent behavior.

Similar expressions can be derived for systems of fluorophores having different rate constants, the details of the mathematics are not important here. A few rules have been derived for mixing of signals from such systems. The modulation and phase for a system of multiple noninteracting fluorophores that are not undergoing excited state reactions can be computed from a sum related to the fractional contribution of the individual fluorophores, the lifetime, and the modulation frequency. The fractional contributions f_i, for the ith fluorophore in a mixture is given by:

$$f_i = \frac{I_{0i}}{\sum_{i=1}^{n} I_{0i}} = \frac{\alpha_i \tau_i}{\sum_{i=1}^{n} \alpha_i \tau_i} \tag{2.8}$$

In this expression, I_{0i} is the fluorescence intensity calculated for the ith fluorophore at a particular wavelength and α_i is the corresponding amplitude of its exponential decay.

For a mixture of n directly excited noninteracting fluorophores, the apparent phase and modulation may be calculated according to Eqs. (2.9–2.11):

$$N = \frac{\sum_{i=1}^{n} \frac{\alpha_i \omega \tau_i^2}{1 + \omega^2 \tau_i^2}}{\sum_{i=1}^{n} \alpha_i \tau_i} \tag{2.9a}$$

$$D = \frac{\sum_{i=1}^{n} \frac{\alpha_i \tau_i}{1 + \omega^2 \tau_i^2}}{\sum_{i=1}^{n} \alpha_i \tau_i} \tag{2.9b}$$

$$\phi = \tan^{-1}\left(\frac{N}{D}\right) \tag{2.10}$$

$$m = \sqrt{N^2 + D^2} \tag{2.11}$$

Practically speaking, these expressions allow the prediction of the phase and modulation for an arbitrary mixture of noninteracting fluorophores and the respective modulation and phase lifetimes.

It is important to note that if a mixture of fluorophores with different fluorescence lifetimes is analyzed, the lifetime computed from the phase is not equivalent to the lifetime computed from the modulation. As a result, the two lifetimes are often referred to as "apparent" lifetimes and should not be confused with the true lifetime of any particular species in the sample. These equations predict a set of phenomena inherent to the frequency domain measurement.

Consequences of Eqs. (2.9–2.11) are:

1. A mixture of noninteracting fluorophores might be observed by spatially variant FRET in a specimen, which is blurred because of optical resolution issues, will result in different lifetimes being measured for τ_m and τ_f and $\tau_m > \tau_f$. In many instances, a single frequency measurement will be insufficient to determine the number of fluorophores or the number of fluorophore environments in a sample.
2. Equations (9–11) are dependent on three parameters: ω, α_i, and τ_i. All may be exploited in the analysis of mixtures of fluorophores; however, ω is the only one that can be systematically varied without altering the sample.
3. Most reported measurements are a tradeoff between speed of image collection, photobleaching, and operator convenience. The limitations of variously designed instruments are consequences of these defining equations.

In the next section of this chapter, we will review a variety of instrumentation approaches to the FLIM experiment. In particular, we describe conventional systems as well as those designed to observe variation in α, and systems designed for the collection of multifrequency data. In this context, we will also look at data collection strategies and the subsequent first pass analysis of the acquired images using Fourier transforms or sinusoidal fitting. The two strategies have slightly different needs for data collection. Sinusoidal fitting will be successful for any number of phase steps, the phase steps need not be equally spaced, and the sum of the phase steps can be quite flexible (e.g., π, 2π, 4π, etc.), with no special conditions placed on the subsequent analysis. Discrete Fourier transform (DFT) processing requires equally spaced phase steps. Depending on the implementation, changing the number of steps and the total distance over which those steps are taken may require special processing and programming.

2.2. Instrumentation

The instrumentation for frequency domain fluorescence lifetime imaging consists of a modulatable light source, the microscope and associated optics, and a modulatable detector. Traditionally, the FLIM light source has been a laser modulated with an acousto-optic modulator (AOM) [15, 28, 29], although pulsed laser sources have also been used in frequency domain systems [30]. Recently, LEDs have appeared as an alternative to lasers in frequency domain FLIM [31–34]. There are a variety of advantages of the LED over the laser. The cost of LEDs and repair of LED equipment is substantially less than that of lasers. Various publications have described the use of LEDs and advances in LED sources have produced devices with high output across wide ranges of the UV–vis spectrum. Integrated LED systems are available commercially for the determination of lifetimes. [35] A disadvantage of LEDs is the more extended nature of the LED as a source. This tends to lead to a lower flux reaching the sample. In comparison, traditional systems using AOMs and ion lasers will be more costly. A few noteworthy points about light sources for the frequency domain lifetime measurement:

1. Although LEDs are emerging as a likely dominant light source for FLIM, there are some regions of the spectrum

where lasers are still the best choice. Since the technologies are advancing rapidly, before purchasing components for the UV portion of the spectrum, a quick review of the literature should be done.

2. In the UV, standard LEDs are still relatively untested and either the UV lines of the Ar+ ion laser or directly modulatable LED lasers are likely the component of choice at time of writing.
3. Use of a laser that is not directly modulatable will require the use of a suitable modulator. Although a number of technologies are available for FLIM, the only one in widespread use is the AOM. The AOM must be on resonance and finding a suitable frequency is facilitated by access to a high-speed photodetector. AOMs must be temperature controlled or the modulation depth and phase will drift (perhaps rapidly). The modulated signal from an AOM is at twice the driving frequency and is found in the zero-order position of the AOM output. The zero- and higher-order beams exiting the AOM diverge at relatively low angles and require roughly 1–2 meters to separate. As a result, the AOM modulated system is likely to require more space, an optical bench, and occasional beam adjustment.
4. High-power argon ion or similar lasers require cooling, high voltage, and skilled technical assistance. These points should be taken into account when computing the cost of a system.

Over the next 1–5 years, it is likely that nearly all frequency domain FLIM will migrate to LED light sources. Unless a user has a specialty application requiring a traditional high-power laser, the added cost will likely be unjustified. Specialty applications that might justify the laser are very high speed—the best LEDs do not, at present, bring as much light to the object plane of a microscope as a laser—or UV-excitation in the range required to excite the intrinsic fluorescence of proteins or molecules with similar UV-excitation requirements.

2.3. Instrumentation: Frequency domain FLIM

FLIM systems can be purchased as an add-on for a standard fluorescence microscope. Such a system will consist of a CCD camera coupled to a modulatable image intensifier, an LED light source, and driver electronics. This system will modulate the LED and image intensifier while shifting the phase between them as it takes a series of images (Fig. 2.1).

An alternative to purchasing a turnkey system is to construct a similar system from components. CCD cameras are available from a wide range of suppliers. For FLIM applications, the CCD need not have particularly high QE or exceptionally low noise.

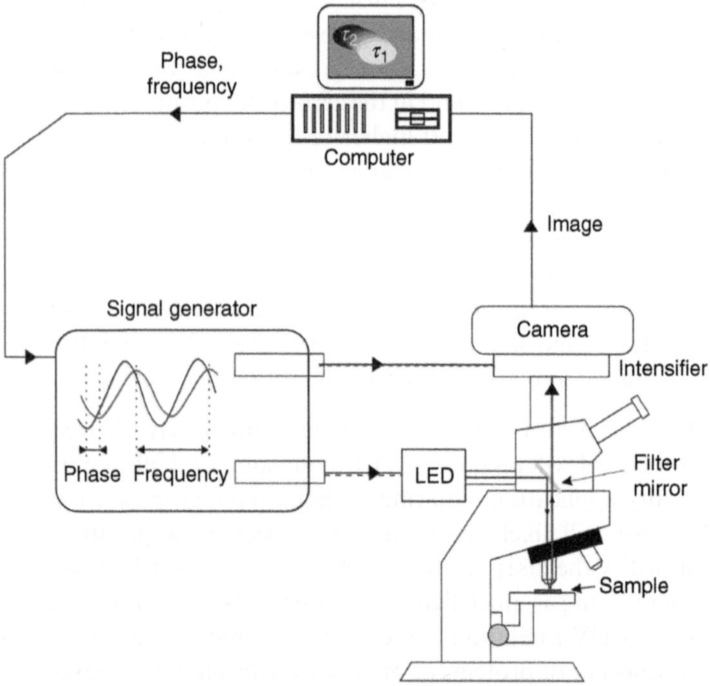

Fig. 2.1. Frequency domain FLIM system.

The image intensifier will define the overall noise characteristics and QE of the system. The CCD camera should be chosen so as to have a good response at the emission wavelengths of the phosphor in the intensifier and, if there is a demand for high frame rates, the CCD should be chosen to support this. Image intensifiers suitable for frequency domain FLIM are available from Hamamatsu, Kentech, and Lambert Instruments. There is wide variation in cost and flexibility between these suppliers. If camera and intensifier are purchased separately, thought should be given to coupling the intensifier to the camera. A pair of standard high-quality camera lenses can be used to accomplish this task. Usually this will require c-mount adapters and a small amount of machining. Some manufacturers provide a unit consisting of a suitable intensifier precoupled to a CCD camera.

The intensifier manufacturers listed above provide electronics to drive the intensifier at the frequency of an input signal. Generation of the input signal requires a precisely controlled frequency synthesizer. Two such synthesizers will be needed for many applications to independently control the intensifier and the light source. For example, when modulating an AOM, the frequency must be half the desired modulation frequency and may need to be at higher power than the signal driving the image intensifier. In the literature, units made by Marconi [3, 22], Hewlett Packard, and Rhode & Schwarz [29] are often cited; however, some of these brands are currently available. The Marconi synthesizers are currently sold by Aeroflex under the IFR Systems name. The test and measurement portion of Hewlett Packard has been reorganized under the Agilent name.

A similar arrangement may be used for driving an AOM to modulate a laser. An AOM and amplifier may be purchased together. The main difference between driving an LED and an AOM from an operational point of view is that the AOM is driven at half the frequency of modulation. As noted earlier, the AOM does not support continuous frequency modulation. A frequency must be chosen close to the desired frequency at the position of a node in the AOM response. Typically, these are regularly spaced about 100–200 kHz apart. This makes finding a convenient node relatively straightforward; however, the position of the AOM nodes must be found and tabulated and the temperature of operation matched to that of the test conditions.

2.4. Systems for measuring lifetimes at multiple frequencies

Examination of Eqs. (2.9–2.11) suggests that having frequency domain lifetimes measured at a variety of frequencies is desirable, as it will allow a mixture of fluorophores to be determined. With this in mind, two approaches may be taken to obtain multifrequency results. The first of these is simply to make a series of FLIM measurements while stepping through a predetermined set of frequencies. In practice, this is of limited utility for biological systems because of photo-induced damage to the specimen.

A second multiplexed approach is to use a nonsinusoidal excitation source, by employing pulsed light sources or multiple AOMs and a nonsinusoidal modulation of the detector, which leads to the presence of multiple harmonics in the FLIM signal. These harmonics can be separated using Fourier methods, allowing measurement at several frequencies. This approach appears to have been reported only once by Squire *et al.* where he used two AOMs placed in series, and modulated an image intensifier in a block-wave fashion [36]. From a theoretical point of view, there appear to be few disadvantages of the multiplexed approach to obtaining multiple frequencies. The barrier to more widespread application is, at present, the complexity of the hardware.

2.5. Spectral FLIM

Spectral FLIM involves measuring the apparent lifetimes in a preparation at many wavelengths with the assistance of a spectrograph or a series of filters (see also Chapter 4, Figs. 4.7 and 4.8 depicting hyperspectral FLIM in the time domain). The goal of the measurement is similar to that of the multifrequency approach:

determining a mixture of lifetimes with the assistance of Eqs. (2.9–2.11), except exploiting spectral variations in α rather than using variations in frequency. Provided the fluorophores are spectrally dissimilar enough, variation in α is of similar value to variation in frequency. However, variations in α in a sample or a set of dyes tend to be fortuitous rather than systematically adjustable like the modulation frequency.

Instrumentally, spectral FLIM generates a spectrally resolved set of lifetimes by either introducing filters to provide spectral resolution or a spectrograph between the sample and image intensifier. The first such system was created for looking at the long lifetimes of lanthanide dyes [37]. Later, a spectral FLIM system was described for measuring from a two-dimensional (2D) area of a microscope field [38]. Simpler systems have also been described and adopted by others [39]. Introducing the spectrograph is relatively straightforward compared with the difficulty of assembling and programming a FLIM system and may be completed at reasonable cost (Fig. 2.2).

There are a number of practical issues associated with introduction of a spectrograph. A spectrograph disperses the light from a single point into a spectrum, which is measured as several points. As a result, the total light in the spectrum will be equal to the light from a single point in a standard FLIM image. The more spectral elements the system has, the less signal will be observed in each of

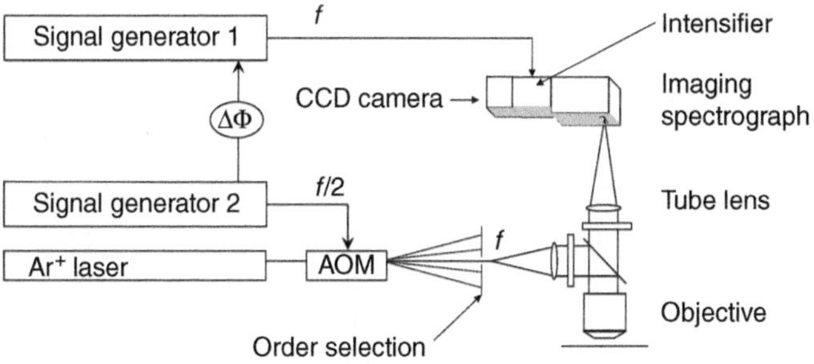

Fig. 2.2. Imaging Spectroscopic FLIM system.

them. As a result, users should plan on taking longer to acquire data, should be realistic about the number of spectral segments they wish to collect, and should bin the camera in the spectral direction. The only alternative to longer exposures and binning is to increase the slit width of the spectrograph. Increasing the slit width is best avoided, since microscope optics can produce features smaller than the slit width if it is increased too much. Often this will not be a problem; however, it can result in small shifts in the position of spectral features in the resulting images. To avoid this, a slit width should be selected that samples the image plane of the microscope well. A slit around 10 μm will work well for many cases.

2.6. Data acquisition strategies

A serious practical limitation on lifetime measurements of all types is photodestruction of the sample. Photodamage can take many forms. The primary fluorophore of interest may simply bleach making the sample unusable. This is perhaps the least worrying because it is readily apparent. Other mechanisms of light-induced change may be more subtle. An acceptor may bleach, leaving the donor apparently unaffected but the measurement altered. Fluorescent proteins have been shown to undergo photoconversion between different forms with different lifetimes. This may be sufficient to alter a result. Most of the time this photon-induced change is undesirable.

A variety of strategies to mitigate sample photodamage have been described. The most widely used is a procedure in which the data is collected twice while reversing the direction of the phase steps. This approach works for minimal photobleaching. Photobleaching is a kinetic process similar to the first-order kinetics discussed earlier in the chapter and exhibits an exponential decrease with time. Over a short time period, the exponential decay because of photobleaching may be treated as approximately linear. By summing up the two measurements at a particular phase shift in the acquisition sequence, approximately linear photobleaching and related processes can be corrected.

The disadvantage of this procedure is that it doubles the amount of time required to acquire a data set, doubles the size of the data set, and exposes the sample to twice as much excitation light. More recently, an approach has been described in which phase step acquisition is randomized so that there is no trend in photobleaching or related effects in the analysis [29]. This approach is relatively new, is reported to work very well, and should see more widespread use in the future.

2.7. Calibration and measurement validation

Frequency domain requires careful calibration and spending time to test and validate the results is worth the effort. Users should be aware that an error in the standard lifetime, using the standard lifetime, or transferring a calibration inappropriately does not propagate linearly to the resulting measured lifetimes. Frequency domain measurements are relative measurements insofar as the lifetime of the sample is made relative to the lifetime of a reference standard of known lifetime. Three methods have been described for calibrating the measurement. Each one seeks to determine the modulation depth and the zero-phase position of the excitation light source. The phase of a sample is the difference between the raw measured phase of the sample and the zero-phase position of the light source standard. Similarly, the corrected modulation depth of the sample is the raw measured value divided by the modulation depth of the excitation light. All three of these methods attempt to determine the modulation and zero-phase position of the light source by using a sample of known "lifetime."

2.8. Calibration by comparison with a scattering solution

In this case, the excitation and emission filters and dichroic mirror used with the sample are removed and replaced with a beam splitter [3, 36]. A scattering solution is placed on the microscope and a measurement series is collected. Scattering of light is a fast process and as such the "lifetime" of the scattering solution is assumed to be 0 ns. The computed phase and modulation depths obtained are therefore equivalent to the position of zero phase and modulation depth of the light source. This method has the advantage of providing a stable reference value and the scattering solution will mimic fluorescence better than a reflecting surface.

This procedure should be used cautiously as image intensifiers can be damaged by too much light. When the standard microscope filters are replaced by the beam splitter, a neutral density filter should be inserted to protect the intensifier.

2.9. Calibration by use of reflecting surfaces

This approach works similar to the scattering solution. The filters are replaced with a beam splitter and a mirrored surface is placed in the position of the sample [36, 40]. Similar to the scattering solutions, the "lifetime" of the reflection process is assumed to be 0 ns. The remaining aspects of the approach are the same as for the scatterer.

There is a variation on this approach in which a specialized filter cube is constructed consisting of a true mirror rotated 90 ° from the way a dichroic mirror is typically installed such that the incident beam is directed to the camera without passing through the objective [41]. This gives a reference phase for a path length that does not include the trip through the objective to the sample and back. The phase delays and demodulation factors for this additional distance are calibrated independently for each objective. Reflecting surfaces are advantageous because they are readily available, provide a robust mimic of a very short lifetime, and will be stable over time.

Both versions of this procedure should be done with care to avoid damaging the image intensifier.

2.10. Calibration by use of fluorophores of known lifetime

This procedure involves selecting a fluorophore of known lifetime and placing it in the microscope and measuring the phase and modulation depth [11]. Rearranging Eqs. (2.5 and 2.6) allows the expected phase and modulation to be predicted. These may then be used to compute the position of zero phase and the modulation depth of the light source. An advantage of the method is that it may be done under conditions exactly matching those of a sample.

2.11. Comparison of calibration methods

All three methods work and have yielded good results when done properly, but none is ideal. Deciding which is best for a particular purpose will depend mostly on personal preference and the traditions of the group doing the measurement. There are a few points to note in each case, which are worth keeping in mind.

1. When implementing methods based on reflection and scattering from the object plane, reflecting surfaces inside the microscope must be kept to a minimum. Phase rings inside an objective can cause spurious reflections, which may cause small changes in phase and modulation depth. Some objectives have polished metal surfaces, which can generate a signal at the intensifier, which interferes with the signal from the object plane.
2. The calibration should be done for the specific objective in use. Some objectives have been shown to be interchangeable with minimal effect on the modulation depth and phase; however, this is not easily predicted in advance and should never be assumed to be negligible.
3. Introduction of metalized neutral density filters in FLIM systems should be done cautiously as artifacts have been observed due to single or multiple reflections between pairs of ND filters.

4. When using fluorophores of known lifetime, it is important to validate the lifetime used. Fluorescence lifetimes can be sensitive to concentration, temperature, pH, and other environmental variables. Fluorophores from different suppliers can have variable purity. As a result, one should not assume that a value reported in the literature will be exactly transferable to other labs and conditions. Users of the method should be particularly careful to use low concentrations of fluorophore ($<10\ \mu M$) to avoid a variety of processes which can perturb lifetimes in solution. There are a limited number of well characterized fluorophores. If one is not available for a particular wavelength this will require a change of filters leaving the method with nothing to recommend it over reflection and scatter.
5. The reflection calibration method with the specialized filter block has the advantage that it does not require the sample to be moved to recalibrate. As a result it might be particularly useful for long time scale time series data.
6. Most of the calibration methods described in the literature have been on systems using laser excitation and AOM modulation. There is much reason to believe that directly modulated LEDs are more stable; however, the base of experience with LEDs is currently less.
7. On the RF time scale, the transit times of electrons in long coaxial cables and the time of flight of photons in optical paths as short as a few centimeters are significant. These effects become more pronounced as the modulation frequency increases. Even simple changes made to a system will affect the resulting measurements.

2.12. Validation after calibration

Once familiar with methods for calibrating the FLIM system, it is worthwhile to verify the range over which a given FLIM system performs well. This is particularly useful for persons new to the method to build confidence and an intuition for conditions likely to cause problems. One approach to doing this is with a set of quenched solutions. The use of Rhodamine 6G solutions quenched with variable amounts of iodide has been well investigated, both in cuvettes and in imaging arrangements [11, 42]. Detailed approaches to this may be found in the literature and it has been used successfully in a number of labs.

2.13. First pass analysis—data to modulation depth and phase shift

To analyze frequency domain FLIM data, first the phase shift and demodulation of the fluorescence light with respect to the excitation light are estimated. In the case of single frequency data, this reduces the FLIM data to only three parameters: phase shift, demodulation, and total intensity. This step can be done in various ways as described in the following sections. From these parameters, the lifetimes can be estimated either by Eqs. (2.6 and 2.7), or by more elaborate approaches as described below.

2.14. Fourier methods for estimating phase and modulation

When analyzing a data set using Fourier methods to estimate the phase shift and demodulation, a stack of images is transformed along the stack direction using a DFT. When using DFTs, there is some computational advantage of using a radix-2 type Fast Fourier transform (FFT); however, for practical work on the data lengths typical of FLIM, other considerations may outweigh any computational advantage these might have. For example, a user may find that collection of eight images does not give the desired results but not wish to expose the sample to a full 16. Twelve images may be processed instead using a DFT.

When using a DFT, the images in the stack must be collected at equal phase intervals and it is convenient to restrict the full sampling to multiples of 2π. For proper sampling of the sinusoidally modulated waveform, the highest frequency in the data set must be sampled greater than or equal to twice per period; a condition dictated by the Nyquist limit.

Processing of the output of one of these algorithms allows recovery of phase and modulation depth from the Fourier coefficients. As an example, assume that a series of N images has been collected. Each point in an image can be designated by an indexing system (x, y) where x and y are the pixel positions in the image in the horizontal and vertical directions, respectively. If the data are measured as a function of wavelength, as has been done in some specialized applications, the data may be indexed further to include (x, y, λ). For the transform, the raw data are treated as a function $g(x, y, n)$, where x and y represent the pixel position in the image, and n is the index into the N evenly spaced phase samples. The data are transformed over n to obtain the discrete Fourier coefficients $G(x, y, \omega)$, where ω is the frequency corresponding to particular component of the sinusoidal driving function:

$$G(x, y, \omega) = \sum_{n=0}^{N-1} g(x, y, n) e^{i\omega n/N} \tag{2.12}$$

It should be recognized that the discrete Fourier coefficients $G(x, y, \omega)$ are represented by complex numbers. The real part $\mathrm{Re}(G(x, y, \omega))$ of the complex number represents the amplitude of the cosine part of the sinusoidal function and the imaginary part $\mathrm{Im}(G(x, y, \omega))$ represents the amplitude of the sine wave.

DFT methods are valuable for determining the magnitude and phase of a complex mixture of frequency components simultaneously such as might be encountered in the multiplexed systems for collection of several frequencies. Once the discrete Fourier coefficients have been computed the uncorrected values of m and ϕ may be computed for every pixel in the sample image:

$$m(x, y) = \frac{| G(x, y, \omega_0) |}{| G(x, y, 0) |} \tag{2.13}$$

$$\phi(x, y) = \tan^{-1}\left[\frac{\mathrm{Im}\Big(G(x, y, \omega_0)\Big)}{\mathrm{Re}\Big(G(x, y, \omega_0)\Big)}\right] \qquad (2.14)$$

Re(·) and Im(·) refer to the real and imaginary parts of discrete Fourier coefficients $G(x, y, \omega)$. $G(x, y, \omega_0)$ and $G(x, y, 0)$ refer to the Fourier coefficients corresponding to the frequency of excitation and to zero frequency, respectively, and $|G(x, y, \omega)| = \sqrt{\mathrm{Re}\Big(G(x, y, \omega)\Big)^2 + \mathrm{Im}\Big(G(x, y, \omega)\Big)^2}$. A variety of methods for using Fourier methods have been presented in the literature, which are optimized for specific purposes. For the purposes of this chapter, the discussion will be limited to the treatment given. Users desiring assistance with particular methods and optimizations are encouraged to consult the original literature.

2.15. Sine fitting methods for estimating phase and modulation

The Fourier method is not a requirement, and direct sinusoidal fitting procedures are also used to fit the data from a set of images. A number of specialized procedures have been described over the years and it is worth noting that extracting the amplitude and phase may be done as a simple extension to conventional linear regression.

Linear least squares fitting of sine and cosine data may be accomplished using a similar indexing system to those used for the Fourier methods. Suppose that a data set consisting of N images may be indexed as $g(x, y, n)$, where x and y represent the pixel position in the image, and n is the index into the N phase samples. If the frequency of modulation and the sampling is known (as is usually the case) this becomes a standard multilinear regression problem. Of the form:

$$f(x, y, n) = A(x, y)\cos(\omega n) + B(x, y)\sin(\omega n) + C(x, y) \qquad (2.15)$$

Presented in this manner, the analysis may proceed similarly to the treatment obtained from the Fourier analysis. C is the zero frequency component of the fit and A and B may be treated as the real and imaginary parts of the complex number.

$$m(x, y) = \frac{\sqrt{A(x, y)^2 + B(x, y)^2}}{C(x, y)} \tag{2.16}$$

$$\phi(x, y) = \tan^{-1}\left[\frac{B(x, y, \lambda)}{A(x, y, \lambda)}\right] \tag{2.17}$$

Sinusoidal fitting is more flexible than Fourier methods, as it does not require evenly spaced phase steps. There is no special convenience associated with sampling of an angle of 2π and estimation of errors in parameters is somewhat more straightforward.

2.16. *Two-component analysis of FLIM data*

With some further assumptions, it is possible to use single frequency FLIM data to fit a two-component model, and calculate the relative concentration of each species, in each pixel [16]. To simplify the analysis, we will assume that in each pixel of the sample we have a mixture of two components with single exponential decay kinetics. We assume that the unknown fluorescence lifetimes, τ_1 and τ_2, are invariant in the sample. In each pixel, the relative concentrations of species may be different and are unknown. We first seek to estimate the two spatially invariant lifetimes, τ_1 and τ_2. We make a transformation of the estimated phase-shifts and demodulations as follows:

$$N_i = m_i \sin(\Delta\phi_i) \tag{2.18a}$$

$$D_i = m_i \cos(\Delta\phi_i) \tag{2.18b}$$

We have added a subscript i, to indicate the number of the pixel that is being considered. From Eq. (2.9) we get:

$$N_i = \frac{f_i \omega \tau_1}{1 + \omega^2 \tau_1^2} + \frac{(1 - f_i)\omega \tau_2}{1 + \omega^2 \tau_2^2} \tag{2.19a}$$

$$D_i = \frac{f_i}{1 + \omega^2 \tau_1^2} + \frac{1 - f_i}{1 + \omega^2 \tau_2^2} \tag{2.19b}$$

where f_i is the relative contribution of the first component to the steady-state fluorescence in pixel i. Eliminating f_i from Eq. (2.19), we find a linear relation between N_i and D_i:

$$N_i = u + vD_i \tag{2.20}$$

where

$$u = \frac{1}{\omega(\tau_1 + \tau_2)} \tag{2.21a}$$

and

$$v = \frac{\omega^2 \tau_1 \tau_2 - 1}{\omega(\tau_1 + \tau_2)} \tag{2.21a}$$

The only unknowns in these equations are the two fluorescence lifetimes, which considerably reduces the complexity of the problem. Figure 2.3 shows a plot of N versus D for all possible monoexponential decays, and for all possible mixtures of two monoexponential species with lifetimes equal to 2.5 and 1 ns. The half-circle though (0,0) and (0,1) represents the values of N and D that correspond to all possible monoexponential decay kinetics [13, 16, 43]. All the values of N_i and D_i for a mixture of two species lie on a straight line connecting the two points on the half-circle that correspond to the lifetimes of the two species. The offset and the slope of this straight line are given by Eq. (2.21).

Figure 2.3 suggests a simple strategy to recover the fluorescence lifetimes τ_1 and τ_2: Given estimated phase shifts and demodulations,

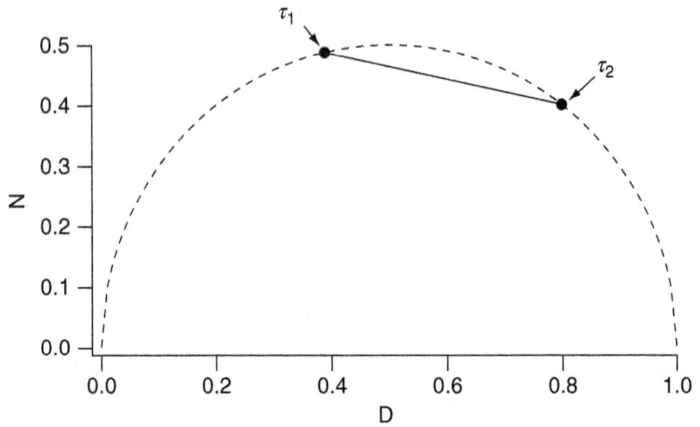

Fig. 2.3. Plot of N versus D for a mixture of two monoexponential species, with fluorescent lifetimes equal to τ_1 and τ_2. Single exponential lifetimes are found on the half-circle passing trough the points (0,0) and (0,1). All possible fluorescent lifetimes of a mixture of free donor and complex are found on the straight line connecting the two points on the half-circle that correspond to the lifetimes of both species.

we calculate the corresponding N_i and D_i values and fit a straight line through them to obtain the slope v and the offset u [16]. The estimated fluorescence lifetimes of the two species are then found by inverting Eq. (2.21):

$$\tau_{1,2} = \frac{1 \pm \sqrt{1 - 4u(u + v)}}{2\omega u} \tag{2.22}$$

Knowing the values of the two lifetimes, the fractions f_i can be recovered in each pixel, by solving Eq. (2.19) in a least squares sense:

$$f_i = \frac{N_i + \omega\tau_1 - \omega D_i\tau_1 - \omega(D_i + \omega N_i\tau_1)\tau_2}{\omega(\tau_1 - \tau_2)} \tag{2.23}$$

The fraction f_i is the fractional contribution of the first species to the total fluorescence in pixel i. The *molar fraction* c_i of the first species in each pixel can be derived from it by dividing by its quantum yield, and renormalizing, which for a two-component system yields:

$$c_i = \frac{f_i Q_2}{Q_1 + (Q_2 - Q_1) f_i} \tag{2.24}$$

where Q_1 and Q_2 are the quantum yields of the first and the second species, respectively. To apply Eq. (2.24), the quantum yields of the two species, or at least their relative magnitudes, must be known. This is straightforward for the case where the two species are the same donor fluorophore in the presence and absence of fluorescent resonance energy transfer. In this case, $Q_1 \propto \tau_1$ and $Q_2 \propto \tau_2$ were given by the lifetimes in the presence and absence of FRET.

This approach, derived first by Clayton *et al.* [16] and subsequently developed further by others [9, 10, 13, 17–19] is conceptually very simple and gives analytical solutions for the lifetimes of the two species in the mixture and the relative concentrations in each pixel. To be able to fit a straight line, it is crucial that the lifetimes (and thus u and v) are invariant over all pixels, and that there is sufficient variation in f_i (and thereby in N_i and D_i). These requirements of invariance in some parameters and of sufficient variation in at least one other parameter also form the basis of the so-called global analysis methods [44, 45] that were applied earlier to fit FLIM images [20]. In these approaches, nonlinear least squares fitting were used to estimate the lifetimes and molar fractions of each species. The relation of these global analysis methods to the approach described above can be seen, if we introduce error weighting. Both N_i and D_i are distorted by errors, and any proper fit should include those in the estimation of the lifetimes. However, since both N_i and D_i have errors, standard error weighting methods for linear fitting methods cannot be applied. Instead, we directly formulate a least square criterion that we minimize:

$$\chi^2(u, v) = \sum_i \frac{(N_i - u - vD_i)^2}{\sigma_{N,i}^2 + v^2 \sigma_{D,i}^2} \tag{2.25}$$

where $\sigma_{N,i}$ and $\sigma_{D,i}$ are the standard deviations of N_i and D_i, respectively. They can be found by propagation of the errors found for the estimated phase and modulation. This least squares criterion can be minimized to find u and v, which is a nonlinear problem, that is, however, not computationally complicated since only two parameters need to be estimated. In practice, we use a different function that is obtained by substituting Eq. (2.21) into (2.25), and minimize directly for τ_1 and τ_2:

$$\chi^2(\tau_1,\tau_2) = \sum_i \frac{\left((\omega^2\tau_1\tau_2 - 1)D_i - \omega(\tau_1+\tau_2)N_i + 1\right)^2}{\omega^2(\tau_1+\tau_2)^2\sigma_{N,i}^2 + (\omega^2\tau_1\tau_2 - 1)\sigma_{D,i}^2} \tag{2.26}$$

This function is identical to the one that was derived earlier using global analysis methods [46].

2.17. Application: Semi-quantitative FRET analysis

As an example of the usefulness of simple estimations of the lifetime from phase and modulation, we consider the case of donor quenching by FRET. Figure 2.4 shows the results of FLIM measurements on epidermal growth factor receptor tagged with GFP or YFP (EGFR-GFP and EGFR-YFP) [47]. Cells were stimulated with epidermal growth factor for 1 min, and then fixed, permeabilized, and incubated with PY72, a generic antibody against phosphorylation. The antibody was tagged with Cy3 (in the case of EGFR-GFP) or Cy3.5 (in the case of EGFR-YFP). Binding of the antibody to phosphorylated EGFR can be detected specifically by FRET from the donor-tagged receptor to the acceptor-tagged antibody. This can be observed in intact cells by measuring the fluorescence lifetime of the donor. Figure 2.4a shows the result for EGFR-GFP. Both the phase and modulation lifetimes of EGFR-GFP are lower in cells that are incubated with PY72-Cy3, near the plasma membrane of the cells. Similar results are shown for EGFR-YPF in Fig. 2.4b.

To summarize the results of multiple cells, 2D histograms of the phase and modulation lifetimes can be used (Fig. 2.4c). Such 2D

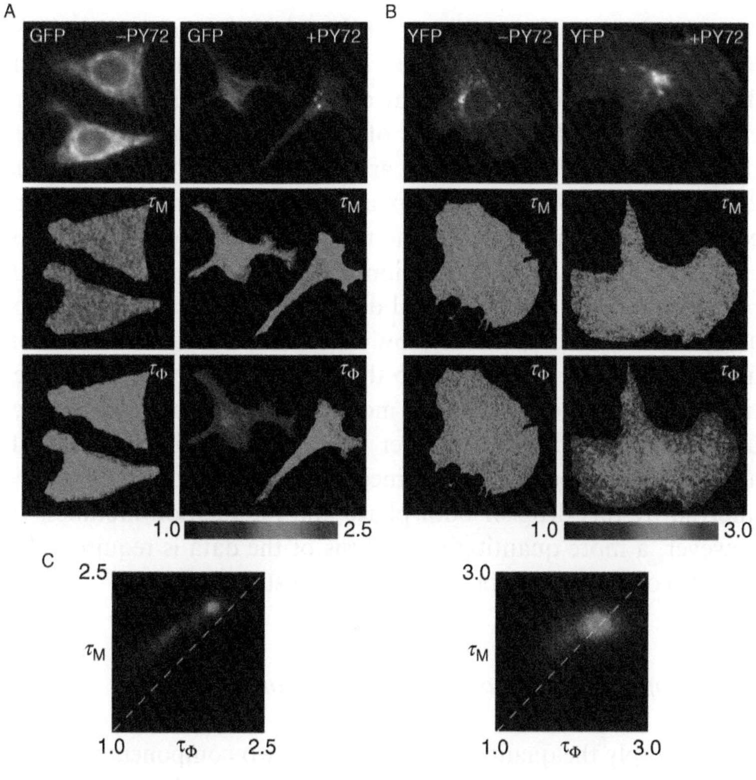

Fig. 2.4. (A) FLIM measurements of EGFR-GFP in the absence (left panels) and presence (right panels) of PY72-Cy3. (B) FLIM measurements of EGFR-YFP in the absence (left panels) and presence (right panels) of PY72-Cy3.5. The scaling of the pseudocolored lifetime images in panel (A) and (B) are indicated with the color bar in ns. Top panels: GFP intensity; Middle: modulation lifetimes; Bottom: phase lifetimes. (C) 2D histograms of τ_ϕ versus τ_M for EGFR-GFP (left) and EGFR-YFP (right). Red: samples incubated with PY72-Cy3 or PY72-Cy3.5, respectively; Green: control samples without PY72. (See Color Insert.)

histograms can be calculated from many images and represent a joint distribution of the phase and modulation lifetimes. These 2D histograms demonstrate some general properties of the GFP and YFP kinetics in the absence and presence of FRET. For EGFR-GFP, the values of the lifetimes in the absence of acceptor do not center on the diagonal of the histogram. This indicates that EGFR-GFP does not have monoexponential decay kinetics. However, in the case of EGFR-YPF, the values of the phase and modulation lifetimes, in the absence of FRET, are located on the diagonal of the 2D histogram showing that the phase and modulations lifetimes are equal within the precision of the measurement. From this we can conclude that within the precision of the instrument EGFR-YFP has essentially monoexponential decay kinetics. In the presence of acceptor, the 2D histograms show that both phase and modulation lifetimes are lower compared to the control, where the phase lifetime is generally lower than the modulation lifetime. In this particular analysis, it does not matter much that the lifetime of GFP alone is not monoexponential, since the occurrence of FRET can be inferred by this drop of both phase and modulation lifetimes. If, however, a more quantitative analysis of the data is required, this issue becomes more important as we will show below.

2.18. Application: Quantitative FRET analysis

We can apply the quantitative analysis of two-component mixtures to the FRET data from Fig. 2.4 [20]. In this case, the two lifetimes are equal to the lifetimes of the donor in the absence and presence of acceptor. Figure 2.5a shows the results for EGFR-GFP. It can be seen that in the control sample, the relative concentrations of the short-lifetime component are not equal to zero, as would be expected for a monoexponential donor in the absence of FRET. Indeed, if we look at the plot of N_i versus D_i for a subset of the pixels of the EGFR-GFP data, we see that the data for the control are not centered on a point on half-circle of monoexponential decays. In contrast, if we look at the results of EGFR-YFP (Fig. 2.5b), we find that the relative concentrations of the short-lifetime component are indeed much closer to zero. Indeed, the plot of N_i versus D_i shows that the data for the control are centered on a

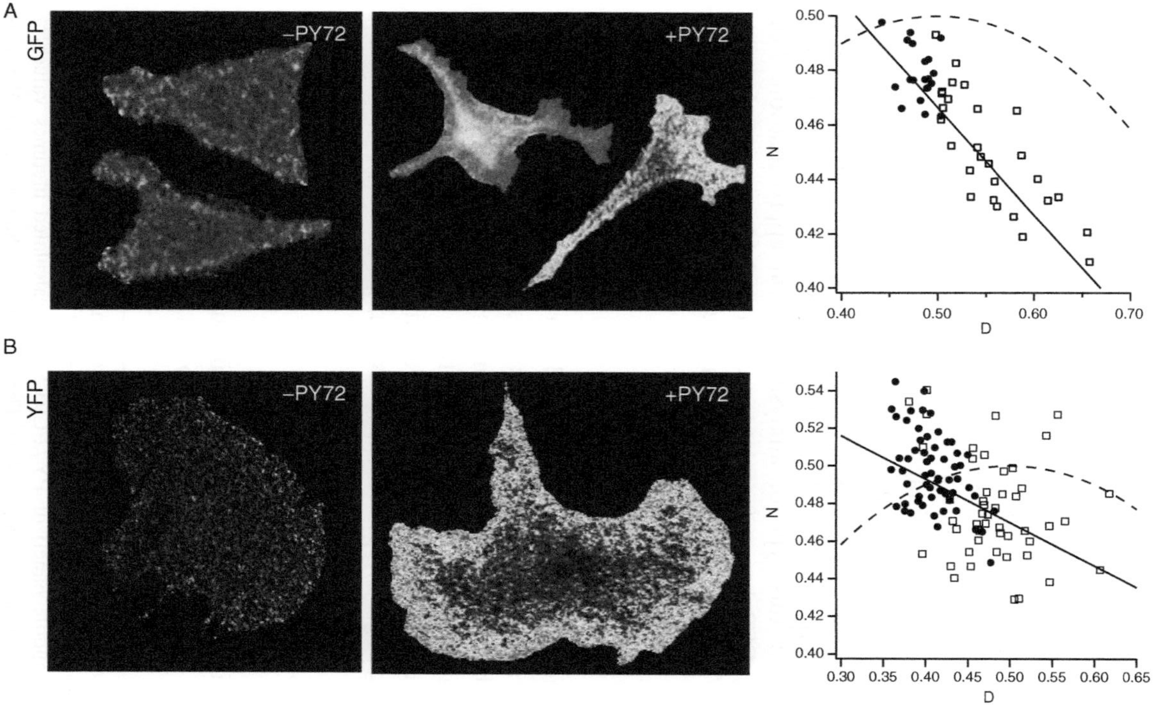

Fig. 2.5. (A) The relative concentrations of the short lifetime component, in samples expressing EGFR-GFP in the absence (left panel) and presence (middle panel) of PY72-Cy3. The calculated lifetimes values of the two species

point on the half-circle of monoexponential decays. This confirms our earlier observations that EGFR-YFP has a monoexponential decay and is therefore more suitable for quantitative analysis than EGFR-GFP. For EGFR-GFP, a systematic error must be expected, but as shown in Fig. 2.5a, the results may still be useful. For biological experiments where a relatively big error is acceptable, this approach can still be used successfully if GFP is used as a donor tag. Indeed GFP tagged donors have been used successfully in several biological applications [8, 48, 49], although YFP tagged donors have proven more reliable for the purpose of quantitative analysis [50]. It should be stressed that not all fluorophores are suitable for this type of analysis, even in approximation. For instance, CFP has a far more pronounced biexponential decay than GFP and is unsuitable for this type of quantitative analysis.

2.19. Emerging techniques

A few methods have been described worth following over the next few years as they represent potentially significant advances. Most of these have to do with improvements in the modulated intensifiers or attempts to remove them altogether. At present, none of these have a sufficient base of users to evaluate their significance but all are intriguing concepts in FLIM instrumentation.

2.20. Segmented intensifiers

One limitation of frequency domain and other FLIM systems is the time of acquisition. For example, a frequency domain FLIM

---

were 0.7 and 2.2 ns. (B) The relative concentrations of the short lifetime component, in samples expressing EGFR-YFP in the absence (left panel) and presence (middle panel) of PY72-Cy3.5. The calculated lifetimes values of the two species were 1.0 and 2.4 ns. The right panels show a plot of N_i versus D_i for a subset of the pixels from the samples shown. (See Color Insert.)

system requiring four or eight phase steps to generate a lifetime image typically takes a second or more to collect. During the data acquisition cycle, the specimen can move or bleach. Earlier in the chapter, the notion of randomizing the phase steps was introduced as a method to limit the impact of photobleaching. Given that the images at the different phase steps are not done coincident in time, randomization is perhaps the best approach; however, this only assures that the computed results are not systematically affected.

One approach to rectify this problem has been to divide an intensifier into segments [51]. Each segment is provided with a different time delay. In the current arrangement, the system has been described with four segments and was demonstrated for use with time domain measurements, however, the approach is significant for frequency domain FLIM. To use the segmented intensifier, the image arriving from the microscope is put into a beam splitter dividing the light into four separate images, each of which is directed to a different segment of the intensifier. This allows four time delays to be collected simultaneously. This seems to be a useful technology as the other approaches to FLIM require multiple exposures to be taken at different times.

2.21. Directly modulated detection schemes

Another problem of widely used instrumentation for FLIM has been the intensifier. In general, the intensifier is a necessary evil required to make the FLIM experiment work. A typical image intensifier can resolve around 12.5 line pairs/mm (about 80 μm). This is a very significant degradation of image quality relative to direct measurements with a CCD camera. The overall effect can be summarized as a process in which a good image goes in one side of an intensifier and a bad image emerges out the other side. Replacing the modulated image intensifier with an alternative with less degradation (and perhaps lower cost) is worth considerable effort.

Two approaches to intensifier-free detection have been described but neither has been used extensively. In one approach, the CCD is modulated by the application of a high-frequency signal to the device and using it to shuffle charge in and out of light sensitive areas of the CCD [33, 52]. This remains an intriguing approach and as CCDs advance this approach may provide users of frequency domain methods with a viable alternative to the intensifier. As with the segmented image intensifier, the pioneering work on this instrument has been done on time domain lifetime measurements; however, the approach holds promise for the frequency domain as well.

A related technology is a so-called "time-of-flight" imager that was originally designed for 3D-vision applications [40]. This can be thought of as a "lock-in" approach to the measurement of a signal at a particular frequency. The time-of-flight imager has the property that a pixel has two gates, which collect charge generated by light striking the detector depending on the phase of the signal applied to the gate. The two gates operate 180 ° out of phase from each other. The approach has some advantages in that the light effectively rejected by an image intensifier would be collected in the out-of-phase image. Although this technology is at an early stage of development, it holds promise for the future.

Acknowledgments

The authors thank Bert van Geest of Lambert Instruments for Fig. 2.1 and permission to use it in this publication.

References

[1] Bastiaens, P. I. H. and Squire, A. (1999). Fluorescence lifetime imaging microscopy: Spatial resolution of biochemical processes in the cell. Trends Cell Biol. *9*, 48–52.

[2] Gadella, T. W. J. and Jovin, T. M. (1995). Oligomerization of epidermal growth-factor receptors on A431 cells studied by time-resolved fluorescence imaging microscopy—a stereochemical model for tyrosine kinase receptor activation. J. Cell Biol. *129*, 1543–58.

[3] Gadella, T. W. J., Jr. Jovin, T. M. and Clegg, R. M. (1993). Fluorescence lifetime imaging microscopy (Flim)—spatial-resolution of microstructures on the nanosecond time-scale. Biophys. Chem. *48*, 221–39.

[4] Lakowicz, J., Szmacinski, H., Nowaczyk, K., Berndt, K. and Johnson, M. (1992). Fluorescence lifetime imaging. Anal. Biochem. *202*, 316–30.

[5] Lakowicz, J. R. and Berndt, K. W. (1991). Lifetime-selective fluorescence imaging using an Rf phase-sensitive camera. Rev. Sci. Instrum. *62*, 1727–34.

[6] Ng, T., Shima, D. S., Hansra, G., Bornancin, F., Prevostel, C., Hanby, A., Harris, W., Barnes, D., Schmidt, S., Mellor, H., Bastiaens, P. I. H. and Parker, P. J. (1999). PKC alpha regulates beta 1 integrin-dependent cell motility through association and control of integrin traffic. EMBO J. *18*, 3909–23.

[7] Ng, T., Squire, A., Hansra, G., Bornancin, F., Prevostel, C., Hanby, A., Harris, W., Barnes, D., Schmidt, S., Mellor, H., Bastiaens, P. I. H. and Parker, P. J. (1999). Imaging protein kinase Cα activation in cells. Science *283*, 2085–9.

[8] Verveer, P. J., Wouters, F. S., Reynolds, A. R. and Bastiaens, P. I. H. (2000). Quantitative imaging of lateral ErbB1 receptor signalling propagation in the plasma membrane. Science *290*, 1567–70.

[9] Colyer, R. A., Lee, C. and Gratton, E. (2008). A novel fluorescence lifetime imaging system that optimizes photon efficiency. Microsc. Res. Tech. *71*, 201–13.

[10] Digman, M. A., Caiolfa, V. R., Zamai, M. and Gratton, E. (2008). The phasor approach to fluorescence lifetime imaging analysis. Biophys. J. *94*, L14–L16.

[11] Hanley, Q. S., Subramaniam, V., Arndt-Jovin, D. J. and Jovin, T. M. (2001). Fluorescence lifetime imaging: Multi-point calibration, minimum resolvable differences, and artifact suppression. Cytometry *43*, 248–60.

[12] Kremers, G. J., Van Munster, E. B., Goedhart, J. and Gadella, T. W. (2008). Quantitative lifetime unmixing of multi-exponentially decaying fluorophores using single-frequency FLIM. Biophys. J. *95*, 378–89.

[13] Redford, G. I. and Clegg, R. M. (2005). Polar plot representation for frequency-domain analysis of fluorescence lifetimes. J. Fluoresc. *15*, 805–15.

[14] Redford, G. I., Majumdar, Z. K., Sutin, J. D. B. and Clegg, R. M. (2005). Properties of microfluidic turbulent mixing revealed by fluorescence lifetime imaging. J. Chem. Phys. *123*, 224504.
[15] Schneider, P. C. and Clegg, R. M. (1997). Rapid acquisition, analysis, and display of fluorescence lifetime-resolved images for real-time applications. Rev. Sci. Instrum. *68*, 4107–19.
[16] Clayton, A. H. A., Hanley, Q. S. and Verveer, P. J. (2004). Graphical representation and multicomponent analysis of single-frequency fluorescence lifetime imaging microscopy data. J. Microsc. *213*, 1–5.
[17] Esposito, A., Gerritsen, H. C. and Wouters, F. S. (2005). Fluorescence lifetime heterogeneity resolution in the frequency domain by lifetime moments analysis. Biophys. J. *89*, 4286–99.
[18] Forde, T. and Hanley, Q. S. (2006). Spectrally resolved frequency domain analysis of multi-fluorophore systems undergoing energy transfer. Appl. Spectrosc. *60*, 1442–52.
[19] Hanley, Q. S. and Clayton, A. H. A. (2005). AB-plot assisted determination of fluorophore mixtures in a fluorescence lifetime microscope using spectra or quenchers. J. Microsc. *218*, 62–7.
[20] Verveer, P. J., Squire, A. and Bastiaens, P. I. H. (2000). Global analysis of fluorescence lifetime imaging microscopy data. Biophys. J. *78*, 2127–37.
[21] Carlsson, K. and Liljeborg, A. (1998). Simultaneous confocal lifetime imaging of multiple fluorophores using the intensity-modulated multiple-wavelength scanning (IMS) technique. J. Microsc. *191*, 119–27.
[22] Hanley, Q. S., Lidke, K. A., Heintzmann, R., Arndt-Jovin, D. J. and Jovin, T. M. (2005). Fluorescence lifetime imaging in an optically sectioning programmable array microscope (PAM). Cytometry A *67A*, 112–8.
[23] van Munster, E. B., Goedhart, J., Kremers, G. J., Manders, E. M. M. and Gadella, T. W. J. (2007). Combination of a spinning disc confocal unit with frequency-domain fluorescence lifetime imaging microscopy. Cytometry A *71A*, 207–14.
[24] Gaviola, E. (1927). Ein fluorometer. Zeitschrift fur Physik *42*, 853–61.
[25] Spencer, R. D. and Weber, G. (1969). Measurement of subnanosecond fluorescence lifetimes with a cross-correlation phase fluorometer. Ann. N. Y. Acad. Sci. *158*, 361–76.
[26] Lakowicz, J. R. (1999). Principles of fluorescence spectroscopy. Kluwer/Plenum,** New York.
[27] Lakowicz, J. R. and Balter, A. (1982). Theory of phase-modulation fluorescence spectroscopy for excited-state processes. Biophys. Chem. *16*, 99–115.
[28] Marriott, G., Clegg, R. M., Arndtjovin, D. J. and Jovin, T. M. (1991). Time resolved imaging microscopy—phosphorescence and delayed fluorescence imaging. Biophys. J. *60*, 1374–87.

[29] van Munster, E. B. and Gadella, T. W. J. (2004). Suppression of photobleaching-induced artifacts in frequency-domain FLIM by permutation of the recording order. Cytometry A *58A*, 185–94.
[30] Hanson, K. M., Behne, M. J., Barry, N. P., Mauro, T. M., Gratton, E. and Clegg, R. M. (2002). Two-photon fluorescence lifetime imaging of the skin stratum corneum pH gradient. Biophys. J. *83*, 1682–90.
[31] Dinish, U. S., Fu, C. Y., Chao, Z. X., Seah, L. K., Murukeshan, V. M. and Ng, B. K. (2006). Subnanosecond-resolution phase-resolved fluorescence imaging technique for biomedical applications. Appl. Opt. *45*, 5020–6.
[32] Elder, A. D., Matthews, S. M., Swartling, J., Yunus, K., Frank, J. H., Brennan, C. M., Fisher, A. C. and Kaminski, C. F. (2006). The application of frequency-domain fluorescence lifetime imaging microscopy as a quantitative analytical tool for microfluidic devices. Opt. Express *14*, 5456–67.
[33] Mitchell, A. C., Wall, J. E., Murray, J. G. and Morgan, C. G. (2002). Direct modulation of the effective sensitivity of a CCD detector: A new approach to time-resolved fluorescence imaging. J. Microsc. *206*, 225–32.
[34] Moser, C., Mayr, T. and Klimant, I. (2006). Filter cubes with built-in ultrabright light-emitting diodes as exchangeable excitation light sources in fluorescence microscopy. J. Microsc. *222*, 135–40.
[35] Anonymous. (2003). LIFA system for fluorescence lifetime imaging microscopy (FLIM). J. Fluoresc. *13*, 365–7.
[36] Squire, A., Verveer, P. J. and Bastiaens, P. I. H. (2000). Multiple frequency fluorescence lifetime imaging microscopy. J. Microsc. *197*, 136–49.
[37] Vereb, G., Jares-Erijman, E., Selvin, P. R. and Jovin, T. M. (1998). Temporally and spectrally resolved imaging microscopy of lanthanide chelates. Biophys. J. *74*, 2210–22.
[38] Hanley, Q. S., Arndt-Jovin, D. J. and Jovin, T. M. (2002). Spectrally resolved fluorescence lifetime imaging microscopy. Appl. Spectrosc. *56*, 155–66.
[39] Hanley, Q. S. and Ramkumar, V. (2005). An internal standardization procedure for spectrally resolved fluorescence lifetime imaging. Appl. Spectrosc. *59*, 261–6.
[40] Esposito, A., Oggier, T., Gerritsen, H. C., Lustenberger, F. and Wouters, F. S. (2005). All-solid-state lock-in imaging for wide-field fluorescence lifetime sensing. Opt. Express *13*, 9812–21.
[41] Van Munster, E. B. and Gadella, T. W. J. (2004). phi FLIM: A new method to avoid aliasing in frequency-domain fluorescence lifetime imaging microscopy. J. Microsc. *213*, 29–38.
[42] Harris, J. M. and Lytle, F. E. (1977). Measurement of subnanosecond fluorescence decays by sampled single-photon detection. Rev. Sci. Instrum. *48*, 1469–76.

[43] Jameson, D. M., Gratton, E. and Hall, R. D. (1984). The measurement and analysis of heterogeneous emissions by multifrequency phase and modulation fluorometry. Appl. Spectrosc. Rev. *20*, 55–106.
[44] Beechem, J. M. (1992). Global analysis of biochemical and biophysical data. Methods Enzymol. *210*, 37–54.
[45] Beechem, J. M., Knutson, J. R., Ross, B. A., Turner, B. W. and Brand, L. (1983). Global resolution of heterogeneous decay by phase/modulation fluorometry: Mixtures and proteins. Biochemistry *22*, 6054–8.
[46] Verveer, P. J. and Bastiaens, P. I. H. (2003). Evaluation of global analysis algorithms for single frequency fluorescence lifetime imaging microscopy data. J. Microsc. *209*, 1–7.
[47] Wouters, F. S. and Bastiaens, P. I. H. (1999). Fluorescence lifetime imaging of receptor tyrosine kinase activity in cells. Curr. Biol. *9*, 1127–30.
[48] Ng, T., Parsons, M., Hughes, W. E., Monypenny, J., Zicha, D., Gautreau, A., Arpin, M., Gschmeissner, S., Verveer, P. J. Bastiaens, P. I. H. *et al.* (2001). Ezrin is a downstream effector of trafficking PKC-integrin complexes involved in the control of cell motility. EMBO J. *20*, 2723–41.
[49] Reynolds, A. R., Tischer, C., Verveer, P. J., Rocks, O. and Bastiaens, P. I. H. (2003). EGFR activation coupled to inhibition of tyrosine phosphatases causes lateral signal propagation. Nat. Cell. Biol. *5*, 447–53.
[50] Rocks, O., Peyker, A., Kahms, M., Verveer, P. J., Koerner, C., Lumbierres, M., Kuhlmann, J., Waldmann, H., Wittinghofer, A. and Bastiaens, P. I. H. (2005). An acylation cycle regulates localization and activity of palmitoylated Ras isoforms. Science *307*, 1746–52.
[51] Elson, D. S., Munro, I., Requejo-Isidro, J., McGinty, J., Dunsby, C., Galletly, N., Stamp, G. W., Neil, M. A. A., Lever, M. J. Kellett, P. A. *et al.* (2004). Real-time time-domain fluorescence lifetime imaging including single-shot acquisition with a segmented optical image intensifier. New J. Phys. *6*. 1–13.
[52] Mitchell, A. C., Wall, J. E., Murray, J. G. and Morgan, C. G. (2002). Measurement of nanosecond time-resolved fluorescence with a directly gated interline CCD camera. J. Microsc. *206*, 233–8.

Laboratory Techniques in Biochemistry and Molecular Biology, Volume 33
FRET and FLIM Techniques
T. W. J. Gadella (Editor)

CHAPTER 3

Time domain FLIM: Theory, instrumentation, and data analysis

H. C. Gerritsen,[1] A. V. Agronskaia,[1] A. N. Bader,[1] and A. Esposito[1,2]

[1]*Molecular Biophysics Group, Debye Institute, Utrecht University, NL 3508 TA, Utrecht, The Netherlands*
[2]*Laser Analytics Group, Department of Chemical Engineering and Biotechnology, University of Cambridge, Cambridge, UK*

The lifetime of the excited state of fluorophores may be altered by physical and biochemical properties of its environment. Fluorescence lifetime imaging microscopy (FLIM) is thus a powerful analytical tool for the quantitative mapping of fluorescent molecules that reports, for instance, on local ion concentration, pH, and viscosity. the fluorescence lifetime of a donor fluorophore, Förster resonance energy transfer can be also imaged by FLIM. This provides a robust method for mapping protein–protein interactions and for probing the complexity of molecular interaction networks.

Quantitative fluorescence imaging techniques and FLIM in particular are becoming increasingly important in biological and biomedical sciences. Knowledge of instrumentation and data analysis is required to avoid misinterpretation of the experimental results and to exploit the wealth of information provided by these techniques.

DOI: 10.1016/S0075-7535(08)00003-X

In this chapter, instrumentation and methods of analysis for FLIM in the time domain will be described. Advantages and limitations of time-correlated single photon counting (TCSPC) and time-gated imaging techniques will be discussed together with general issues on photon efficiency of detection schemes, data analysis, and practical experimental examples.

3.1. Introduction

Because of the underlying photophysics, fluorescence lifetimes are intrinsically short, usually on the order of a few nanoseconds. Detection systems with a high timing resolution are thus required to resolve and quantify the fluorescence decays. Developments in electronics and detector technology have resulted in sophisticated and easy to use equipment with a high time resolution. Fluorescence lifetime spectroscopy has become a popular tool in the past decades, and reliable commercial instrumentation is readily available.

At present, two main streams of techniques exist for the measurement of fluorescence lifetimes, time domain based methods, and frequency domain methods. In the frequency domain, the fluorescence lifetime is derived from the phase shift and demodulation of the fluorescent light with respect to the phase and the modulation depth of a modulated excitation source. Measurements in the time domain are generally performed by recording the fluorescence intensity decay after exciting the specimen with a short excitation pulse.

At the end of the 1980s and early 1990s, first experiments were carried out to combine fluorescence lifetime measurements with imaging using both time domain [1–4] and frequency domain [5–7] based approaches. This chapter will deal exclusively with time domain based fluorescence lifetime imaging methods. For the frequency domain based methods, refer Chapter 2.

Lifetime imaging can be implemented both in wide field and in scanning microscopes such as confocal microscopes and two-photon excitation microscopes. The most common implementations in time-domain fluorescence lifetime imaging microscopy (FLIM) are based on TCSPC [8, 9] and time-gating (TG) [2, 10].

3.2. Lifetime detection methods

3.2.1. Time-correlated single photon counting

In TCSPC, the fluorescent molecules are excited with very short light pulses. Typically picosecond and femtosecond excitation pulses are used, and the time delays between the excitation pulse and the detection of a single-photon is recorded. By repeating this procedure many times, the probability distribution for the emission of a single photon, and thus the fluorescence decay curve, is obtained. TCSPC is characterized by a high overall time resolution of 25–300 ps and a wide dynamic lifetime range. The TCSPC electronics often have a very high timing accuracy (<1–200 ps). In almost all cases, the time resolution with which photons are being detected is limited by the timing jitter of the detector (25–300 ps). At present, plug-in boards for PCs are commercially available that contain all the TCSPC electronics [8].

A schematic diagram of a typical TCSPC setup is shown in Fig. 3.1. The sample is excited by a short excitation pulse usually provided by a pulsed laser. A trigger signal that is synchronized with the excitation light pulse is used to start an accurate timing device such as a time-to-amplitude converter (TAC). The fluorescence emitted by the specimen is then selected by an emission filter and detected by a fast detector (e.g., a photomultiplier tube (PMT) or an avalanche photodiode) that is able to detect single photons (single photon counting (SPC)). The output pulses of the detector are sent through a discriminator and used to stop the TAC.

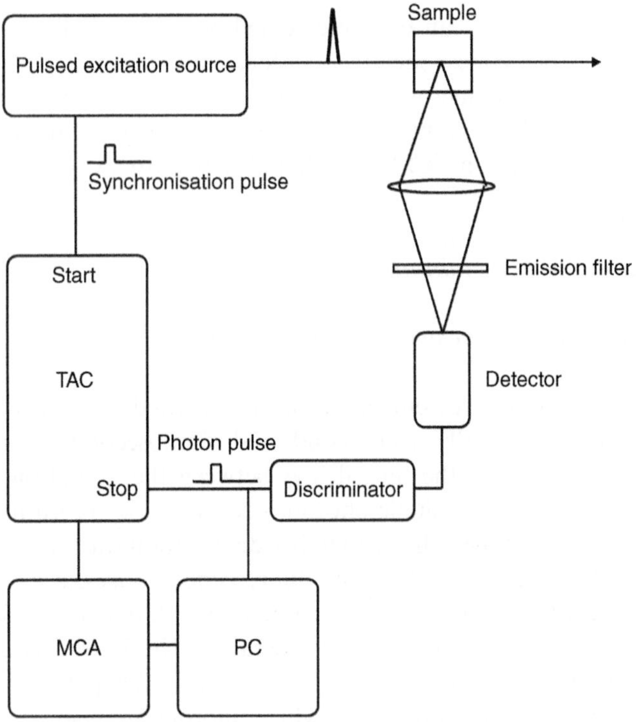

Fig. 3.1. Schematic diagram of a TCSPC setup. Using a fast timing device (e.g., time-to-amplitude-converter) the time is measured between the excitation pulse and the detection of a photon. By repeating this procedure many times a decay curve is measured. TAC: time to amplitude convertor, MCA: multi channel analyzer, PC: personal computer.

The output from the TAC is an analog signal that is proportional to the time difference between the start and stop pulses. The next step consists of digitizing the TAC output and storing the event in a multichannel analyzer (MCA). After repeating this process many times, a histogram of the arrival times of photons is accumulated in the memory of the MCA. In fluorescence lifetime spectroscopy the histogram usually contains 512–2048 channels and represents the fluorescence decay curve of the specimen (see Fig. 3.2). Note that the recorded decay curve is convoluted with the total timing response of the instrument, the instrument response function (IRF). Nowadays, there are other timing devices used for TCSPC such as time to digital converters (TDCs) [11]. These devices directly convert the timing difference between start and stop signals into a digital word. To this end, a series of solid state delay lines is used and, in practice, TDCs have similar timing accuracies and limitations as TACs.

Importantly, the dead-time of TACs and TDCs is comparatively long, typically 125–350 ns. When a photon arrives within this time interval after the detection of a photon, it will not be observed. Therefore, care must be taken that the count rate of the experiment is sufficiently low to prevent this pulse-pileup. TACs and TDCs usually operate in reversed start–stop geometry. Here, the TAC is started by the fluorescence signal and stopped by the laser trigger.

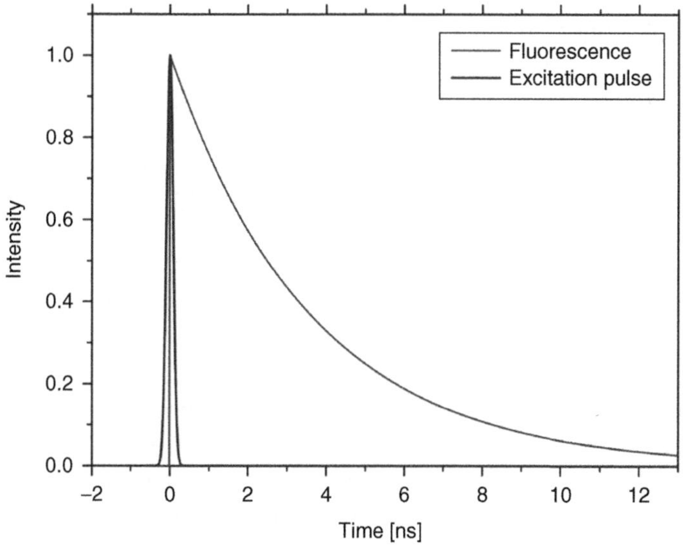

Fig. 3.2. A decay curve of a fluorescent dye and an IRF.

In this way the TAC is only triggered by usable events, and not by laser trigger pulses that do not result in a detected fluorescence photon. This mode of operation suffers less from dead-time effects. If two photons arrive within a period equal to the dead-time of the system, pulse-pileup occurs and the second photon cannot be detected. In the reversed start–stop geometry, pile-up is minimized by reducing the excitation intensity to about 1–5 detected photons per 100 excitation pulses. Furthermore, in spectroscopy applications excitation frequencies not exceeding 10 MHz are used to ensure that the fluorescence decay signal from one excitation pulse is not affected by the tail of the fluorescence decay produced by other excitation pulses. The maximum count rate employed in conventional spectroscopy applications of TCSPC is less than 100 kHz. The time required to access the histogramming memory and to transfer the decay curve from the histogramming memory to the computer system can be substantial in particular for TCSPC electronics designed for spectroscopic application.

In general the decay curves recorded by TCSPC are fitted to a (multi)exponential decay employing an iterative deconvolution technique to account for the time response of the instrument [9]. This requires the recording of the timing response of the system (see Fig. 3.2) and can be done by, for instance, the recording of excitation light from a scattering sample or fluorescence from a fast decaying dye such as Rose Bengal (τ ~80 ps) [12, 13].

3.2.2. Time gating

In TG methods, the fluorescence emission is detected in two or more time-gates each delayed by a different time relative to the excitation pulse (see Fig. 3.3). In the case of a detection scheme equipped with two time-gates, the ratio of the signals acquired in the two time-gates is a measure of the fluorescence lifetime. For a decay that exhibits only a single exponent, the fluorescence lifetime is given by:

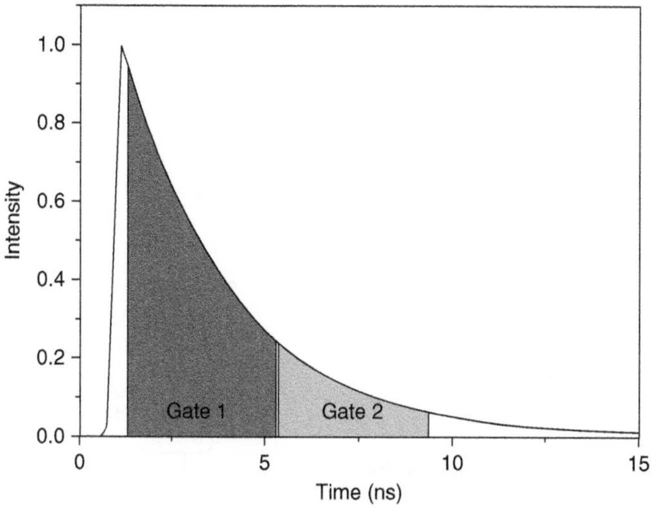

Fig. 3.3. Principle of time gating (TG). After exciting the specimen with a short light pulse, the fluorescence is detected in a number of time gates that open after a specific delay with respect to the excitation pulse.

$$\tau = \Delta T / \ln(I_1/I_2) \tag{3.1}$$

where ΔT is the time-offset between the start of the two time-gates and I_1 and I_2 are the corresponding fluorescence intensities accumulated in the gates. In this "rapid lifetime determination method," the assumption is made that the two time-gates are of equal width [14]. In the case of a multiexponential fluorescence decay (Eq. (3.1)), yields only an "average" fluorescence lifetime. This limitation can be circumvented by increasing the number of time-gates enabling the recording of multiexponential decays (de Grauw and Gerritsen, 2001; [15, 16]. Increasing the number of gates requires more sophisticated data analyses approaches like fitting the decay to (multi)exponential functions.

Practical implementation of TG requires careful synchronization of the opening of the gates with respect to the laser pulses, see

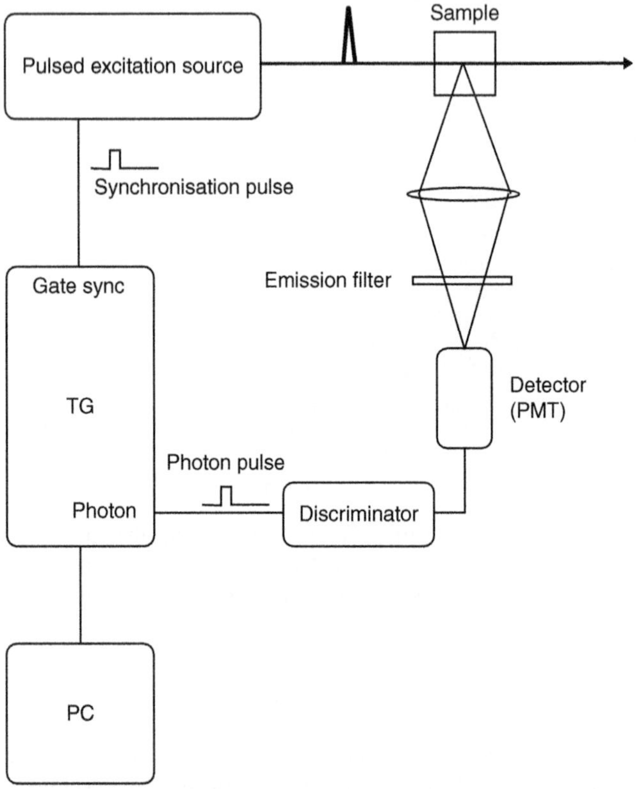

Fig. 3.4. Schematic diagram of a TG setup. The TG electronics need careful synchronization with the excitation pulse. Here, time-gated single photon counting is shown.

Fig. 3.4. In addition, if TG is implemented using SPC, a discriminator is required to separate the photon signal from background noise.

Time-gated detection offers the possibility to suppress background signals correlated with the excitation pulse. Direct and multiple scattered excitation light as well as Raman scattering reaches the detector at $t \approx 0$, and can be effectively suppressed by opening the first gate a few hundred picoseconds after $t = 0$.

This can improve the signal-to-background ratio of the images without a significant loss of signal. Furthermore, TG can be employed to discriminate autofluorescence in biological specimens. Often, autofluorescence has a comparatively short fluorescence lifetime and the signal-to-background ratio of the images can be improved by offsetting the first gate with respect to the excitation pulse [17].

In Fig. 3.5A a comparison between time-gated detection and TCSPC is shown. The time-gated detection system was based on four 2 ns wide gates. The first gate opened about 0.5 ns after the peak of the excitation pulse from a pulsed diode laser. The TCSPC trace was recorded using 1024 channels of 34.5 ps width. The specimen consisted of a piece of fluorescent plastic with a lifetime of about 3.8 ns. In order to compare the results, approximately 1700–1800 counts were recorded in both experiments. The lifetimes obtained with TG and TCSPC amounted to 3.85 ± 0.2 ns and 3.80 ± 0.2 ns respectively, see Fig. 3.5B. Both techniques yield comparable lifetime estimations and statistical errors.

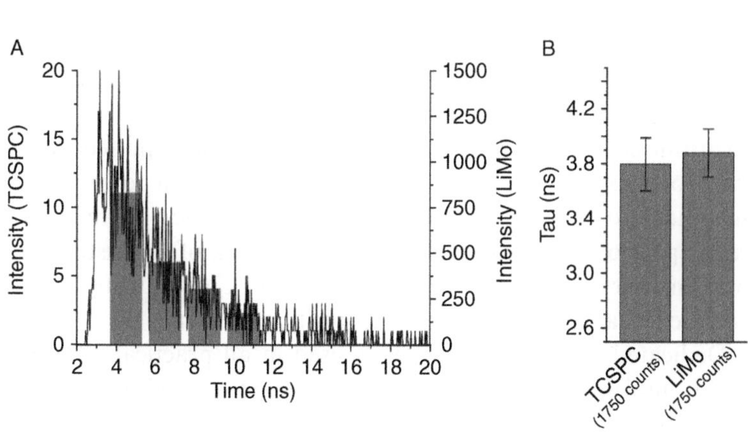

Fig. 3.5. A comparison between TG and TCSPC using the same number of detected photons. (A) The distribution of photons over the time bins. (B) Bar plot of the lifetimes including errors ($n = 4$).

3.3. Point scanning time domain FLIM implementations

Implementation of time domain FLIM methods is comparatively straightforward in laser scanning microscopes (LSMs). Here, point-scanning is used so that single channel lifetime detection suffices. In principle, standard fluorescence lifetime detection equipment developed for spectroscopy can be used in combination with point-scanning systems and a pulsed laser.

3.3.1. Point scanning TCSPC-based FLIM

Conventional TCSPC equipment has been successfully employed in LSM for fluorescence spectroscopy on discrete microscopic volumes [18, 19] and for fluorescence lifetime imaging at a low acquisition speed [1]. The use of conventional TCSPC equipment for imaging results in very long acquisition times, several to many minutes per (time-resolved) image. Importantly, operating the TCSPC detection system at too high detection rates, above 5% of the excitation frequency, results in distortion of the recorded decay curve [20].

At present dedicated TCSPC FLIM boards are commercially available. They are compatible with most LSMS and are easily synchronized with the scanning microscope and pulsed laser. These boards, often plug-in cards for PCs, have a lower dead-time than do the conventional TCSPC electronics intended for use in spectroscopy and the memory bottle neck of the histogramming memory has been removed [21, 22]. Consequently, these dedicated boards provide higher acquisition speeds.

Dedicated TCSPC electronics is used in all practical TCSPC–FLIM implementations [21, 22]. There are several issues that should be noted. First of all, the lifetime acquisition has to be synchronized with the scanning of the confocal or multiphoton microscope. To this end, the pixel clock and often the line and frame synchronization signals of the scanning microscope are used.

After each and every pixel clock pulse, the MCA memory is transferred to buffer memory (either on the TCSPC board or on the PC memory), the MCA memory is reset and accumulation is (re) started. The repetition rate of the laser is typically in the range 20–80 MHz, much higher than common pixel clock rates in lifetime imaging of 10^4–10^5 Hz. Therefore, no synchronization is required between the pulse train coming out of the laser and the pixel clock. The frame synchronization signal can be used to provide an overall start signal for the acquisition and the line synchronization signal can be used to stop acquisition during the retrace of the laser beam.

In TCSPC imaging, the number of time channels is usually restricted to 32–128. In general, a higher number of channels do not provide additional information because only a limited number of detected photons are accumulated per pixel, often several hundred to a few thousand.

3.3.2. Point scanning TG-based FLIM

TG-based FLIM in LSMs is usually implemented using SPC. Here, fast and efficient detection schemes can be employed where the detected photons are counted in the time-gates that are opened sequentially after each and every laser pulse (de Grauw and Gerritsen, 2001; [23]). Similarly to TCSPC, the fluorescence signal from hundreds to thousands of excitation pulses is accumulated at each pixel to obtain sufficient signal level for reliable lifetime analysis. Also TG data acquisition needs to be synchronized with the pixel clock of the scanning microscope. Again, frame and line synchronization signals are used as overall start and line start triggers, respectively.

For the simple TG scheme with only two time-gates, the optimum gate-width amounts to $2.5\,\tau$. Consequently, the total integration time per pulse amounts to $5\,\tau$ and approximately 99% of all photons in the decay are detected. The detected fraction decreases when an offset is applied between the laser pulse and the opening of the first gate. Such an offset is often used to avoid detection of the signal corresponding with the tail of the excitation pulse.

In time-gated photon counting, comparatively high photon count rates can be employed; count rates as high as 10 MHz are often used. TG has the advantage of virtually no dead-time of the detection electronics ($\sim$1 ns), whereas the dead-time of the TCSPC electronics is usually on the order of 125–350 ns. This causes loss of detected photons, and a reduced actual photon economy of TCSPC at high count rates.

3.3.3. Detectors for single photon counting

Detector properties determine to a great extent the performance of TCSPC and TG experiments. In TCSPC the timing resolution of the electronics is in general much better than that of the detector. Often, fast PMTs are used. Here, the timing resolution is limited by the timing jitter in the arrival of electrical pulses at the output of the PMT. This transit time spread (TTS) of fast PMTs is in the order of 25–300 ps, much smaller than the width of the pulses coming out of the PMT, which are usually on the order of $<$1–3 ns. We note that the standard PMTs used in commercial confocal microscopes are not suitable for the detection of fluorescence lifetimes. These PMTs are selected for sensitivity and not for timing resolution. Often they exhibit a TTS of several nanoseconds, which makes them useless for fluorescence lifetime imaging.

Another important property of PMTs is the pulse height distribution. The amplification of individual photoelectrons by the PMT is a stochastic process that causes variations in the gain of individual photoelectrons. As a result significant jitter in the amplitude of the output pulses is observed, see Fig. 3.6. These pulse height variations can be more than a factor of 10. The lowest pulse heights mainly consist of (thermal) noise, indicated by the dashed line in Fig. 3.6. The pulse height distribution exhibits a peak corresponding to detected photons. The threshold level of the

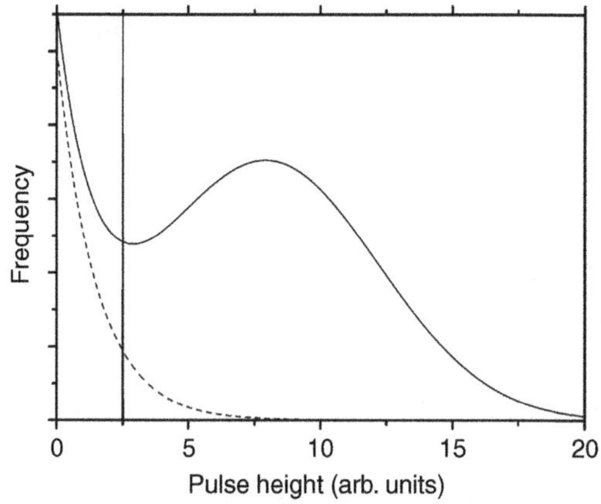

Fig. 3.6. The pulses produced by PMTs show a distribution in pulse heights. The lowest pulses are caused by noise (dashed line) and the higher pulses are due to detected photons. The vertical line indicates the position of the valley in the pulse height distribution. This position would correspond to the optimal setting for the discriminator.

discriminator should be set at the appropriate level to suppress the background signal. Setting the threshold level at the position of the valley between background pulses and the photon pulses is a good compromise between sensitivity and noise suppression (vertical line in Fig. 3.6).

Special "constant fraction" discriminators (CFDs) are being used to realize the highest timing accuracy. This type of discriminator determines the position of the maximum of the pulse coming out of the detector. Conventional discriminators trigger at a constant level of the detector pulses. Because of pulse height variations, constant level triggering results in timing jitter as high as 1–2 ns. In contrast, CFDs yield timing jitter, depending on the type of detector, of 25 to a few hundred ps.

After the detection of a single photon, PMTs need a specific recovery time before they are sensitive again. During this dead-time, no photons can be detected. Because of the stochastic nature of the fluorescence signals, the dead-time reduces the detection probability when the count rate goes up. The relative detection efficiency for a system with a (overall) dead-time t_d at an incident count rate C_i amounts to:

$$f = 1/(1 + t_d C_i) \tag{3.2}$$

At low count rates $C_i \ll 1/t_d$, the detection sensitivity is not affected by the dead-time ($f = 1$). However, at a count rate of $C_i = 1/t_d$ the detection sensitivity is reduced to 50% of its sensitivity at low count rates, see Fig. 3.7. Typical values for dead-times of PMTs are on the order of 50–100 ns.

Not only PMTs and other detectors such as avalanche photodiodes suffer from dead-time effects also the detection electronics may have significant dead-times. Typical dead-times of TCSPC electronics are in the range 125–350 ns. This may seriously impair the efficiency of detection at high count rates. The dead-time effects of the electronics in time-gated single photon detection are usually negligible.

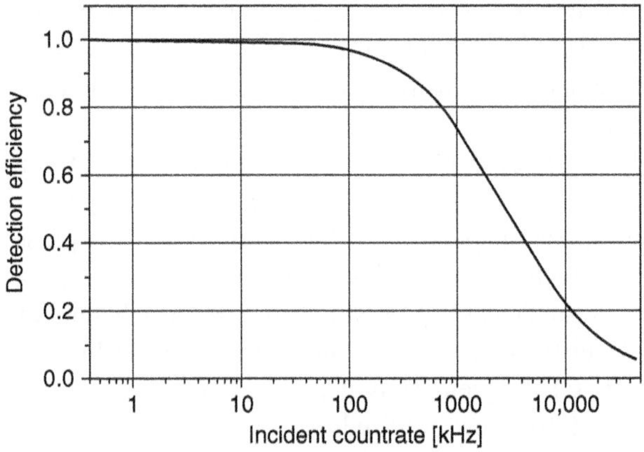

Fig. 3.7. The detection efficiency of a system with a dead-time 350 ns as a function of the incident count rate. At high count rates the detection efficiency reduces due to pileup effects.

Therefore, the throughput of current systems based on time-gated SPC is somewhat higher than in TCSPC-based systems.

Both TCSPC and TG benefit from operation in SPC mode. SPC results in little or no noise and a high photon-economy [10]. Therefore, TCSPC and TG are ideal for high spatial and lifetime resolution imaging [24]. Both techniques offer high image contrast also on dim samples. However, the dead-time of the detectors and the point scanning character limit the throughput of these systems.

Imaging in biology is often affected by other uncertainties than the instrumental sensitivity and precision. Therefore, detectors with very low dead-times, for example PMTs capable of counting at high count rates ($\sim 10^7$ Hz), may be preferred to achieve higher throughputs at the cost of higher timing jitter.

Often, experiments are carried out on specimens that emit only very weak fluorescence. For these cases, the most sensitive detectors should be used, for instance fast avalanche photodiodes or high quantum yield PMTs. These detectors may have somewhat longer dead-times causing longer exposure times but maximal sensitivity.

3.4. Wide field time-domain FLIM implementations

In wide field microscopy, spatial information of the entire image is acquired simultaneously thus providing comparatively short acquisition times compared with scanning microscopy implementations. Combining TCSPC with wide field microscopy is not straightforward. However, a four quadrant anode multichannel plate (MCP) has been used for time- and space-correlated SPC experiments [25, 26]. This detector has excellent timing properties that make it very suitable for FLIM. Unfortunately, it can be operated only at low count-rates ($\sim 10^5$–10^6 Hz); therefore, it requires comparatively long acquisition times (minutes).

A more common approach to time domain wide field FLIM is based on a time-gated image intensifier MCP in combination with a CCD camera (see Fig. 3.8). After every excitation pulse the gated

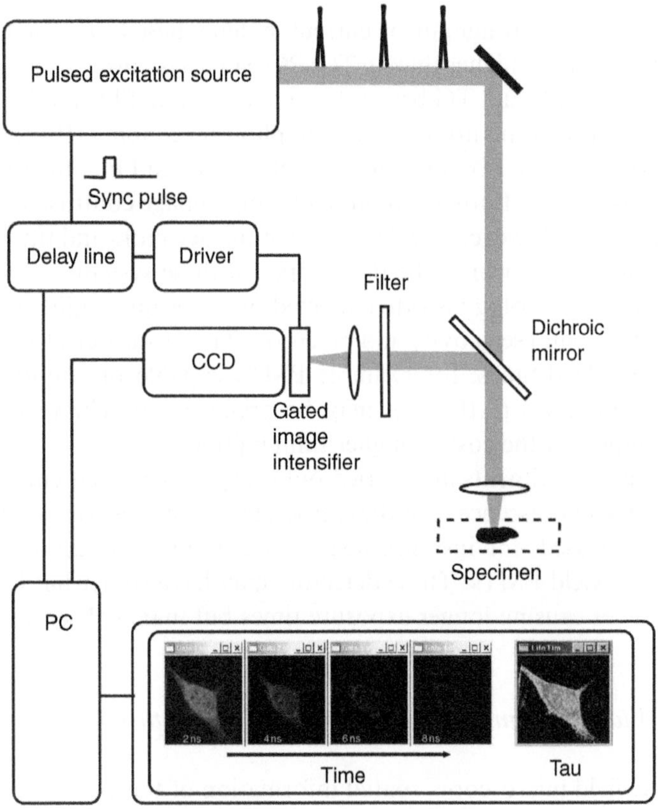

Fig. 3.8. Wide field time-gated FLIM based on a gated image intensifier in combination with a CCD camera.

image intensifier is triggered and its output detected by the CCD camera. After a large number of excitation pulses sufficient signal is detected by the CCD chip and the time-gated image is transferred to the computer. This procedure is repeated for every gate setting. Finally, the lifetime image is calculated from the series of time-gated images. The gate width is determined by the voltage pulse that is applied to the photocathode of the image intensifier. Furthermore, the time offset between the excitation pulse and the opening of the gate is set by a delay line. Data acquisition and gate settings are controlled by the computer.

In contrast to point scanning implementation of TG discussed earlier, wide field TG is an analog detection technique. The sensitivity of analog detection methods is in general lower than that of SPC-based methods. Both the intensifier and the CCD camera introduce noise that reduces the sensitivity of the system. The image intensifier in particular strongly deteriorates the performance of the system. The quantum efficiency and noise properties of the intensifier determine to a great extend the sensitivity; furthermore, the rise time is a critical parameter for the timing properties of the system.

In most implementations of wide field time-gated detection, multiple images are recorded sequentially at different time offsets with respect to the excitation pulse. At present, only few custom-built systems [27, 28] offer the combined advantage of spatial and TG parallelization. This can be implemented by using image splitters and offers the possibility of high speed lifetime imaging at acquisition speeds of hundreds of hertz without artifacts due to sequential opening of gates. In Fig. 3.9, the principle of such a parallelized approach is shown [27]. The fluorescence image generated by pulsed excitation is split into two images by means of a

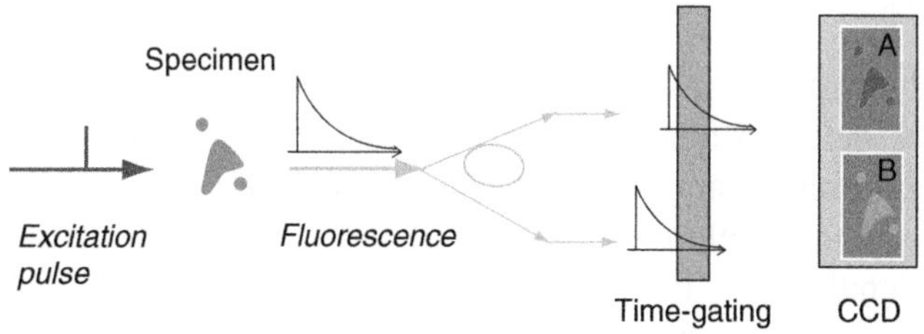

Fig. 3.9. Principle of a wide field FLIM system with simultaneous detection of two time gates. The fluorescence image is split into two images and one of the images is optically delayed with respect to the other. Both images are detected simultaneously with the same time-gated detector.

beam splitter. One of the images is (optically) delayed with respect to the other and both images are projected onto the same gated image intensifier. The gating process results in the simultaneous recording of two time-gated images. The images have the same gate width but a different time offset with respect to the excitation pulse. Based on this approach, fluorescence lifetime images could be recorded at a rate of 100 Hz.

Although new emerging technologies may provide more efficient applications in the future [29, 30], so far in all wide field TG-based FLIM systems, the gating process results in the loss of photons and a consequent reduction of the sensitivity (photon-economy).

In contrast to point scanning TG, wide field TG is comparatively inefficient; only a (small) fraction of the decay is recorded per time-gate acquisition. Nevertheless, this approach generally results in acquisition times that are significantly shorter than in LSM-based FLIM (0.1–10 s).

3.5. Signal considerations and limitations

3.5.1. Noise

There are many sources of noise that affect the performance of fluorescence imaging systems. The sources of noise can be classified into [31]:

(i) *Intrinsic noise* caused by statistics related to the number of detected quanta. This type of noise is often referred to as Poissonian noise or shot-noise. For specific number of detected photons, N_e, the standard deviation of the detected signal will be never less than $N_e^{1/2}$. Consequently, the highest achievable signal-to-noise ratio (SNR) equals $N_e/N_e^{1/2} = N_e^{1/2}$.

(ii) *Subtractive noise* caused by the loss of photons because of inefficiencies in the collection and detection of the fluorescence emission. If the efficiency of the total system equals η, the number of detected photons will be $N_d = \eta N_e$ and the SNR ratio relative to the ideal case will be reduced to $(\eta N_e)^{1/2}$. An important cause for subtractive noise is the limited solid angle (α) of microscope objectives. A high numerical aperture objective may have an α as high as 72 °. This results in detected fraction of ~30% of all photons. Other contributions to subtractive noise include inefficiencies of the optics (e.g., band pass filters) and the limited quantum efficiency of detectors.

(iii) *Additive noise* caused by background light, dark counts (current) of the detector and noise of the electronics. Additive noise only increases the noise and not the signal, therefore reducing the SNR of the system.

(iv) *Multiplicative noise* caused by the uncertainty in the gain of detectors, which results in the increase of noise by a factor a; a will be between 1 and 2. This noise is usually absent in SPC and dominates in systems that employ analog detection.

(v) *Digitization noise* is caused by the digitization of the detector output. Photon counting is not affected by digitization noise due to the discrete nature of the detection.

Noise can be also introduced by *biochemical heterogeneity* of the specimen. This can be a major cause of uncertainty in biological imaging. The high (three-dimensional) spatial resolution of fluorescence microscopy results in low numbers of fluorophores in the detection volume. In a typical biological sample, the number of fluorophores in the detection volume can be as low as 2–3 fluorophores for a confocal microscope equipped with a high NA objective at a fluorescent dye concentration of 100 nM. This introduces another source of noise for imaging applications, chemical or *molecular noise*, related to the inherent randomness of diffusion and the interaction of molecules.

There are several other sources of noise that are specific to lifetime imaging. In particular, noise related to the *timing jitter* of detector and electronics. This causes uncertainties in the time-resolved detection of photons. In absence of timing jitter, the IRF equals the laser pulse shape. The timing jitter of detector and electronics broadens the IRF and deteriorates the lifetime sensitivity and resolution, in particular for short lifetimes.

3.5.2. Photon economy

The performance of a lifetime detection system can be conveniently quantified by a figure-of-merit F, which is defined as the ratio of the SNR in a lifetime measurement and the SNR in an intensity measurement both carried out with the same number of photons. Based on this definition, F can be written as:

$$F = \frac{\Delta\tau}{\tau}\left(\frac{\Delta N}{N}\right)^{-1} = \frac{\sigma_\tau}{\tau}\sqrt{N} \tag{3.3}$$

where $N^{1/2}$ is the intrinsic Poissonian noise for N detected photons. When $F = 1$, lifetime estimations will exhibit the minimal possible noise, whereas $F > 1$ implies a reduced SNR for the lifetime estimation compared to a perfect system. An increase in F can be compensated for by collecting F^2-fold more photons. This will result in the same SNR at the expense of F^2-fold longer exposure times (or higher excitation intensities). Therefore, F^{-2} is a measure for the efficiency with which information is collected and used by an imaging application; F^{-2} can be defined as the photon-economy of a system.

The photon-economy depends on extrinsic sources of noise, the characteristics and settings of the instrument and also on the analysis method. Usually, the photon-economy depends on the lifetime; therefore it is instructive to construct graphs of F as a function of the lifetime. The photon-economy of time-domain techniques has been extensively characterized [10, 14, 32, 33].

SPC techniques are hardly affected by additive noise and multiplicative noise is absent. However, subtractive noise due to the collection efficiency and transmission of optics and the quantum efficiency of the detector do play a role. In addition, at high count rates, the efficiency goes down due to pileup effects.

$$F_{\mathrm{SPC}} = EF = \sqrt{\frac{1 + C_{\mathrm{i}} t_{\mathrm{d}}}{\eta}} \frac{\sigma_\tau}{\tau} \sqrt{N} \tag{3.4}$$

E represents the combined collection and detection efficiency of the system and F the intrinsic photon-economy of the technique. The factor η accounts for the subtractive noise, t_{d} is the dead-time of the detector and C_{i} the count rate of the system.

Methods based on analog detection are usually configured to not suffer from rate-dependent (saturation) effects. Therefore, such effects only occur at very high signal intensities, much higher than used in SPC. Multiplicative (a) and additive noise do make a contribution to the photon economy in analog detection. The effect of additive noise is not straightforward to include in the photon economy description and is here ignored. Inclusion of multiplicative noise in F results in:

$$F_{\mathrm{analog}} > \frac{a}{\sqrt{\eta}} \frac{\sigma_\tau}{\tau} \sqrt{N} = EF \tag{3.5}$$

a is usually in the range 1–2.

An analytical description of the photon-economy and additive noise could be carried out by the estimation of the Fisher-information matrix of the used estimators [34].

Also the relationship between the F-value and imaging parameters such as the number of time-bins, the width of the time-bins and the repetition rate have been studied in detail (de Grauw and Gerritsen, 2001); [10, 14, 32, 33, 35, 36].

When high repetition rate pulsed laser (40–80 MHz) are employed, the fluorescence decay can be resolved over the 12–25 ns between adjacent excitation pulses with high time-resolution. Commercial electronics for TCSPC can digitize the photon arrival time with (sub) picosecond time resolution and the overall time resolution is determined by the detector. Because of the high time resolution and the large number of time-bins, TCSPC exhibits excellent F-values, close to one when the excitation frequency is sufficiently low [33]. TG with SPC can achieve equivalent efficiencies. However, TG is typically implemented with a low number of time-gates (2–8). When only two gates (I_0, I_1) of equal width are employed the rapid lifetime determination algorithm (Eq. 3.2) can be used to calculate the (average) lifetime of the fluorescence decay.

The optimum gate width ΔT for a specific lifetime amounts to 2.5τ. In Fig. 3.10 a typical F–τ curve is shown for time domain lifetime detection with a variable number of time bins and a total detection window (sum of all the time bins/gate widths) of 10 ns (de Grauw and Gerritsen, 2001). The curves are representative for both TCSPC and TG operating in a high excitation frequency mode of

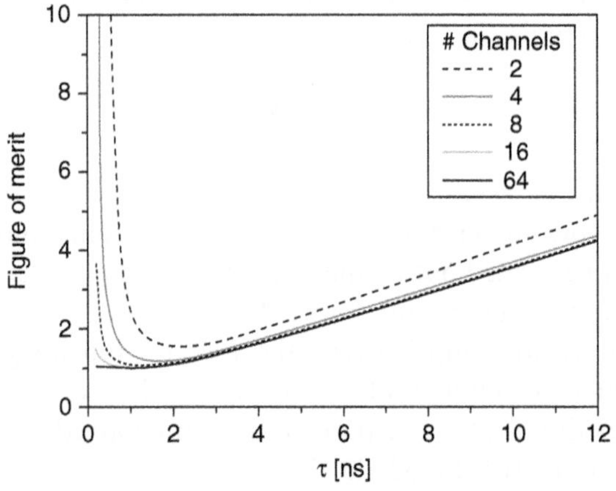

Fig. 3.10. Figure of merit curves for time domain lifetime detection with a variable number of gates. For all the curves the total detection window, the sum of all the gate widths, is 10 ns.

operation. The curves were calculated assuming that no pulse-pileup occurs and that all the TG gates are opened sequentially after the excitation pulse. Moreover, only the intrinsic photon-economy of the technique is taken into account and no additional noise contributions. Also, the effect of the IRF is not taken into account.

The two-gate curve has its minimum ($F \sim 1.5$) at 2 ns and below 1 ns the F-value rapidly increases. Interestingly, the minimum F-values for four and more channels hardly differ at 2 ns. The value is 1.18 and 1.10 for 4 and 64 gates respectively. Only for (very) short lifetimes <500 ps the advantage of a large number of (narrow) gates/time bins shows up. Interestingly, in the case of long detection windows ($\gg\tau$), F converges to 1 for high numbers of gates.

3.5.3. Calibration and accuracy

The accuracy with which a system can measure lifetimes depends on a number of different factors including: calibration of the instrument, the number of detected photons and also the efficiency of the analysis routines. In addition, sources of background and scattered light should be eliminated. Emission filters should be chosen with great care to make sure that no scattered laser light reaches the detector. Detection of scattered excitation light results in a spurious fast component in the decay and complicates the interpretation of the data. The choice of emission filters is much more critical in FLIM than in conventional fluorescence intensity imaging methods.

The time-domain IRF can be comparatively broad and skewed functions. The IRF needs to be taken into account in the data acquisition procedure and analyses to minimize systematic errors in the lifetime determination, in particular if the lifetimes are short.

TCSPC is inherently self-referenced and therefore, with the exception of the regular recording of the IRF, a TCSPC system requires practically no day-to-day calibration. In TCSPC the recording of the decay is started well before the rise of the fluorescence signal and the high timing resolution allows measuring the entire fluorescence rise and decay of the fluorophore. This makes TCSPC insensitive to long-term drift of reference timing signals from, for instance, the lasers source. A (slow) drift in the timing is automatically accounted for in the data analyses procedure.

TG-based microscopes make use of a comparatively low number of gates and only a part of the decay is sampled. The opening of the first gate should be carefully chosen to start after the initial rise of the fluorescence emission. Opening the first gate before or during the rise of the fluorescence will cause a reduction of counts in the first gate relatively to the second and therefore a bias in the lifetime. This needs to be taken into account in the analyses. A further delay of the opening of the first gate can be used for the suppression of (fast) background lifetime contributions, but at the expense of loss of detected photons. Due to the scale-invariant properties of exponentials, a time-shift in the opening of the gates will not alter the measured lifetimes.

A common cause of inaccuracy in SPC-based time domain detection is pulse-pileup, that is, the arrival of photons during the dead-time of the detection system. Because the higher probability of emission (and detection) in the earlier part of the decay, pulse-pileup is more probable in this part of the decay. Consequently, the decay will be distorted and the lifetime will be biased towards higher values. Moreover, pulse-pileup will also result in a reduction of the detection efficiency (see Fig. 3.7 and Eq. (3.4)). Therefore, care should be taken to avoid excitation rates too close to the efficacy count rate (i.e., the inverse of the dead-time) in order to minimize these effects.

For comparatively high repetition-rates (period $T < 5\tau$) fluorescence decays could also overlap between adjacent pulses. Thanks to the scale-invariant properties of the exponentials, no error is introduced when the decay is a pure single-exponential. Conversely, the preexponential factors can be altered when multiple lifetime decays are present, causing bias in the estimation of the (average) lifetimes of the sample.

Finally, it is interesting to note that biases can be introduced by data fitting at low counts even with the use of ordinarily unbiased estimators like the maximum likelihood estimator [37].

Day-to-day drifts of instrumental characteristics, differences in temperature and sample preparation may affect the recorded lifetimes. For instance, differences in temperature can affect both the excited state lifetime of the fluorophore and instrument properties like noise or delays of wirings and electronics.

Importantly, the intrinsic heterogeneity of biological samples can cause lifetime differences between different preparations [38]. Although the broadness of estimated lifetime distributions of single-exponential fluorophores can be narrowed by collecting higher numbers of photons, in biological imaging there will be often broader lifetime distribution because of biochemical intracellular heterogeneity and differences among cells and preparations.

It is recommended to characterize these errors in order to estimate the statistical relevance of the measurement. Importantly, relative estimates are usually less prone to errors and may offer higher sensitivities. For example, in FRET–FLIM experiments the ratio of the donor lifetime in the absence and presence of an acceptor is measured. This offers a higher precision than absolute lifetime values.

3.5.4. Lifetime resolution and lifetime heterogeneity

The lifetime resolution is the smallest variation in lifetime that can be detected. If external noise sources are ignored, the lifetime resolution depends essentially on the photon-economy of the system. For instance, if a 2 ns lifetime is measured with a 4 gate TG single-photon counting FLIM ($F = 1.3$) and 1000 photons, variations of about 80 ps can be resolved. However, for reasons discussed earlier, in biological samples these values could be higher.

To optimize resolution in lifetime-based assays, a comparison of relative estimates is always favorable. If the FLIM experiment is carried out in an environment where temperature cannot be tightly controlled, it is also convenient to cycle between different samples during the same experimental session, in order to average out thermal and other instrumental drifts. When applicable, this practice may be useful to suppress any nonrandom variation in the detection.

Lifetime heterogeneity can be analyzed by fitting the fluorescence decays with appropriate model function (e.g., multiexponential, stretched exponential, and power-like models) [39]. This, however, always requires the use of additional fitting parameters and a significantly higher number of photons should be collected to obtain meaningful results. For instance, two lifetime decays with time constants of 2 ns, 4 ns and a fractional contribution of the fast component of 10%, requires about 400,000 photons to be resolved at 5% confidence [33].

3.5.5. Shortest lifetimes

The shortest lifetime that an instrument can measure is mainly determined by the instrument characteristics and for sufficiently short excitation pulses it is limited by the timing-jitter of detector and electronics. The most accurate detectors are MCP–PMTs that exhibit TTSs as low as 25 ps but other commonly used detectors may have values as high as 300 ps. At very high count numbers, the lifetime resolution can be virtually infinitely high; however, the timing jitter of the system is a practical detection limit for the shortest measurable lifetime at any realistic SNR. Moreover, the IRF needs to be taken into account to measure lifetimes that are on the order of the width of the IRF or shorter. Contrary to time-resolved spectroscopy, measuring lifetimes on the order of the width of the IRF is challenging in lifetime imaging due to the significantly lower numbers of photons per pixels that are typically detected.

3.5.6. Acquisition throughput

The acquisition throughput of a microscope is often determined by photon statistics, but depends also on many parameters including instrumental limitations, for example, the read-out and dead-time of the detector and electronics [40].

For instance, a time-gated SPC microscope with four gates exhibits a better maximal ($F_{TG4} = 1.3$) photon-economy than a two-gated system ($F_{TG2} = 1.5$). The latter setup will thus require $(F_{TG2}/F_{TG4})^2 = 1.7$-fold longer acquisition time than the former because of photon-statistics alone.

Throughput also depends on the degree of parallelization of the detection system. In wide-field imaging the fluorescence emission needs to be acquired and gated sequentially by a MCP or simultaneously by the use of image splitters. In both the cases, however, only a fraction of photons equal to the inverse of the number of gates is collected because of the gating process. Based on the sequential acquisition setup, a two gate system will be $(4/2)(F_{TG4}/F_{TG2})^2 \sim 1.5$ times faster than a four-gate system. On the other hand, the importance of parallelization lies in the minimization of read-out time-lags, photobleaching, and motion artefacts, which could affect wide-field imaging when performed by the sequential acquisition of the images. Such systems proved capable of acquisition rates in excess of 100 Hz [27, 41]. TCSPC can be also performed using wide-field detectors. These detectors cannot, however, sense more than one photon at a time and are inherently slow [26]. Therefore, at present wide-field TCSPC systems do not take advantage of the parallelization in terms of acquisition speed.

SPC techniques offer the advantage of low noise detection, providing F-values of 1–2 times lower than in analog detection. Although this can in principle result in four times faster acquisition speed this gain in speed is not realized in practice. In SPC, the comparatively high dead-times of detectors and electronics limits the acquisition speed. SPC system should be operated at count rates below the inverse of the dead-time of the system (electronics or detector). Above this count rate the lifetime may be distorted. Moreover, the efficiency of the system goes down because of pileup effects (see Fig. 3.7)

Because of the low timing-jitter (down to 25 ps) TCSPC-based systems are often equipped with a MCP-PMT at detriment of acquisition speed ($<10^6$ counts per second). On the other hand, a TG-SPC system equipped with four gates and a fast PMT (10 MHz) could be slower than a TCSPC at low count-rates (<100 kHz), because of a lower photon-economy. However, already at 1 MHz, the former would be almost three times faster and more the one order of magnitude faster at 10 MHz.

The above limitations are implementation dependent and no intrinsic limitations. The throughput of TCSPC can for instance be improved by the use of multiple detectors, and multiple TCSPC boards [42]. The photon-economy of TGSPC could be optimized somewhat by increasing the number of gates.

3.6. Data analysis

Time-domain detection results in histograms of photon arrival times. In time-correlated SPC, the time-bins correspond to the analog-to-digital conversion levels, whereas in time-gated SPC, the time-bins are the time-windows during which the photon-counting gate is activated.

The analysis of the histograms of photon arrival times is equivalent in both cases and relies on fitting appropriate model functions to the measured decay. The selection of the fitting model depends on the investigated system and on practical considerations such as noise. For instance, when a cyan fluorescent protein (CFP) is used, a multi-exponential decay is expected; furthermore, when CFP is used in FRET experiments more components should be considered for molecules exhibiting FRET. Several thousands of photons per pixel would be required to separate just two unknown fluorescent decay times; a signal level that can often not be realized in biological imaging.

On the basis of a priori knowledge of the system, the number of fit parameters can be sometimes constrained. In the case of a FRET imaging experiment, the lifetime component corresponding to the donor molecules that do not exhibit FRET can be assumed constant. Now, the fitting would require one fit parameter less and consequently fewer detected photons per pixel are required for a reliable fit.

Lifetime heterogeneity itself can be the target of the measurement. In this case, high photon counts and alternative model functions like stretched exponentials and power-distribution-based models can be used [39, 43]. These provide information on the degree of heterogeneity of the sample with the addition of only one fit parameter compared with single exponential fits.

It is not uncommon to detect only a few hundred photons per pixel or less. Therefore, spatial binning of the data may be necessary to obtain the sufficient signal for a reliable fit. At low counts, also rebinning of the time histograms may be beneficial to avoid "empty bins" and increase the efficiency of the fit. TD-FLIM can be implemented with only two time-gates. By using Eq. (3.1)—the rapid lifetime determination method—the average lifetime of the sample can be estimated without any fitting. This method offers high speed in both acquisition and analysis, but a comparatively low photon-economy and accuracy. The latter is partly due to lack of background correction.

Generally, inaccuracies can also be expected at low photon counts ($N < 100$). Besides comparatively large statistical fluctuations, also a bias in lifetime is introduced by the data fitting procedure [37].

As a criterion for the quality of the fit usually the reduced χ^2 is used by the fitting algorithm. This function is defined as:

$$\chi^2 = \sum_{i=1}^{N} \frac{1}{\sigma_i^2} \left(y_i - f(x_i) \right)^2 \tag{3.6}$$

Here, N is the number of data points (time bins), y_i the measured intensity in time bin i, σ_i^2 the measurement error (variance) for y_i, x_i the time position of bin i, and f the theoretical function describing the decay.

The fitting algorithm minimizes the χ^2, or another goodness-of-fit function [44], to minimize the difference between the experimental data and the fit model. In spectroscopic measurements, high values of the reduced χ^2 (>1.4) indicate that the model may be not a good representation of the experimental data and a different fit function may have to be selected. However, in lifetime imaging experiments even values higher than 2 may be acceptable because systematic errors cannot be always excluded in FLIM data sets [20]. On the other hand, a reduced χ^2 close to one implies only that the model fits the data with high accuracy within the experimental errors. This does not necessarily mean that the model itself is a realistic model. If the model has too many fit parameters, the fitting function will yield low χ^2 values without reliable values for the estimated parameters.

Furthermore, at very low photon counts in general low values of the reduced χ^2 ($\approx$1) are obtained because of the high noise level in the data. Therefore, the reduced χ^2 and other goodness of parameters should be used with caution.

Also global fitting techniques, where the space invariance (or any other invariance property) of one or more fitting parameters is exploited, have been successfully used to analyze fluorescence lifetime images [45, 46]. When applicable, global analysis techniques provide more homogeneous SNRs and reduce the number of fitted parameters.

In the analyses of the FLIM data, it is important to set an appropriate threshold on the number of counted photons above which the data is fitted. If a pixel contains a number of counts below the threshold, the fit parameters are usually set to 0, that is, those pixels are masked out.

When for a single exponential fit an error of <5% is required, the threshold should be set at 400–700 counts, depending on the

F-value of the system. The exact number of total counts per (eventually binned) pixel can be calculated from the F-value using Eq. (3.3). For precise measurements, care should be taken that the contribution of autofluorescence to the measurement is negligible. This may mean that the threshold has to be raised to comparatively high values.

The inspection of the fit residuals, that is, the (normalized) differences between the experimental and fitted data point, is a reliable tool to check for deviations from the fitted model. Residuals should be statistically noncorrelated and randomly distributed around zero. For example, if a bi-exponential decay is fitted to a single exponential function, the residuals will show systematic errors. Therefore, correlations in the residuals may indicate that another fit model should be used.

Recently, a method used for the analysis of frequency-domain data has been proposed for the analysis of time-domain images. The AB-plot or phasor plot provides a useful graphical representation of lifetime data that can be used for the segmentation of the images prior to data fitting [47, 48]. With this method, data fitting may be avoided in many instances.

3.7. FRET–FLIM example application

An important application of fluorescence lifetime imaging is the imaging of molecular colocalization based on FRET imaging, see Chapters 1 and 2). The energy transfer from a fluorescent donor molecule to a matched acceptor introduces an additional deactivation pathway for the donor's excited state. As a result, the donor's fluorescence lifetime is reduced. Therefore, the occurrence of FRET can be imaged using FLIM. Figure 3.11, shows an example of a FRET-FLIM experiment in which two lipid raft markers, GPI-GFP (donor) and CTB-Alexa594 (acceptor), are present in the plasma membrane of NIH 3T3 cells. The lifetime images were recorded using a confocal microscope equipped with a four channel

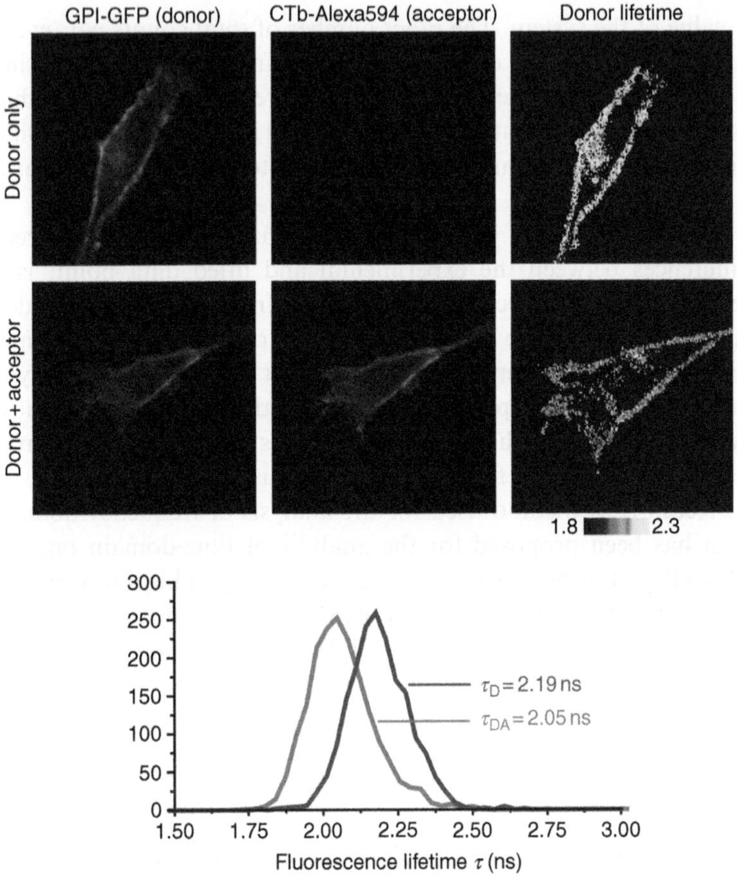

Fig. 3.11. FRET–FLIM experiment to study colocalization of two lipid raft markers, GPI-GFP and CTB-Alexa594. The rows of images show intensity and lifetime images of donor-labeled and donor + acceptor-labeled cells. The histogram shows the lifetime distribution of the whole cells. The FRET efficiency is ~6%. (See Color Insert.)

time-gated photon counting module with 2-ns wide time gates (LIMO) [23]. A low power pulsed 440 nm laser diode was used with a pulse width of about 100 ps and the power at the specimen was approximately 50 μW. Because of the low excitation power the acquisition time was comparatively long, about 60 s.

The top series of images in Fig. 3.11 was recorded on a specimen containing donor only. Left and middle images show the intensity images of donor and acceptor channel respectively. As expected no signal is present in the acceptor channel and the donor channel shows clear labeling of the plasma membrane. On the right, the donor-only FLIM image is shown. Here, a threshold was set at 900 counts and the maximum number of counts amounted to about 2500 counts. Only above the threshold the lifetimes were calculated. The donor-only image shows a fairly constant lifetime over the whole image. From the lifetime histograms of the image, an average lifetime over the whole cell of 2.19 ns was found.

The presence of the acceptor, lower row of images, results in a clear reduction in the lifetime to about 2.05 ns. The reduction corresponds to a 6% FRET efficiency. Experiments on more cells ($N = 4$, not shown) confirms that this reduction is indeed significant and that the two lipid raft markers colocalize in the plasma membrane.

A closer look at the data shows the lifetime distributions are comparatively broad, about 0.25 ns for both distributions. This is in fact much broader than what one would expect from photon statistics alone. Based on realistic F-values (1.2–1.5) lifetime images recorded with this many counts are expected to yield distributions with widths on the order of 0.1 ns. The broadening is therefore not because of photon statistics. Variations in the microenvironment of the GFP are the most likely source of the lifetime heterogeneities. Importantly, such sensitivity for local microenvironment may be the source of apparent FRET signals. In this particular FRET–FLIM experiment, we found that the presence of CTB itself without the acceptor dye already introduced a noticeable shift of the donor lifetime. Therefore, in this experiment the donor-only lifetime image was recorded after unlabeled CTB was added to the cells. The low FRET efficiency and broadened lifetime distribution call for careful control experiments and repeatability checks.

For this particular experiment, we also carried out additional photobleaching measurements. Here, the acceptor was photo bleached and it was found that the donor lifetime returned to the donor only lifetime of about 2.19 ns. This confirms that the observed lifetime reduction is indeed due to FRET [49].

References

[1] Bugiel, I., König, K. and Wabnitz, H. (1989). Investigation of cells by fluorescence laser scanning microscopy with subnanosecond time resolution. Laser Life Sci. *3*, 1–7.

[2] Buurman, E. P., Sanders, R., Draaijer, A., Gerritsen, H. C., Van Veen, J. J. F., Houpt, P. M. and Levine, Y. K. (1992). Fluorescence lifetime imaging using a confocal laser scanning microscope. Scanning *14*, 155–9.

[3] Schneckenburger, H. and Konig, K. (1992). Fluorescence decay kinetics and imaging of Nad(P)H and flavins as metabolic indicators. Opt. Eng. *31*, 1447–51.

[4] Wang, X. F., Uchida, T., Coleman, D. M. and Minami, S. (1991). A 2-dimensional fluorescence lifetime imaging-system using a gated image intensifier. Appl. Spectrosc. *45*, 360–6.

[5] Gadella, T. W., Jovin, T. M. and Clegg, R. M. (1993). Fluorescence lifetime imaging microscopy (FLIM) - spatial resolutions of microstructures on the nanosecond time scale. Biophys. Chem. *48*, 221–39.

[6] Lakowicz, J. R. and Berndt, K. W. (1991). Lifetime-selective fluorescence imaging using an Rf phase- sensitive camera. Rev. Sci. Instrum. *62*, 1727–34.

[7] Morgan, C. G., Mitchell, A. C. and Murray, J. G. (1990). Nano-second time-resolved fluorescence microscopy: principles and practice. Trans. R. Microsc. Soc. *1*, 463–6.

[8] Becker, W., Bergmann, A., Hink, M. A., Konig, K., Benndorf, K. and Biskup, C. (2004b). Fluorescence lifetime imaging by time-correlated single-photon counting. Microsc. Res. Tech. *63*, 58–66.

[9] O'Connor, D. V. and Phillips, D. (1984). Time-Correlated Single Photon Counting. Academic press, London.

[10] Gerritsen, H. C., Asselbergs, M. A., Agronskaia, A. V. and Van Sark, W. G. (2002a). Fluorescence lifetime imaging in scanning micro-

scopes: Acquisition speed, photon economy and lifetime resolution. J. Microsc. *206*, 218–24.

[11] Bohmer, M., Pampaloni, F., Wahl, M., Rahn, H. J., Erdmann, R. and Enderlein, J. (2001). Time-resolved confocal scanning device for ultrasensitive fluorescence detection. Rev. Sci. Instrum. *72*, 4145–52.

[12] Esposito, A., Gerritsen, H. C., Wouters, F. S. and Resch-Genger, U. (2007b). Fluorescence lifetime imaging microscopy: Quality assessment and standards. In: Standardization in Fluorometry: State of the Art and Future Challenges (Wolfbeis, O. S., ed.). Springer, Berlin Heidelberg New York.

[13] Stiel, H., Teuchner, K., Paul, A., Leupold, D. and Kochevar, I. E. (1996). Quantitative comparison of excited state properties and intensity- dependent photosensitization by rose bengal. J. Photoch. Photobiol. B *33*, 245–54.

[14] Ballew, R. M. and Demas, J. N. (1989). An error analysis of the rapid lifetime determination method for the evaluation of single exponential decays. Anal. Chem. *61*, 30–3.

[15] Scully, A. D., Ostler, R. B., Phillips, D., O'Neill, P. O., Townsend, K. M., Parker, A. W. and MacRobert, A. J. (1997). Application of fluorescence lifetime imaging microscopy to the investigation of intracellular PDT mechanisms. Bioimaging *5*, 9–18.

[16] Siegel, J., Elson, D. S., Webb, S. E. D., Lee, K. C. B., Vlanclas, A., Gambaruto, G. L., Leveque-Fort, S., Lever, M. J., Tadrous, P. J., Stamp, G. W. H., Wallace, A. L., Sandison, A., *et al.* (2003). Studying biological tissue with fluorescence lifetime imaging: microscopy, endoscopy, and complex decay profiles. Appl. Optics *42*, 2995–3004.

[17] Vroom, J. M., De Grauw, K. J., Gerritsen, H. C., Bradshaw, D. J., Marsh, P. D., Watson, G. K., Birmingham, J. J. and Allison, C. (1999). Depth penetration and detection of pH gradients in biofilms by two-photon excitation microscopy. Appl. Environ. Microb. *65*, 3502–11.

[18] Ghiggino, K. P., Harris, M. R. and Spizzirri, P. G. (1992). Fluorescence lifetime measurements using a novel fiberoptic laser scanning confocal microscope. Rev. Sci. Instrum. *63*, 2999–3002.

[19] Van der Oord, C. J. R., Gerritsen, H. C., Rommerts, F. F. G., Shaw, D. A., Munro, I. H. and Levine, Y. K. (1995). Microvolume time-resolved fluorescence spectroscopy using a confocal synchrotron-radiation microscope. Appl. Spectrosc. *49*, 1469–73.

[20] Lakowicz, J. R. (1999). Principles of Fluorescence Spectroscopy. Plenum press, New York.

[21] Becker, W., Bergmann, A., Biskup, C., Kelbauskas, L., Zimmer, T., Klöcker, N. and Benndorf, K. (2003). High resolution TCSPC lifetime imaging. Proc. SPIE *4963*, 175–84.
[22] Kwak, E. S. and Vanden Bout, D. A. (2003). Fully time-resolved near-field scanning optical microscopy fluorescence imaging. Anal. Chim. Acta *496*, 259–66.
[23] Van der Oord, C. J., de Grauw, C. J. and Gerritsen, H. C. (2001). Fluorescence lifetime imaging module LIMO for CLSM. Proc. SPIE *4252*, 119–23.
[24] Gratton, E., Breusegem, S., Sutin, J., Ruan, Q. and Barry, N. (2003). Fluorescence lifetime imaging for the two-photon microscope: time-domain and frequency-domain methods. J. Biomed. Opt. *8*, 381–90.
[25] Jose, M., Nair, D. K., Reissner, C., Hartig, R. and Zuschratter, W. (2007). Photophysics of clomeleon by FLIM: Discriminating excited state reactions along neuronal development. Biophys. J. *92*, 2237–54.
[26] Kemnitz, K., Pfeifer, L. and Ainbund, M. R. (1997). Detector for multi-channel spectroscopy and fluorescence lifetime imaging on the picosecond timescale. Nucl. Instrum. Meth. Phy. Res. A *387*, 86–7.
[27] Agronskaia, A. V., Tertoolen, L. and Gerritsen, H. C. (2003). High frame rate fluorescence lifetime imaging. J. Phys. D: Appl. Phys. *36*, 1655–62.
[28] Elson, D. S., Munro, I., Requejo-Isidro, J., McGinty, J., Dunsby, C., Galletly, N., Stamp, G. W., Neil, M. A. A., Lever, M. J., Kellett, P. A., Dymoke-Bradshaw, A., Hares, J., *et al.* (2004). Real-time time-domain fluorescence lifetime imaging including single-shot acquisition with a segmented optical image intensifier. New J. Phys. *6*, 180–93.
[29] Mosconi, D., Stoppa, D., Pancheri, L., Gonzo, L. and Simoni, A. (2006). CMOS single-photon avalanche diode array for time-resolved fluorescence detection. IEEE ESSCIRC. 564–67.
[30] Niclass, C., Gersbach, M., Henderson, R., Grant, L. and Charbon, E. (2007). A single photon avalanche diode implemented in 130-nm CMOS technology. IEEE J. Sel. Top. Quant. Electron *13*, 863–9.
[31] Pawley, J. (2006). Handbook of Biological Confocal Microscopy, 3rd edition. Springer Verlag, New York.
[32] Gerritsen, H. C., Draaijer, A., van den Heuvel, D. J. and Agronskaia, A. V. (2006). Fluorescence lifetime imaging in scanning microscopy. In: Handbook of Biological Confocal Microscopy (Pawley, J., ed.). Plenum Press, New York, pp. 516–33.
[33] Kollner, M. and Wolfrum, J. (1992). How many photons are necessary for fluorescence-lifetime measurements? Chem. Phys. Lett. *200*, 199–204.

[34] Watkins, L. P. and Yang, H. (2004). Information bounds and optimal analysis of dynamic single molecule measurements. Biophys. J. *86*, 4015–29.

[35] Ballew, R. M. and Demas, J. N. (1991). Error analysis of the rapid lifetime determination method for single exponential decays with a non-zero base-line. Anal. Chim. Acta *245*, 121–7.

[36] Moore, C., Chan, S. P., Demas, J. N. and DeGraff, B. A. (2004). Comparison of methods for rapid evaluation of lifetimes of exponential decays. Appl. Spectrosc. *58*, 603–7.

[37] Tellinghuisen, J. and Wilkerson, C. W. (1993). Bias and precision in the estimation of exponential decay parameters from sparse data. Anal. Chem. *65*, 1240–6.

[38] Grailhe, R., Merola, F., Ridard, J., Couvignou, S., Le Poupon, C., Changeux, J. P. and Laguitton-Pasquier, H. (2006). Monitoring protein interactions in the living cell through the fluorescence decays of the cyan fluorescent protein. Chemphyschem *7*, 1442–54.

[39] Lee, K. C. B., Siegel, J., Webb, S. E. D., Leveque-Fort, S., Cole, M. J., Jones, R., Dowling, K., Lever, M. J. and French, P. M. W. (2001). Application of the stretched exponential function to fluorescence lifetime imaging. Biophys. J. *81*, 1265–74.

[40] Esposito, A., Gerritsen, H. C. and Wouters, F. S. (2007a). Optimizing frequency-domain fluorescence lifetime sensing for high-throughput applications: photon economy and acquisition speed. J. Opt. Soc. Am. *24*, 3261–73.

[41] Agronskaia, A. V., Tertoolen, L. and Gerritsen, H. C. (2004). Fast fluorescence lifetime imaging of calcium in living cells. J. Biomed. Opt. *9*, 1230–7.

[42] Becker, W., Bergmann, A., Biscotti, G., König, K., Riemann, I., Kelbauskas, L. and Biskup, C. (2004a). High-speed FLIM data acquisition by time-correlated single photon counting. Proc. SPIE *5223*, 1–9.

[43] Wlodarczyk, J. and Kierdaszuk, B. (2003). Interpretation of fluorescence decays using a power-like model. Biophys. J. *85*, 589–98.

[44] Awaya, T. (1979). New method for curve fitting to the data with low statistics not using the Chi-2-method. Nucl. Instrum. Methods *165*, 317–23.

[45] Barber, P. R., Ameer-Beg, S. M., Gilbey, J. D., Edens, R. J., Ezike, I. and Vojnovic, B. (2005). Global and pixel kinetic data analysis for FRET detection by multi-photon time-domain FLIM. *In* Multiphoton Microscopy in the Biomedical Sciences V.Vol. 5700. SPIE, San Jose, CA, USA, pp. 171–81.

[46] Pelet, S., Previte, M. J., Laiho, L. H. and So, P. T. (2004). A fast global fitting algorithm for fluorescence lifetime imaging microscopy based on image segmentation. Biophys. J. *87*, 2807–17.

[47] Clayton, A. H., Hanley, Q. S. and Verveer, P. J. (2004). Graphical representation and multicomponent analysis of single-frequency fluorescence lifetime imaging microscopy data. J. Microsc. *213*, 1–5.

[48] Digman, M., Caiolfa, V. R., Zamai, M. and Gratton, E. (2007). The Phasor approach to fluorescence lifetime imaging analysis. Biophys. J. *94*, L14–16.

[49] Barzda, V., de Grauw, C. J., Vroom, J., Kleima, F. J., van, G. R., van, A. H. and Gerritsen, H. C. (2001). Fluorescence lifetime heterogeneity in aggregates of LHCII revealed by time-resolved microscopy. Biophys. J. *81*, 538–46.

[50] de Grauw, C. J. and Gerritsen, H. C. (2001). Multiple time-gate module for fluorescence lifetime imaging. Appl. Spectro. *55*(6), 670-78.

Laboratory Techniques in Biochemistry and Molecular Biology, Volume 33
FRET and FLIM Techniques
T. W. J. Gadella (Editor)

CHAPTER 4

Multidimensional fluorescence imaging

James McGinty, Christopher Dunsby, Egidijus Auksorius, Richard K. P. Benninger, Pieter De Beule, Daniel S. Elson, Neil Galletly, David Grant, Oliver Hofmann, Gordon Kennedy, Sunil Kumar, Peter M. P. Lanigan, Hugh Manning, Ian Munro, Björn Önfelt, Dylan Owen, Jose Requejo-Isidro, Klaus Suhling, Clifford B. Talbot, P. Soutter, M. John Lever, Andrew J. deMello, Gordon S. Stamp, Mark A. A. Neil, and Paul M. W. French

Photonics Group, Department of Physics, Imperial College London, South Kensington Campus, London SW7 2AZ, UK

Fluorescence measurement techniques are being ever more widely applied in biology, medicine, and numerous other fields, utilizing both exogenous fluorescent labels and naturally occurring fluorophores to provide information concerning different molecular species, states, and environments. To extract the desired information, the fluorescence signal can be resolved with respect to a number of dimensions (e.g., spatial, spectral, temporal, polarization). Such measurements were initially undertaken in single-channel cuvette-based instruments but the increasing demand for molecular spatiotemporal information to study complex and dynamic systems, such as signal pathways in cell biology, has stimulated the development of multidimensional fluorescence imaging instrumentation. The work described

DOI: 10.1016/S0075-7535(08)00004-1

in this chapter begins with wide-field, time-gated fluorescence lifetime imaging (FLIM), which is a core technology for such instrumentation. This approach to FLIM reduces the total acquisition time compared with photon counting techniques and the combination with wide-field optical sectioning techniques, such as Nipkow disc microscopy, provides a powerful means to study live cells and other dynamic systems. The extension to spectrally resolved FLIM is then described, including excitation-resolved FLIM, using a supercontinuum source to provide tunable pulsed excitation, and hyperspectral FLIM that is implemented in a line-scanning microscope. The rich content of such spectro-temporal data sets is illustrated using tissue autofluorescence and a complex multiply stained sample. Finally, the combination of polarization-resolved and time-gated imaging is shown to map both fluorescence lifetime and rotational correlation time, as demonstrated by an application to microfluidic devices.

4.1. Introduction

As discussed elsewhere in this volume, fluorescence provides optical molecular contrast for applications in biology, medicine, and other fields. Fluorophores can be used as "labels" to tag specific molecules of interest or the autofluorescence of target molecules themselves can be exploited to provide label-free intrinsic contrast. Furthermore, the sensitivity of the fluorescence process to the local fluorophore environment can provide a sensing function. Fluorescence-based measurements are made in a wide range of instruments, including cuvette-based systems, microscopes, endoscopes, and multiwell plate readers. In principle, fluorescence may be analyzed with respect to the excitation and emission spectra, the quantum efficiency, the polarization response, and the fluorescence decay profile [1]. Conventionally, however, fluorescence imaging has mainly been concerned with mapping the distribution of fluorophores, deriving information concerning localization from intensity measurements, while experiments resolving excitation, emission, lifetime, and polarization information have typically been undertaken using cuvette-based instruments such as spectrophotometers and spectrofluorometers. The limited information obtained from fluorescence imaging has been partly due to the relatively weak fluorescence signal available from biological samples and partly due to instrumentation limitations—particularly with respect to the sparse availability of suitable light sources for excitation- and lifetime-resolved imaging.

The last decade has seen tremendous changes in both imaging technology and user aspirations. The commercial availability of user-friendly tunable femtosecond Ti:Sapphire lasers stimulated the wide-spread uptake of multiphoton microscopy [2] and subsequently of fluorescence lifetime imaging (FLIM), particularly when implemented as a relatively straightforward and inexpensive "add-on" using the time-correlated single photon counting (TCSPC) method. FLIM [3, 4] provides a robust means to interrogate fluorescence signals, its inherently ratiometric nature provids reduced sensitivity to fluorophore concentration, optical path length, sample absorption, scattering, and background fluorescence, which often affect intensity measurements. The dependence of the fluorescence decay time on both the radiative and nonradiative decay rates provides the means to quantitate changes in local fluorophore environment as well as to contrast different fluorescent molecular species. The sensitivity to the decay rate also makes FLIM an attractive approach to image intermolecular and intramolecular interactions using Förster resonant energy transfer (FRET) [5, 6].

In our laboratory, FLIM has been applied to a range of applications, from the study of protein–protein interactions for molecular biology and drug discovery, through the label-free diagnosis of diseased tissue to the assessment of organic semiconductor devices. For most current applications, however, we do not consider fluorescence lifetime in isolation but are developing multidimensional fluorescence imaging (MDFI) instrumentation to also provide spectral or polarization resolution, as well as optical sectioning. We note that for any FLIM experiment there is always some selection (often constrained) of the excitation and the detection spectral properties and, in order to optimize this selection, it is useful to know the spectrotemporal properties of the fluorescence signal. For fluorescence imaging in general, spectral resolution often provides more information when maximizing or investigating the contrast between different fluorophores or fluorophore environments. To achieve further specificity, excitation–emission matrices can provide a superior separation of different fluorophores than do consideration of the emission or excitation spectra alone. Thus the ability to multiplex different fluorescent labels, to discriminate between different endogenous fluorophores, and to probe fluorophore environments may be enhanced by resolving fluorescence lifetime and emission and/or excitation spectra. An important consideration, however, is that most samples will only emit a limited number of fluorescence photons before the onset of photobleaching or photodamage. There is therefore a limited photon "budget" that must be "spent" carefully to maximize the information that can be obtained from a sample. This photon budget may be reduced further for many biological experiments (particular those undertaken *in vivo*) that require imaging on a certain time scale or that can only tolerate limited excitation irradiance. As the photon budget is allocated over multiple measurement dimensions, the signal-to-noise (S/N) ratios will be reduced and so it is important to maximize the photon efficiency and to judiciously sample the measurement dimensions such that the information per photon is also maximized. In practice, the optimum approach will, of course, vary with the application depending on a priori knowledge of the fluorescence properties and the aims of the investigator.

This chapter describes the development and first applications of MDFI instrumentation. It will first discuss FLIM, with an emphasis on rapid wide-field time-gated imaging, including application to molecular biology, FLIM of tissue autofluorescence, and high-speed optically sectioned FLIM for live cell imaging. Optical sectioning is important to enhance image contrast and to minimize unwanted mixing of signals from axially separate fluorophores.

The extension to spectral FLIM is then discussed, including (emission-resolved) hyperspectral FLIM implemented using line-scanning microscopy and excitation-resolved imaging and FLIM utilizing supercontinuum generation to provide excitation throughout the visible spectrum. The combination of polarization-resolved and time-resolved fluorescence imaging is then described, mapping both lifetime and rotational correlation time as illustrated by an application to microfluidic devices. Finally, there is a discussion of future trends.

4.2. FLIM and wide-field time-gated imaging

Our laboratory has a background in ultrafast laser technology and we have focused on time-domain FLIM instrumentation, initially based on wide-field time-gated imaging and more recently also TCSPC implemented in single photon confocal and multiphoton scanning microscopes. These techniques are discussed in Chapter 3 and elsewhere, see also Refs. [3, 7]. In general, TCSPC produces high-quality FLIM data provided sufficient photons are detected—the S/N ratio improves indefinitely as the number of detected photons increases and the decay profiles are typically sampled with a large number of time bins. This is particularly useful when imaging fluorophores with complex exponential decays of which there is little or no a priori knowledge. TCSPC is usually implemented in laser scanning microscopes that provide excellent image quality, but for which the serial pixel acquisition limits the image acquisition speed. The fundamental limit is due to nonlinear photobleaching and photodamage effects that occur when the excitation power is increased in order to increase the imaging rate. Further limits are imposed by the maximum count rate of the TCSPC electronics but, in practice, this limit is rarely reached before the onset of photobleaching in most biological samples. The image acquisition rate can be further reduced for multiphoton microscopy due to the reduced excitation cross-sections and increased nonlinear photobleaching/photodamage in the focal plane. We therefore tend to use single photon excitation with TCSPC in confocal microscopes for FLIM of cells. We note, however, that multiphoton microscopy is particularly useful for imaging fluorophores that would otherwise require ultraviolet excitation and we often apply this to imaging tissue autofluorescence. Figure 4.1 shows an example where multiphoton TCSPC FLIM is preeminent: the samples were fresh (~200 μm thickness) unstained sections from a human cervical biopsy and this figure illustrates the intrinsic lifetime contrast available from tissue autofluorescence indicating the potential for clinical diagnosis [8]. The multiphoton excitation is convenient for UV absorbing endogenous fluorophores and limits unwanted fluorescence from outside the focal plane that could degrade the FLIM images in such a thick and heterogeneous sample.

For high-speed FLIM applications, including real-time imaging of biological tissue using endoscopes or macroscopes, as well as for imaging live cell dynamics and microfluidic systems, we have

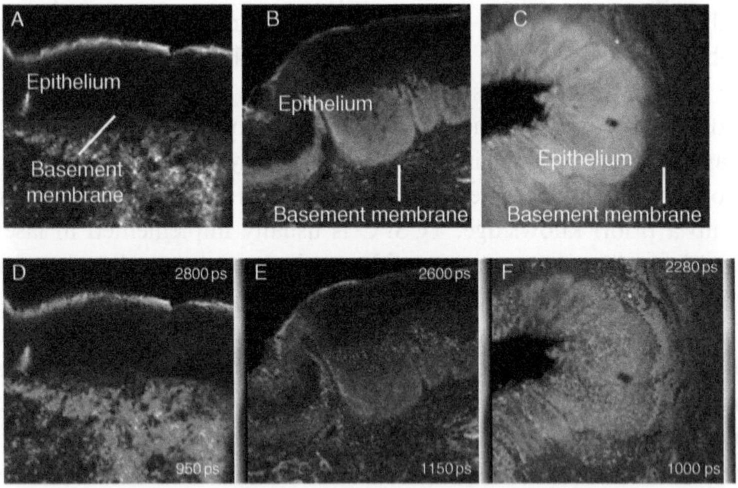

Fig. 4.1. Multiphoton fluorescence intensity (A–C) and TCSPC fluorescence lifetime images (D–F) of fresh unstained sections of human cervical biopsy excited at 740 nm and imaged between 385 and 600 nm. The individual acquisition times were 600 s. Adapted from Fig. 22.11 of Ref. [8]. (See Color Insert.)

developed wide-field time-gated imaging systems. While time-gated imaging is a sampling technique that cannot match the photon economy of TCSPC, the massive parallelism of wide-field detection confers significantly improved imaging speed and reduced photobleaching compared with scanning microscopy. A schematic of a typical wide-field time-gated FLIM system is shown in Chapter 3 Fig. 3.8, for which the key components are the pulsed excitation source, the gated image intensifier, the CCD camera, and the delay-control electronics used to position the time gates with respect to the excitation pulses. In our early work we combined a (home-built) ultrafast solid-state laser oscillator–amplifier system operating at ~10 kHz with a gated image intensifier (model GOI, Kentech Instruments Ltd.), which provided sub-100 ps time gates by ultrafast gating of a mesh in front of the photocathode, but was limited to sub-MHz repetition rates [9]. This system was shown to be able to image lifetime differences of less than 10 ps and was applied to FLIM of tissue autofluorescence [10].

With the development of high repetition rate gated image intensifiers operating at up to 1 GHz, we moved to more convenient commercially available mode-locked laser oscillators for excitation, such as femtosecond Ti:Sapphire lasers, although the minimum gate width available from the high repetition rate gated image intensifiers was ~200 ps. For FLIM of tissue autofluorescence, we typically used many (~16) time gates in order to fully characterize the complex autofluorescence decay profiles, demonstrating label-free contrast between different tissue constituents (e.g., collagen, elastin, keratin) and between healthy and diseased tissue [11, 12]. Figure 4.2 shows some examples of time-gated FLIM applied to tissue autofluorescence of fixed sections of atherosclerotic plaque and of freshly resected colon tissue presenting a dysplastic growth (precancerous state). Using many subnanosecond time gates to sample fluorescence decay profiles that typically exhibit mean lifetimes of several nanoseconds, however, is relatively photon inefficient and this approach required acquisition times of ten's of seconds. The prospect of developing clinical instrumentation

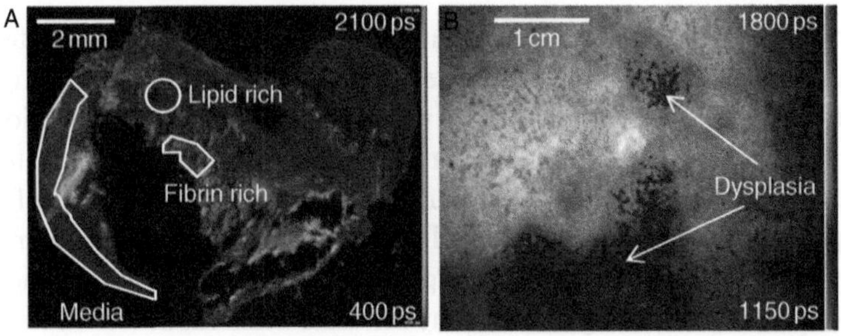

Fig. 4.2. Time-gated FLIM of (A) fixed section of human artery presenting atherosclerotic plaque, excited at 400 nm and imaged through a 435 nm long pass filter, and (B) freshly resected human colon tissue presenting dysplasia (precancerous state), excited at 355 nm and imaged through a 375 nm long pass filter. (See Color Insert.)

flagged up the need for real-time FLIM and we worked to optimize high-speed time-gated imaging.

There is inevitably a trade-off between the information that can be extracted from a fluorescence signal and the total acquisition time (i.e., the number of photons detected). Typically, a few hundred photons are required to permit the lifetime of a single exponential decay profile to be calculated with errors of less than ~10% [13, 14]. To fit a double exponential decay profile and obtain the two lifetime components and their respective magnitudes to the same accuracy with no a priori knowledge, however, requires $\sim 10^4$ photons [14]. For high-speed FLIM, we limit ourselves to the assumption of single exponential decay profiles and can then avoid the need for nonlinear fitting. Instead, we sample the decay profiles with two or more time gates and use rapid lifetime determination (RLD) [15] algorithms to immediately calculate our fluorescence lifetime images [16–18]. The use of gate widths that are comparable with the fluorophore lifetimes increases the photon efficiency significantly and we typically use gates of 1 ns width or longer. The photon efficiency is helped by the fast rise time (~100 ps) of the gated image intensifier, which permits the first

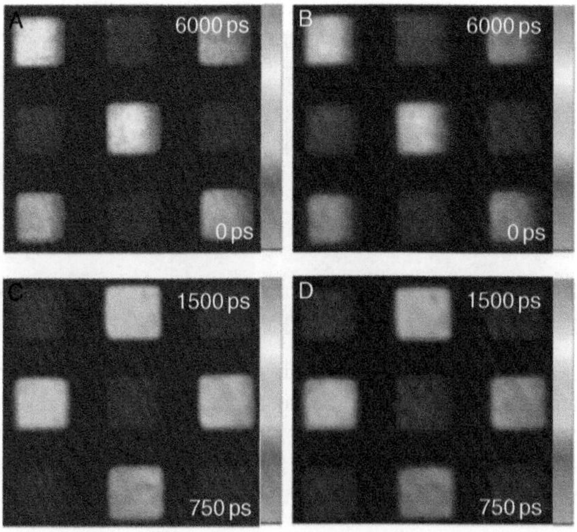

Fig. 4.3. FLIM of multiwell plate with alternate wells containing (A, B) Stilbene_1 and Coumarin_314 and (C, D) Stilbene_1 and Stilbene_3. Figures (A, C) were acquired using 8 time gates (400 ps gate width, ~10 s total acquisition time) followed by WNLLS fitting; figures (B, D) were calculated using RLD from two time-gated images (gate width 2.4 ns, acquired at 16 frames per second). All acquisition excited at 355 nm and imaged through (A, B) 435 nm and (C, D) 375 nm long pass filters respectively. Adapted from Fig. 2 of Ref. [18]. (See Color Insert.)

time gate to be positioned almost immediately after the excitation pulse. Figure 4.3 shows a comparison between fluorescence lifetime images obtained using the "conventional" approach, with a ~10 s acquisition using eight time gates of 400 ps width followed by weighted nonlinear least squares (WNLLS) fitting, and an RLD acquisition using two time gates of 2.4 ns width, with the lifetime images being calculated from time-gated images acquired at 16 frames per second. Fluorescence lifetime images calculated using RLD may be displayed in real-time, which is important for potential clinical applications. Figure 4.4A shows an RLD fluorescence

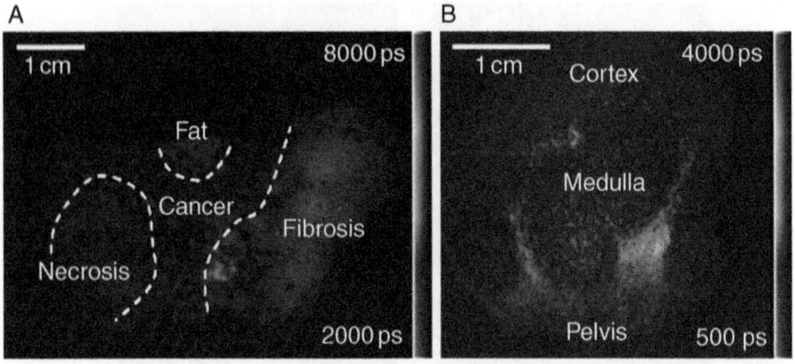

Fig. 4.4. RLD FLIM of (A) unstained freshly resected human pancreas imaged though a macroscope at 7.7 frames per second and (B) unstained sheep's kidney imaged through a rigid endoscope at 7 frames per second. Both samples illuminated at excitation wavelength of 355 nm and fluorescence imaged through a 375 nm long pass filter. Adapted from Fig. 3 of Ref. [18]. (See Color Insert.)

lifetime image of unstained freshly resected human pancreas imaged with a low magnification macroscope, acquiring time-gated images at 7.7 frames per second while Fig. 4.4B shows an RLD fluorescence lifetime image of an unstained sheep's kidney acquired through a rigid endoscope at 7 frames per second. We note that comparable imaging speeds can be achieved using the frequency domain approach to wide-field FLIM [19].

These wide-field time-gated RLD fluorescence lifetime images do not benefit from optical sectioning and do not take into account the complex decay profiles of tissue autofluorescence, but they do provide contrast that may be useful for some clinical applications where FLIM may be used as a "red flag" technique to indicate potential regions of tissue abnormality. Similarly, this high-speed FLIM approach may be used for rapid (high throughput) FRET imaging, where wide-field microscope images of mean fluorescence lifetime may be used to indicate the presence of FRET or for imaging dynamics on fast timescales. We are particularly interested in the application of FLIM and MDFI to characterize fluorescence-based experiments in microfluidic devices and to provide online read-outs. We have applied RLD FLIM to image fluid dynamics in a microfluidic mixer in real time (12.3 frames per second), using the solvent-viscosity dependent fluorescence lifetime of a dye, DASPI, to distinguish water and a 50% glycerol solution [20].

For many applications, it is desirable to have both the optical sectioning capabilities of confocal microscopy and the imaging speed of wide-field detection. This may be realized combining wide-field FLIM with a wide-field optical sectioning technique such as structured illumination [21, 22], multibeam multiphoton microscopy [23–26], or Nipkow disc microscopy [27]. Of these techniques, structured illumination is perhaps the least useful for FLIM because the calculation required to obtain the sectioned image reduces the S/N ratio [28], reducing the dynamic range for the final lifetime calculation. Multibeam multiphoton microscopy provides effective optical sectioning and reduced out of focus photobleaching as well as straightforward excitation of UV excited fluorophores. Nipkow disc microscopy, however, generally permits faster imaging with parallel single photon excitation of a large number of pixels, albeit with a small (~4%) degree of crosstalk. To date, (quasi) wide-field optically sectioned FLIM has been implemented with multibeam multiphoton microscopy [29], structured illumination [30], and Nipkow disc microscopy [31, 32]. The latter approach is a focus of our current work, directed towards high-speed imaging of live cells, in which we obtain FLIM images of comparable S/N as confocal TCSPC FLIM microscopy but in approximately a tenth of the acquisition time. Figure 4.5 compares optically sectioned FLIM images of live COS cells, transfected throughout with EGFP, acquired using time-gated imaging in a Nipkow disc microscope, or TCSPC in a confocal microscope [33]. For the same excitation wavelength of 488 nm, provided by a frequency doubled Ti:Sapphire laser, and a one second acquisition time, the wide-field FLIM returned an average lifetime of 2.6 ns while the TCSPC FLIM retuned an average value of 3.0 ns. For 10 s acquisition time, both systems returned an average lifetime of 2.6 ns.

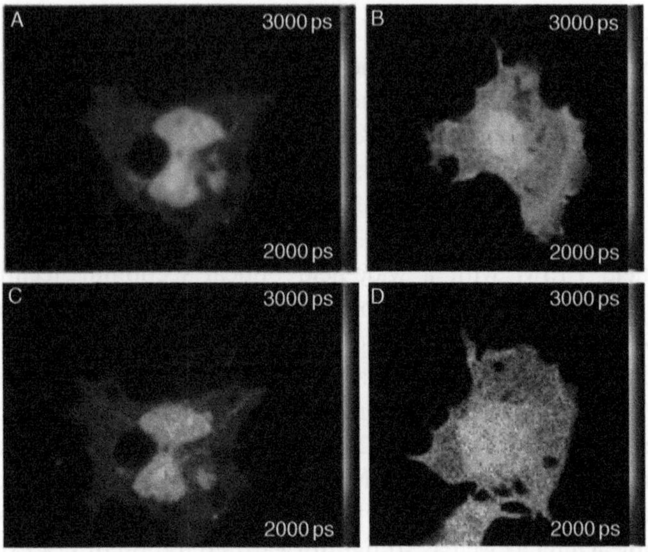

Fig. 4.5. FLIM images of live COS cells transfected with EGFP acquired in (A, B) 10 s and (C, D) 1 s using (A, C) wide-field time-gated Nipkow microscope (10 time-gates of 1 ns gate width, calculated using WNLLS fitting) and (B, D) TCSPC-enabled confocal microscope using 64 time bins with a maximum pixel count rate of 2 million photons per second. Excitation wavelength 488 nm, provided by a frequency doubled Ti:Sapphire laser, imaged through 500–550 nm emission band filter. Adapted from Fig. 2 of Ref. [33]. (See Color Insert.)

4.3. Spectrally resolved FLIM

To investigate and optimize fluorescence contrast, it is desirable to resolve spectral properties as well as fluorescence lifetime. In practice, for any fluorescence imaging experiment, spectral discrimination is applied through the choice of filters, dichroic beam splitters, and excitation wavelengths. Different choices can obviously result in different fluorescence intensity or fluorescence lifetime images being recorded, particularly when multiple fluorophores are present in the sample. The parameter of prime importance is the excitation wavelength, since this will determine which fluorophores can be imaged.

Historically, this has been the most constrained parameter, particularly for confocal laser scanning microscopes that require spatially coherent sources and so have been typically limited to a few discrete excitation wavelengths, traditionally obtained from gas lasers. Convenient tunable continuous wave (c.w.) excitation for wide-field microscopy was widely available from filtered lamp sources but, for time domain FLIM, the only ultrafast light sources covering the visible spectrum were c.w. mode-locked dye lasers before the advent of ultrafast Ti: Sapphire lasers.

The Ti:Sapphire laser provides a convenient multiphoton excitation source, which is more or less essential when imaging in turbid media such as biological tissue, but is often employed with less challenging samples such as cell cultures. Here the convenience of a user-friendly (and often computer-controlled) excitation laser can outweigh the disadvantages of reduced spatial resolution, lower excitation cross-sections and increased nonlinear photobleaching/ photodamage in the focal plane. There are situations, however, where single photon excitation is preferable, for example for more selective excitation of fluorophores (owing to the narrower single-photon excitation spectra), to achieve the best possible spatial resolution, to reduce nonlinear photodamage, and to realize faster imaging for example of live (dynamic) samples or in a high throughput context. Single photon excitation may be provided by frequency-doubled Ti:Sapphire lasers, but these only cover a range of ~350–520 nm, which will not excite many important fluorophores. To address the issue of convenient broadly tunable single-photon excitation sources, we and others [34–36] have exploited visible supercontinuum generation in microstructured ("holey") fibers [37, 38], as previously demonstrated for multiphoton excitation [39, 40].

4.3.1. Excitation-resolved imaging: Tunable continuum source

By launching femtosecond pulses from a mode-locked Ti:Sapphire laser into an appropriate microstructured fiber, one can generate a supercontinuum output spanning from the blue to the near infrared. This radiation—sometimes described as a "white laser"—is spatially coherent, making it suitable for confocal microscopes and a range of other instrumentation, and still comprises a train of ultrashort pulses—although the pulse duration in the continuum extends to a few picoseconds. Figure 4.6A shows a schematic of our first tunable continuum source (TCS), for which spectral selection was conveniently realized using a prism spectrometer arrangement to provide continuously tunable radiation from 435 to 1150 nm with ~70 mW available in visible spectral window from 435 to 700 nm [35]. Using the (motorized) slit, we can readily select an arbitrary spectral region to realize convenient and continuous (electronically controlled) tuning, providing a versatile excitation source for confocal and wide-field microscopy. We note that supercontinuum sources can be pumped using compact, robust, and relatively inexpensive ultrafast fiber lasers [41, 42], to provide spectral coverage spanning from the ultraviolet to the near infrared [43]. Supercontinuum sources are now commercially available and it appears likely that this technology will have a major impact on fluorescence microscopy and FLIM.

We applied our TCS in a confocal fluorescence microscope and obtained optical performance equivalent to that obtained with argon ion laser excitation, as seen in Fig. 4.6B which shows a confocal fluorescence intensity image of human B cells transfected with GPI-GFP [44]. The results are comparable even though the TCS exhibits higher pulse-to-pulse amplitude noise. This is not surprising if one considers the relatively small number of photons detected per pixel for a typical confocal fluorescence microscope image. A simple calculation suggests that shot noise will dominate any contributions from amplitude noise under typical conditions. For example, a 20% standard deviation in the excitation pulse energy would only become significant for photon counts of greater than 10^4 photons per pixel in a confocal microscope (which would correspond to 25 photons/excitation pulse for a pixel dwell time of 5 μs and 80 MHz pulse repetition rate) [35]. On a longer time scale, our TCS typically exhibited pulse energy fluctuations of only 2% RMS over a 10 min measurement period and this has been

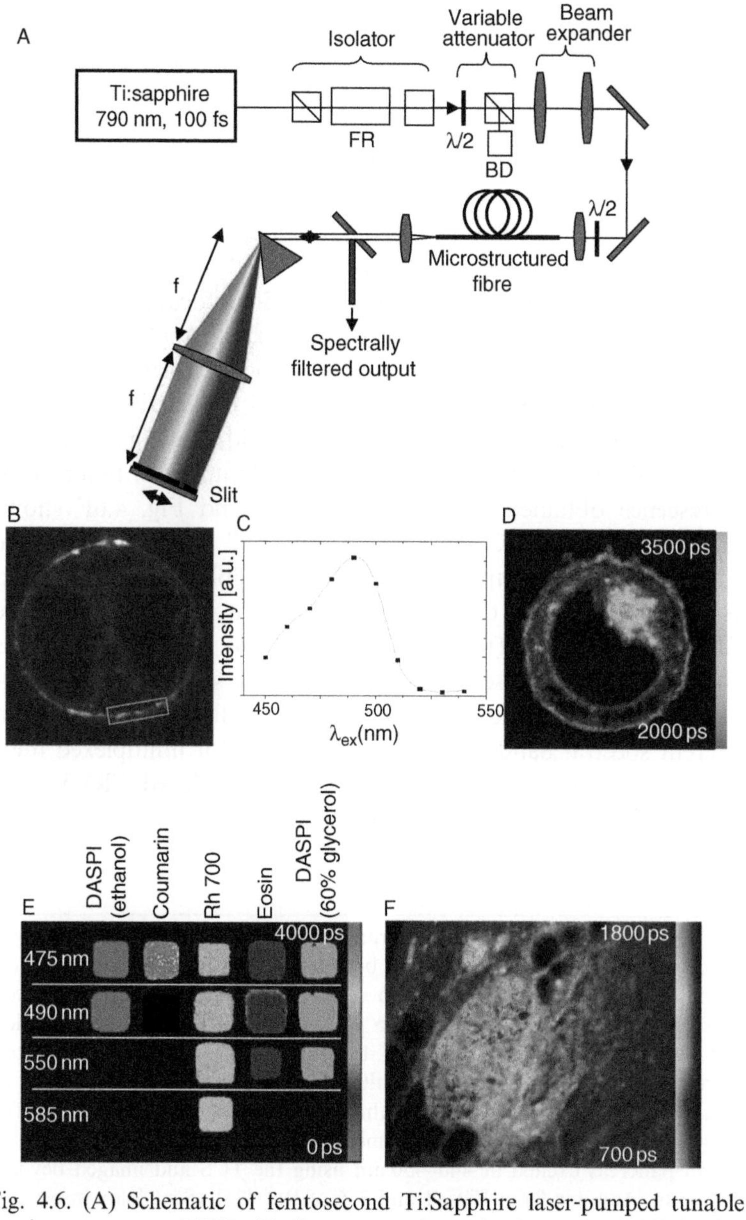

Fig. 4.6. (A) Schematic of femtosecond Ti:Sapphire laser-pumped tunable continuum source (TCS); (B) fluorescence intensity image; and (C) in situ

improved for supercontinuum systems that are now commercially available. Figure 4.6C illustrates the first application of a TCS to single-photon excitation-resolved spectroscopic imaging, measuring the excitation spectrum of the GPI-GFP in situ in the cell. Since the TCS provides ultrafast radiation, it is also straightforward to realize confocal FLIM using TCSPC and Fig. 4.6D shows the first application of a TCS to FLIM, illustrated with a 721.221 cell transfected with HLA-Cw6 linked to CFP.

The TCS has many applications beyond confocal laser scanning microscopy. Being spatially coherent, it is readily applicable to Nipkow disc-based confocal microscopy utilizing microlens arrays [31], indeed the data presented in Fig. 4.5A was obtained in this way. It may also be used for conventional wide-field imaging and FLIM. Figure 4.6E shows an example of a wide-field image of tissue autofluorescence obtained using our first TCS and Fig. 4.6F shows an example of the TCS applied to multiwell plate imaging. The latter illustrates the capability to excite multiple fluorophores and distinguish them—or their local environment—using fluorescence lifetime. Because the TCS does not need to be tuned, but provides tunability via spectra selection, it can provide simultaneous excitation at multiple wavelengths and can be rapidly switched between different spectral bands. This could be useful for multiplexed fluorescence imaging experiments, including multiplexed FRET, and

excitation spectrum of a GPI–GFP transfected human B-cell imaged using a confocal microscope (from region in box) using long pass filter with cut-off 20 nm beyond excitation wavelength. (D) FLIM image of a 721.221 cell transfected with HLA-Cw6 linked to CFP, acquired on a TCSPC-enabled confocal microscope excited using the TCS; (E) wide-field time-gated fluorescence lifetime images of a multiwell plate sample array acquired at different TCS excitation wavelengths through long pass filters (shown on left), and (F) a wide-field fluorescence lifetime image of a fixed unstained section of human pancreas excited at 440–450 nm using the TCS and imaged beyond 470 nm. Adapted from Figs. 1 and 3–6 of Ref. [35]. (See Color Insert.)

should provide a convenient source for stimulated emission depletion (STED) microscopy [45].

4.3.2. Emission-resolved imaging

As well as the excitation wavelength, the outcome of any fluorescence imaging experiment also depends on the spectral discrimination of the detection. It is standard practice to select appropriate filters to optimize detection of a target fluorophore and discriminate against background fluorescence from endogenous fluorophores or other fluorescent labels, for example in FRET experiments. Ratiometric imaging, that is comparing the signal detected in two or more spectral channels, can significantly improve quantitative measurements by avoiding uncertainties associated with fluorophore concentration, illumination and detection geometries and with sample attenuation (the "inner filter" effect). For many experiments, a simple "multispectral" imaging approach, sampling fluorescence emission in a few discrete spectral windows, is sufficient to obtain the desired information [46]. To optimize filter selection, however, for example for spectral unmixing or ratiometric imaging, it is necessary to know the spectral profiles of the fluorophores concerned. Since the spectral properties of a fluorophore can change with variations in their local environment, it is often advisable to characterize fluorescence spectra in the experimental (biological) system under investigation. This entails determining the fluorescence emission spectral profile for each image pixel of a sample, that is implementing "hyperspectral" imaging. Similarly, to optimize a FLIM experiment using, for example tissue autofluorescence, it is necessary to know how the fluorescence lifetime contrast changes with emission wavelength. Hyperspectral imaging—and hyperspectral FLIM—can be invaluable when attempting to contrast different fluorescence signals for which there is no a priori knowledge and can also provide useful information when studying the fluorophore environment.

There are several established approaches for implementing hyperspectral imaging. Usually they require some form of sequential sampling (scanning) since a three dimensional data set (x, y, λ) is to be acquired using, at most, two dimensional detection. In single point scanning fluorescence microscopes, the fluorescence radiation can be dispersed in a spectrometer and the spectral profiles acquired sequentially, pixel by pixel. In wide-field instruments, hyperspectral imaging can be undertaken using manually operated or automatic filter wheels, acousto-optic [47], or liquid crystal [48] tunable filters, to sequentially acquire (x–y) images in different spectral channels. Even with automation, this can be a time-consuming process and is not photon efficient because the "out-of-band" light is rejected. Alternative approaches that are more photon efficient, but still require sequential acquisitions, include Fourier-transform spectroscopic imaging, for example as implemented using a Sagnac interferometer [49], and encoding Hadamard transforms using spatial light modulator technology [50]. It is, in principle, possible to combine spectral and 2-D spatial information such that it can be detected on a single 2-D detector during a single acquisition and subsequently separated in computational post-processing, for example using a complex diffractive optic (computer generated hologram) to mix the spatial and spectral information [51]. In general, any spectrally resolved imaging approach that requires calculations to recover the desired image information can suffer from a reduction in the S/N compared to direct detection, reducing the dynamic range available for FLIM, as for structured illumination. A compromise between photon economy and parallel pixel acquisition is the so-called "push-broom" approach [52] implemented in a line-scanning microscope, for which the emission resulting from a line excitation is imaged to the entrance slit of a spectrograph to produce an (x–λ) "sub-image" that can be recorded on a 2-D detector. Stage scanning along the y-axis then provides the full (x–y–λ) data cube.

4.3.3. Hyperspectral FLIM

To measure the spectral variation of fluorescence lifetime in an image, in order to maximize contrast between fluorophores or to study the local fluorophore environment, FLIM can be combined with hyperspectral imaging. This is readily achieved in scanning microscopes using commercially available TCSPC systems that incorporate a spectrometer and multianode photomultiplier [53–55]. This approach works well, although care has to be taken with calibration and the sequential pixel acquisition, with its concomitant photobleaching/photodamage considerations, which can constrain the imaging speed. To address the latter issue, wide-field (2-D) FLIM technology can be exploited. In our work studying tissue autofluorescence, we initially acquired sequential spectrally resolved fluorescence lifetime images using a filter wheel with wide-field time-gated imaging, but this proved to be a slow, laborious process—made worse by the need to readjust the optical alignment after changing filters. We, therefore, decided to pursue the "push-broom" approach and developed a line-scanning hyperspectral FLIM microscope [56, 57], which also provides "semi-confocal" optical sectioning. Figure 4.7 shows the setup in which the excitation radiation is focused onto an input slit and this line excitation is imaged onto the sample through the microscope objective. The resulting fluorescence line emission is imaged onto the input slit of a compact imaging spectrograph. Using a CCD camera to record the output would provide an (x–λ) "sub-image", but by deploying a gated image intensifier in front of the CCD camera, as is typically done for wide-field FLIM [10], we can record the emission spectrum and fluorescence decay profile for each image pixel in the line (x–λ–τ). Stage-scanning the sample then provides the full 4-D hyperspectral (x–y–λ–τ) data stack. If we exploit the optical sectioning and also scan the sample in z, we can acquire 5-D (x–y–z–λ–τ) fluorescence data stacks. This instrument is particularly intended as a tool to study autofluorescence of unstained tissue samples, for applications in histopathology and research into the molecular origins of disease-related autofluorescence contrast.

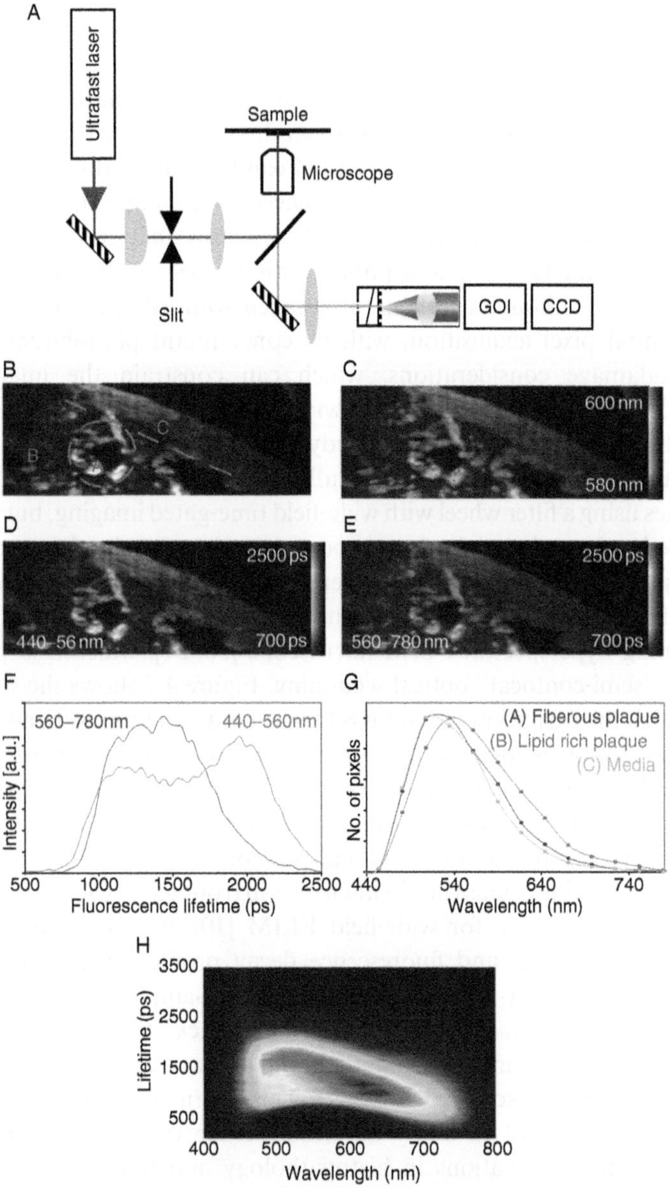

Fig. 4.7. Application of hyperspectral FLIM to an unstained fixed section of human artery excited at 400 nm: (A) schematic of a line-scanning hyperspectral

Figure 4.7 shows a first example, displaying representations of a hyperspectral FLIM data set for an unstained fixed section of human artery. We note that, for sufficiently bright samples, our high-speed wide-field time-gated imaging technology permits such hyperspectral data stacks to be recorded in ~40 s [56], of which ~30 s is taken up by mechanical stage scanning rather than data acquisition, suggesting that these images could be acquired in a little over 10 s.

This MDFI approach can be further extended if we employ the TCS to also resolve the excitation wavelength. As a demonstration of this, we applied the TCS to image a Convallaria slide preparation (Molecular Probes) in the slit-scanning hyperspectral FLIM microscope, varying the excitation wavelength from 390 to 510 nm. Figure 4.8 presents some of the information that can be obtained from such a rich data stack: for each pixel in the field of view, we can obtain the fluorescence excitation–emission–lifetime matrix, subsequently generating "conventional" excitation–emission matrices for any image pixel and then obtaining the fluorescence decay profile for any point in this EEM space. We note that the size and complexity of such hyperspectral FLIM data sets is such that they are almost beyond the scope of ad hoc analysis by human investigators and, in general, sophisticated bioinformatics software tools are now required to "automatically" analyze and present FLIM and MDFI data, for example using approaches to reduce the dimensionality [58], in order that trends and fluorescence "signatures" may be visualized. The combination with image segmentation should provide powerful tools for histopathology and for very high content screening applications.

FLIM microscope; (B) integrated intensity image marked with regions of A-fiberous plaque, B-lipid rich plaque and C-media; (C) central wavelength image; (D, E) fluorescence lifetime images calculated by integrating over emission spectral windows 440-560 and 560-780 nm respectively; (F) fluorescence lifetime histograms comparing these two spectrally integrated FLIM images, (G) time-integrated emission spectra for areas A and B in intensity image and (H) 2-D histogram of fluorescence lifetime and emission wavelength for the whole data set. Adapted from Fig.3 of Ref. [56]. (See Color Insert.)

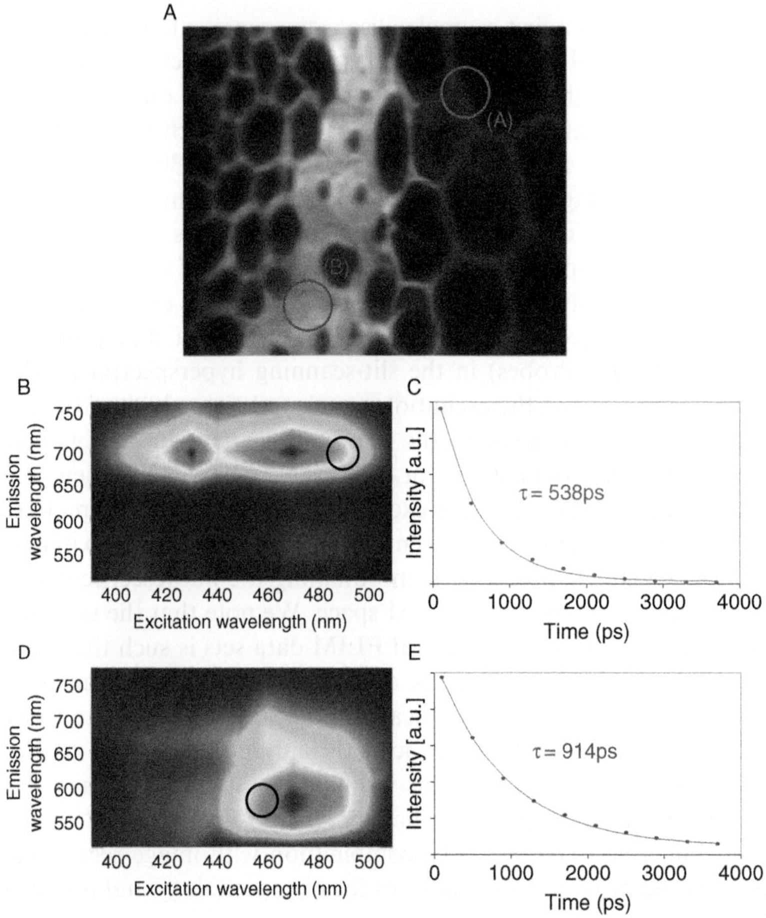

Fig. 4.8. Fluorescence lifetime of a stained section of Convallaria resolved with respect to lifetime, excitation and emission wavelength: (A) intensity image integrated over the time-resolved excitation-emission matrix (EEM); (B, D) time-integrated EEM from areas A and B respectively in (A); (C) fluorescence decay profile for λ_{ex}~490 nm and λ_{em}~700 nm corresponding to area A; (E) fluorescence decay profile for λ_{ex}~460 nm and λ_{em}~570 nm corresponding to area B. (See Color Insert.)

4.4. Polarization-resolved imaging

Polarization provides a further dimension in which fluorescence signals can be analyzed. Steady-state imaging of polarization-resolved fluorescence (i.e., fluorescence anisotropy) is a well-established technique [1] to obtain information concerning fluorophore orientation and can also probe resonant energy transfer and rotational decorrelation, which can be applied to imaging homo-FRET [59] (as discussed in more detail in Chapter 14). This sensitivity of fluorescence anisotropy measurements to both static fluorophore orientation and rotational or FRET dynamics can compromise measurements of either. To image only fluorophore orientation, one can use linear dichroism, that is polarization-resolved absorption, which can be calculated from fluorescence measurements [60]. Conversely, to image rotational mobility, one can combine polarization-resolved imaging with FLIM technology to obtain maps of both fluorescence lifetime and rotational correlation time [61–64]. To minimize errors when imaging fluorescence anisotropy, it is desirable to acquire the polarization-resolved images simultaneously and this has been realized for wide-field [65] and confocal scanning microscopy [66].

Figure 4.9 illustrates time-gated imaging of rotational correlation time. Briefly, excitation by linearly polarized radiation will excite fluorophores with dipole components parallel to the excitation polarization axis and so the fluorescence emission will be anisotropically polarized immediately after excitation, with more emission polarized parallel than perpendicular to the polarization axis (r_0). Subsequently, however, collisions with solvent molecules will tend to randomize the fluorophore orientations and the emission anistropy will decrease with time ($r(t)$). The characteristic timescale over which the fluorescence anisotropy decreases can be described (in the simplest case of a spherical molecule) by an exponential decay with a time constant, θ, which is the rotational correlation time and is approximately proportional to the local solvent viscosity and to the size of the fluorophore. Provided that

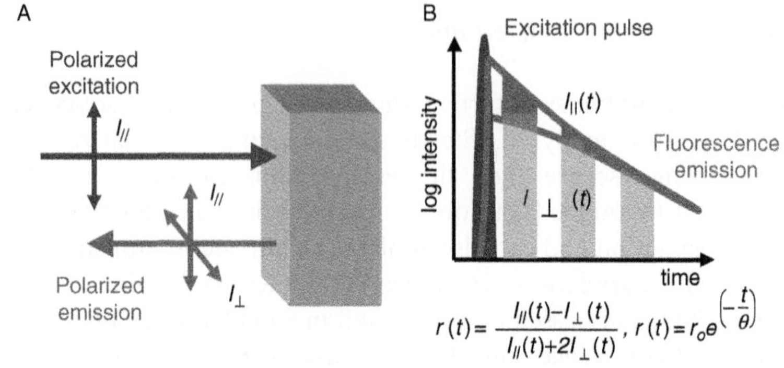

Fig. 4.9. Schematic of time-resolved fluorescence anisotropy: sample is excited with linearly polarized light and time-resolved fluorescence images are acquired with polarization analyzed parallel and perpendicular to excitation polarization. Assuming a spherical fluorophore, the temporal decay of the fluorescence anisotropy, $r(t)$, can be fitted to an exponential decay model from which the rotational correlation time, θ, can be calculated.

the fluorescence decay time is comparable to or greater than the rotational decay time, it is possible to determine θ from time-resolved measurements of fluorescence anisotropy. Figure 4.10 presents an example of data obtained using our wide-field time-gated imaging system combined with a polarization-resolved image splitter to image a multiwell plate array of fluorescein solutions of varying solvent viscosity [64]. Time-resolved fluorescence anisotropy imaging (TR-FAIM) can also be applied to image ligand binding that changes the size of a fluorophore complex [67].

In our laboratory, we are developing time-gated and polarization-resolved fluorescence imaging systems for application to microfluidic devices. By imaging rotational correlation time, we can distinguish between solvents of different viscosity and this can permit us to analyze fluid mixing. Figure 4.11 shows a schematic of an experiment to image the mixing of high and low viscosity solutions of fluorescein in a microfluidic Y-junction, together with the fluorescence lifetime and rotational correlation time images obtained [20]. This experiment was undertaken using a multibeam

A

0	0	63.0	63.0
29.9	29.9	67.4	67.4
38.4	38.4	72.4	72.4
46.9	46.9	75.9	75.9
53.3	53.3	81.0	81.0

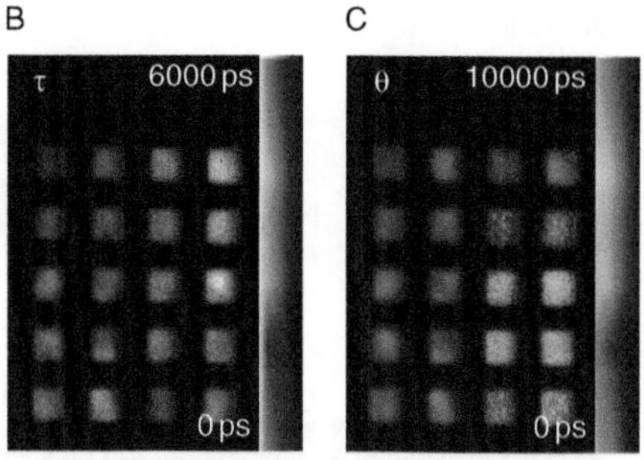

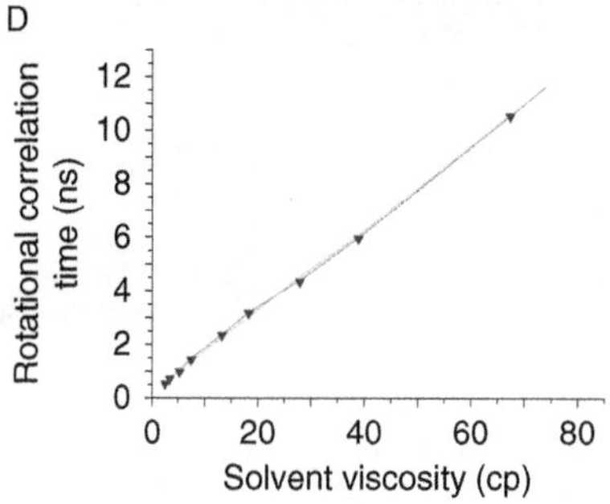

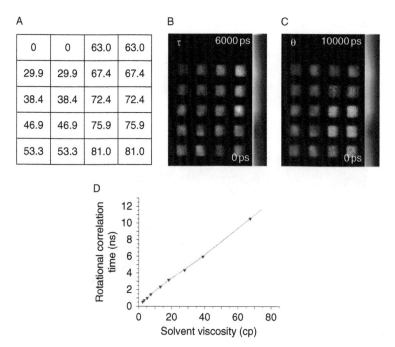

Fig. 4.10. (A) Schematic of percentage weights of glycerol in composite solvents corresponding to array of fluorescein solutions of varying viscosity; (B) fluorescence lifetime; (C) rotational correlation time images of this array; and (D) plot of the rotational correlation time as a function of viscosity for this sample array exited at 470 nm: the straight line fit yields a fluorophore radius of 0.54 nm for fluorescein. Adapted from Fig. 2 of Ref. [64]. (See Color Insert.)

multiphoton microscope that permits us to use our wide-field polarization-resolved time-gated imaging system [20]. As well as optical sectioning, the nonlinear excitation provides an enhanced degree of polarization compared with single photon excitation [68]. Figures 4.11B and 4.11C shows how the rotational correlation time, but not the fluorescence lifetime, reports variations in viscosity while Fig. 4.11D shows how the multiphoton optical sectioning permits 3-D imaging of fluids mixing. As well as providing a tool to optimize microfluidic devices, we anticipate the capability to image

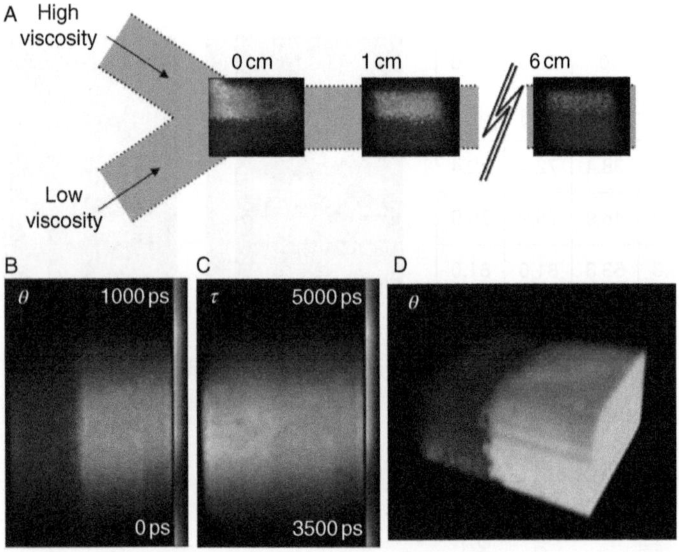

Fig. 4.11. (A) Schematic of microfluidic Y-junction mixer of 50 μm channel width with inset rotational correlation time images (40× magnification) recorded at 0, 1, and 6 cm after the junction, together with 60× magnification images of (B) rotational correlation time, (C) fluorescence lifetime, and (D) 3-D rendered image of rotational correlation time recorded immediately after the junction, excited 780 nm. Adapted from Figs. 5 and 6 [20]. (See Color Insert.)

fluid mixing could be used in flow-based fluorescence assays to separate out fluid dynamics from reaction kinetics, thereby increasing time resolution and accuracy otherwise limited by uncertainties associated with the fluid mixing. We note that we have also reported the application of MDFI to map 3-D temperature distributions in a microfluidic device, utilizing the temperature-dependant fluorescence lifetime of Rhodamine B [69]. This is important for the optimization of microfluidic PCR devices.

A final example of MDFI exploiting polarization resolution is given in Fig. 4.12. This shows the application of optically sectioned TR-FAIM to image ligand binding in a microfluidic reactor [67]. Solutions of a small dye molecule (Hoechst 33258) and a (non-fluorescing) 5.8 kbp DNA plasmid were mixed in a 50-μm wide

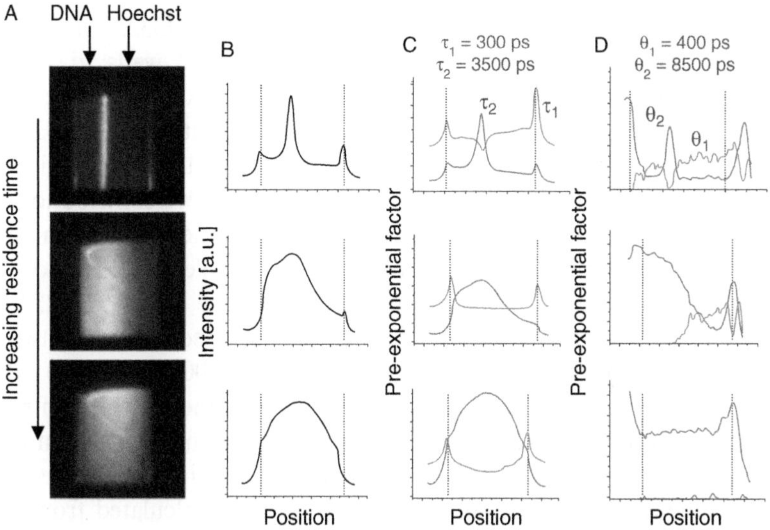

Fig. 4.12. Polarization resolved imaging, excited at 780 nm, applied to ligand binding in a microfluidic device using the fluorescent molecule Hoechst 332258 and a nonfluorescing DNA plasmid (5.8 kbp): (A) integrated intensity images (residence times of 300 ms, 7.2 s, and $t \to \infty$ from top to bottom) and cross-sectional plots (vertically averaged over image) of (B) intensity, (C) preexponential factors resulting from a double exponential fit (lifetimes fixed to 300 and 3500 ps after global analysis) to the decay of fluorescence intensity, and (D) preexponential factors resulting from a double exponential fit (rotational correlation times fixed to 500 and 10,000 ps after global analysis) to decay of the fluorescence anisotropy. The dashed lines show the location of the channel walls. Adapted from Figs. 2–4 of Ref. [67]. (See Color Insert.)

channel and imaged using multibeam multiphoton excitation with wide-field polarization-resolved and time-gated detection. The fluorescence intensity images shown in column (A), arranged top to bottom by increasing residence time (0.3 s, 7.2 s, and $t \to \infty$ respectively) in the microfluidic mixing channel, indicate that the H33258 molecules increase their fluorescence upon binding to the DNA plasmid and that the smaller dye molecule diffuses more rapidly across the channel than the larger DNA plasmid. This is also shown in the transverse line sections taken across the channel, displayed in column (B), which shows an enhancement of the H33258 fluorescence intensity upon interaction with the channel walls (dashed lines). Global analysis was used to fit a double exponential decay model, corresponding to free and bound components of the H33258 dye, with the image data subsequently reanalyzed fixing the lifetime components (300 and 3500 ps) to calculate the relative contributions from free and bound H33258 across the channel, as shown in column (C). Again the rapid diffusion of the smaller H33258 molecules can be seen as well as the appearance of the bound H33258–DNA complex. Similarly, global analysis was used to fix the two components of the rotational correlation time (500 and 10,000 ps) and again subsequent analysis used to calculate the relative contributions from bound and free H33258 across the channel, shown in column (D). Although these data were acquired in a sequence of experiments, they could, in principle, have been calculated from a single experiment and such data could readily be acquired in optical sections to provide a 3-D mapping of ligand binding in the microfluidic device. This can provide an important tool to study and optimize the performance of microfluidic reactors and the instrumentation can also be applied to study ligand binding in cells.

4.5. Conclusions

We hope that this chapter has indicated how FLIM and MDFI can add significant value to microscopy, endoscopy, and assay technology. By resolving the fluorescence signal with respect to multiple dimensions including excitation and emission wavelength, lifetime, and polarization, it may be possible to enhance the fluorescence read-out for an experiment or assay, for example, in terms of improving the separation of multiple fluorophores or imaging variations in the local molecular environment. Generally, it is possible to obtain more information from the fluorescence signal than just fluorophore localization, although localization information—including colocalization for example using FRET—is still extremely valuable when considering the exquisite labeling specificity of genetically expressed fluorophores and other approaches to target proteins. Label-free imaging of autofluorescence can be enhanced by MDFI with obvious applications for clinical imaging and drug discovery. For any application, however, one must consider that a fluorescence sample will only emit a limited number of fluorescence photons at a particular rate before the onset of photobleaching or damage and the limited photon "budget" must therefore be "spent" carefully over the various dimensions of detection to maximize the information that can be obtained from a sample. This inevitably depends on what the investigator expects from the "read-out" and the a priori knowledge available. While limited multispectral imaging and/or RLD FLIM may suffice for a particular high speed application such as diagnostic imaging or a high throughput assay, more sophisticated approaches such as hyperspectral FLIM may be needed to obtain the a priori knowledge or to study unknown samples or complex systems.

Advances in FLIM and MDFI technology have enhanced the prospects for clinical imaging, for drug discovery and for molecular biology research, giving investigators more choices when designing experiments. FLIM–FRET is a powerful technique that is readily applicable to optically sectioned live cell imaging, for example using Nipkow disc microscopy and wide-field FLIM detection. In principle, frequency-domain and time-domain approaches offer comparable performance and speed and both can provide rapid wide-field FLIM—although both approaches can suffer from severe motion artifacts if the samples move between sequential image acquisitions. Such motion artifacts can be avoided using single-shot wide-field FLIM techniques [16, 70]. For some applications, TCSPC FLIM may be more appropriate since it does not suffer so significantly from sample motion; the spatial resolution is degraded, but movement has less impact on the fluorescence lifetime measurements.

The development of tunable (ultrafast) excitation sources based on supercontinuum generation offers convincing alternatives to multiphoton excitation for confocal and other microscopes, permitting faster image acquisition, particularly when only a single optical section through a cell is required. Confocal microscopy becomes less favorable as the number of optical sections in a *z*-stack increases, due to the unwanted excitation of out-of-focus planes during the acquisition of a 3-D image stack. Multiphoton microscopy is advantageous for polarization-resolved imaging, owing to the increased polarization of the nonlinear excitation [68], and is usually preferable when imaging through scattering media such as biological tissue. We note that both single photon confocal and multiphoton excitation can induce nonlinear as well as linear photobleaching and photodamage [71, 72, 73]. The photophysics underlying photodamage and photobleaching depends on the fluorophore being imaged and is the subject of continued research. It is apparent that transitions from long-lived excited states, such as triplet states, and the generation of reactive oxygen species should be minimized. Developing imaging strategies to address these issues is currently an active area of research [74, 75] and it seems likely that quantitative fluorescence imaging will develop still further as a powerful tool for biomedicine.

Acknowledgments

The authors gratefully acknowledge funding from the Biotechnology and Biological Sciences Research Council (BBSRC), the Department of Trade and Industry (DTI), the Engineering and Physical Sciences Research Council (EPSRC), an EU Framework VI Integrated Project (# LSHG-CT-2003-503259), a Joint Infrastructure Fund Award from the Higher Education Funding Council for England (HEFCE JIF), and a Wellcome Trust Showcase Award.

References

[1] Lakowicz, J. R. (1999). Principles of Fluorescence Spectroscopy. Kluwer Academic/Plenum Publishers, New York.

[2] Denk, W., Strickler, J. H. and Webb, W. W. (1990). 2-photon laser scanning fluorescence microscopy. Science *248*, 73–6.

[3] Cubeddu, R., Comelli, D., D'Andrea, C., Taroni, P. and Valentini, G. (2002). Time-resolved fluorescence imaging in biology and medicine. J. Phys. D *35*, R61–R76.
[4] Gadella, T. W. J., Jovin, T. M. and Clegg, R. M. (1993). Fluorescence lifetime imaging microscopy (flim)—spatial-resolution of microstructures on the nanosecond time-scale. Biophys. Chem. *48*, 221–39.
[5] Bastiaens, P. I. H. and Squire, A. (1999). Fluorescence lifetime imaging microscopy: Spatial resolution of biochemical processes in the cell. Trends. Cell. Biol. *9*, 48–52.
[6] Jares-Erijman, E. A. and Jovin, T. M. (2003). FRET imaging. Nat. Biotechnol. *21*, 1387–95.
[7] Suhling, K., French, P. M. W. and Phillips, D. (2005). Time-resolved fluorescence microscopy. Photochem. Photobiol. Sci. *4*, 13–22.
[8] Elson, D. S., Galletly, N., Talbot, C., Requejo-Isidro, J., McGinty, J., Dunsby, C., Lanigan, P. M. P., Munro, I., Benninger, R. K. P.,, De Beule, P., Auksorius, E., Hegyi, L. *et al.* (2006). Multidimensional fluorescence imaging applied to biological tissue. In: Reviews in Fluorescence 2006 (Geddes, C. D. and Lakowicz, J. R., eds.). Springer Science + Business Media Inc, New York, pp. 477–524.
[9] Dowling, K., Hyde, S. C. W., Dainty, J. C., French, P. M. W. and Hares, J. D. (1997). 2-D fluorescence lifetime imaging using a time-gated image intensifier. Opt. Commun. *135*, 27–31.
[10] Dowling, K., Dayel, M. J., Lever, M. J., French, P. M. W., Hares, J. D. and Dymoke-Bradshaw, A. K. L. (1998). Fluorescence lifetime imaging with picosecond resolution for biomedical applications. Opt. Lett. *23*, 810–2.
[11] Elson, D., Requejo-Isidro, J., Munro, I., Reavell, F., Siegel, J., Suhling, K., Tadrous, P., Benninger, R. K. P., Lanigan, P. M. P.,, McGinty, J., Talbot, C. B., Treanor, B. *et al.* (2004a). Time-domain fluorescence lifetime imaging applied to biological tissue. Photochem. Photobiol. Sci. *3*, 795–801.
[12] Tadrous, P. J., Siegel, J., French, P. M. W., Shousha, S., Lalani, E. N. and Stamp, G. W. H. (2003). Fluorescence lifetime imaging of unstained tissues: Early results in human breast cancer. J. Pathol. *199*, 309–17.
[13] Becker, W., Bergmann, A., Hink, M. A., Konig, K., Benndorf, K. and Biskup, C. (2004). Fluorescence lifetime imaging by time-correlated single-photon counting. Microsc. Res. Tech. *63*, 58–66.
[14] Kollner, M. and Wolfrum, J. (1992). How many photons are necessary for fluorescence-lifetime measurements. Chem. Phys. Lett. *200*, 199–204.

[15] Ballew, R. M. and Demas, J. N. (1989). An error analysis of the rapid lifetime determination method for the evaluation of single exponential decays. Anal. Chem. *61*, 30–3.
[16] Elson, D. S., Munro, I., Requejo-Isidro, J., McGinty, J., Dunsby, C., Galletly, N., Stamp, G. W., Neil, M. A. A., Lever, M. J.,, Kellet, P. A., Dymoke-Bradshaw, A., Hares, J. *et al.* (2004b). Real-time time-domain fluorescence lifetime imaging including single-shot acquisition with a segmented optical image intensifier. New. J. Phys. *6*, 180.
[17] Munro, I., McGinty, J., Galletly, N., Requejo-Isidro, J., Lanigan, P. M. P., Elson, D. S., Dunsby, C., Neil, M. A. A., Lever, M. J.,, Stamp, G. W. H. and French, P. M. W. (2005). Toward the clinical application of time-domain fluorescence lifetime imaging. J. Biomed. Opt. *10*, 051403.
[18] Requejo-Isidro, J., McGinty, J., Munro, I., Elson, D. S., Galletly, N. P., Lever, M. J., Neil, M. A. A., Stamp, G. W. H., French, P. M. W., Kellett, P. A., Hares, J. D. and Dymoke-Bradshaw, A. K. L. (2004). High-speed wide-field time-gated endoscopic fluorescence-lifetime imaging. Opt. Lett. *29*, 2249–51.
[19] Holub, O., Seufferheld, M. J., Gohlke, C., Govindjee, R. M. and Clegg, R. M. (2000). Fluorescence lifetime imaging (FLI) in real-time—a new technique in photosynthesis research. Photosynthetica *38*, 581–99.
[20] Benninger, R. K. P., Hofmann, O., McGinty, J., Requejo-Isidro, J., Munro, I., Neil, M. A. A. and French, P. M. W. (2005a). Time-resolved fluorescence imaging of solvent interactions in microfluidic devices. Opt. Express *13*, 6275–85.
[21] Neil, M. A. A., Squire, A., Juskaitis, R., Bastiaens, P. I. H. and Wilson, T. (2000). Wide-field optically sectioning fluorescence microscopy with laser illumination. J. Microsc. *197*, 1–4.
[22] Neil, M. A. A., Juskaitis, R. and Wilson, T. (1997). Method of obtaining optical sectioning by using structured light in a conventional microscope. Opt. Lett. *22*, 1905–7.
[23] Bewersdorf, J., Pick, R. and Hell, S. W. (1998). Multifocal multiphoton microscopy. Opt. Lett. *23*, 655–7.
[24] Buist, A. H., Muller, M., Squier, J. and Brakenhoff, G. J. (1998). Real time two-photon absorption microscopy using multi point excitation. J. Microsc. *192*, 217–26.
[25] Kumar, S., Dunsby, C., De Beule, P. A. A., Owen, D. M., Anand, U., Lanigan, P. M. P., Benninger, R. K. P., Davis, D. M., Neil, M. A. A., Anand, P., Benham, C., Naylor, A. *et al.* (2007). Multifocal multiphoton excitation and time correlated single photon counting detection for 3-D fluorescence lifetime imaging. Opt. Express *15*, 12548–61.

[26] Nielsen, T., Fricke, M., Hellweg, D. and Anderson, P. (2000). High efficiency beam splitter for multifocal multiphoton microscopy. J. Microsc. *201*, 368–76.

[27] Petran, M., Hadravsky, M., Egger, M. D. and Galambos, R. (1968). Tandem-scanning reflected-light microscope. J. Opt. Soc. Am. *58*, 661–4.

[28] Poher, V., Elson, D. S., Lanigan, P. M. P., Dunsby, C., French, P. M. W. and Neil, M. A. A. (2005). A Study of Shot Noise in Wide Field Structured Illumination Microscopy. Poster at Focus on Microscopy, Jena.

[29] Straub, M. and Hell, S. W. (1998). Fluorescence lifetime three-dimensional microscopy with picosecond precision using a multifocal multiphoton microscope. Appl. Phys. Lett. *73*, 1769–71.

[30] Cole, M. J., Siegel, J., Webb, S. E. D., Jones, R., Dowling, K., French, P. M. W., Lever, M. J., Sucharov, L. O. D., Neil, M. A. A.,, Juskaitis, R. and Wilson, T. (2000). Whole-field optically sectioned fluorescence lifetime imaging. Opt. Lett. *25*, 1361–3.

[31] Grant, D. M., Elson, D. S., Schimpf, D., Dunsby, C., Requejo-Isidro, J., Auksorius, E., Munro, I., Neil, M. A. A., French, P. M. W.,, Nye, E., Stamp, G. and Courtney, P. (2005). Optically sectioned fluorescence lifetime imaging using a Nipkow disk microscope and a tunable ultrafast continuum excitation source. Opt. Lett. *30*, 3353–5.

[32] van Munster, E. B., Goedhart, J., Kremers, G. J., Manders, E. M. M. and Gadella, T. W. J. (2007). Combination of a spinning disc confocal unit with frequency-domain fluorescence lifetime imaging microscopy. Cytometry. A. *71A*, 207–14.

[33] Grant, D. M., McGinty, J., McGhee, E. J., Bunney, T. D., Owen, D. M., Talbot, C. B., Zhang, W., Kumar, S., Munro, I., Lanigan, P. M. P., Kennedy, G. T., Dunsby, C. *et al.* (2007). High speed optically sectioned fluorescence lifetime imaging permits study of live cell signaling events. Opt. Express *15*, 15656–73.

[34] Birk, H. and Storz, R. (2001). Illuminating Device and Microscope. United States Patent, Leica Microsystems Heidelberg GmbH.

[35] Dunsby, C., Lanigan, P. M. P., McGinty, J., Elson, D. S., Requejo-Isidro, J., Munro, I., Galletly, N., McCann, F., Treanor, B., Önfelt, B., Davis, D. M., Neil, M. A. A. *et al.* (2004). An electronically tunable ultrafast laser source applied to fluorescence imaging and fluorescence lifetime imaging microscopy. J. Phys. D *37*, 3296–303.

[36] McConnell, G. (2004). Confocal laser scanning fluorescence microscopy with a visible continuum source. Opt. Express *12*, 2844–50.

[37] Knight, J. C., Arriaga, J., Birks, T. A., Ortigosa-Blanch, A., Wadsworth, W. J. and Russell, P. S. (2000). Anomalous dispersion in photonic crystal fiber. IEEE. Photon. Technol. Lett. *12*, 807–9.

[38] Ranka, J. K., Windeler, R. S. and Stentz, A. J. (2000). Visible continuum generation in air-silica microstructure optical fibers with anomalous dispersion at 800 nm. Opt. Lett. *25*, 25–7.

[39] Deguil, N., Mottay, E., Salin, F., Legros, P. and Choquet, D. (2004). Novel diode-pumped infrared tunable laser system for multi-photon microscopy. Microsc. Res. Tech. *63*, 23–6.

[40] Jureller, J. E., Scherer, N. F., Birks, T. A., Wadsworth, W. J. and Russell, P. S. J. (2003). Widely tunable femtosecond pulses from a tapered fiber for ultrafast microscopy and multiphoton applications. In: Ultrafast Phenomena xiii (Miller, R. J. D., ed.). Springer, Berlin, pp. 684–6.

[41] Champert, P. A., Popov, S. V. and Taylor, J. R. (2002). Generation of multiwatt, broadband continua in holey fibers. Opt. Lett. *27*, 122–4.

[42] Schreiber, T., Limpert, J., Zellmer, H., Tunnermann, A. and Hansen, K. P. (2003). High average power supercontinuum generation in photonic crystal fibers. Opt. Commun. *228*, 71–8.

[43] Kudlinski, A., George, A. K., Knight, J. C., Rulkov, A. B., Popov, S. V. and Taylor, J. R. (2006). Zero-dispersion wavelength decreasing photonic crystal fibers for ultraviolet-extended supercontinuum generation. Opt. Express *14*, 5715–22.

[44] Eleme, K., Taner, S. B., Önfelt, B., Collinson, L. M., McCann, F. E., Chalupny, N. J., Cosman, D., Hopkins, C., Magee, A. I. and Davis, D. M. (2004). Cell surface organization of stress-inducible proteins ULBP and MICA that stimulate human NK cells and T cells via NKG2D. J. Exp. Med. *199*, 1005–10.

[45] Hell, S. W. and Wichmann, J. (1994). Breaking the Diffraction Resolution Limit by Stimulated-Emission—Stimulated-Emission-Depletion Fluorescence Microscopy. Opt. Lett. *19*, 780–2.

[46] Neher, R. and Neher, E. (2004). Optimizing imaging parameters for the separation of multiple labels in a fluorescence image. J. Microsc. *213*, 46–62.

[47] Wachman, E. S., Niu, W. H. and Farkas, D. L. (1996). Imaging acousto-optic tunable filter with 0.35-micrometer spatial resolution. Appl. Opt. *35*, 5220–6.

[48] Farkas, D. L., Du, C. W., Fisher, G. W., Lau, C., Niu, W. H., Wachman, E. S. and Levenson, R. M. (1998). Non-invasive image acquisition and advanced processing in optical bioimaging. Comput. Med. Imaging. Graph *22*, 89–102.

[49] Malik, Z., Cabib, D., Buckwald, R. A., Talmi, A., Garini, Y. and Lipson, S. G. (1996). Fourier transform multipixel spectroscopy for quantitative cytology. J. Microsc. *182*, 133–40.

[50] Hanley, Q. S., Verveer, P. J. and Jovin, T. M. (1999). Spectral imaging in a programmable array microscope by hadamard transform fluorescence spectroscopy. Appl. Spectrosc. *53*, 1–10.
[51] Volin, C. E., Ford, B. K., Descour, M. R., Garcia, J. P., Wilson, D. W., Maker, P. D. and Bearman, G. H. (1998). High-speed spectral imager for imaging transient fluorescence phenomena. Appl. Opt. *37*, 8112–9.
[52] Schultz, R. A., Nielsen, T., Zavaleta, J. R., Ruch, R., Wyatt, R. and Garner, H. R. (2001). Hyperspectral imaging: A novel approach for microscopic analysis. Cytometry *43*, 239–47.
[53] Becker, W. (2005). Advanced time-correlated single photon counting techniques. Springer, Berlin Heidelberg New York.
[54] Becker, W., Bergmann, A., Biskup, C., Zimmer, T., Klöcker, N. and Benndorf, K. (2002). Multiwavelength TCSPC lifetime imaging. Proc. SPIE *4620*, 79–84.
[55] Bird, D. K., Eliceiri, K. W., Fan, C. H. and White, J. G. (2004). Simultaneous two-photon spectral and lifetime fluorescence microscopy. Appl. Opt. *43*, 5173–82.
[56] De Beule, P., Owen, D. M., Manning, H. B., Talbot, C. B., Requejo-Isidro, J., Dunsby, C., McGinty, J., Benninger, R. K. P., Elson, D. S.,, Munro, I., Lever, M. J., Anand, P. *et al.* (2007). Rapid hyperspectral fluorescence lifetime imaging. Microsc. Res. Tech. *70*, 481–4.
[57] Owen, D. M., Auksorius, E., Manning, H. B., Talbot, C. B., De Beule, P. A. A., Dunsby, C., Neil, M. A. A. and French, P. M. W. (2007). Excitation-resolved hyperspectral fluorescence lifetime imaging using a UV-extended supercontinuum source. Opt. Lett. *32*, 3408–10.
[58] Lekadir, K., Elson, D. S., Requejo-Isidro, J., Dunsby, C., McGinty, J., Galletly, N., Stamp, G. W., French, P. M. W. and Yang, G. Z. (2006). Tissue characterization using dimensionality reduction and fluorescence imaging. Med. Image Comput. Comput. Assist. Interv. Miccai *4191*, 586–93 Pt 2.
[59] Squire, A., Verveer, P. J., Pocks, O. and Bastiaens, P. I. H. (2004). Red-edge anisotropy microscopy enables dynamic imaging of homo-FRET between green fluorescent proteins in cells. J. Struct. Biol. *147*, 62–9.
[60] Benninger, R. K. P., Önfelt, B., Neil, M. A. A., Davis, D. M. and French, P. M. W. (2005b). Fluorescence imaging of two-photon linear dichroism: Cholesterol depletion disrupts molecular orientation in cell membranes. Biophys. J. *88*, 609–22.
[61] Buehler, C., Dong, C. Y., So, P. T. C., French, T. and Gratton, E. (2000). Time-resolved polarization imaging by pump-probe (stimulated emission) fluorescence microscopy. Biophys. J. *79*, 536–49.

[62] Clayton, A. H. A., Hanley, Q. S., Arndt-Jovin, D. J., Subramaniam, V. and Jovin, T. M. (2002). Dynamic fluorescence anisotropy imaging microscopy in the frequency domain (rflim). Biophys. J. *83*, 1631–49.

[63] Gautier, I., Tramier, M., Durieux, C., Coppey, J., Pansu, R. B., Nicolas, J. C., Kemnitz, K. and Coppey-Moisan, M. (2001). Homo-FRET microscopy in living cells to measure monomer-dimer transition of GFP-tagged proteins. Biophys. J. *80*, 3000–8.

[64] Suhling, K., Siegel, J., Lanigan, P. M. P., Leveque-Fort, S., Webb, S. E. D., Phillips, D., Davis, D. M. and French, P. M. W. (2004). Time-resolved fluorescence anisotropy imaging applied to live cells. Opt. Lett. *29*, 584–6.

[65] Siegel, J., Suhling, K., Leveque-Fort, S., Webb, S. E. D., Davis, D. M., Phillips, D., Sabharwal, Y. and French, P. M. W. (2003). Wide-field time-resolved fluorescence anisotropy imaging (TR-FAIM): Imaging the rotational mobility of a fluorophore. Rev. Sci. Instrum. *74*, 182–92.

[66] Bigelow, C. E., Conover, D. L. and Foster, T. H. (2003). Confocal fluorescence spectroscopy and anisotropy imaging system. Opt. Lett. *28*, 695–7.

[67] Benninger, R. K. P., Hofmann, O., Önfelt, B., Munro, I., Dunsby, C., Davis, D. M., Neil, M. A. A., French, P. M. W. and de Mello, A. J. (2007). Fluorescence-lifetime imaging of DNA-dye interactions within continuous-flow microfluidic systems. Angew. Chem. *46*, 2228–31.

[68] Volkmer, A., Hatrick, D. A. and Birch, D. J. S. (1997). Time-resolved nonlinear fluorescence spectroscopy using femtosecond multiphoton excitation and single-photon timing detection. Meas. Sci. Technol. *8*, 1339–49.

[69] Benninger, R. K. P., Koc, Y., Hofmann, O., Requejo-Isidro, J., Neil, M. A. A., French, P. M. W. and de Mello, A. J. (2006). Quantitative 3D mapping of fluidic temperatures within microchannel networks using fluorescence lifetime imaging. Anal. Chem. *78*, 2272–8.

[70] Agronskaia, A. V., Tertoolen, L. and Gerritsen, H. C. (2003). High frame rate fluorescence lifetime imaging. J. Phys. D. *36*, 1655–62.

[71] Patterson, G. H. and Piston, D. W. (2000). Photobleaching in two-photon excitation microscopy. Biophys. J. *78*, 2159–62.

[72] Drummond, D. R., Carter, N. and Cross, R. A. (2002). Multiphoton versus confocal high resolution z-sectioning of enhanced green fluorescent microtubules: Increased multiphoton photobleaching within the focal plane can be compensated using a Pockels cell and dual widefield detectors. J. Microsc. *206*, 161–9.

[73] White, J. G., Amos, W. B. and Fordham, M. (1987). An Evaluation of Confocal Versus Conventional Imaging of Biological Structures by Fluorescence Light-Microscopy. J. Cell. Biol. *105*, 41–8.

[74] Donnert, G., Eggeling, C. and Hell, S. W. (2007). Major signal increase in fluorescence microscopy through dark-state relaxation. Nat. Methods *4*, 81–6.

[75] Hoebe, R. A., Van Oven, C. H., Gadella, T. W. J., Dhonukshe, P. B., Van Noorden, C. J. F. and Manders, E. M. M. (2007). Controlled light-exposure microscopy reduces photobleaching and phototoxicity in fluorescence live-cell imaging. Nat. Biotechnol. *25*, 249–53.

Laboratory Techniques in Biochemistry and Molecular Biology, Volume 33
FRET and FLIM Techniques
T. W. J. Gadella (Editor)

CHAPTER 5

Visible fluorescent proteins for FRET

Gert-Jan Kremers[1] and Joachim Goedhart[2]

[1]*Department of Molecular Physiology and Biophysics, Vanderbilt University Medical Center, 702 Light Hall, Nashville, Tennessee 37232, USA*
[2]*Section Molecular Cytology and Centre for Advanced Microscopy, Swammerdam Institute for Life Sciences, University of Amsterdam, Kruislaan 316, NL-1098 SM Amsterdam, The Netherlands*

Visible fluorescent proteins (VFPs) are genetically encoded fluorophores that are available in a wide variety of colors, spanning the whole visible spectrum. These probes have revolutionized cell-biology, allowing the visualization of biological processes with high sensitivity and specificity in living cells, tissue, or even whole organs. Besides localization studies, VFPs can be used for studies that use fluorescence resonance energy transfer (FRET) to detect molecular interactions and molecular conformations. This chapter gives an introduction to general features of VFPs. The use of VFPs for FRET studies is discussed, a comprehensive table with Förster radii of VFP pairs is presented and recommendations for choosing the right pairs are made. The most promising pairs are discussed in detail and finally examples of VFPs in FRET studies are highlighted.

DOI: 10.1016/S0075-7535(08)00005-3

5.1. Introduction

During the past two decades, cell-biological and biomedical research has greatly benefited from innovations in fluorescence microscopy. Both the increase in the repertoire of fluorescence staining techniques at the (sub)cellular level and the development of a multitude of novel fluorescence microscopy techniques contributed significantly.

Before the 1990s, labeling of proteins with fluorescent probes inside cells has been mainly achieved by immunocytochemistry. However, to enable fluorescent antibodies to enter cells, these need to be fixed and permeabilized, and therefore, immunolabeling is not suited for the study of proteins in living cells. Direct protein labeling with organic fluorophores is possible. However, this is laborious because it requires protein purification, chemical labeling, and microinjection into cells. Therefore, there has been great urge for methods allowing noninvasive and site-specific labeling of proteins in living cells or tissue. This was satisfied with the revolutionary discovery of the genetically encoded green fluorescent protein (GFP). This protein contains a fluorophore that is synthesized within the protein itself, without the need for additional substrates for generating fluorescence. Today, fluorescent proteins are available in a variety of colors, which are generally named visible fluorescent proteins (VFPs). Specific fluorescent labeling of multiple proteins simultaneously with spectrally different VFPs can now be achieved. Moreover, the availability of spectral variants has enabled the use of VFPs in fluorescence resonance energy transfer (FRET)-based approaches in cell biology. In this chapter we focus on the use of VFPs for measuring FRET. Other genetically encoded labeling strategies (AGT, FlAsH) and FRET to or from organic fluorophores in general will not be discussed, more information can be found in other chapters (Chapters 6 and 12).

We will briefly discuss the origin and properties of several VFPs; for extensive discussion and comparison, the reader is referred to the literature [1–5]. Recommendations for choosing the right pair of VFPs for FRET are made and finally some examples of the application of VFPs in FRET-based imaging are highlighted.

5.2. Fluorescent proteins

5.2.1. Discovery of the fluorescent proteins

The first reported observation of GFP fluorescence dates back to the early 1960s. A protein exhibiting a greenish glow was observed as a by-product after purification of the chemiluminescent protein aequorin from the jellyfish *Aequorea victoria* [6]. In the following decades, GFPs were also discovered in several other light emitting marine species, such as the sea pansy *Renilla reniformis* [7].

The first major breakthrough for the application of fluorescent proteins occurred 30 years after the initial observation, when the gene encoding *A. victoria* GFP (avGFP) was isolated [8], followed 2 years later by the discovery that GFP could be expressed and also become fluorescent in organisms other than *A. victoria* [9, 10]. This provided evidence that avGFP is an autofluorescent protein and requires no additional (jellyfish-specific) substrates or enzymes, in contrast to the well-known chemiluminescent proteins aequorin and luciferase [11, 12]. The impact of this work has been so significant, that in 2008 the Nobel Prize for Chemistry was awarded to Shimomura, Chalfie and Tsien for their efforts in isolating, identifying and applying GFP.

The second major breakthrough for the application of fluorescent proteins was the isolation of the red fluorescent protein (RFP) drFP583 or DsRed from the *Anthozoa* and *Discosoma* sp., a mushroom-shaped anemone found in the warm waters of the Indo-Pacific ocean [13]. The breakthrough was not only the discovery of the first true RFP, but equally important was the fact that it was discovered in a nonbioluminescent species and that the gene was cloned immediately.

5.2.2. Evolution of fluorescent proteins

The isolation of fluorescent proteins from nonbioluminescent species has led to the discovery of a super family of GFP-like proteins [1, 14]. Recently, six additional GFP-like proteins were isolated from *A. victoria*-related jellyfish [14–16]. Furthermore, a large number of GFP-like proteins have been isolated from *Anthozoa* species, ranging in fluorescence from green to orange-red, as well as nonfluorescent purple-blue chromoproteins [17–19].

Hydrozoa, such as *A. victoria*, and *Anthozoa* both belong to the phylum Cnidaria. In addition, fluorescent proteins have been isolated from planktonic Copepods, which belong to the evolutionary distant phylum Arthropodia [14]. This wide phylogenetic distribution of GFP-like proteins might implicate that these proteins developed early in evolution and hence that almost every animal taxon can potentially contain GFP homologs [1].

5.2.3. Biological function of fluorescent proteins

The biological function of fluorescent proteins is still unclear. In *A. victoria* and other bioluminescent species harboring GFP-like proteins, GFP is bound to aequorin, where it converts the blue light emitted by aequorin into green light in a process called bioluminescence resonance energy transfer (BRET) [4]. The advantage might be that green light provides higher contrast in the bluish marine environment.

Although this might be one of the functions of fluorescent proteins, it is certainly not the only function. The association of GFP with aequorin is (most likely) a special case, since most bioluminescent species apparently lack fluorescent proteins and in addition, the majority of species containing GFP-like proteins are not bioluminescent [1]. Furthermore, the existence of highly absorbing but nonfluorescent chromoproteins, together with the fact that many of the organisms containing fluorescent proteins have been found in subtropical climates, suggests a possible role for fluorescent proteins in protection against UV-radiation from the sun [20]. Another function proposed for GFP-like proteins in Copepods is that they are involved in visual mate-recognition [1, 21, 22].

5.2.4. Protein structure of VFPs

The crystal structure of avGFP was first solved in 1996, independently by two groups [23, 24]. The structure revealed a cylindrical protein, consisting of 11 β-strands (Fig. 5.1), which was named 11-stranded β-barrel. A single α-helix runs along the axis inside the β-barrel and was found to contain the chromophore, the source

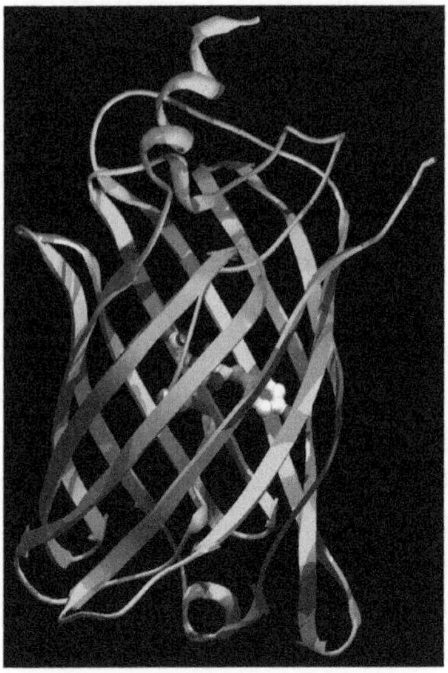

Fig. 5.1. Ribbon diagram of a fluorescent protein (citrine, PDB entry 1HUI) crystal structure. The chromophore is buried in the protein's interior and shown in balls and sticks representation. (See Color Insert.)

of fluorescence. The chromophore is placed almost exactly in the center of the protein and is therefore buried deep inside the GFP protein. This protective shell effectively shields the chromophore from the environment.

Remarkably, although there is little sequence homology between the members of the GFP super family (DsRed and avGFP share less then 30% sequence homology), their crystal structures are highly similar [25–28]. The β-barrel structure is a feature common to all members of the GFP super family for which the crystal structure has been solved. However, whereas avGFP is present mainly as a monomer, many other VFPs form obligate di- or tetramers.

5.2.5. Chromophore formation in avGFP and DsRed

The chromophore of avGFP is formed by amino acids Ser65, Tyr66, and Gly67 by a series of autocatalytic reactions (Fig. 5.2A). Although the precise sequence of events is still under debate [29, 30], it is generally accepted that chromophore formation involves cyclization of the tripeptide by nucleophilic attack of the amide nitrogen of Gly67 on the carbonyl carbon of Ser65 [8, 31–33]. This results in ring closure and the release of one H_2O molecule. The mature chromophore is formed upon oxidation of the Tyr66 C_α–C_β bond by molecular oxygen, causing the formation of a *p*-hydroxybenzylidene–imidazolidinone derivative with a large conjugated π-electron system. Since the chromophore is created by autocatalytic reactions within the protein itself, no cofactors are required for fluorescence, except for a single O_2 molecule during chromophore maturation [34].

The red-fluorescent chromophore of DsRed is formed through a green intermediate formed by the tripeptide Gln66-Tyr67-Gly68, analogous to avGFP (Fig. 5.2B). Red fluorescence develops only after additional oxidation of the bond between the α-carbon and the nitrogen of Gln66, which extents the conjugation of the

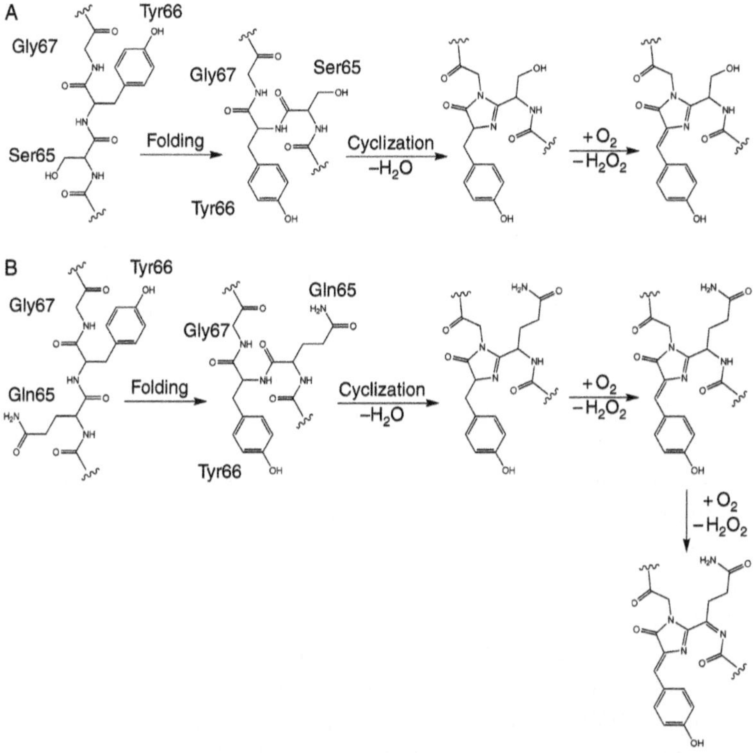

Fig. 5.2. Chromophore formation in avGFP and DsRed. Chromophore formation in avGFP (A) requires folding of the tripeptide into the right conformation in order to enable cyclization and oxidation to form the mature green chromophore. In DsRed (B) chromophore formation follows the same path as for avGFP but requires an additional oxidation step to extend the conjugation of the chromophore.

chromophore [28, 35, 36]. The second oxidation step requires an additional O_2 molecule and occurs rather slow and incomplete. This results in residual green chromophore. Importantly, since DsRed is an obligate and closely packed tetramer, as soon as a single chromophore in the tetramer is fully matured (red), then the residual green fluorescence is efficiently converted to red via FRET.

5.3. Variants of avGFP

Use of wild-type avGFP is limited by inefficient folding at temperatures above 20 °C. Therefore, much effort has been put into optimizing protein folding at elevated temperatures to facilitate GFP expression at 37 °C for applications in mammalian cells. Furthermore, a range of mutations has been found, which alter the spectral properties of avGFP. This has resulted in a variety of color variants of avGFP with fluorescence ranging from blue to greenish-yellow.

5.3.1. Mutations for improving expression of avGFP

The *A. victoria* codon usage has been changed into an optimized mammalian codon usage, thereby eliminating several rare codons [37, 38]. The altered codon usage also removed a cryptic splice site that prevented expression of avGFP in some types of plant cells [39]. Codon optimization has been shown to increase levels of protein expression fourfold at 37 °C and GFP variants with such optimized codon usage are generally named enhanced fluorescent proteins or EVFPs (enhanced visible fluorescent proteins) [37, 40]. Furthermore, random mutagenesis has yielded several mutations that do not increase the intrinsic brightness of fluorescence, but rather increase the amount of expressed protein that reaches a fluorescent state. Some of these mutations are present close to the chromophore (Phe64Leu and Ser72Ala [41, 42]), whereas others are located more distant and can be buried inside the protein (Val163Ala [43]) or exposed to the outside (Met153Thr, Ser175Gly, Ala206Lys [44–46]). Although the exact mechanism of most of these mutations is unknown, mutations close to the chromophore are presumed to directly improve chromophore formation, whereas the more distant mutations Val163Ala and Ser175Gly are likely to prevent protein misfolding at elevated temperatures [47]. Mutation

Met153Thr presumably improves solubility, by decreasing the surface hydrophobicity. Mutation Ala206Lys was found to abolish the tendency of GFP to dimerize, because the charged lysine disrupts the hydrophobic interphase required for dimerization [46]. Although the affinity for dimerization is low ($k_d = 0.11$ mM), it can induce artifacts when using GFP fusion proteins to study intermolecular interactions [46]. This is especially important if the VFPs are targeted to subcellular locations or to membranes. Introducing the whole set of mutations was shown to improve expression of all color variants derived from avGFP [48–50].

In general, VFPs are preferred with optimized codon usage and containing folding mutations to facilitate expression and optimize maturation.

5.3.2. Wild-type avGFP and variants with neutral chromophore

The chromophore in avGFP can exist in two states, protonated and deprotonated, with different spectroscopic features. In wild-type avGFP, the chromophore is mainly present in a protonated neutral form and a minor fraction is in a deprotonated anionic state [51]. Wild-type avGFP has a major excitation peak at 395 nm arising from the protonated chromophore and a minor excitation peak at 475 nm corresponding to the deprotonated anionic chromophore [4]. Excitation at either excitation maximum gives rise to similar but not identical fluorescence with peak emission at approximately 505 nm. The similar fluorescence is the result of excited-state proton transfer upon excitation of the neutral chromophore [52]. T-Sapphire is a GFP variant containing several folding mutations in addition to mutation Tyr203Ile [53]. Mutation Thr203Ile abolishes the minor excitation peak at 475 nm; therefore, it can be detected separately from avGFP variants with a stabilized anionic chromophore (discussed in the following section).

5.3.3. avGFP variants with anionic phenolate chromophore

For cell biological applications, excitation at 475 nm is more favorable than excitation at 395 nm, because the risk of (near-UV) radiation damage is decreased and the levels of cellular autofluorescence are less. Random mutagenesis of wild-type avGFP has yielded several mutations that efficiently abolish chromophore protonation, resulting in increased excitation around 475 nm at the expense of excitation at 395 nm. Key mutations include changing Ser65 into Thr, Gly, Ala, Leu, or Cys [41, 42, 54]. These mutations generally act by disrupting the hydrogen bond network between the chromophore and Glu222. Gly, Ala, and Leu cannot donate hydrogen bonds and Cys is too large to adopt the right conformation [4], therefore Glu222 cannot transfer a proton to the chromophore [51]. The most commonly used mutation to prevent chromophore protonation is Ser65Thr [54]. Compared with serine, threonine contains an additional methyl group and the increased size causes steric hindrance, preventing hydrogen bonding to Glu222 [51].

In addition to disrupting the hydrogen bond network between the chromophore and Glu222, chromophore protonation can be prevented by direct mutagenesis of Glu222. Replacing Glu222 with Gly or Gln has been shown to effectively abolish excitation at 395 nm [48, 55–57]. However, Glu222Gln seems to reduce the efficiency of chromophore maturation [57].

The most commonly used avGFP variant today still is enhanced GFP (EGFP) [37, 41]. EGFP has an optimized codon usage for expression in mammalian cells and contains the mutations, Phe46-Leu and Ser65Thr. The anionic state of the chromophore (Fig. 5.3A) results in an excitation maximum at 488 nm and emission maxima at 507 nm, therefore EGFP is compatible with the standard fluorescence filter sets used for green fluorescent dyes, for example, fluorescein. SGFP2 is a recently developed GFP variant containing several additional folding mutations, which improve protein expression in prokaryotic and eukaryotic cells [48]. SGFP2 also has an increased fluorescence lifetime and photostability. These

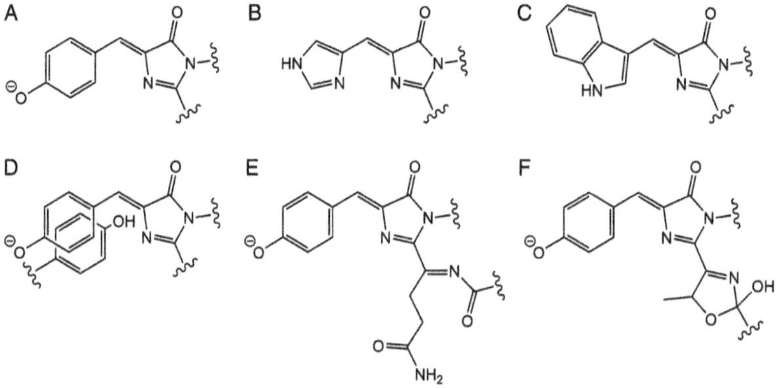

Fig. 5.3. Chromophores of different colored VFPs. (A) EGFP, (B) EBFP, (C) ECFP, (D) EYFP, (E) DsRed, and (F) mOrange.

properties make SGFP2 better suited as a donor fluorophore for FRET studies. The chromophore in SGFP2 is partially protonated resulting in a secondary excitation peak at 400 nm. SGFP2 is therefore less suitable as a FRET acceptor in combination with a blue fluorescent protein donor if sensitized emission based FRET detection is used (see Chapters 7 and 8). Excitation at 400 nm might be beneficial for FRET applications using two-photon excitation.

5.3.4. Blue fluorescent variants of avGFP

One of the first color variants of avGFP was a blue fluorescent protein (BFP). BFP has gained interest because the blue-shifted fluorescence permitted, for the first time, dual-color imaging together with GFP [40, 44, 58]. Replacing Tyr66 for His in the chromophore (Fig. 5.3B) results in blue-shifted fluorescence with excitation and emission maxima at 380 and 450 nm, respectively [34]. Optimization of the codon usage and incorporation of the additional mutation Tyr145Phe (EBFP) [40] have been shown to improve expression and enhance the brightness; nevertheless, fluorescence remains dim and susceptible to photobleaching [44, 59]. Recently two new brighter BFP variants have been developed, SBFP2 and Azurite [48, 60]. Both BFP variants display an increased quantum yield and improved photostability. SBFP2 is mammalian codon optimized and appears slightly brighter than Azurite. These bright BFPs are expected to gain renewed interest for FRET applications with BFP as donor fluorophore.

5.3.5. Cyan fluorescent variants of avGFP

Cyan fluorescent proteins (CFPs) have blue-shifted excitation and emission spectra, because of the mutation Tyr66Trp inside the chromophore (Fig. 5.3C) [34]. CFP fluorescence (Ex 435 nm/Em 474 nm) is less blue-shifted than for EBFP and CFP excitation is intermediate to the excitation of the neutral and anionic chromophores of avGFP [4]. CFPs are widely used for dual-color imaging and FRET applications together with yellow fluorescent proteins (YFP, Section 3.6).

ECFP is a widely used CFP variant with mammalian optimized codon usage and contains the additional mutations Phe64Leu, Ser65Thr, Met153Thr, and Val163Ala. Although not as bright as EGFP, ECFP fluorescence is readily monitored because of the high photostability and the high intensity output of a pressured mercury lamp at 436 nm.

Recently, several novel CFP variants, including CyPet, Cerulean, and SCFP3A, have been developed [49, 61, 62]. Cerulean, or better mCerulean (contains mutation A206K), is rapidly taking over the role of preferred donor fluorophore for FRET applications. CyPet contains several novel mutations and was presented as an improved donor for FRET assays. However, the improved FRET performance might well be the result of increased interaction between CyPet and YPet [63]. In addition, CyPet has been found to express relatively poorly at 37 °C [3, 61]. Cerulean and SCFP3A are brighter fluorescent variants than ECFP mainly because of an increased quantum yield as a result of mutation His148Asp. Most CFPs show complex multiexponential fluorescence decays. Purified Cerulean protein was found to have a fluorescence lifetime best fitted by a monoexponential decay based on time-domain FLIM measurements [62]. Frequency-domain FLIM measurements of mammalian cells expressing Cerulean, however, indicated a multiexponential fluorescence decay, with lifetime values similar to ECFP [49]. SCFP3A is closely related to Cerulean but has a higher quantum yield (see Table 5.1). SCFP3A also exhibits a complex multiexponential fluorescence decay, but the lifetime values are increased compared with ECFP and mCerulean. Therefore SCFP3A is a better donor for FRET-FLIM applications.

5.3.6. Yellow fluorescent variants of avGFP

avGFP was also the basis for the creation of red-shifted GFP variants. These fluorescent proteins are called YFPs because of the yellowish appearance of their fluorescence (Ex 514 nm/Em 526 nm) [4]. Although all other color variants of avGFP were initially found by random mutagenesis strategies, YFP was rationally designed, based on the crystal structure of GFP [23]. Ormö and coworkers hypothesized that the phenolic ring of a tyrosine residue at position 203 might result in π–π stacking with the phenolic ring of the chromophore and hence would reduce the excited state energy. This hypothesis was later confirmed by the crystal structure of YFP (Fig. 5.3D) [64]. YFPs have found wide application, because for several years YFP together with CFP was the best combination of fluorescent proteins available for dual-color imaging and for FRET applications.

Use of early YFP variants, for example EYFP, suffered from several disadvantages, since they did not express well at 37 °C, were sensitive to photobleaching and to environmental factors, for example pH and Cl^- concentration [65–69]. The recently developed optimized YFP variants Venus and SYFP2 have overcome most of the drawbacks of early YFP variants [50]. Venus and SYFP2 are both fast maturing fluorescent proteins that express well at 37 °C and are less sensitive to changes in pH and insensitive to the Cl^- concentration; however, they have been implied to be somewhat less photostable than EYFP [3, 49].

5.4. Variants of DsRed

DsRed (Fig. 5.3E) is a bright RFP with excitation and emission maxima at 558 and 583 nm, respectively. Despite the bright red fluorescence, application of DsRed has been restricted, because of slow and inefficient maturation and its tetrameric structure [70, 71]. The poor maturation efficiency has been overcome by random mutagenesis, which resulted in the fast maturing variant DsRedT1 [72]. However, DsRedT1 remains tetrameric.

DsRedT1 has served as the basis for further mutagenesis by directed evolution to finally yield a monomeric RFP, mRFP1 [73]. An additional feature of mRFP1 is a further red-shift in excitation and emission of approximately 25 nm, resulting in excitation and emission maxima at 584 and 607 nm, respectively. An undesirable side effect in mRFP1 is a significant reduction in brightness. Therefore, much effort has been put forth to develop brighter monomeric RFPs. This has resulted in the generation of a whole series of bright monomeric fluorescent proteins [74, 75]. These fluorescent proteins range in color from green and orange to red and far-red. The red variant mCherry shows fast and efficient maturation and high photostability [74]. The excitation and emission spectra of mCherry are 3 nm further red-shifted, with maxima at 587 and 610 nm, respectively. Another red variant mStrawberry is even brighter but the fluorescence is somewhat blue-shifted compared with mCherry. Unfortunately, it is also less photostabile and requires more time for maturation. Among the monomeric fluorescent proteins in the Fruit-series derived from DsRed and mRFP1 is an orange fluorescent protein, mOrange. The blue-shift from red to orange fluorescence (excitation and emission maxima at 548 and 562 nm, respectively) is caused by a unique covalent modification around the chromophore, which reduces conjugation within the chromophore (Fig. 5.3F) [76].

5.5. Variants of eqFP578 and eqFP611

DsRed is an obligate tetramer, and much effort was put into its monomerization. Unfortunately, the resulting monomeric fluorescent proteins often have substantially lower brightness and therefore alternative RFPs, which have less tendency to oligomerize are desirable. Two fluorescent proteins, eqFP578, and eqFP611, have been isolated from the sea anemone *Entacmaea quadricolor* and served as the basis for the development of several novel mRFPs.

EqFP578 is a dimeric bright RFP and was used to develop a faster maturing variant, turboRFP [77]. TurboFP is a bright fluorescent protein with an extinction coefficient of 92,000 M^{-1} cm^{-1} and a quantum yield of 0.67, but is still a dimer. Both site-directed mutagenesis at the dimer interface and random mutagenesis were used to monomerize the protein. The resulting dim variants were subjected to several rounds of semirandom and random mutagenesis to finally yield tagRFP. This orange-red-fluorescent protein has excitation and emission maxima at 553 and 574 nm, respectively. With an extinction coefficient of 100,000 M^{-1} cm^{-1} and a quantum yield of 0.48, it is almost threefold brighter than mCherry [77]. The photostability is somewhat lower than mCherry, but it was reported that a single mutation increases the photostability of tagRFP around one order of magnitude [78], without altering other spectroscopic parameters.

TurboFP was also used as a basis for far-red fluorescent proteins (fRFPs). Residues surrounding the chromophore were mutagenized to create a library, which was subsequently subjected to random mutagenesis. A bright far-red variant with excitation and emission maxima at 588 and 635 nm, respectively was isolated and named Katushka [79]. This fast-maturing protein has an extinction coefficient of 65,000 M^{-1} cm^{-1} and a quantum yield of 0.34. Hence, Katushka is the brightest far-red protein. To develop a monomeric variant that can be used for protein tagging, the mutations surrounding the chromophore of Katushka were introduced into tagRFP. This resulted in a monomeric fRFP, mKate, which is spectrally similar to Katushka, with an extinction coefficient of 45,000 M^{-1} cm^{-1} and a quantum yield of 0.33. mKate is remarkably photostable, but shows complex photobleaching kinetics under laser excitation.

The development of fRFPs is of great interest. One reason is the expansion of the color palette of fluorescent proteins; however, equally important is the potential of fRFP for use in tissues and whole-body mapping [1, 80]. The most red-shifted native fluorescent proteins at present are derived from eqFP611, which has its emission maximum at 611 nm [81]. Although eqFP611 is brightly fluorescent and has a low tendency to oligomerize, its expression is restricted to temperatures below 30 °C. EqFP611 has been subjected to extensive mutagenesis to generate better folding and further red-shifted variants. Better folding at elevated temperatures was enabled by introducing mutations that facilitate dimerization [82]. Even further red-shifted variants were generated by introducing mutations that cause trans–cis isomerization of the chromophore [83]. The most far-red-fluorescent variant has an emission maximum at 639 nm and is efficiently expressed at 37 °C as a dimer.

5.6. VFP variants from other sources

Several other fluorescent proteins have been isolated from a variety of species. We will briefly discuss the variants with the most optimal spectroscopic properties. A true orange fluorescent protein was isolated from the stony coral *Fungia concinna*, with good absorbance and a quantum yield of 0.45. Targeted mutagenesis was used to monomerize the protein, yielding mKO with an absorbance of 51,600 M^{-1} cm^{-1} and a quantum yield of 0.74 [84, 85]. The crystal structure of mKO has not been resolved yet, but the high similarity in absorbance and fluorescence between mKO and mOrange suggests a similar chromophore structure. The lower pK_a, higher quantum yield, relative high fluorescence lifetime, and higher photostability make mKO better suited for FRET applications than do mOrange [3, 84]. In addition, the high fluorescence lifetime makes mKO especially suited for FLIM applications.

A bright cyan-green fluorescent protein was isolated from *Clavularia* coral [86]. Since one of the intermediates displayed fast bleaching, a screen for more photostable variants was performed. The optimized monomeric variant was named teal fluorescent protein 1 (mTFP1). It has an excitation and emission maximum at 462 and 492 nm, respectively, so this protein is spectrally located in between CFP and GFP. With an extinction coefficient of 64,000 M^{-1} cm^{-1} and a quantum yield of 0.85 mTFP1 is a very bright fluorescent protein.

5.7. Chromoproteins and fluorescent derivatives

Nonfluorescent chromoproteins constitute a surprising alternative source of fRFPs. These GFP homologs have highly absorbing chromophores, however, without the ability to fluoresce. It was shown that a single amino acid mutation was sufficient to render these proteins fluorescent. Whereas GFP has a histidine at position 148 and VFPs from *Anthozoa* have a serine at the corresponding position, the chromoproteins have an Ala, Cys, or Asn residue [17, 87]. Several chromoprotein mutants containing 148Ser display (far-) red fluorescence with emission maxima beyond 600 nm, probably due to alteration of the chromophore environment, resulting in a coplanar chromphore structure [5]. Currently, the most far-red-shifted fluorescent protein is the tetrameric protein AQ143 with excitation and emission maxima at 595 and 655 nm, respectively, and fluorescence extending beyond 750 nm [88]. Until recently, fRFPs were only weakly fluorescent, due to a low quantum yield (0.04 for AQ143). However, because of the high molar extinction coefficient (90,000 M^{-1} cm^{-1} for AQ143) and the low autofluorescence in this spectral region, far-red fluorescence can be readily measured. Although these proteins are weakly fluorescent and usually oligomeric, monomeric versions could in principle be interesting for use as FRET acceptors, because of their high absorbance. Moreover, the nonfluorescent chromoproteins can be used for FLIM-based FRET measurements, as has been demonstrated for a nonfluorescent YFP-based acceptor [89].

5.8. Optical highlighter fluorescent proteins

Among the large number of fluorescent proteins known today several proteins have been identified which contain unique properties, for example the ability to change the color of their fluorescence [1, 90]. These fluorescent proteins can be turned on at will, where, and when you want.

Photoactivatable visible fluorescent proteins (PA-VFPs) become fluorescent upon intense illumination with near-UV light around 400 nm. At present, three PA-VFPs are known: PAGFP, PS-CFP2, and PAmRFP1. PAGFP was created by inserting mutation Thr203His into avGFP [91]. This mutation prevents chromophore deprotonation. Therefore, before activation PAGFP does not fluoresce when excited with 488 nm light. Upon irradiation with intense violet light, the chromophore is converted into its anionic deprotonated state, thus enabling excitation at 488 nm and producing a 100-fold increase in fluorescence. Photoactivation has been shown to result in decarboxylation of Glu222, making photoactivation irreversible [92, 93].

PAmRFP1 is a variant of the RFP DsRed and mRFP1 [94]. Upon irradiation with 380 nm light, PAmRFP1 displays a 70-fold increase in red fluorescence. However, use of PAmRFP1 is limited, due to its dim red fluorescence.

PS-CFP2 is actually a photoconvertable fluorescent protein and emits cyan fluorescence when excited with low intensity violet light

[95]. Upon illumination with intense near-UV light, PS-CFP2 fluorescence converts to green, resulting in a more than 400-fold increase in green fluorescence and over 2000-fold change in the ratio green/cyan fluorescence. The mechanism of photoconversion is irreversible and thought to be similar to that of PAGFP.

Kaede-like photoconvertable visible fluorescent proteins (PC-VFPs) change the color of fluorescence from green to red upon illumination with intense violet light. An advantage of Kaede-like PC-VFPs is that the excitation wavelength required for fluorescence does not cause photoconversion. In Kaede [96], the green-fluorescent chromophore before photoswitching consists of His65-Tyr66-Gly67 (amino acid numbering according to avGFP). Photoconversion involves cleavage of the protein backbone between the Nα and Cα of His65 and the subsequent formation of a double bond between Cα and Cβ of His65. This extends the conjugated π-electron system of the chromophore to the imidazole ring of His65 and results in red-shifted fluorescence. At present, four Kaede-like PC-VFPs have been isolated and all except one are obligate tetramers. Kaede and EosFP [97] are native PC-VFPs, whereas KiKGr [98] was rationally engineered. A monomeric variant of EosFP (mEosFP; [97] has been created; however, expression of mEosFP is limited to temperatures under 30 °C. Dendra is a monomeric photoswitchable fluorescent protein, which can be photoconverted with near-UV as well as high intensity 488 nm light [99].

Photoswitchable fluorescent proteins (PS-VFPs) have the ability to be repeatedly turned on and off. Only two PS-VFPs, KFP1, and Dronpa have been studied in detail [100, 101]. Dronpa is a monomeric GFP and is switched on by near-UV irradiation and turned off by intense blue light. KFP1 is a tetrameric RFP that is switched on by illumination with green light and is switched off again by blue light. Photoactivation of KFP1 can be made irreversible by using high-intensity green illumination. The mechanism of photoswitching has been proposed to involve a cis–trans isomerization of the chromophore, with fluorescence associated to the cis-conformation [27, 90, 102].

5.9. Choosing the right VFP FRET pair

With the large number of spectral classes and several variants within each spectral class, choosing the right VFP FRET pair for FRET can be a daunting task. Due to space limitations, it is impossible to discuss technical considerations such as the microscope set-ups (including available excitation and emission wavelengths, detection methods, etc.) and the nature of the sample (thickness, autofluorescence, fixed, or live cells). Instead, we will discuss biochemical and biophysical properties of VFPs that are relevant in FRET measurements, including R_0, oligomerization, pH dependence, and photostability, before focusing on the most promising pairs in detail.

Some of the requirements for the VFP pair are dependent on the approaches used to measure FRET. We will only briefly mention these issues as more comprehensive discussions can be found elsewhere in this book. Although many options exist to measure FRET using a microscope [103, 104], the most popular methods are: (1) measuring the excited-state lifetime (Chapters 2–4), (2) donor dequenching upon acceptor photobleaching (Chapter 7), (3) ratio-imaging (Chapters 6 and 7), or (4) measuring sensitized emission (Chapters 7 and 8). In the case where only (a change in) donor fluorescence is detected (methods 1 and 2), it is important to have a highly fluorescent donor, whereas fluorescence of the acceptor is unimportant. Moreover, even a nonfluorescent acceptor, that is, a chromoprotein, can be used because a larger part of the donor emission spectrum can be detected. However, in this case, the presence of the dark acceptor cannot be detected directly. For methods that measure sensitized emission (3 and 4), the direct excitation of the acceptor should be minimal and the absorbance and quantum yield of the acceptor are preferably high, since this increases the intensity of the sensitized emission.

5.9.1. Förster radius

The FRET efficiency (E) is highly dependent on the distance between donor and acceptor and is defined by Försters theory [105]:

$$E = \frac{1}{1 + \left(r_{DA}/R_0\right)^6} \tag{5.1}$$

Here, r_{DA} is the distance between donor and acceptor and R_0 is the Förster radius for the donor–acceptor pair. The Förster radius is the key quality measure for a VFP-FRET pair. R_0 is defined as the distance at which 50% FRET occurs and can be calculated from the following equation [105, 106]:

$$\mathrm{R}_0^6 = \mathrm{c}\kappa^2\eta^{-4}\mathrm{Q}_D\mathrm{J}(\lambda) \tag{5.2}$$

where c is 8.786 10^{-11} mol l^{-1} cm nm^2, κ^2 is the orientation factor of the interacting dipoles, η the refractive index of the medium separating donor and acceptor chromophore, Q_D the quantum yield of the donor, and $J(\lambda)$ (in M^{-1} cm^{-1} nm^4), the overlap integral which is defined as (Chapters 1 and 12):

$$J(\lambda) = \frac{\int_0^\infty F_D(\lambda)\varepsilon_A(\lambda)\lambda^4 d\lambda}{\int_0^\infty F_D(\lambda)d\lambda} \tag{5.3}$$

$F_D(\lambda)$ is the fluorescence emission spectrum of the donor, $\varepsilon_A(\lambda)$ is the molar absorbance spectrum of the acceptor, and λ is the wavelength. Generally, κ^2 is set to 2/3, which is true only if donor and acceptor are rapidly randomly orientated [107–109]. The maximum distance at which FRET can still be detected is in general $1.5R_0$, therefore choosing a pair with a large Förster radius is recommended. For example, changing the pair from ECFP–EYFP ($R_0 = 4.72$ nm) to mKO–mCherry ($R_0 = 6.37$ nm), assuming similar conditions, increases the distance over which reasonable FRET can be detected from 7.1 to 9.4 nm (Fig. 5.4). When applying VFPs to measure FRET, the intrinsic size of the fluorophore should be kept in mind. All VFPs have the fluorophore buried inside a β-barrel, implying that the distance from the fluorophore to the surface of the protein is at least 1 nm. Hence, the donor and acceptor chromophores cannot approach each other closer than 2 nm.

In Table 5.1, the R_0 values of several VFP pairs are presented. From the table, several conclusions can be drawn: First, a clear trend towards higher R_0 values is observed for red-shifted FRET pairs. Second, when BFP and CFP are considered as donors, relative high R_0 values are observed in combination with red acceptors, such as mStrawberry. Third, the R_0 for pairs consisting of identical VFPs can be significant, with the highest R_0 for SYFP2-SYFP2 (5.33 nm),

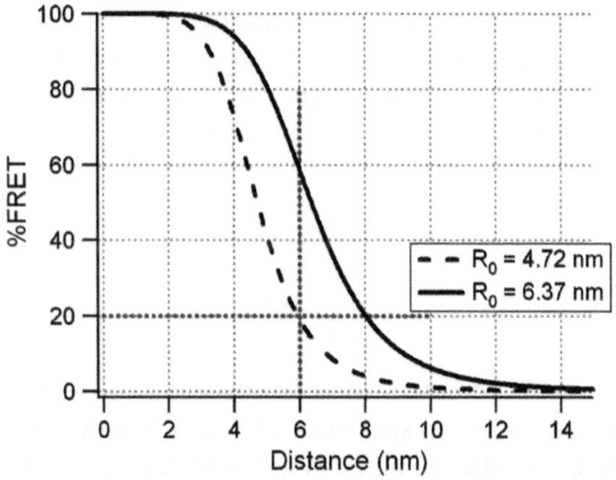

Fig. 5.4. Correlation between R_0 and FRET efficiency. Exchanging ECFP/EYFP ($R_0 = 4.72$ nm, dashed line) for the red-shifted VFP FRET-pair mKO/mCherry ($R_0 = 6.37$ nm, solid line) will increase the measured FRET efficiency, since the distance between donor and acceptor is expected to remain unchanged. A FRET pair with increased R_0 yields detectable FRET over longer distances and can be used to measure protein–protein interaction between larger proteins.

TABLE 5.1
Spectroscopic parameters[a] and Förster radii of monomeric VFP pairs

	EB	SB2	EC	mCer	SC3A	EG	SG2	EY	mVenus	SY2	mOr	mKO	tagRFP	mStrawb	mCherry	mKate
ϵ_{max} (x10^3)	31	34	28	33	30	63	46	72	105	101	71	52	100	90	72	45
QY	0.15	0.47	0.36	0.49	0.56	0.65	0.70	0.76	0.64	0.68	0.69	0.74	0.48	0.29	0.22	0.33
pKa	6.3	5.5	4.8	< 4.5	< 4.5	5.8	5.9	5.8	5.5	6.0	6.5	5.0	< 4	< 4.5	<4.5	6.0
EB	1.92	1.99	3.41	3.48	3.42	3.94	3.75	3.60	3.79	3.78	3.78	3.31	3.78	3.66	3.35	3.01
SB2	2.28	2.35	4.12	4.20	4.13	4.78	4.54	4.38	4.61	4.59	4.58	4.02	4.57	4.44	4.06	3.65
EC			3.17	3.14	3.06	4.74	4.58	4.72	5.00	4.97	4.87	4.44	4.94	4.88	4.52	3.98
mCer			3.30	3.26	3.17	4.98	4.82	4.98	5.27	5.25	5.12	4.69	5.20	5.14	4.78	4.19
SC3A			3.43	3.40	3.31	5.11	4.94	5.08	5.38	5.35	5.24	4.77	5.32	5.23	4.85	4.28
EG						4.76	4.85	5.51	5.85	5.82	5.70	5.27	5.83	5.73	5.31	4.66
SG2						4.63	4.77	5.59	5.94	5.91	5.81	5.38	5.97	5.87	5.44	4.78
EY						3.67	4.11	5.12	5.51	5.47	6.00	5.59	6.33	6.42	5.90	5.26
mVenus						3.48	3.90	4.92	5.30	5.27	5.83	5.44	6.15	6.24	5.75	5.11
SY2						3.52	3.95	4.98	5.36	5.33	5.89	5.50	6.21	6.30	5.81	5.16
mOr											5.21	4.89	6.15	6.46	6.27	5.75
mKO											5.36	5.17	6.25	6.59	6.37	5.85
tagRFP													5.06	5.75	5.81	5.43
mStrawb													4.32	5.05	5.27	4.98
mCherry													3.52	4.23	4.34	4.53
mKate													3.33	3.74	4.21	4.24

Förster radii (in nm) are calculated based on the spectroscopic parameters (molar absorbance in M^{-1} cm^{-1} and quantum yield) indicated in the top rows and the assumption that $k^2 = 2/3$and $n = 1.33$. EB = EBFP; SB2 = SBFP2; EC = ECFP; mCer = mCerulean; SC3A = SCFP3A; EG = EGFP; SG2 = SGFP2; EY = EYFP(Q69K); SY2 = SYFP2; mOr = mOrange; mStrawb = mStrawberry. R_0 values for pairs consisting of identical VFPs have a gray background.
[a]Spectroscopic data has been cited from [48, 49, 74, 77, 79, 84, 85].

indicating that homo-FRET can occur. Several VFP–FRET pairs and their R_0 value will be discussed in more detail.

5.9.2. Oligomeric state of fluorescent proteins

Besides the R_0 values, other factors have to be taken into account when choosing the optimal VFP FRET pair. Maybe most important is the oligomeric state of the fluorescent protein. Many fluorescent proteins, especially the red fluorescent variants, exist as obligate dimers or tetramers. Fusions of such fluorescent protein with your protein of interest will result in oligomerization of your protein of interest. This can easily result in artefacts in localization and function of the protein of interest. Therefore, only monomeric proteins are included in Table 5.1. Some monomeric VFPs, including the enhanced VFPs (EGFP, EBFP, ECFP, and EYFP) derived from *A. victoria* GFP, have a tendency to form dimers at high concentration [4, 46]. In cells, such high concentrations can be achieved when the fluorescent protein is compartmentalized, for example at the plasma membrane [46, 110]. Dimerization of VFPs should of course be prevented at all cost during protein–protein interaction studies as this can yield false positive protein–protein interactions. Much effort has been put into developing monomeric VFP variants and nowadays the whole visible spectral window is covered by monomeric VFPs. For fluorescent proteins derived from *A. Victoria*, GFP a single point mutation Ala206Lys abolishes the tendency to dimerize [46]. We advise to only use such "nonsticky" GFP variants containing the Ala206Lys mutation. An exception to the rule of only using nondimerizing VFPs, is the application in so-called "biosensors" (discussed in Section 8.3). Since these constructs are based on intramolecular FRET changes, dimerizing VFPs might be instrumental for decreasing the distance between donor and acceptor chromophores, thereby increasing the FRET efficiency in one of the two states and thus effectively increasing changes in FRET efficiency.

5.9.3. pH dependence

The fluorescence intensity of fluorescent proteins is pH dependent and most fluorescent proteins are less fluorescent at lower pH mainly because of a reduction in absorbance. Since the absorbance of the acceptor determines the FRET efficiency, changes in the acceptor absorbance spectrum due to pH variations can be wrongly interpreted as changes in FRET efficiency. Thus, a pK_a well below physiological pH is recommended to prevent artifacts due to pH changes inside cells. This is especially challenging if the fluorescent proteins are to be targeted to acid cellular compartments, for example, endosomes, lysosomes, or plant vacuoles.

5.9.4. Photostability

Photostability is an important parameter for the quality of VFPs for FRET applications as well. Bleaching will limit the time a fluorophore can be imaged. In case of ratio-imaging, ratio-changes that are not related to changes in FRET may arise from the fact that one of the two VFPs bleaches faster than the other. Thus in general one would prefer the most photostable VFPs available. When employing acceptor-bleaching experiments, however, a less photostable acceptor can be favorable especially if the intensity of the bleach light-source is limited.

Related to photobleaching is photoconversion, a process in which the spectral properties of a fluorophore change upon (intense) illumination. Photoconversion of an acceptor VFP into a species resembling the donor VFP or vice versa can change the apparent FRET efficiency and in this way lead to artifacts. Generation of fluorescence in VFPs is complex in nature and photoconversion has been demonstrated for several fluorescent proteins, including green, yellow, and red fluorescent VFPs [59, 91, 94, 111]. Furthermore, there is a debate going whether YFP variants can photoconvert into a CFP-like species [112–115]. In our lab, we have observed photoconversion of mOrange into a red-shifted species upon intense illumination with 488 nm laser-light (unpublished data), and photoconversion of mKO into a green-fluorescent species under intense illumination with 436 nm light [84].

The nuisance of photoconversion can be turned into a strategy for measuring FRET. When the absorbance of a VFP increases upon photoconversion, it can be used as an activatable FRET acceptor. Such a strategy was reported, in which FRET was measured, as well as mobility [116]. The donor was a CFP, and the acceptor photoactivatable GFP. An even more sophisticated FRET experiment, called photochromic FRET, can be performed when the acceptor can be switched reversibly between an absorbing and a nonabsorbing state. The use of photochromic FRET can result in a substantial improvement in the detection sensitivity to less than 1% FRET efficiency [117]. At present only organic probes have been used as reversible switching FRET acceptors [117, 118]. Nevertheless, with the recent discovery of reversible photoswitching VFPs, like Dronpa and KFP1, fully VFP-based photochromic FRET is expected to be feasible in the near future.

5.10. VFP-based FRET pairs

In this section, several VFP FRET pairs will be specifically discussed. We will start with the most blue-shifted pair, BFP-GFP, and then move towards the red part of the spectrum, subsequently highlighting CFP–YFP, GFP–RFP, YFP–RFP, and OFP–RFP pairs. For each of the pairs the excitation/emission spectra are plotted including the overlap integral (Fig. 5.5) as well as an emission spectrum of the FRET pair in a FRET and non-FRET situation (Fig. 5.6). These graphs will be useful for choosing the optimal VFP pair for a FRET study.

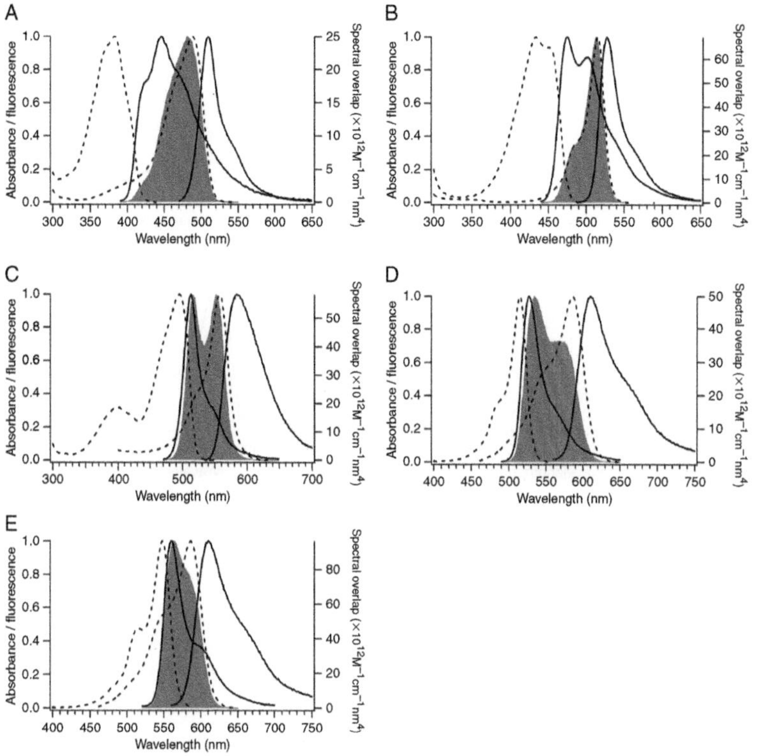

Fig. 5.5. Spectral overlap between VFP-based FRET pairs. Shown are the normalized absorbance spectra (dashed) and emission spectra (solid), and the λ^4-weighted spectral overlap (grey) of SBFP2–EGFP (A), SCFP3A–SYFP2 (B), SGFP2–tagRFP (C), SYFP2–mStrawberry (D), and mKO–mCherry (E).

5.10.1. BFP/GFP

The first FRET-based biosensors employing fluorescent proteins were developed over 10 years ago. These protease sensors consisted of a BFP donor fused to a GFP acceptor by a protease-sensitive linker [44, 119]. BFP and GFP have well separated emission spectra, resulting in little fluorescence bleed-through (Figs. 5.5A and 5.6A). This facilitates data analysis for FRET ratio imaging

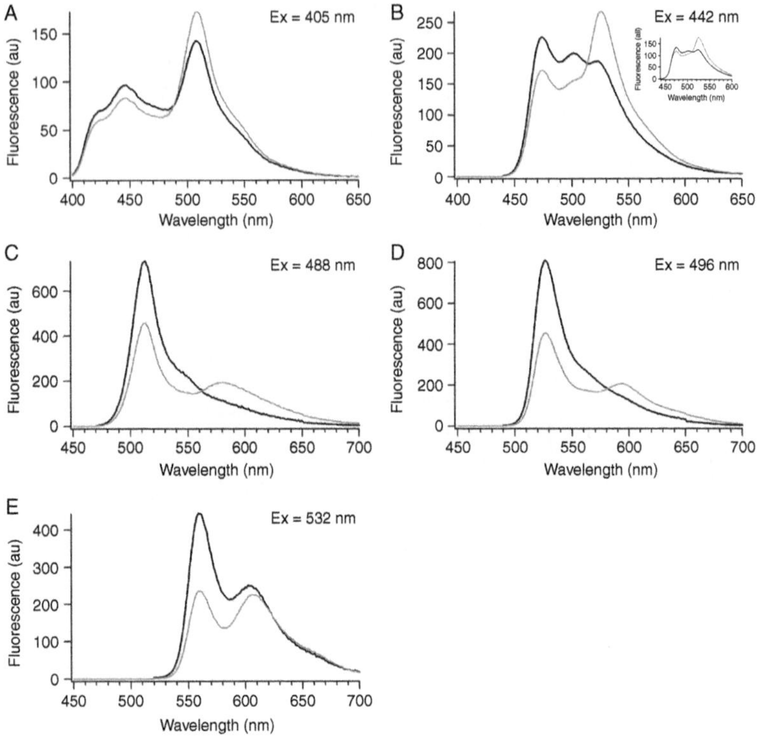

Fig. 5.6. Simulated emission spectra of VFP-based FRET pairs in a non-FRET situation (black line) and a FRET situation (grey line). Shown are SBFP2–EGFP, $E = 14\%$ (A), SCFP3A–SYFP2, $E = 24\%$ (inset shows ECFP–EYFP, $E = 12\%$) (B), SGFP2–tagRFP $E = 38\%$ (C), SYFP2–mStrawberry $E = 44\%$ (D), mKO–mCherry $E = 47\%$ (E). The FRET situation assumes a distance of $r_{DA} = 6.5$ nm between donor and acceptor and uses a 1:1 stoichiometry between donor and acceptor. Excitation at the indicated wavelength, other parameters from Table 5.1.

techniques and also allows collection of most of the BFP fluorescence. It is important to take into account cellular autofluorescence when using BFPs.

The use of BFPs for FRET applications have however been limited due to the poor extinction coefficients, quantum yields, and photostability of the early BFP variants. The recent development of brighter and more photostable BFP variants however might provide a comeback for the BFP/GFP FRET pair [48, 60]. The increased quantum yield of SBFP results in larger R_0 for BFP/GFP FRET pairs and the calculated R_0 for SBFP2/EGFP is 4.78 nm. This is similar to the R_0 for ECFP/EYFP (4.72 nm) thus the efficiency of energy transfer of BFP/GFP-based FRET applications is expected to be similar as when using ECFP/EYFP.

The requirement of near-UV excitation for BFPs can be a limiting factor. A possible solution would be to use two-photon excitation; alternatively a 405 nm diode laser could be used for excitation of BFP. A unique advantage of the BFP/GFP FRET-pair in principle is that it can be used in combination with a red-shifted VFP FRET pair. This enables the simultaneous expression of two different FRET-based biosensors (see Chapter 6).

5.10.2. CFP/YFP

CFP and YFP are currently the most widely used VFP FRET pair, mainly due to the fact that for a long time YFP was the most red-shifted monomeric VFP available. As a result, CFP and YFP are among the best-characterized and therefore the most reliable VFPs available at present. CFP and YFP were the first two VFPs bright and photostable enough for prolonged dual-color imaging and with considerable spectral overlap to enable FRET. Instrumentation for imaging CFP and YFP has become relatively standard. Excitation of CFP benefits from the high intensity 436 nm spectral line of a high-pressure mercury lamp, the 458 nm line of an argon laser, 440 nm diode lasers and blue LEDs. The argon laser also has, in addition to the standard 488 nm line, a laser line at 514 nm, which is excellent for YFP excitation. Filter sets for detecting CFP and YFP fluorescence and sensitized emission are available. FRET methods based on sensitized emission benefit from the relative high-quantum yield of YFP.

FRET applications employing CFP and YFP are complicated due to considerable bleed-through between CFP and YFP fluorescence (Figs. 5.5B and 5.6B). Direct excitation of YFP and bleed-through of CFP fluorescence into the YFP detection channel have to be corrected for as shown in Chapters 7 and 8. The multiexponential fluorescence decay of all CFP variants complicates the quantification of FRET by donor lifetime methods. Altogether these factors make quantitative analysis of the FRET efficiency relatively difficult.

5.10.3. GFP/RFP

In theory, red-shifted donor–acceptor pairs have increased Förster radii, due to a fourth power contribution of wavelength to the overlap integral $J(\lambda)$ (see Eq. (5.3)). Another advantage is the reduction in cellular autofluorescence because of the longer excitation wavelengths. Before the development of monomeric RFPs, YFP has been used as a FRET acceptor for GFP [120]. The main advantage being the high R_0 of 5.51 nm, however, due to the largely overlapping emission spectra (separation of the maxima is less than 20 nm) detection of FRET is complicated. Several solutions to this problem have been presented. First, changing the GFP to the optimized GFP variant GFP^2, which has improved brightness and an excitation maximum around 400 nm similar to wild-type avGFP [121]. At this excitation wavelength YFP is not or hardly excited, thus effectively decreasing acceptor fluorescence due to direct excitation. This is especially useful for FRET measurements that are based on measuring the sensitized emission and this pair has indeed been used to construct a ratiometric FRET-based sensor with a relatively high dynamic range [122].

Second, FRET between GFP and YFP can be measured by determining the lifetime of the FRET-pair together. The lifetime of the combined fluorescence emission of GFP and YFP is increased in case of FRET due to a process known as "acceptor in-growth". It was shown that cleavage of the GFP-YFP construct decreased the detected lifetime due to the loss of FRET [120].

Finally, a nonfluorescent YFP variant has been constructed for use as a FRET acceptor for GFP [89]. This allows the detection of the complete emission spectrum of GFP, while the FRET efficiency is high ($R_0 = 5.9$ nm) due to strong overlap of the GFP fluorescence and YFP absorbance band. The occurrence of FRET was detected by a reduction in the excited state lifetime of the GFP by FLIM. The main disadvantage is that the presence of the acceptor cannot be detected in living cells.

An important step towards useful red FRET acceptors was the development of the first mRFP1 [73], as this variant solved the problems of tetramerization and slow/incomplete maturation of its parent, DsRed [13]. Unfortunately the EGFP–mRFP1 FRET pair has only a slightly increased R_0 relative to CFP–YFP, due to the low extinction coefficient of mRFP1 and the poor spectral overlap of the donor emission and acceptor absorbance. Further optimization of mRFP1 has yielded novel RFPs, such as mCherry and mStrawberry, which have increased extinction coefficients and photostability, thereby rendering mRFP1 obsolete [74]. In addition, a GFP variant, SGFP2, with an improved quantum yield and photostability has been reported [48]. The combination of the higher quantum yield of SGFP2, the high extinction coefficient of mStrawberry and the increased spectral overlap due to a 10 nm blueshift in the absorbance of mStrawberry increased the R_0 of the SGFP2–mStrawberry pair to 5.87 nm. At present, based on the R_0 value and quantum yield, the best acceptor for GFPs is tagRFP (R_0 6.0 nm, Table 5.1). Moreover, the relative high quantum yield should be advantageous for FRET measurements based on sensitized emission (Figs. 5.5C and 5.6C).

5.10.4. YFP/RFP

Bright YFPs are available that can be used as donor. Two variants, SYFP2 and Venus, have been optimized for folding and have quantum yields up to 0.7 [49, 50]. Since the fluorescence emission spectrum of YFP is red-shifted compared with GFP the spectral overlap with the RFP is greater (Figs. 5.5D and 5.6D). This translates directly into higher R_0 values. On the other hand, the red-shifted YFP fluorescence does complicate the separation of donor and acceptor fluorescence. We have recently measured FRET between fusion proteins consisting of SYFP2 and mRFP1, mStrawberry or mCherry, both by FLIM and acceptor photobleaching [84]. The FRET efficiency measured for SYFP2-mStrawberry and SYFP2–mCherry was significantly higher than measured for SYFP2–mRFP1, reflecting the increased extinction coefficient of these novel RFP. It is of note that cell-to-cell variation in the FRET efficiency was observed for the SYFP2-mStrawberry pair one day after transfection, which was attributed to a slower folding/maturation of mStrawberry in mammalian cells grown at 37 °C. This heterogeneity was not observed when mCherry was used as the acceptor, or when cells expressing SYFP2–mStrawberry were examined at least 2 days after transfection. In plant protoplasts, the YFP–mStrawberry pair was used to detect protein–protein interactions without heterogeneity in the FRET efficiency [123]. Probably the lower incubation temperature for the plant cells (20 °C) is beneficial for folding/maturation of the protein. In our lab we now routinely use the SYFP2–mCherry and SYFP2–mStrawberry pairs to detect FRET in living cells using either FRET-based FLIM or acceptor photobleaching.

5.10.5. Orange/RFP

Two orange fluorescing proteins are available with an emission maximum around 560 nm. As their quantum yields are high (>0.69) and their emission spectra overlap strongly with the absorbance spectrum of the RFP, high R_0 values are calculated (see Table 5.1). However, the spectral separation becomes increasingly more difficult, demanding careful design of the experiment. The highest R_0 of 6.59 nm is calculated for mKO–mStrawberry pair, but application of this pair is not recommended for methods in which the donor fluorescence needs to be detected without bleed-through from acceptor fluorescence. Therefore, the only application we envision for this pair is in FRET-based sensors that are examined by ratio-imaging or spectral unmixing (see also Chapters 7 and 8). On the other hand, mCherry is sufficiently red-shifted to enable its use as a FRET acceptor for orange fluorescent proteins (Fig. 5.5E), although the relatively low quantum yield of mCherry, limits the sensitized emission (Fig. 5.6E). The higher quantum yield of mKate would be beneficial for sensitized emission based methods. The mOrange–mCherry and mKO–mCherry pair have R_0 values of 6.27 and 6.37 nm, respectively, which is a substantial improvement over the R_0 of the widely used ECFP–EYFP pair. High FRET efficiencies have been observed between orange fluorescent proteins and mCherry in single living cells expressing tandem fusions [84]. The occurrence of FRET was detected by FLIM, exciting the orange fluorescence at 514 nm and detecting the donor fluorescence with a bandpass filter of 530–560 nm. Unfortunately, a large fraction of the orange fluorescence (75%) cannot be detected due to overlap with the red acceptor fluorescence. Still, sufficient signal remained to acquire the excited state lifetime of the donor. From these experiments was concluded that mKO is the best donor since it has a high quantum yield and a high lifetime, increasing the lifetime contrast in FLIM-based FRET studies. Moreover, it was shown that the mKO–mCherry pair detected homodimerization of a NF-κB subunit with a twofold higher FRET efficiency than with the ECFP–EYFP couple.

5.11. Applications

The three main applications of VFPs in FRET studies are (1) the study of protein–protein interactions in living cells, (2) the study of conformational changes within a protein, and the (3) use of biosensors, which are designed to undergo a conformational change, indicating a posttranslational change or binding of a (small) molecule. The latter type includes proteolysis sensors that have a cleavable linker to measure protease activity. In this section we will highlight examples for each of the three applications.

5.11.1. Protein–protein interactions

In addition to studying protein dynamics, visualization of protein–protein interactions is a key aspect for understanding the regulation of protein-function (especially in signal transduction). Because of the limited spatial resolution of conventional light microscopy (200 nm), it is impossible to distinguish between two differently labeled proteins that actually interact or merely co-localize. To detect protein–protein interactions in living cells, methods are required which are sensitive to changes in the nanometer range. Recently, several strategies to increase the resolution in fluorescence microscopy by almost an order of magnitude have been reported, that is, STED [124], STORM [125], PALM [126], however, at this moment these methods still do not have sufficient resolution to detect protein–protein interactions.

Detection of FRET between fluorescently labeled proteins has become an extremely powerful tool for studying protein–protein interaction in living cells. Since FRET is restricted to distances <9 nm, it provides resolution on the nanometer scale [127–130]. In practice, a number of issues complicate measurements of protein–protein interaction by FRET, and can lead to misleading or even meaningless results [131, 132]. One major consideration is that the donor:acceptor stoichiometry needs to be within the range of 10:1–1:10 [133]. For FRET measurements of protein–protein interactions, where one partner might be in excess, the main problem is to measure a small amount of FRET in a background of fluorescent labels that are not undergoing FRET. Many possible protein–protein interaction experiments that fall into this category are simply unsuitable for examination by FRET. Nevertheless, FRET has been used to visualize interaction between a variety of proteins, including oligomerization of receptors [134–137], complex formation between heterotrimeric G-proteins [123, 138, 139], transcription factor interactions [140–142], and many others. An elegant study in which protein–protein interaction was demonstrated by FRET was recently published by Eberle and coworkers [143]. They demonstrated oligomerization of CFP/YFP-labeled CD95 death receptors upon stimulation of cells with different apoptosis inducing treatments. Furthermore they found that oligomerization in the cytosol is required for translocation of the receptor complex to the plasma membrane. In this study, ECFP was excited at 405 nm. Although this wavelength in suboptimal for excitation, it minimized direct acceptor excitation and hence facilitated the detection of sensitized emission due to FRET. Minimizing direct acceptor excitation is of key importance for protein–protein interaction studies when sensitized emission is used as the read-out, because donor- and acceptor-labeled proteins are not necessarily expressed at equal concentrations. Direct acceptor excitation in a cell overexpressing acceptor can be mistaken for FRET.

Importantly, in most applications the measured (change in) FRET efficiency cannot be translated directly into an average distance between donor and acceptor fluorophore because the fraction of donor molecules involved in FRET is unknown (i.e., all molecules display 25% FRET or 50% of the molecules display 50% FRET), and the orientation factor (κ^2) is unknown (see also Chapter 7).

5.11.2. Conformational change

Since FRET is sensitive to the dipole orientation of the donor and acceptor, changes in orientation of either the donor fluorophore or the acceptor fluorophore may lead to changes in FRET efficiency. Therefore conformational changes within a protein can be detected when the protein of interest is tagged with both a donor and an acceptor fluorophore. This is also the basis for the FRET-based sensors that will be described in the next section. Several examples of papers reporting on conformational change using intramolecular FRET in living cells can be found [144, 145]. An elegant example of this approach is reported by Vilardaga and coworkers [146], reporting on a G-protein coupled receptor that has been tagged with CFP and YFP. A G-protein coupled receptor is located at the plasma membrane in eukaryotic cells and has seven transmembrane helices. It is known, mainly from studies on purified proteins, that upon binding of a ligand, the orientation of helices changes. In order to study this process in single living cells receptors were tagged with CFP and YFP. YFP was added to the N-terminus of the receptor that resides in the cytosol, while the CFP was inserted into one of the cytosolic loops. It is of note that the receptors were truncated to allow monitoring of the conformation. Furthermore the presence of the two large VFPs hampered downstream signaling. Nonetheless, the authors of the paper demonstrated that the kinetics of receptor activation depend on the receptor type. Subsequently, this approach has been improved and extended to several other receptors [147–149].

5.11.3. Biosensors

The first straightforward application of a biosensor was to detect cleavage of the fusion protein [44, 119]. This approach is of interest for the detection of protease activity in single living cells, and has been applied to monitor caspase activity [150, 151]. More sophisticated biosensors are based on a FRET change due to a structural rearrangement, changing the distance, and/or the spatial orientation of the fluorophores. In almost all cases the underlying molecular mechanism is unknown. The prototypical FRET based biosensor is termed "cameleon." The approach was published at the same time by two different groups [152, 153]. The sensor for calcium consists of three main parts, a donor, a sensor, and an acceptor. (It is of note that the sensor in its turn may consist of multiple parts). The first version of cameleon showed only a modest change in FRET efficiency when calcium levels were elevated. Nevertheless, the fact that it is genetically encoded has tremendous advantages, since it can be easily tuned for a certain calcium affinity [152], it can be targeted to any cell organelle [68], and has potential for use in whole animal imaging [154]. The cameleon has undergone several rounds of improvement to increase the dynamic range of the FRET change, including the replacement of the YFP by a version with a lower pK_a [68] or enhanced folding [155], optimization of the sensor part [156, 157], and changing the relative orientation of the donor and acceptor fluorophore [158]. The latter improvement has been achieved by using circular permutated YFP variants, in which the C- and N-terminus have been replaced by a new C- and N-terminus at another position in the protein [159]. One of the circular permutated variants yields a cameleon with a tremendously increased dynamic range. Permutation of the YFP is a generally applicable approach and we expect that the circular permutated variants will be used to optimize other CFP–YFP-based FRET sensors. Moreover, circular permutation is not restricted to YFP and could therefore be applied to all GFP-mutants and probably to all fluorescent proteins since the VFPs are structurally similar.

An alternative approach was used to optimize a FRET-based glutamate sensor [160]. A glutamate binding protein was sandwiched between CFP and YFP. To optimize the FRET change, a linker truncation library was screened, by systematically removing single residues from the N- and/or C-terminus of the glutatamate-sensitive domain. There was a weak trend towards higher FRET contrast upon truncation of the N- and C-terminus, but no prediction could be made as to which of the combinations would produce the optimal FRET change and only a single variant turned out to be superior. These results are in line with a previous study in the FRET efficiency of several concatenated CFP–YFP fusions was analyzed [161]. Again the dependence of FRET efficiency on the linker length was irregular. Together, these results indicate that the removal of amino acid residues changes both the distance and the orientation between the donor and acceptor, and although the effects on the FRET efficiency are unpredictable, this provides another strategy for optimization of FRET sensors. Both the use of circularly permutated VFPs and the deletion of amino acids increase the number of possible combinations substantially. Therefore, it will be essential to rapidly produce constructs and screen them for the response of interest.

Since the first report of cameleon, sensors based on the same approach have been designed to detect small molecules (including cAMP [162, 163], cGMP [164], Ins(1,4,5)P3 [165, 166], lipids [167, 168], and sugars [169]) and the activity of a wide variety of kinases [122, 168, 170, 171].

5.12. Conclusion

VFPs have revolutionized the way cell-biological research is approached and have revived many fluorescence microscopy techniques including TIRF, FRAP, and not least the application of FRET. Nowadays, many combinations of VFPs are available for application in FRET studies. In this chapter we have summarized the properties of several VFP pairs. The most widely used CFP/YFP pair is still a good choice when nonsticky and optimized variants are used (e.g., SCFP3A–SYFP2), although we think that the future holds great promise for novel red-shifted pairs to study protein–protein interactions, mainly because of the increased Förster radius [84]. Papers in which the red-shifted pairs are used start to emerge and it will be exciting to see whether these pairs will become as successful as the CFP/YFP pair. Application of red-shifted pairs in biosensors is still less attractive since the biosensors are often analyzed by measuring the sensitized emission, which is hampered by a relative poor quantum yield of the red acceptors. The recent development of red variants with improved quantum yields, tagRFP, and mKate, is an encouraging step towards bright RFP. Finally, we anticipate an increase in studies employing two or more VFP-based FRET pairs in single cells [172, 173]. This kind of multiparameter approaches will help to elucidate the molecular mechanisms that underlie cellular physiology [174].

Acknowledgments

We like to thank Dmitriy Chudakov (Institute of Bioorganic Chemistry, Moscow, Russia) for sharing data on RFP prior to publication and Dorus Gadella for help with creating Table 5.1 and emission plots. This work is supported by NIH Grant GM72048 (GJK) and the EU integrated project on "Molecular Imaging" LSHG-CT-2003-503259 (JG).

References

[1] Chudakov, D. M., Lukyanov, S. and Lukyanov, K. A. (2005). Fluorescent proteins as a toolkit for in vivo imaging. Trends Biotechnol. *23*, 605–13.

[2] Shaner, N. C., Patterson, G. H. and Davidson, M. W. (2007). Advances in fluorescent protein technology. J. Cell Sci. *120*, 4247–60.

[3] Shaner, N. C., Steinbach, P. A. and Tsien, R. Y. (2005). A guide to choosing fluorescent proteins. Nat. Methods *2*, 905–9.

[4] Tsien, R. Y. (1998). The green fluorescent protein. Annu. Rev. Biochem. *67*, 509–44.

[5] Verkhusha, V. V. and Lukyanov, K. A. (2004). The molecular properties and applications of *Anthozoa* fluorescent proteins and chromoproteins. Nat. Biotechnol. *22*, 289–96.

[6] Shimomura, O., Johnson, F. H. and Saiga, Y. (1962). Extraction, purification and properties of aequorin, a bioluminescent protein from the luminous hydromedusan, Aequorea. J. Cell. Comp. Physiol. *59*, 223–39.

[7] Morin, J. G. and Hastings, J. W. (1971). Energy transfer in a bioluminescent system. J. Cell Physiol. *77*, 313–8.

[8] Prasher, D. C., Eckenrode, V. K., Ward, W. W., Prendergast, F. G. and Cormier, M. J. (1992). Primary structure of the *Aequorea victoria* green-fluorescent protein. Gene *111*, 229–33.

[9] Chalfie, M., Tu, Y., Euskirchen, G., Ward, W. W. and Prasher, D. C. (1994). Green fluorescent protein as a marker for gene expression. Science *263*, 802–5.
[10] Inouye, S. and Tsuji, F. I. (1994). Aequorea green fluorescent protein. Expression of the gene and fluorescence characteristics of the recombinant protein. FEBS Lett. *341*, 277–80.
[11] Greer, L. F., III and Szalay, A. A. (2002). Imaging of light emission from the expression of luciferases in living cells and organisms: A review. Luminescence *17*, 43–74.
[12] Inouye, S. and Shimomura, O. (1997). The use of Renilla luciferase, Oplophorus luciferase, and apoaequorin as bioluminescent reporter protein in the presence of coelenterazine analogues as substrate. Biochem. Biophys. Res. Commun. *233*, 349–53.
[13] Matz, M. V., Fradkov, A. F., Labas, Y. A., Savitsky, A. P., Zaraisky, A. G., Markelov, M. L. and Lukyanov, S. A. (1999). Fluorescent proteins from nonbioluminescent *Anthozoa* species. Nat. Biotechnol. *17*, 969–73.
[14] Shagin, D. A., Barsova, E. V., Yanushevich, Y. G., Fradkov, A. F., Lukyanov, K. A., Labas, Y. A., Semenova, T. N., Ugalde, J. A., Meyers, A. Nunez, J. M. *et al.* (2004). GFP-like proteins as ubiquitous metazoan superfamily: Evolution of functional features and structural complexity. Mol. Biol. Evol. *21*, 841–50.
[15] Gurskaya, N. G., Fradkov, A. F., Pounkova, N. I., Staroverov, D. B., Bulina, M. E., Yanushevich, Y. G., Labas, Y. A., Lukyanov, S. and Lukyanov, K. A. (2003). A colourless green fluorescent protein homologue from the non-fluorescent hydromedusa *Aequorea coerulescens* and its fluorescent mutants. Biochem. J. *373*, 403–8.
[16] Xia, N. S., Luo, W. X., Zhang, J., Xie, X. Y., Yang, H. J., Li, S. W., Chen, M. and Ng, M. H. (2002). Bioluminescence of *Aequorea macrodactyla*, a common jellyfish species in the East China Sea. Mar. Biotechnol. (NY) *4*, 155–62.
[17] Gurskaya, N. G., Fradkov, A. F., Terskikh, A., Matz, M. V., Labas, Y. A., Martynov, V. I., Yanushevich, Y. G., Lukyanov, K. A. and Lukyanov, S. A. (2001). GFP-like chromoproteins as a source of far-red fluorescent proteins. FEBS Lett. *507*, 16–20.
[18] Labas, Y. A., Gurskaya, N. G., Yanushevich, Y. G., Fradkov, A. F., Lukyanov, K. A., Lukyanov, S. A. and Matz, M. V. (2002). Diversity and evolution of the green fluorescent protein family. Proc. Natl. Acad. Sci. USA *99*, 4256–61.
[19] Matz, M. V., Lukyanov, K. A. and Lukyanov, S. A. (2002). Family of the green fluorescent protein: journey to the end of the rainbow. Bioessays *24*, 953–9.

[20] Salih, A., Larkum, A., Cox, G., Kuhl, M. and Hoegh-Guldberg, O. (2000). Fluorescent pigments in corals are photoprotective. Nature *408*, 850–3.

[21] Marshall, J. and Oberwinkler, J. (1999). The colourful world of the mantis shrimp. Nature *401*, 873–4.

[22] Mazel, C. H., Cronin, T. W., Caldwell, R. L. and Marshall, N. J. (2004). Fluorescent enhancement of signaling in a mantis shrimp. Science *303*, 51.

[23] Ormo, M., Cubitt, A. B., Kallio, K., Gross, L. A., Tsien, R. Y. and Remington, S. J. (1996). Crystal structure of the *Aequorea victoria* green-fluorescent protein. Science *273*, 1392–5.

[24] Yang, F., Moss, L. G. and Phillips, G. N., Jr. (1996). The molecular structure of green fluorescent protein. Nat. Biotechnol. *14*, 1246–51.

[25] Petersen, J., Wilmann, P. G., Beddoe, T., Oakley, A. J., Devenish, R. J., Prescott, M. and Rossjohn, J. (2003). The 2.0-A crystal structure of eqFP611, a far red fluorescent protein from the sea anemone *Entacmaea quadricolor*. J. Biol. Chem. *278*, 44626–31.

[26] Wall, M. A., Socolich, M. and Ranganathan, R. (2000). The structural basis for red fluorescence in the tetrameric GFP homolog DsRed. Nat. Struct. Biol. *7*, 1133–8.

[27] Wilmann, P. G., Petersen, J., Pettikiriarachchi, A., Buckle, A. M., Smith, S. C., Olsen, S., Perugini, M. A., Devenish, R. J., Prescott, M. and Rossjohn, J. (2005). The 2.1A crystal structure of the far-red fluorescent protein HcRed: inherent conformational flexibility of the chromophore. J. Mol. Biol. *349*, 223–37.

[28] Yarbrough, D., Wachter, R. M., Kallio, K., Matz, M. V. and Remington, S. J. (2001). Refined crystal structure of DsRed, a red fluorescent protein from coral, at 2.0-A resolution. Proc. Natl. Acad. Sci. USA *98*, 462–7.

[29] Barondeau, D. P., Kassmann, C. J., Tainer, J. A. and Getzoff, E. D. (2005). Understanding GFP chromophore biosynthesis: Controlling backbone cyclization and modifying post-translational chemistry. Biochemistry *44*, 1960–70.

[30] Rosenow, M. A., Huffman, H. A., Phail, M. E. and Wachter, R. M. (2004). The crystal structure of the Y66L variant of green fluorescent protein supports a cyclization-oxidation-dehydration mechanism for chromophore maturation. Biochemistry *43*, 4464–72.

[31] Cody, C. W., Prasher, D. C., Westler, W. M., Prendergast, F. G. and Ward, W. W. (1993). Chemical structure of the hexapeptide chromophore of the *Aequorea* green-fluorescent protein. Biochemistry *32*, 1212–8.

[32] Cubitt, A. B., Heim, R., Adams, S. R., Boyd, A. E., Gross, L. A. and Tsien, R. Y. (1995). Understanding, improving and using green fluorescent proteins. Trends Biochem. Sci. *20*, 448–55.
[33] Reid, B. G. and Flynn, G. C. (1997). Chromophore formation in green fluorescent protein. Biochemistry *36*, 6786–91.
[34] Heim, R., Prasher, D. C. and Tsien, R. Y. (1994). Wavelength mutations and posttranslational autoxidation of green fluorescent protein. Proc. Natl. Acad. Sci. USA *91*, 12501–4.
[35] Baird, G. S., Zacharias, D. A. and Tsien, R. Y. (2000). Biochemistry, mutagenesis, and oligomerization of DsRed, a red fluorescent protein from coral. Proc. Natl. Acad. Sci. USA *97*, 11984–9.
[36] Gross, L. A., Baird, G. S., Hoffman, R. C., Baldridge, K. K. and Tsien, R. Y. (2000). The structure of the chromophore within DsRed, a red fluorescent protein from coral. Proc. Natl. Acad. Sci. USA *97*, 11990–5.
[37] Yang, T. T., Cheng, L. and Kain, S. R. (1996). Optimized codon usage and chromophore mutations provide enhanced sensitivity with the green fluorescent protein. Nucleic Acids Res. *24*, 4592–3.
[38] Zolotukhin, S., Potter, M., Hauswirth, W. W., Guy, J. and Muzyczka, N. (1996). A "humanized" green fluorescent protein cDNA adapted for high-level expression in mammalian cells. J. Virol. *70*, 4646–54.
[39] Haseloff, J., Siemering, K. R., Prasher, D. C. and Hodge, S. (1997). Removal of a cryptic intron and subcellular localization of green fluorescent protein are required to mark transgenic *Arabidopsis* plants brightly. Proc. Natl. Acad. Sci. USA *94*, 2122–7.
[40] Yang, T. T., Sinai, P., Green, G., Kitts, P. A., Chen, Y. T., Lybarger, L., Chervenak, R., Patterson, G. H., Piston, D. W. and Kain, S. R. (1998). Improved fluorescence and dual color detection with enhanced blue and green variants of the green fluorescent protein. J. Biol. Chem. *273*, 8212–6.
[41] Cormack, B. P., Valdivia, R. H. and Falkow, S. (1996). FACS-optimized mutants of the green fluorescent protein (GFP). Gene *173*, 33–8.
[42] Delagrave, S., Hawtin, R. E., Silva, C. M., Yang, M. M. and Youvan, D. C. (1995). Red-shifted excitation mutants of the green fluorescent protein. Biotechnology (NY) *13*, 151–4.
[43] Crameri, A., Whitehorn, E. A., Tate, E. and Stemmer, W. P. (1996). Improved green fluorescent protein by molecular evolution using DNA shuffling. Nat. Biotechnol. *14*, 315–9.
[44] Heim, R. and Tsien, R. Y. (1996). Engineering green fluorescent protein for improved brightness, longer wavelengths and fluorescence resonance energy transfer. Curr. Biol. *6*, 178–82.

[45] Siemering, K. R., Golbik, R., Sever, R. and Haseloff, J. (1996). Mutations that suppress the thermosensitivity of green fluorescent protein. Curr. Biol. *6*, 1653–63.
[46] Zacharias, D. A., Violin, J. D., Newton, A. C. and Tsien, R. Y. (2002). Partitioning of lipid-modified monomeric GFPs into membrane microdomains of live cells. Science *296*, 913–6.
[47] Cubitt, A. B., Woollenweber, L. A. and Heim, R. (1999). Understanding structure-function relationships in the *Aequorea victoria* green fluorescent protein. Methods Cell. Biol. *58*, 19–30.
[48] Kremers, G. J., Goedhart, J., van den Heuvel, D. J., Gerritsen, H. C. and Gadella, T. W., Jr. (2007). Improved green and blue fluorescent proteins for expression in bacteria and mammalian cells. Biochemistry *46*, 3775–83.
[49] Kremers, G. J., Goedhart, J., van Munster, E. B. and Gadella, T. W., Jr. (2006). Cyan and yellow super fluorescent proteins with improved brightness, protein folding, and FRET Forster radius. Biochemistry *45*, 6570–80.
[50] Nagai, T., Ibata, K., Park, E. S., Kubota, M., Mikoshiba, K. and Miyawaki, A. (2002). A variant of yellow fluorescent protein with fast and efficient maturation for cell-biological applications. Nat. Biotechnol. *20*, 87–90.
[51] Brejc, K., Sixma, T. K., Kitts, P. A., Kain, S. R., Tsien, R. Y., Ormo, M. and Remington, S. J. (1997). Structural basis for dual excitation and photoisomerization of the *Aequorea victoria* green fluorescent protein. Proc. Natl. Acad. Sci. USA *94*, 2306–11.
[52] Chattoraj, M., King, B. A., Bublitz, G. U. and Boxer, S. G. (1996). Ultrafast excited state dynamics in green fluorescent protein: Multiple states and proton transfer. Proc. Natl. Acad. Sci. USA *93*, 8362–7.
[53] Zapata-Hommer, O. and Griesbeck, O. (2003). Efficiently folding and circularly permuted variants of the Sapphire mutant of GFP. BMC Biotechnol. *3*, 5.
[54] Heim, R., Cubitt, A. B. and Tsien, R. Y. (1995). Improved green fluorescence. Nature *373*, 663–4.
[55] Ehrig, T., O'Kane, D. J. and Prendergast, F. G. (1995). Green-fluorescent protein mutants with altered fluorescence excitation spectra. FEBS Lett. *367*, 163–6.
[56] Jung, G., Wiehler, J. and Zumbusch, A. (2005). The photophysics of green fluorescent protein: Influence of the key amino acids at positions 65, 203, and 222. Biophys. J. *88*, 1932–47.
[57] Sniegowski, J. A., Phail, M. E. and Wachter, R. M. (2005). Maturation efficiency, trypsin sensitivity, and optical properties of Arg96, Glu222,

and Gly67 variants of green fluorescent protein. Biochem. Biophys. Res. Commun. *332*, 657–63.
[58] Rizzuto, R., Brini, M., De Giorgi, F., Rossi, R., Heim, R., Tsien, R. Y. and Pozzan, T. (1996). Double labelling of subcellular structures with organelle-targeted GFP mutants *in vivo*. Curr. Biol. *6*, 183–8.
[59] Patterson, G. H., Knobel, S. M., Sharif, W. D., Kain, S. R. and Piston, D. W. (1997). Use of the green fluorescent protein and its mutants in quantitative fluorescence microscopy. Biophys. J. *73*, 2782–90.
[60] Mena, M. A., Treynor, T. P., Mayo, S. L. and Daugherty, P. S. (2006). Blue fluorescent proteins with enhanced brightness and photostability from a structurally targeted library. Nat. Biotechnol. *24*, 1569–71.
[61] Nguyen, A. W. and Daugherty, P. S. (2005). Evolutionary optimization of fluorescent proteins for intracellular FRET. Nat. Biotechnol. *23*, 355–60.
[62] Rizzo, M. A., Springer, G. H., Granada, B. and Piston, D. W. (2004). An improved cyan fluorescent protein variant useful for FRET. Nat. Biotechnol. *22*, 445–9.
[63] Ohashi, T., Galiacy, S. D., Briscoe, G. and Erickson, H. P. (2007). An experimental study of GFP-based FRET, with application to intrinsically unstructured proteins. Protein Sci. *16*, 1429–38.
[64] Wachter, R. M., Elsliger, M. A., Kallio, K., Hanson, G. T. and Remington, S. J. (1998). Structural basis of spectral shifts in the yellow-emission variants of green fluorescent protein. Structure *6*, 1267–77.
[65] Elsliger, M. A., Wachter, R. M., Hanson, G. T., Kallio, K. and Remington, S. J. (1999). Structural and spectral response of green fluorescent protein variants to changes in pH. Biochemistry *38*, 5296–301.
[66] Griesbeck, O., Baird, G. S., Campbell, R. E., Zacharias, D. A. and Tsien, R. Y. (2001). Reducing the environmental sensitivity of yellow fluorescent protein. Mechanism and applications. J. Biol. Chem. *276*, 29188–94.
[67] Jayaraman, S., Haggie, P., Wachter, R. M., Remington, S. J. and Verkman, A. S. (2000). Mechanism and cellular applications of a green fluorescent protein-based halide sensor. J. Biol. Chem. *275*, 6047–50.
[68] Llopis, J., McCaffery, J. M., Miyawaki, A., Farquhar, M. G. and Tsien, R. Y. (1998). Measurement of cytosolic, mitochondrial, and Golgi pH in single living cells with green fluorescent proteins. Proc. Natl. Acad. Sci. USA *95*, 6803–8.
[69] McAnaney, T. B., Zeng, W., Doe, C. F., Bhanji, N., Wakelin, S., Pearson, D. S., Abbyad, P., Shi, X., Boxer, S. G. and Bagshaw, C. R. (2005). Protonation, photobleaching, and photoactivation of yellow fluo-

rescent protein (YFP 10C): a unifying mechanism. Biochemistry *44*, 5510–24.

[70] Lauf, U., Lopez, P. and Falk, M. M. (2001). Expression of fluorescently tagged connexins: A novel approach to rescue function of oligomeric DsRed-tagged proteins. FEBS Lett. *498*, 11–5.

[71] Mizuno, H., Sawano, A., Eli, P., Hama, H. and Miyawaki, A. (2001). Red fluorescent protein from Discosoma as a fusion tag and a partner for fluorescence resonance energy transfer. Biochemistry *40*, 2502–10.

[72] Bevis, B. J. and Glick, B. S. (2002). Rapidly maturing variants of the Discosoma red fluorescent protein (DsRed). Nat. Biotechnol. *20*, 83–7.

[73] Campbell, R. E., Tour, O., Palmer, A. E., Steinbach, P. A., Baird, G. S., Zacharias, D. A. and Tsien, R. Y. (2002). A monomeric red fluorescent protein. Proc. Natl. Acad. Sci. USA *99*, 7877–82.

[74] Shaner, N. C., Campbell, R. E., Steinbach, P. A., Giepmans, B. N., Palmer, A. E. and Tsien, R. Y. (2004). Improved monomeric red, orange and yellow fluorescent proteins derived from *Discosoma* sp. red fluorescent protein. Nat. Biotechnol. *22*, 1567–72.

[75] Wang, L., Jackson, W. C., Steinbach, P. A. and Tsien, R. Y. (2004). Evolution of new nonantibody proteins via iterative somatic hypermutation. Proc. Natl. Acad. Sci. USA *101*, 16745–9.

[76] Shu, X., Shaner, N. C., Yarbrough, C. A., Tsien, R. Y. and Remington, S. J. (2006). Novel chromophores and buried charges control color in mfruits. Biochemistry *45*, 9639–47.

[77] Merzlyak, E. M., Goedhart, J., Shcherbo, D., Bulina, M. E., Shcheglov, A. S., Fradkov, A. F., Gaintzeva, A., Lukyanov, K. A., Lukyanov, S. Gadella, T. W. *et al.* (2007). Bright monomeric red fluorescent protein with an extended fluorescence lifetime. Nat. Methods *4*, 555–7.

[78] Shaner, N. C., Lin, M. Z., McKeown, M. R., Steinbach, P. A., Hazelwood, K. L., Davidson, M. W. and Tsien, R. Y. (2008). Improving the photostability of bright monomeric orange and red fluorescent proteins. Nat. Methods *5*, 545–51.

[79] Shcherbo, D., Merzlyak, E. M., Chepurnykh, T. V., Fradkov, A. F., Ermakova, G. V., Solovieva, E. A., Lukyanov, K. A., Bogdanova, E. A., Zaraisky, A. G. Lukyanov, S. *et al.* (2007). Bright far-red fluorescent protein for whole-body imaging. Nat. Methods *4*, 741–6.

[80] Weissleder, R. and Ntziachristos, V. (2003). Shedding light onto live molecular targets. Nat. Med. *9*, 123–8.

[81] Wiedenmann, J., Schenk, A., Rocker, C., Girod, A., Spindler, K. D. and Nienhaus, G. U. (2002). A far-red fluorescent protein with fast maturation

and reduced oligomerization tendency from *Entacmaea quadricolor* (*Anthozoa, Actinaria*). Proc. Natl. Acad. Sci. USA *99*, 11646–51.

[82] Wiedenmann, J., Vallone, B., Renzi, F., Nienhaus, K., Ivanchenko, S., Rocker, C. and Nienhaus, G. U. (2005). Red fluorescent protein eqFP611 and its genetically engineered dimeric variants. J. Biomed. Opt. *10*, 14003.

[83] Kredel, S., Nienhaus, K., Oswald, F., Wolff, M., Ivanchenko, S., Cymer, F., Jeromin, A., Michel, F. J., Spindler, K. D. Heilker, R. *et al.* (2008). Optimized and far-red-emitting variants of fluorescent protein eqFP611. Chem. Biol. *15*, 224–33.

[84] Goedhart, J., Vermeer, J. E., Adjobo-Hermans, M. J., van Weeren, L. and Gadella, T. W., Jr. (2007). Sensitive detection of p65 homodimers using red-shifted and fluorescent protein-based FRET couples. PLoS ONE *2*, e1011.

[85] Karasawa, S., Araki, T., Nagai, T., Mizuno, H. and Miyawaki, A. (2004). Cyan-emitting and orange-emitting fluorescent proteins as a donor/acceptor pair for fluorescence resonance energy transfer. Biochem. J. *381*, 307–12.

[86] Ai, H. W., Henderson, J. N., Remington, S. J. and Campbell, R. E. (2006). Directed evolution of a monomeric, bright and photostable version of *Clavularia cyan* fluorescent protein: Structural characterization and applications in fluorescence imaging. Biochem. J. *400*, 531–40.

[87] Bulina, M. E., Chudakov, D. M., Mudrik, N. N. and Lukyanov, K. A. (2002). Interconversion of *Anthozoa* GFP-like fluorescent and non-fluorescent proteins by mutagenesis. BMC Biochem. *3*, 7.

[88] Shkrob, M. A., Yanushevich, Y. G., Chudakov, D. M., Gurskaya, N. G., Labas, Y. A., Poponov, S. Y., Mudrik, N. N., Lukyanov, S. and Lukyanov, K. A. (2005). Far-red fluorescent proteins evolved from a blue chromoprotein from *Actinia equina*. Biochem. J. *392*, 649–54.

[89] Ganesan, S., Ameer-Beg, S. M., Ng, T. T., Vojnovic, B. and Wouters, F. S. (2006). A dark yellow fluorescent protein (YFP)-based Resonance Energy-Accepting Chromoprotein (REACh) for Forster resonance energy transfer with GFP. Proc. Natl. Acad. Sci. USA *103*, 4089–94.

[90] Lukyanov, K. A., Chudakov, D. M., Lukyanov, S. and Verkhusha, V. V. (2005). Innovation: Photoactivatable fluorescent proteins. Nat. Rev. Mol. Cell Biol. *6*, 885–91.

[91] Patterson, G. H. and Lippincott-Schwartz, J. (2002). A photoactivatable GFP for selective photolabeling of proteins and cells. Science *297*, 1873–7.

[92] Bell, A. F., Stoner-Ma, D., Wachter, R. M. and Tonge, P. J. (2003). Light-driven decarboxylation of wild-type green fluorescent protein. J. Am. Chem. Soc. *125*, 6919–26.

[93] van Thor, J. J., Gensch, T., Hellingwerf, K. J. and Johnson, L. N. (2002). Phototransformation of green fluorescent protein with UV and visible light leads to decarboxylation of glutamate 222. Nat. Struct. Biol. *9*, 37–41.
[94] Verkhusha, V. V. and Sorkin, A. (2005). Conversion of the monomeric red fluorescent protein into a photoactivatable probe. Chem. Biol. *12*, 279–85.
[95] Chudakov, D. M., Verkhusha, V. V., Staroverov, D. B., Souslova, E. A., Lukyanov, S. and Lukyanov, K. A. (2004). Photoswitchable cyan fluorescent protein for protein tracking. Nat. Biotechnol. *22*, 1435–9.
[96] Ando, R., Hama, H., Yamamoto-Hino, M., Mizuno, H. and Miyawaki, A. (2002). An optical marker based on the UV-induced green-to-red photoconversion of a fluorescent protein. Proc. Natl. Acad. Sci. USA *99*, 12651–6.
[97] Wiedenmann, J., Ivanchenko, S., Oswald, F., Schmitt, F., Rocker, C., Salih, A., Spindler, K. D. and Nienhaus, G. U. (2004). EosFP, a fluorescent marker protein with UV-inducible green-to-red fluorescence conversion. Proc. Natl. Acad. Sci. USA *101*, 15905–10.
[98] Tsutsui, H., Karasawa, S., Shimizu, H., Nukina, N. and Miyawaki, A. (2005). Semi-rational engineering of a coral fluorescent protein into an efficient highlighter. EMBO Rep. *6*, 233–8.
[99] Gurskaya, N. G., Verkhusha, V. V., Shcheglov, A. S., Staroverov, D. B., Chepurnykh, T. V., Fradkov, A. F., Lukyanov, S. and Lukyanov, K. A. (2006). Engineering of a monomeric green-to-red photoactivatable fluorescent protein induced by blue light. Nat. Biotechnol. *24*, 461–5.
[100] Chudakov, D. M., Feofanov, A. V., Mudrik, N. N., Lukyanov, S. and Lukyanov, K. A. (2003). Chromophore environment provides clue to "kindling fluorescent protein" riddle. J. Biol. Chem. *278*, 7215–9.
[101] Habuchi, S., Ando, R., Dedecker, P., Verheijen, W., Mizuno, H., Miyawaki, A. and Hofkens, J. (2005). Reversible single-molecule photoswitching in the GFP-like fluorescent protein Dronpa. Proc. Natl. Acad. Sci. USA *102*, 9511–6.
[102] Quillin, M. L., Anstrom, D. M., Shu, X., O'Leary, S., Kallio, K., Chudakov, D. M. and Remington, S. J. (2005). Kindling fluorescent protein from *Anemonia sulcata*: Dark-state structure at 1.38 A resolution. Biochemistry *44*, 5774–87.
[103] Jares-Erijman, E. A. and Jovin, T. M. (2003). FRET imaging. Nat. Biotechnol. *21*, 1387–95.
[104] Jares-Erijman, E. A. and Jovin, T. M. (2006). Imaging molecular interactions in living cells by FRET microscopy. Curr. Opin. Chem. Biol. *10*, 409–16.

[105] Förster, T. (1948). Zwischenmolekulare energiewanderung und fluoreszenz. Annalen der Physik *2*, 55–75.

[106] Patterson, G. H., Piston, D. W. and Barisas, B. G. (2000). Forster distances between green fluorescent protein pairs. Anal. Biochem. *284*, 438–40.

[107] Clegg, R. M. (1996). Fluorescence resonance energy transfer. In: "Fluorescence Imaging Spectroscopy and Microscopy" (Wang, X. F. and Herman, B., eds.). Wiley, New York, pp. 179–252.

[108] Stryer, L. (1978). Fluorescence energy transfer as a spectroscopic ruler. Annu. Rev. Biochem. *47*, 819–46.

[109] van der Meer, B. W. (2002). Kappa-squared: From nuisance to new sense. J. Biotechnol. *82*, 181–96.

[110] Vermeer, J. E., Van Munster, E. B., Vischer, N. O. and Gadella, T. W., Jr. (2004). Probing plasma membrane microdomains in cowpea protoplasts using lipidated GFP-fusion proteins and multimode FRET microscopy. J. Microsc. *214*, 190–200.

[111] Sinnecker, D., Voigt, P., Hellwig, N. and Schaefer, M. (2005). Reversible photobleaching of enhanced green fluorescent proteins. Biochemistry *44*, 7085–94.

[112] Kirber, M. T., Chen, K. and Keaney, J. F., Jr. (2007). YFP photoconversion revisited: Confirmation of the CFP-like species. Nat. Methods *4*, 767–8.

[113] Thaler, C., Vogel, S. S., Ikeda, S. R. and Chen, H. (2006). Photobleaching of YFP does not produce a CFP-like species that affects FRET measurements. Nat. Methods *3*, 491; author reply 4921–13.**

[114] Valentin, G., Verheggen, C., Piolot, T., Neel, H., Coppey-Moisan, M. and Bertrand, E. (2005). Photoconversion of YFP into a CFP-like species during acceptor photobleaching FRET experiments. Nat. Methods *2*, 801.

[115] Verrier, S. E. and Soling, H. D. (2006). Photobleaching of YFP does not produce a CFP-like species that affects FRET measurements. Nat. Methods *3*, 492; author reply 492–3.**

[116] Demarco, I. A., Periasamy, A., Booker, C. F. and Day, R. N. (2006). Monitoring dynamic protein interactions with photoquenching FRET. Nat. Methods *3*, 519–24.

[117] Mao, S., Benninger, R. K., Yan, Y., Petchprayoon, C., Jackson, D., Easley, C. J., Piston, D. W. and Marriott, G. (2008). Optical lock-in detection of FRET using synthetic and genetically encoded optical switches. Biophys. J. *94*, 4515–24.

[118] Giordano, L., Jovin, T. M., Irie, M. and Jares-Erijman, E. A. (2002). Diheteroarylethenes as thermally stable photoswitchable acceptors in

photochromic fluorescence resonance energy transfer (pcFRET). J. Am. Chem. Soc. *124*, 7481–9.

[119] Mitra, R. D., Silva, C. M. and Youvan, D. C. (1996). Fluorescence resonance energy transfer between blue-emitting and red-shifted excitation derivatives of the green fluorescent protein. Gene *173*, 13–7.

[120] Harpur, A. G., Wouters, F. S. and Bastiaens, P. I. (2001). Imaging FRET between spectrally similar GFP molecules in single cells. Nat. Biotechnol. *19*, 167–9.

[121] Zimmermann, T., Rietdorf, J., Girod, A., Georget, V. and Pepperkok, R. (2002). Spectral imaging and linear un-mixing enables improved FRET efficiency with a novel GFP2-YFP FRET pair. FEBS Lett. *531*, 245–9.

[122] Schleifenbaum, A., Stier, G., Gasch, A., Sattler, M. and Schultz, C. (2004). Genetically encoded FRET probe for PKC activity based on pleckstrin. J. Am. Chem. Soc. *126*, 11786–7.

[123] Adjobo-Hermans, M. J., Goedhart, J. and Gadella, T. W., Jr. (2006). Plant G protein heterotrimers require dual lipidation motifs of G(alpha) and G(gamma) and do not dissociate upon activation. J. Cell Sci. *119*, 5087–97.

[124] Willig, K. I., Kellner, R. R., Medda, R., Hein, B., Jakobs, S. and Hell, S. W. (2006). Nanoscale resolution in GFP-based microscopy. Nat. Methods *3*, 721–3.

[125] Rust, M. J., Bates, M. and Zhuang, X. (2006). Sub-diffraction-limit imaging by stochastic optical reconstruction microscopy (STORM). Nat. Methods *3*, 793–5.

[126] Betzig, E., Patterson, G. H., Sougrat, R., Lindwasser, O. W., Olenych, S., Bonifacino, J. S., Davidson, M. W., Lippincott-Schwartz, J. and Hess, H. F. (2006). Imaging intracellular fluorescent proteins at nanometer resolution. Science *313*, 1642–5.

[127] Bastiaens, P. I. and Squire, A. (1999). Fluorescence lifetime imaging microscopy: Spatial resolution of biochemical processes in the cell. Trends Cell Biol. *9*, 48–52.

[128] Gadella, T. W., Jr., van der Krogt, G. N. and Bisseling, T. (1999). GFP-based FRET microscopy in living plant cells. Trends. Plant Sci. *4*, 287–91.

[129] Truong, K. and Ikura, M. (2001). The use of FRET imaging microscopy to detect protein–protein interactions and protein conformational changes *in vivo*. Curr. Opin. Struct. Biol. *11*, 573–8.

[130] Zhang, J., Campbell, R. E., Ting, A. Y. and Tsien, R. Y. (2002). Creating new fluorescent probes for cell biology. Nat. Rev. Mol. Cell. Biol. *3*, 906–18.

[131] Piston, D. W. and Kremers, G. J. (2007). Fluorescent protein FRET: The good, the bad and the ugly. Trends Biochem. Sci. *32*, 407–14.

[132] Vogel, S. S., Thaler, C. and Koushik, S. V. (2006). Fanciful FRET. Sci STKE 2006, re2 **.

[133] Chen, H., Puhl, H. L., III, Koushik, S. V., Vogel, S. S. and Ikeda, S. R. (2006). Measurement of FRET efficiency and ratio of donor to acceptor concentration in living cells. Biophys. J. *91*, L39–41.

[134] Gadella, T. W., Jr. and Jovin, T. M. (1995). Oligomerization of epidermal growth factor receptors on A431 cells studied by time-resolved fluorescence imaging microscopy. A stereochemical model for tyrosine kinase receptor activation. J. Cell Biol. *129*, 1543–58.

[135] Overton, M. C. and Blumer, K. J. (2000). G-protein-coupled receptors function as oligomers *in vivo*. Curr. Biol. *10*, 341–4.

[136] Patel, R. C., Lange, D. C. and Patel, Y. C. (2002). Photobleaching fluorescence resonance energy transfer reveals ligand-induced oligomer formation of human somatostatin receptor subtypes. Methods *27*, 340–8.

[137] Tertoolen, L. G., Blanchetot, C., Jiang, G., Overvoorde, J., Gadella, T. W., Jr., Hunter, T. and den Hertog, J. (2001). Dimerization of receptor protein-tyrosine phosphatase alpha in living cells. BMC Cell Biol. *2*, 8.

[138] Ruiz-Velasco, V. and Ikeda, S. R. (2001). Functional expression and FRET analysis of green fluorescent proteins fused to G-protein subunits in rat sympathetic neurons. J. Physiol. *537*, 679–92.

[139] Zhou, J. Y., Toth, P. T. and Miller, R. J. (2003). Direct interactions between the heterotrimeric G protein subunit G beta 5 and the G protein gamma subunit-like domain-containing regulator of G protein signaling 11: Gain of function of cyan fluorescent protein-tagged G gamma 3. J. Pharmacol. Exp. Ther. *305*, 460–6.

[140] Day, R. N. (1998). Visualization of Pit-1 transcription factor interactions in the living cell nucleus by fluorescence resonance energy transfer microscopy. Mol. Endocrinol. *12*, 1410–9.

[141] Immink, R. G., Gadella, T. W., Jr., Ferrario, S., Busscher, M. and Angenent, G. C. (2002). Analysis of MADS box protein–protein interactions in living plant cells. Proc. Natl. Acad. Sci. USA *99*, 2416–21.

[142] Tonaco, I. A., Borst, J. W., de Vries, S. C., Angenent, G. C. and Immink, R. G. (2006). *In vivo* imaging of MADS-box transcription factor interactions. J. Exp. Bot. *57*, 33–42.

[143] Eberle, A., Reinehr, R., Becker, S., Keitel, V. and Haussinger, D. (2007). CD95 tyrosine phosphorylation is required for CD95 oligomerization. Apoptosis *12*, 719–29.

[144] Niethammer, P., Bastiaens, P. and Karsenti, E. (2004). Stathmin–tubulin interaction gradients in motile and mitotic cells. Science *303*, 1862–6.
[145] Takao, K., Okamoto, K., Nakagawa, T., Neve, R. L., Nagai, T., Miyawaki, A., Hashikawa, T., Kobayashi, S. and Hayashi, Y. (2005). Visualization of synaptic Ca^{2+}/calmodulin-dependent protein kinase II activity in living neurons. J. Neurosci. *25*, 3107–12.
[146] Vilardaga, J. P., Bunemann, M., Krasel, C., Castro, M. and Lohse, M. J. (2003). Measurement of the millisecond activation switch of G protein-coupled receptors in living cells. Nat. Biotechnol. *21*, 807–12.
[147] Chachisvilis, M., Zhang, Y. L. and Frangos, J. A. (2006). G protein-coupled receptors sense fluid shear stress in endothelial cells. Proc. Natl. Acad. Sci. USA *103*, 15463–8.
[148] Hoffmann, C., Gaietta, G., Bunemann, M., Adams, S. R., Oberdorff-Maass, S., Behr, B., Vilardaga, J. P., Tsien, R. Y., Ellisman, M. H. and Lohse, M. J. (2005). A FlAsH-based FRET approach to determine G protein-coupled receptor activation in living cells. Nat. Methods *2*, 171–6.
[149] Nikolaev, V. O., Hoffmann, C., Bunemann, M., Lohse, M. J. and Vilardaga, J. P. (2006). Molecular basis of partial agonism at the neurotransmitter alpha2A-adrenergic receptor and Gi-protein heterotrimer. J. Biol. Chem. *281*, 24506–11.
[150] Rehm, M., Dussmann, H., Janicke, R. U., Tavare, J. M., Kogel, D. and Prehn, J. H. (2002). Single-cell fluorescence resonance energy transfer analysis demonstrates that caspase activation during apoptosis is a rapid process. Role of caspase-3. J. Biol. Chem. *277*, 24506–14.
[151] Takemoto, K., Nagai, T., Miyawaki, A. and Miura, M. (2003). Spatio-temporal activation of caspase revealed by indicator that is insensitive to environmental effects. J. Cell Biol. *160*, 235–43.
[152] Miyawaki, A., Llopis, J., Heim, R., McCaffery, J. M., Adams, J. A., Ikura, M. and Tsien, R. Y. (1997). Fluorescent indicators for Ca^{2+} based on green fluorescent proteins and calmodulin. Nature *388*, 882–7.
[153] Romoser, V. A., Hinkle, P. M. and Persechini, A. (1997). Detection in living cells of Ca^{2+}-dependent changes in the fluorescence emission of an indicator composed of two green fluorescent protein variants linked by a calmodulin-binding sequence. A new class of fluorescent indicators. J. Biol. Chem. *272*, 13270–4.
[154] Hasan, M. T., Friedrich, R. W., Euler, T., Larkum, M. E., Giese, G., Both, M., Duebel, J., Waters, J., Bujard, H. Griesbeck, O. *et al.* (2004). Functional fluorescent Ca^{2+} indicator proteins in transgenic mice under TET control. PLoS Biol. *2*, e163.

[155] Evanko, D. S. and Haydon, P. G. (2005). Elimination of environmental sensitivity in a cameleon FRET-based calcium sensor via replacement of the acceptor with Venus. Cell Calcium *37*, 341–8.

[156] Palmer, A. E., Giacomello, M., Kortemme, T., Hires, S. A., Lev-Ram, V., Baker, D. and Tsien, R. Y. (2006). Ca^{2+} indicators based on computationally redesigned calmodulin-peptide pairs. Chem. Biol. *13*, 521–30.

[157] Truong, K., Sawano, A., Mizuno, H., Hama, H., Tong, K. I., Mal, T. K., Miyawaki, A. and Ikura, M. (2001). FRET-based *in vivo* Ca^{2+} imaging by a new calmodulin-GFP fusion molecule. Nat. Struct. Biol. *8*, 1069–73.

[158] Nagai, T., Yamada, S., Tominaga, T., Ichikawa, M. and Miyawaki, A. (2004). Expanded dynamic range of fluorescent indicators for Ca(2+) by circularly permuted yellow fluorescent proteins. Proc. Natl. Acad. Sci. USA *101*, 10554–9.

[159] Baird, G. S., Zacharias, D. A. and Tsien, R. Y. (1999). Circular permutation and receptor insertion within green fluorescent proteins. Proc. Natl. Acad. Sci. USA *96*, 11241–6.

[160] Hires, S. A., Zhu, Y. and Tsien, R. Y. (2008). Optical measurement of synaptic glutamate spillover and reuptake by linker optimized glutamate-sensitive fluorescent reporters. Proc. Natl. Acad. Sci. USA *105*, 4411–6.

[161] Shimozono, S., Hosoi, H., Mizuno, H., Fukano, T., Tahara, T. and Miyawaki, A. (2006). Concatenation of cyan and yellow fluorescent proteins for efficient resonance energy transfer. Biochemistry *45*, 6267–71.

[162] Ponsioen, B., Zhao, J., Riedl, J., Zwartkruis, F., van der Krogt, G., Zaccolo, M., Moolenaar, W. H., Bos, J. L. and Jalink, K. (2004). Detecting cAMP-induced Epac activation by fluorescence resonance energy transfer: Epac as a novel cAMP indicator. EMBO Rep. *5*, 1176–80.

[163] Zaccolo, M., De Giorgi, F., Cho, C. Y., Feng, L., Knapp, T., Negulescu, P. A., Taylor, S. S., Tsien, R. Y. and Pozzan, T. (2000). A genetically encoded, fluorescent indicator for cyclic AMP in living cells. Nat. Cell Biol. *2*, 25–9.

[164] Nikolaev, V. O., Gambaryan, S. and Lohse, M. J. (2006). Fluorescent sensors for rapid monitoring of intracellular cGMP. Nat. Methods *3*, 23–5.

[165] Remus, T. P., Zima, A. V., Bossuyt, J., Bare, D. J., Martin, J. L., Blatter, L. A., Bers, D. M. and Mignery, G. A. (2006). Biosensors to

measure inositol 1,4,5-trisphosphate concentration in living cells with spatiotemporal resolution. J. Biol. Chem. *281*, 608–16.
[166] Tanimura, A., Nezu, A., Morita, T., Turner, R. J. and Tojyo, Y. (2004). Fluorescent biosensor for quantitative real-time measurements of inositol 1,4,5-trisphosphate in single living cells. J. Biol. Chem. *279*, 38095–8.
[167] Ananthanarayanan, B., Ni, Q. and Zhang, J. (2005). Signal propagation from membrane messengers to nuclear effectors revealed by reporters of phosphoinositide dynamics and Akt activity. Proc. Natl. Acad. Sci. USA *102*, 15081–6.
[168] Violin, J. D., Zhang, J., Tsien, R. Y. and Newton, A. C. (2003). A genetically encoded fluorescent reporter reveals oscillatory phosphorylation by protein kinase C. J. Cell Biol. *161*, 899–909.
[169] Fehr, M., Okumoto, S., Deuschle, K., Lager, I., Looger, L. L., Persson, J., Kozhukh, L., Lalonde, S. and Frommer, W. B. (2005). Development and use of fluorescent nanosensors for metabolite imaging in living cells. Biochem. Soc. Trans. *33*, 287–90.
[170] Ting, A. Y., Kain, K. H., Klemke, R. L. and Tsien, R. Y. (2001). Genetically encoded fluorescent reporters of protein tyrosine kinase activities in living cells. Proc. Natl. Acad. Sci. USA *98*, 15003–8.
[171] Zhang, J., Ma, Y., Taylor, S. S. and Tsien, R. Y. (2001). Genetically encoded reporters of protein kinase A activity reveal impact of substrate tethering. Proc. Natl. Acad. Sci. USA *98*, 14997–5002.
[172] Ai, H. W., Hazelwood, K. L., Davidson, M. W. and Campbell, R. E. (2008). Fluorescent protein FRET pairs for ratiometric imaging of dual biosensors. Nat. Methods *5*, 401–3.
[173] Piljic, A. and Schultz, C. (2008). Simultaneous recording of multiple cellular events by FRET. ACS Chem. Biol. *3*, 156–60.
[174] Schultz, C., Schleifenbaum, A., Goedhart, J. and Gadella, T. W., Jr. (2005). Multiparameter imaging for the analysis of intracellular signaling. Chembiochem *6*, 1323–30.

Laboratory Techniques in Biochemistry and Molecular Biology, Volume 33
FRET and FLIM Techniques
T. W. J. Gadella (Editor)

CHAPTER 6

Small molecule-based FRET probes

Amanda Cobos Correa and Carsten Schultz

European Molecular Biology Laboratory, Meyerhofstr. 1, 69117 Heidelberg, Germany

When imaging of enzyme activity is performed in tissue or when large changes in fluorescence resonance energy transfer are required, almost always reporter molecules based on small molecules are instrumental. In this chapter, we will describe suitable fluorophores and building blocks for the design of FRET probes. In examples, we will describe the strategy and the application of small molecule-based FRET probes for the detection of enzymes involved in bond formation and hydrolysis including proteases and lipases, protein–substrate interaction, as well as sensors based on conformational and environmental changes.

6.1. Introduction

Fluorescence or Förster resonance energy transfer (FRET) is widely accepted as being one of the most useful methods to observe biochemical events in vitro and in living cells. Generally, there are two forms of FRET sensors: those based on a pair of genetically encoded fluorophores, usually employing fluorescent proteins from jellyfish or corals, or those based on small molecules that make use of small organic fluorophores.

DOI: 10.1016/S0075-7535(08)00006-5

FRET applications in entire organisms have rarely been reported, mostly because ratiometric measurements in deep tissue are difficult to combine with two-photon excitation of the FRET donor. In addition, experiments in tissue regularly require the preparation of transgenic animals when genetically encoded sensors are used. Small molecule sensors potentially have the advantage of direct delivery to cells and tissue provided that the molecules are able to pass membranes. However, the main caveat of small sensors cannot be neglected: we have to rely on chemical synthesis to prepare these probes, a notoriously difficult and time-consuming process. Nevertheless, a significant number of fluorogenic probes have been synthesized. The vast majority of them were applied to in vitro problems. Some of them, however, were shown to be useful in cells. These sensors were instrumental to novel approaches for screening receptor antagonists or to determine the onset of enzyme activity during the embryonic development of an organism. This chapter will focus on sensors and reporters based on small molecules, specifically on those bearing non-protein fluorophores and excluding those requiring special labeling techniques such as with AGT, FlAsH, NTA or the introduction of quantum dots (see Chap. 12). The main building blocks and the different strategies for probe design will be described. We will also discuss potential technical problems in their application.

6.2. Fluorophores and quenchers

When compared to fluorescent proteins, fluorophores and quenchers of fluorescence (short: quenchers) are small molecules with sizes varying from 1 to 10 Å. They are the main building blocks for constructing small molecule FRET probes. As molecular entities, they might influence the performance of the probe to a great extent. Their fluorescent properties will determine the sensitivity and dynamic range of the sensor. The success of the probe for a specific application will depend on the selection of the right fluorophores and for this reason we will give here an overview of the most commonly used types of fluorophores and quenchers. In addition, the optical properties and some of the synthetic needs to prepare fluorescent sensors will be discussed with special attention to practical considerations.

6.2.1. Fluorescence and properties of dyes

Fluorescence is a process that occurs after excitation of a molecule with light. It involves transitions of the outermost electrons between different electronic states of the molecule, resulting in emission of a photon of lower energy than the previously absorbed photon. This is represented in the Jablonski diagram (see Fig. 6.1). As every molecule has different energy levels, the fluorescent properties vary from one fluorophore to the other. The main characteristics of a fluorescent dye are absorption and emission wavelengths, extinction

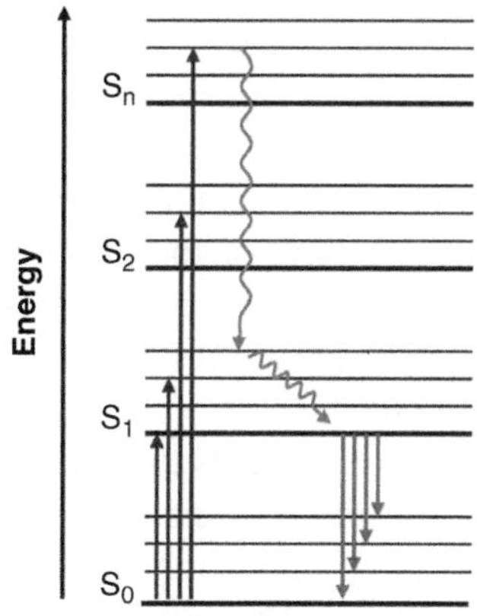

Fig. 6.1. Jablonski diagram, representing electron energy levels of fluorophores and transitions after photon excitation. S = electronic state, different lines within each state represent different vibrational levels. Blue arrows represent absorption events, green arrows depict internal conversion or heat dissipation, and orange arrows indicate fluorescence emission. Intersystem crossing into triplet states has been omitted for simplicity (see also Chaps. 1 and 12). (See Color Insert.)

coefficient, quantum yield, photostability, and Stokes shift [1, 2]. It is important to consider all these properties of the fluorophores before designing a fluorescent probe adequate for a particular experiment.

The absorption and emission spectra of a fluorophore are bands spread over a range of wavelengths with at least one peak of maximal absorbance and emission that corresponds to the S_0–S_1 and S_1–S_0 transitions, respectively. There are several vibrational levels within an electronic state and transitions from one electronic to several vibrational states are potentially possible. This determines that the spectra are not sharp but consist of broad bands. The emission spectrum is independent of the excitation wavelength. The energy used to excite the fluorophore to higher electronic and vibrational levels is very rapidly dissipated, sending the fluorophore to the lowest vibrational level of the first electronic excited state (S_1) from where the main fluorescent transition occurs [3] (see Fig. 6.1).

The amount of fluorescence emitted by a fluorophore is determined by the efficiencies of absorption and emission of photons, processes that are described by the extinction coefficient and the quantum yield. The extinction coefficient (ε/M^{-1} cm^{-1}) is a measure of the probability for a fluorophore to absorb light. It is unique for every molecule under certain environmental conditions, and depends, among other factors, on the molecule cross section. In general, the bigger the π-system of the fluorophore, the greater is the probability that the photon hitting the fluorophore is absorbed. Common extinction coefficient values of fluorophores range from 25,000 to 200,000 M^{-1} cm^{-1} [4].

The quantum yield (Q) represents the ratio between the number of photons absorbed and photons emitted as fluorescence. It is a measure of brightness of the fluorophore and represents the efficiency of the emission process. The determination of absolute quantum yield for a fluorophore is experimentally difficult. Therefore, usually relative quantum yield values are determined. To measure the relative quantum yield of a fluorophore, the sample is compared to a standard fluorophore with an established quantum yield that does not show variations in the excitation wavelength [5, 6].

Photostability refers to the rate at which molecules degrade when UV or visible light is applied and it is directly related to the number of excitation-emission cycles a fluorophore may undergo before being photodestroyed. The number of cycles was quantified in aqueous media and determined to be between 5,000 and 200,000. Coumarin and its derivatives are the least stable and rhodamine and cyanine dyes the most photostable dyes [2, 7]. Photodestruction of a fluorophore by light is referred to as photobleaching. In the excited state, different chemical reactions might occur that lead to a loss of excitability or changes of the emission properties. Light may also excite inner shell electrons creating molecules in a triplet state that are highly reactive radicals with significant cytotoxicity [2]. Photostability is an important issue, especially in microscopy where experiments require high light intensities.

Another important feature of fluorophores is the amount of vibrational energy lost in the excited state. The difference between emission and excitation maxima gives a readout in this respect and is referred to as the Stokes shift. In many sensors, a small Stokes shift is unfavorable for FRET ratio measurements due to overlap of emission spectra.

The polarity of the solvent may affect the absorption and emission spectra of a fluorophore. This effect is more pronounced in fluorophores that have a large excited state dipole moment. Since the excited state is usually more polar than the ground state, polar solvents can arrange better around the excited molecule to minimize the energy, that way stabilizing the excited state. Transitions from a stabilized excited state to the ground state are less energy-rich and this is translated into a bathochromic or red shift in the emission wavelength. Polarity affects the brightness of a fluorophore as well, increasing or decreasing the probability of fluorescent decay. A good example is the small dye NBD, which is very fluorescent in apolar solvents and emits only weakly in polar environments [8].

Chemically, most fluorophores are formed by a coplanar conjugated aromatic ring system, with some exceptions like parinaric acid. In addition, a combination of an electron-pushing and an electron-pulling group at the opposite sides of the aromatic system is usually favorable for fluorescence. The larger the conjugate system, the higher is the emission wavelength and the better the efficiency, measured as fluorescent quantum yield. For example, fluorescein composed by a three member ring system, emits at 520 nm and has a quantum yield of 0.9 whereas methoxycoumarin, formed by a two member ring system, has its maximum emission at 390 nm and a quantum yield that varies from 0.3 to 0.4 [9]. Substitutions on the ring system affect the fluorescence as well. Electron-donating groups increase the quantum yield whereas electron-withdrawing groups decrease it. Halogen atoms produce a decrease in the quantum yield following the order I > Br > Cl, where the presence of iodine quenches the fluorescence almost completely [10]. As an example we can compare some of the derivatives of fluorescein. While fluorescein has a quantum yield of 0.9, its tetrabromo-derivative, eosin, has one of 0.2 [11] and the tetra-iodo derivative, erythrosine, 10 times less (0.02) [12]. On the other hand, fluorine is successfully introduced as a substituent in some derivatives resulting in higher photostability, for example in Oregon Green® 488 [13].

6.2.2. Families of fluorophores

Fluorophores are relative small molecules that, with some exceptions, are not naturally occurring and have to be synthesized chemically. There has been a large development in the synthesis of fluorescent molecules and nowadays there is a vast range of alternatives including dyes with improved photochemical properties, solubility or modified reactivity that allow for conjugation to other molecules of interest and the synthesis and application of fluorescent sensors [10, 13]. Although a lot is known about the physics of fluorescence and a lot of information is available about the properties of dyes, their prediction from the chemical structures cannot be accurately done. For this reason, there has been a tendency to synthesize derivatives of well known fluorophores rather than to discover new fluorescent structures [14, 15]. Here we describe the families of fluorophores that have been more commonly employed in the design of FRET probes (Table 6.1).

6.2.2.1. Fluorescein derivatives

Fluorescein is one of the most common dyes used for labeling probes. It is formed by an upper xanthene ring that provides it with its fluorescent nature, which is substituted with a hydroxyl group at positions 6 and an oxygen atom at position 3. The upper ring system is attached to benzoic acid at position 9 (see Fig. 6.2). The benzene ring does not participate in the conjugate system but the introduction of different functionalities into this moiety influences the fluorescent properties of the molecule by intramolecular photo-induced electron transfer [16, 17]. Further modifications at positions 5 or 6 of the lower benzyl ring create derivatives with different functionalities to couple them to a molecule of interest [10].

Fluorescein is excited at 494 nm, which fits to the argon-ion laser line at 488 nm, a very convenient feature for many microscopy experiments. It emits at 520 nm and the emission band is far from being sharp. The broad fluorescence emission spectrum varies with pH [18]. The advantageous photochemical properties of fluorescein are its high absorption ($\varepsilon_{max} = 79{,}000 M^{-1}cm^{-1}$) and quantum

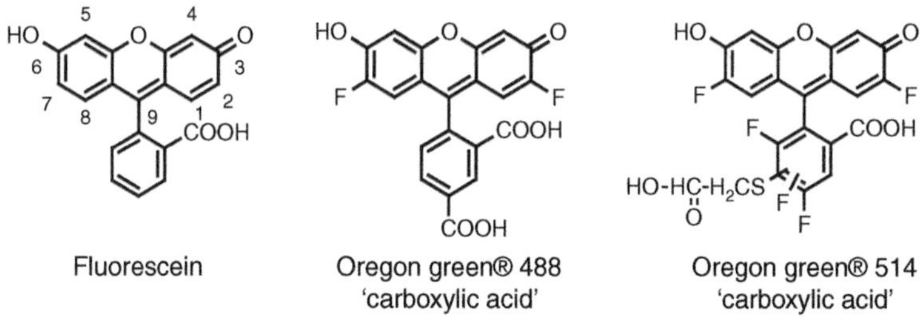

Fig. 6.2. Chemical structures of fluorescein and some commonly used derivatives. The carboxylic acid groups are used for forming conjugates with other molecules.

yield ($Q = 0.9$) and one of its main drawbacks is its high rate of photobleaching [19]. Its solubility in water is good but its fluorescence is very sensitive to pH [13], mainly because the protonation of the phenolate oxygen will decrease the electron withdrawing effect. In acidic pH the lower ring carboxylic acid lactonizes thereby breaking the conjugation of the xanthene system and yielding a non-fluorescent species.

There are some derivatives of fluorescein available with improved photostability. The most common is the analog fluorinated at positions 2 and 7 of the xanthene ring, Oregon green® 488 (see Fig. 6.2). The fluorines increase the photostability of this dye and reduce the pH sensitivity making it more suitable for microscopy experiments especially when continuous excitation may cause photobleaching. For this reason, it is also more reliable for quantitative measurements. A longer wavelength emission analog is Oregon Green® 514, chemically similar to Oregon Green® 488 but with the lower ring fluorinated, too [13].

For experiments in cells it is important to note that fluorescein derivatives are permanently negatively charged at physiological pH and therefore tend to leave cells via unspecific anion channels following the driving force of the negative membrane potential. Negative charges, on the other hand, usually prohibit interaction with polynucleotides and storage in intracellular compartments, therefore leaving the dye evenly distributed in the cytosol [20]. Fluorescein is not membrane-permeant and subsequently derivatives such as the uncharged non-fluorescent fluorescein diacetate have to be used to induce permeability. This derivative can enter cells and becomes fluorescent after esterase-mediated deacetylation [21].

6.2.2.2. Rhodamine derivatives

Due to their longer wavelength fluorescence and photostability, rhodamine and its derivatives have often been employed for the labeling of probes tested in living cells. Like fluorescein, the chemical structure of rhodamine consists of an upper xanthene ring and a lower benzene ring. In this case, the xanthene ring is substituted

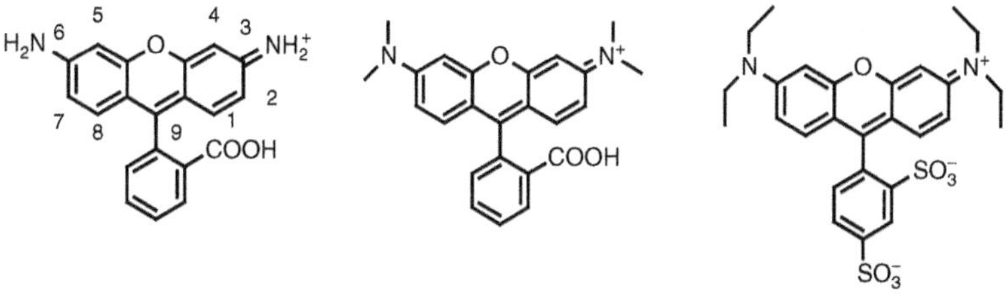

Fig. 6.3. Chemical structures of rhodamine and some derivatives. TAMRA = *N,N,N,N*-tetramethylrhodamine. Lissamine rhodamine = 3,5-disulfonyl-*N,N, N,N*-tetramethylrhodamine.

with nitrogen atoms replacing the oxygens of fluorescein (see Fig. 6.3). Substitutions at the positions 5 and 6 of the lower ring allow the introduction of reactive groups to attach the dye to different molecules [10]. For modification of its fluorescent properties, substitutions are introduced into the upper ring system generating analogs with more stability or different spectral properties. The excitation of rhodamine varies from 520 to 580 nm depending on the derivative and its emission from 550 to 620 nm. Rhodamine derivatives have lower quantum yields than the fluoresceins ($Q \sim 0.2$) [22] and are generally less soluble in water. However, their photostability is higher [23]. Among the derivatives of rhodamine, the most common ones are tetramethylrhodamine (TAMRA) and sulforhodamine B. The first one has two methyl groups attached to each nitrogen atom of the upper ring. Sulforhodamine, also called Lissamine rhodamine, has two ethyl groups attached to each nitrogen atom of the upper ring and two sulfonates in positions 3 and 5 of the lower ring (see Fig. 6.3). Molecular Probes has patented a substitute of fluorescein, Alexa Fluor 488, based on the rhodamine basic structure with sulfonates at positions 4 and 5 of the upper ring. These substitutions result in increased photostability to the dye.

Dyes which are half fluorescein, half rhodamine are called rhodols. Their spectral properties are intermediate with respect to excitation and emission wavelength. Generally, rhodol fluorophores are more photostable and less pH sensitive than fluorescein, and have higher quantum yields than rhodamine [24].

6.2.2.3. Coumarin derivatives

The coumarin moiety is found in nature and some of its derivatives possess biomedical applications [25, 26]. Apart from that, coumarins are commonly used as fluorescent labels in biology. The basic structure of coumarin is a benzo-α-pyrone (see Fig. 6.4) that has been used as a template to develop hundreds of derivatives with different fluorescent properties or with reactive groups that permit the coupling to relevant biological moieties. Most of the coumarin derivatives absorb in the region between 300 and 500 nm and emit fluorescence between 400 and 550 nm. Main modifications are introduced in positions 3, 4, 6–8 and influence the fluorescent properties to a great extent. Several libraries of coumarin derivatives have been created trying to define the structure–fluorescent properties relationship of this family of dyes. Apparently, substitutions with electron-donating groups, such as $-NR_2$, –OH, and –OMe, in position 4, 6, or 7 enhance the fluorescence intensity and shift the emission maxima to longer wavelengths. Especially, when the nitrogen atom is rotationally locked, as in coumarin 343 (see Fig. 6.4), the system shows an increased red-shift. In the same three positions, substitutions with alkyl groups reduce the quantum yield and shift the emission maxima slightly to shorter wavelengths [27, 28].

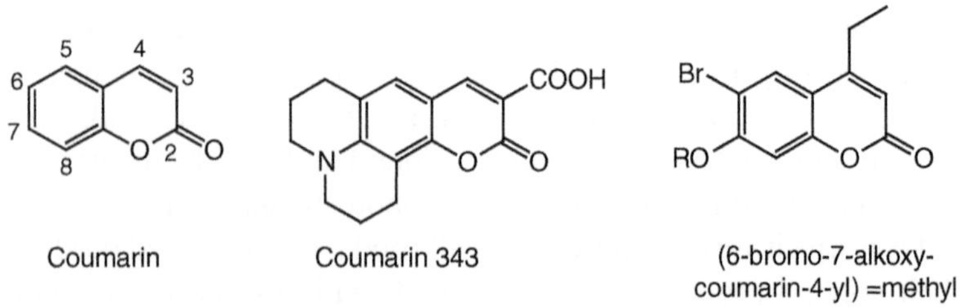

Fig. 6.4. Chemical structures of coumarin and some of its fluorescent derivatives.

Substitutions in position 3 with electron-withdrawing groups such as –CN, –COR, –COOR, and $-NO_2$ or with alkyl groups increase the fluorescence intensity of the basic coumarin structure [29, 30].

Another example of fluorescence intensity modulation in coumarins is the 3-azido substitution that quenches the fluorescence completely. These compounds are used as starting material for the synthesis of fluorescent triazolocoumarins by click chemistry [31].

Interestingly, the fluorescence of some coumarins depends strongly on the solvent. This is the case for 7-alkoxycoumarins that have been used as probes for microenvironments [32], 7-hydroxycoumarin that is pH sensitive, and 7-NR_2 substituted coumarins such as coumarin 120 whose quantum yield is reduced in nonpolar solvents due to a change in the 3D structure [33].

The peculiarity of 4-ylmethyl coumarin derivatives (see Fig. 6.4) is photolability. This feature is currently increasingly employed to create phototriggers or photoactivatable (caged) compounds [34, 35]. Another interesting coumarin derivative can be generated as a co-product after photolysis of ethyl 3,5-dibromo-2,4-dihydroxycinnamate. The two-photon illumination of this caging compound gives the fluorescent 6,8-dibromo-7-hydroxycoumarin allowing quantification of the uncaging rate [36].

The wide variety of absorption and emission wavelengths has allowed the use of different coumarin derivatives as FRET pairs [37]. Contrary to most fluorophores, coumarin is uncharged which makes it intrinsically membrane-permeant. To induce water solubility polar groups are frequently introduced to the basic structure [35].

6.2.2.4. BODIPY derivatives

BODIPY is a short for 4,4-difluoro-4-bora-3a,4a-diaza-s-indacene, the basic structure of this type of fluorophore (see Fig. 6.5). Derivatives of this dye have been created by modification of positions 1, 3, 5, 7, and 8, generating an array of fluorophores with very distinct excitation and emission properties [38]. Molecular Probes has synthesized a wide number of BODIPY dyes whose excitation

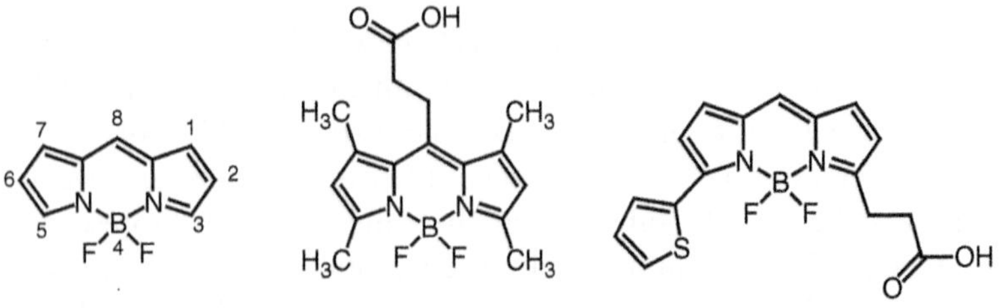

Fig. 6.5. BODIPY and some derivatives. BODIPY = 4,4-difluoro-4bora-3a, 4a-diaza-s-indacene.

and emission spectra cover a broad region of the spectra [13]. Most of them are characteristic for having high extinction coefficients (~80,000), quantum yields over 0.8, insensitivity to pH and solvent polarity and small Stokes shifts in the order of 10–20 nm. The last property might be a problem in FRET measurements due to direct excitation of the acceptor. BODIPY dyes are not suitable for solid phase peptide synthesis (SPPS) when acidic conditions are required during cleavage from the resin since the pyrrole rings tend to open when the conjugate system is disturbed. One BODIPY dye successfully used in our lab in conventional Fmoc SPPS is the 493/503 derivative with positions 1, 3, 5, and 7 occupied with methyl groups which render the opening of the ring impossible even under strong acidic conditions. Another characteristic of BODIPY dyes is the formation of non-fluorescent dimers that can be used for the design of FRET probes [39].

6.2.2.5. Cyanine derivatives

Cyanine dyes have the common molecular formula $R_2N[CH{=}CH]_nCH{=}N^+R_2$ where the mono- or poly-methine chain is flanked by two nitrogen atoms that form part of a heterocyclic system (see Fig. 6.6). The most common cyanine derivatives used for labeling molecules are Cy3 and Cy5. These are symmetrical

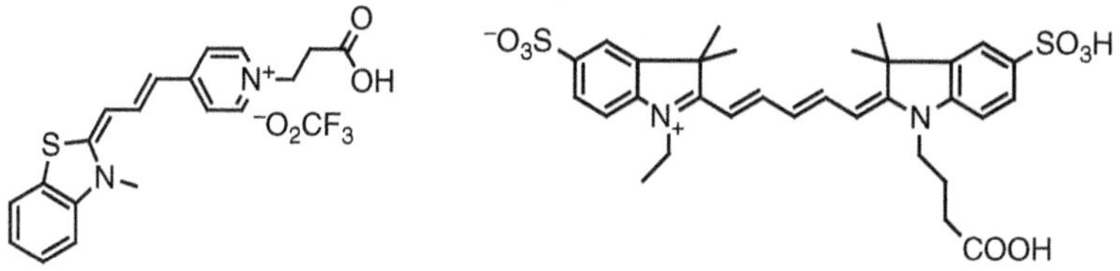

Fig. 6.6. Structure of an asymmetrical cyanine dye, BO-3, and a symmetrical cyanine dye, Cy5.

cyanine dyes because two identical heterocycles are flanking the polymethine chain. Variations in the length of the chain affect the fluorescent properties of these dyes. The larger the conjugated system, the longer is the excitation wavelength. Cyanine dyes have large extinction coefficients (10^5 M^{-1} cm^{-1}) [40], but poor quantum yields ($Q = 0.1$).[41] However, their broad use in probes for cell biology is mainly due to their long excitation wavelength, well separated from cell autofluorescence and resulting in less cell damage. Cyanines form aggregates in aqueous solutions with a high degree of quenching and this characteristic has been used for the design of fluorogenic probes [42].

Unsymmetrical cyanines are relatively new dyes. They are characterized by having two different heterocycles attached to the polymethine chain. New synthetic strategies to synthesize asymmetric cyanines have been developed in the last years [43]. Interestingly, their quantum yield is very sensitive to environmental conditions, being close to 0.001 when they are in fluid solutions and around 0.1 in viscous solvents or attached to molecules, possibly due to restriction of the heterocycles' rotation [44]. This was used to build fluorogenic sensors that report environmental changes [45, 46]. Since most cyanines are cationic compounds, they tend to concentrate in mitochondria due to the electrical potential gradient that exists across the membrane of this compartment [47, 48]. Widely spread is the use of carbocyanines as probes for membrane potential since they accumulate into hyperpolarized membranes where they aggregate and quench the fluorescence. Upon cell depolarization, carbocyanine dyes are released to the surrounding media and fluorescence is recovered [49].

6.2.2.6. Molecular rotors

Molecular rotors are fluorophores characteristic for having a fluorescent quantum yield that strongly depends on the viscosity of the solvent [50]. This property relies on the ability to resume a twisted conformation in the excited state (twisted intramolecular charge transfer or TICT state) that has a lower energy than the planar conformation. The de-excitation from the twisted conformation happens via a non-radiative pathway. Since the formation of the TICT state is favored in viscous solvents or at low temperature, the probability of fluorescence emission is reduced under those conditions [51]. Molecular rotors have been used as viscosity and flow sensors for biological applications [52]. Modifications on their structure have introduced new reactivity that might increase the diversity of their use in the future [53] (see Fig. 6.7).

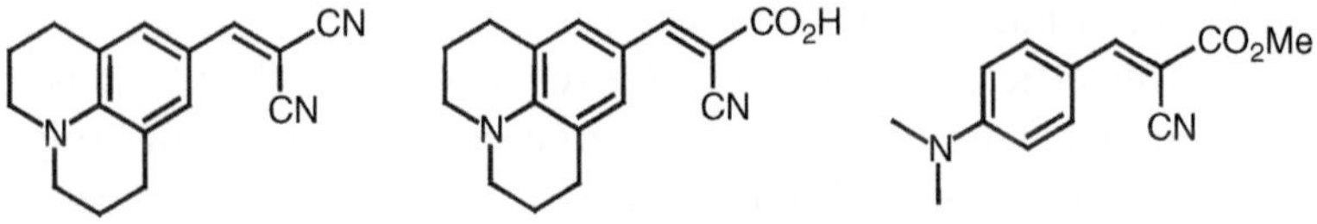

DCVJ CCVJ CMAM

Fig. 6.7. Viscosity sensitive fluorophores: molecular rotors. DCVJ = 9-(dicyanovinyl)-julolidine, CCVJ = 9-(carboxy-2-cyano)vinyl julolidine, CMAM = 2-cyano-3-(p-dimethyl-aminophenyl)acrylic acid, methyl ester.

6.2.2.7. Other fluorophores

6.2.2.7.1. Pyrene. Pyrene and its derivatives emit in the UV region of the spectra. They have two excitation peaks, at 327 and 346 nm, and an emission maximum at 378 nm. The intensity of the first excitation maxima is ~80% of the second and this is useful for FRET studies because exciting at lower wavelength produces less cross-talk. Pyrenes have high absorption values but the most interesting property of these dyes is their ability to form excimers when two molecules are situated in close proximity. An excimer (excite state dimer) is a complex formed by two molecules of the same type in the excited state. Instead of quenching the fluorescence of each other, this complex retains fluorescence and changes its properties, for instance, shifting and substantially broadening the emission spectrum to 490 nm [54]. This property has been used to monitor, for example, membrane fusion by labeling several fatty acids with pyrene molecules as well as protein conformational changes [55, 56].

6.2.2.7.2. Linear polyenes. The measurement of membrane processes requires fluorophores inserted into cell membranes. Suitable fluorophores are polyene compounds such as diphenylhexatriene (DPH) (see Fig. 6.8) or parinaric acid, used in several occasions as

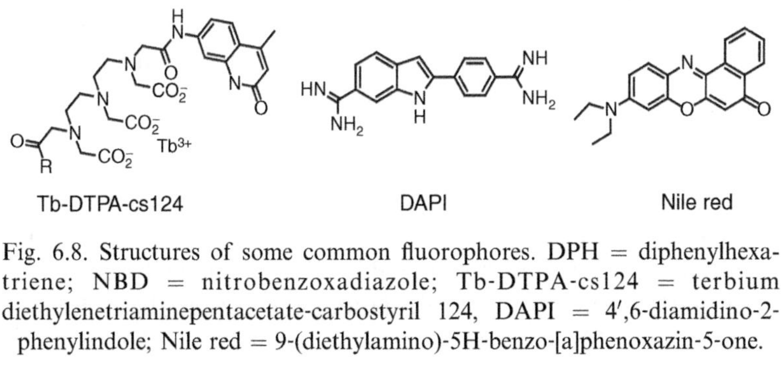

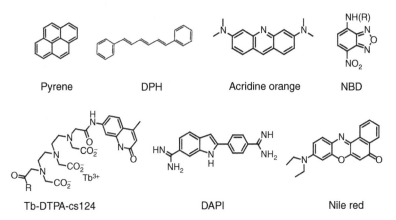

Fig. 6.8. Structures of some common fluorophores. DPH = diphenylhexatriene; NBD = nitrobenzoxadiazole; Tb-DTPA-cs124 = terbium diethylenetriaminepentacetate-carbostyril 124, DAPI = 4′,6-diamidino-2-phenylindole; Nile red = 9-(diethylamino)-5H-benzo-[a]phenoxazin-5-one.

membrane probes [57–59]. They fluoresce in the near UV or visible region. As a consequence of their linearity, they have unhindered structures. DPH has a very high quantum yield in several organic solvents and an extinction coefficient around 85,000 M^{-1} cm^{-1} [60]. Very recently, polyene lipids were introduced to analyze membrane phase partitioning [61]. The lipids integrated well into membranes and showed superior properties compared to NBD- or BODIPY-tagged lipids.

6.2.2.7.3. Lanthanide complexes. Lanthanide complexes emit in the visible part of the spectrum with a number of sharp bands. The complexes consist of metal ions, such as terbium or europium, and a chelating ligand that usually has a chromophore as sensitizer. Fluorescence occurs through energy transfer from the ligand to the central metal ion. The complexes have excellent fluorescent properties and their main characteristics are long fluorescence lifetimes in the range of microseconds, large Stokes shifts, since they are normally excited in the UV and emit at wavelengths longer than 500 nm, multiple sharp emission peaks, and insensitivity to pH [62, 63]. Lanthanide complexes have been used for building small molecule sensors for the detection of hydrolyzing enzymes in vitro. The lanthanide-based resonance energy transfer (LRET) was used with several fluorescent acceptors and quenchers [64]. LRET has a high efficiency of energy transfer when lanthanides complexes are used as donors which are independent of the orientation factor [65]. The use of probes formed by lanthanide complexes in cells has been recently applied to the detection of zinc ions or to selectively stain the nucleoli, to cite some examples [66, 67].

6.2.2.7.4. Nitrobenzoxadiazole. NBD derivatives are environmentally sensitive. In apolar solvents, they are highly fluorescent. When dissolved in polar solvents, the fluorescence intensity decreases significantly. Semi-empirical calculations can explain these changes of quantum yields in terms of decreased energy gaps between the excited state and a triplet state in polar solvents, which increase the probability of intersystem crossing and quenching [68].

6.2.2.7.5. Nile red. An increasingly used red fluorophore is Nile red. It is very attractive in terms of excitation wavelength but its biological use is restricted by its low solubility in aqueous media and its environment-dependent fluorescence [69]. In non-polar solvents, Nile red has a high quantum yield and emission maximum at 530 nm whereas in water it is hardly fluorescent and the emission maximum is shifted to 640 nm [70]. In alcohols, that is methanol, the excitation maximum lies around 540 nm and the emission maximum at 620 nm. Interestingly, the same values are found when the dye is incorporated in cellular membranes. Lipid sensors have been created making use of the polarity-dependent fluorescence of Nile red [71]. In addition, Nile red 2-*O*-butyric acid can be used as a competition sensor for lipid binding to proteins [72, 73]. Recently, the dicarboxylic acid and sulphonic/carboxylic acid derivatives were synthesized. They show higher water solubility while retaining the bathochromic shift and an increased quantum yield [74]. In addition, bromination of nile red or its relative phenothiazine yields derivatives with a significant redshift [72]. However, these derivatives bleach faster and seem to be more cytotoxic. Nevertheless, the larger range of derivatives will broaden the application of Nile red as biological markers.

6.2.3. Quenching and quenchers

Quenching is the reduction in fluorescence intensity and can be caused by various processes. It occurs either during the lifetime of the excited state or in the ground state. Quenching processes that happen in the excited state are collisional quenching, charge transfer reactions, or energy transfer. The latter is the basis for FRET probes but the other events happen as well under certain conditions and it is important to consider them.

TABLE 6.1
Fluorescent properties of some fluorophores commonly used for the construction of fluorescent probes

Fluorophore	ε ($M^{-1}cm^{-1}$)	Excitation wavelength (nm)	Emission wavelength (nm)	Solvent	Quantum yield
Pyrene [60]	54,000	335	390	Cyclohexane	0.32
7-Hydroxycoumarin [28]	29,000	386	448	pH 10	0.7
7-Methoxycoumarin [28]	20,000	336	402	pH 9	0.18
Coumarin 343 [173]	44,300	445	490	Ethanol	0.63
Fluorescein [174]	92,300	482	518	EtOH	0.9
Oregon green 488 [13]	85,000	492	518	pH 9	0.9
TAMRA [13]	91,000	542	568	MeOH	0.2
TEXAS RED [13]	84,000	588	601	$CHCl_3$	0.9
Lissamine rhodamine [175]	88,000	568	583	MeOH	0.3
BODIPY 493/503 [13]	89,000	495	504	MeOH	0.9
BODIPY 558/568 [13]	91,000	558	569	MeOH	0.9
Cy3 [176]	133,000	544	570	MeOH	0.07
Cy5 [176]	200,000	646	660	MeOH	0.4
NBD [177]	19,700	460	539	EtOH	0.36
Nile red [179]	38,000	519	590	Dioxane	0.7

Collisional quenching, also called dynamic quenching, is due to the formation of a complex between the excited fluorophore and surrounding molecules promoting non-radiative relaxation of the fluorophore to the ground state. Some common quenchers are paramagnetic species, such as O_2, and heavy atoms, such as iodine. Dynamic quenching reduces the fluorescence lifetime and quantum yield of the fluorophore in a concentration-dependent manner as is described by the Stern–Volmer equation [75].

Charge transfer reactions occur by interaction of a fluorophore with electron donating or electron accepting molecules and lead to quenching of fluorescence by generation of radical states. The mechanisms of charge transfer reactions are electron-exchange and photo-induced electron transfer (PET). Many fluorophores are quenched by proteins due to charge transfer reactions with aromatic amino acid residues. Typical examples are NBD, fluorescein, and BODIPY dyes [76, 77].

Energy transfer, as described by Förster [78], requires a long range dipole–dipole interaction between the donor and the acceptor fluorophore. This energy transfer is possible at distances between 2 and 10 nm. Contrary to what happens in collisional quenching, there is no need for physical contact between the two molecules.

A different kind of quenching is the one that occurs when the fluorophore is in the ground state (see Fig. 6.9). It requires the formation of a stable complex between the fluorophore and another molecule, possibly of the same class, and brought together by electrostatic or hydrophobic effects. This type of quenching is called static quenching. Fluorescent lifetimes are not affected by the formation of ground state complexes. However, the absorbance spectrum is changed and the extinction coefficient is reduced, resulting in lower fluorescence intensities. Complexes that can be formed between identical dyes are parallel H-dimers and head-to-tail J-dimers. They are formed by strong dipole–dipole interactions where there is alignment of the two dyes with the radius vector that connects them perpendicular to their transition dipoles. These dimers are characteristic for a shift of the absorption spectra and

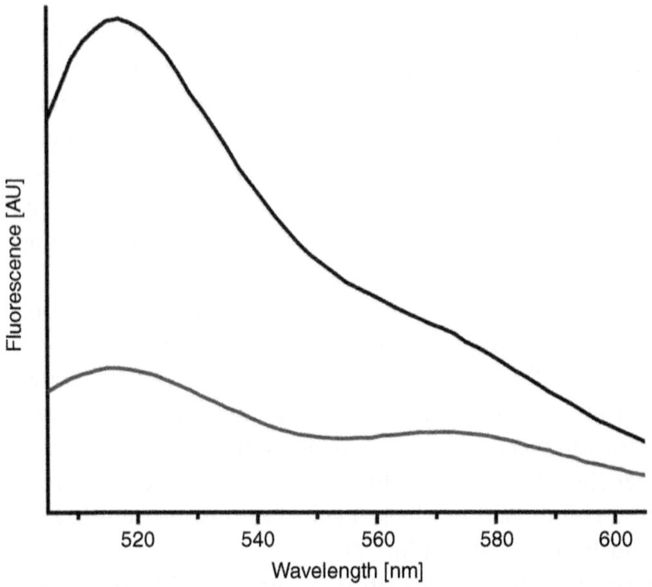

Fig. 6.9. Typical fluorescent spectra of the cleavage of a protease sensor labeled with the FRET pair fluorescein-TAMRA. These dyes tend to aggregate when placed in close proximity in aqueous solutions forming a ground state complex with reduced fluorescence intensity (gray line). After protease cleavage (black line) the dyes separate and the observed effect is not only loss of FRET, as indicated by the strong increase in donor fluorescence, but also an increase of fluorescence intensity of the acceptor due to the break in dye contacts.

changes in quantum yield [79]. The tendency of a dye to form dimers is depending on its charge, symmetry, and dipole strength. Solutions of rhodamines, which have a positive charge in the xanthene ring, form dimers from concentrations of 10^{-5} M or when they are attached to the same molecule in close proximity. The formation of these dimers leads to a blue shift in their absorption spectra from 546 to 518 nm in the case of TAMRA. For fluorescein and its derivatives, which are negatively charged at physiological pH, the formation of these dimers has not been observed unless much higher concentrations are reached [80]. Although the ability of xanthene fluorophores to form dimers can cause trouble in some applications, their introduction into probes has been advantageously used for measuring protease activity [81].

Strictly speaking, all acceptors in FRET probes are quenchers but the term "quencher" is specifically used for those molecules that absorb the fluorescence energy emitted by a dye and do not emit themselves. They are widely employed to create FRET probes in combination with a fluorophore that acts as a donor. The major advantage of this type of probes is that the direct excitation of the acceptor is eliminated and the drawback is that they do not allow ratiometric measurements. The energy transfer can be easily monitored with fluorescence lifetime imaging (FLIM). The most used quenchers and their properties are summarized in Table 6.2. Dinitrophenyl is frequently used to quench pyrene and the fluorescence of tryptophan residues in proteins [82]. Dabcyl [4-((4-(dimethylamino)-phenyl)azo)benzoic acid] quenches in a wide area of the green region of the spectra and Dabsyl is identical but has a sulphonic acid chloride moiety instead of the carboxylic acid group. This helps forming chemically stable conjugates. A variety of QXL quenchers and QSY quenchers cover the full visible spectrum and can be used in combination with the most commonly used fluorescent donors.

TABLE 6.2
The most commonly used quenchers and their properties

Quencher	Absorption maximum (nm)	ε (M^{-1} cm^{-1})
Dabcyl	453	32,000
Dabsyl	466	33,000
QXL	490–680	–
QSY	470–660	23,000–92,000
ATTO580Q	542	10,500
Iowa black FQ	531	38,216
BHQ 3	670	–

All data are from manufacturers specifications.

QSY dyes have high extinction coefficients and very low quantum yields (<0.001) that prevents them from fluorescing.

Iowa black FQ quenches in the green and yellow region and Iowa black RQ in the orange red region. Finally, ATTO quenchers have very sharp absorption spectrum which makes them more selective but less generally used (see Fig. 6.10).

6.2.4. Selection of FRET pairs

A FRET pair is constituted by two chromophores, one acting as energy donor and the other as acceptor. The efficiency of the energy transfer process is described as:

$$E = \left[1 + (r_{DA}/R_0)^6\right]^{-1} \tag{6.1}$$

where r_{DA} is the distance between the two fluorophores and R_0 is the Förster distance defined as the distance between the fluorophores when the transfer of energy is 50%. R_0 can be calculated with the following equation:

$$R_0 = c_0 \kappa^2 J n^{-4} Q_D \tag{6.2}$$

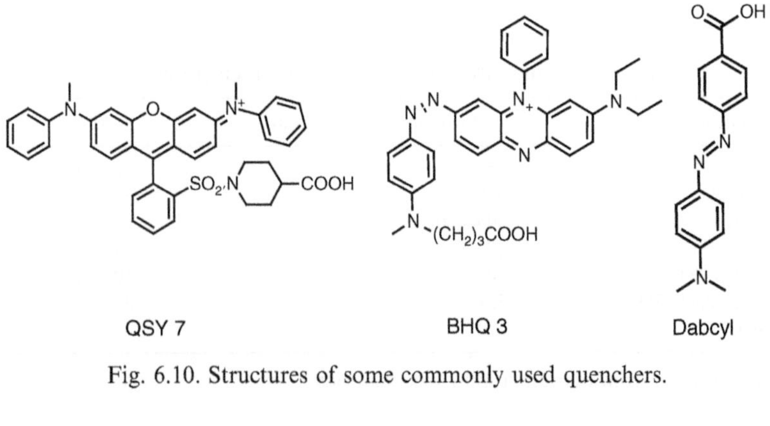

Fig. 6.10. Structures of some commonly used quenchers.

where c_0 is a constant with value 8.8×10^{-28} for distances measured in nm, κ^2 is a dipole orientation factor, J is the overlap integral of donor emission and acceptor excitation, n is the refractive index of the medium, and Q_D the quantum yield of the donor [83, 84].

Following the formulae shown above, it can be deduced that an efficient transfer of energy strongly depends on the following factors [85]:

- Maximal overlap between the fluorescent emission spectra of the donor and the fluorescent absorption spectra of the acceptor is important for efficient FRET. Since the donor can be directly excited, even fluorophores with mediocre quantum yield may be used. However, the success of a ratiometric FRET approach will strongly depend on the quantum yield of the acceptor. While an acceptor with a lousy quantum yield may still be a good quencher of the donor fluorescence, it will not yield acceptor fluorescence to an appreciable extent. When calculating the ratio between donor and acceptor emission values, acceptors with a good quantum yield will therefore be preferred.
- The rotation of the fluorophores is a factor that affects the energy transfer. Only maximal rotational freedom will permit r_{DA} estimation. There is no way to predict this factor. Therefore the dynamic averaged value of κ^2 is considered 2/3. This prediction induces a certain error in the calculation of distances (see Chap. 1).
- The range of distance that allows reliable determination is derived from the Förster equation (Eq. 6.1) as $R_0 \pm R_0/2$. Some R_0 values (assuming $\kappa^2 = 2/3$) for different FRET pairs are shown in Table 6.3.

The pairs of dyes more commonly used to generate FRET have been extensively reviewed [86–88] and their conjugation to different types of molecules has been facilitated by the incorporation of a range of reactive groups. Some examples are activated esters and isothiocyanates for reaction with amino groups of proteins or peptides,

TABLE 6.3
Most commonly used FRET pairs for the design of small molecule FRET sensors

Donor	Acceptor	λ abs/nm	λ em/nm	Ro (Å)	Reference
Tryptophan	Dansyl	280	520	21–24	[169]
EDANS	Dabcyl	335	Quencher	33	[93]
BODIPY 493/503	Cy5	500	667	42	Okamura and Watanabe (2006)
Fluorescein	TAMRA	492	576	49–55	[94]
Pyrene	7-Hydroxy coumarin	325	478	39	[95]
FAM	Texas Red	494	615	51	[171]
NBD	Nile Red	458	640	–	[89]
7-Diethylamino coumarin	Fluorescein	420	520	52	[178]
Naphthalene	Dansyl	280	525	22	[96]
NBD	Sulforhodamine	467	584	40–74	[97]

iodoacetyl derivatives, and maleimide to react with sulfhydryl groups such as the side chain of cysteine or carbonyldiimidazole to activate hydroxyl compounds [10]. Table 6.3 shows the properties of some of the most frequently used FRET pairs for the design of sensors.

6.3. Concepts and design of FRET sensors

As we have seen above, FRET is a technique that provides precise information about distances between 10 and 100 Å which is in the range of the size of biological molecules and proteins. Researchers have taken advantage of this feature and developed different strategies to synthesize FRET sensors that are able to follow in real time and with high sensitivity very diverse processes such as enzymatic activity, conformational change, or molecule-molecule interaction. The design of these FRET sensors is described below.

6.3.1. *Substrates for hydrolytic enzymes*

Generally, a FRET sensor based on a small molecule requires a FRET pair flanking the centre of reactivity. In small molecule sensors, the latter is usually a substrate site for hydrolytic enzymes such as a protease, lipase, esterase, amidase, lactamase, nuclease, phosphodiesterase, or phosphatase. Hydrolytic enzymes are the most common enzymes monitored with FRET probes due to the straightforward design of the sensor [89–92, 177]. The FRET pair can be formed by two fluorophores or by a donor fluorophore and a quencher molecule functioning as the FRET acceptor. Accordingly, two different types of sensors can be designed: ratiometric and quenched probes (see Fig. 6.11). Generally, probes that permit ratiometric measurements have several advantages over more simple donor-quenched probes. By determination of the emission ratio R upon exciting near the donor absorbance maximum $\left(R = I\left(\lambda_{\mathrm{em},D}\right)/I\left(\lambda_{\mathrm{em},A}\right)\right)$ a unit-free set of values is generated that is independent of fluorophore concentration. Since the ratiometric probe usually contains one donor and one acceptor, thus providing a stable stoichiometry of the fluorophores prior to cleavage, calculation of enzyme activities is directly reflected by the change in emission ratio (see Fig. 6.12). Since the fluorescence might be influenced by factors such as pH, changes in the environmental polarity or excitation intensity, or by photobleaching, quenched probes are easily generating artifacts. This is much less the case with ratiometric probes where the fluorescence emission of the acceptor serves as internal reference. For this reason, less calibration is needed and quantification is easier. However, when FLIM is used, both probes types are evenly suitable, because only changes in the donor lifetime will be monitored [99].

An important feature that influences the performance of the probe to a large extent is the selection of the FRET pair. Generally, fluorophores with high excitation wavelengths are preferred over blue fluorophores in single photon microscopy for applications in tissue since there is no autofluorescence in this region and the ratio

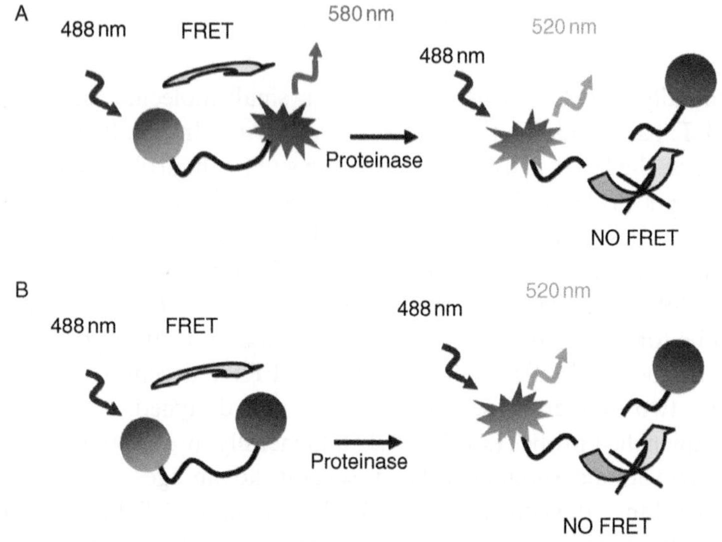

Fig. 6.11. Two types of FRET probes. (A) Ratiometric probes are formed by two fluorescent molecules that allow determination of emission ratio. (B) Quenched probes feature a donor fluorophore and a quencher. The emission increase of the donor after release of the acceptor is detected. Both types are frequently used to build proteinases probes. (See Color Insert.)

signal/background is increased. The spectroscopic characteristics of the donor/acceptor pairs that we have discussed in the previous section will determine the detection limits. The sensitivity of the probe will be established by the extent of energy transfer between the two dyes and their quantum yields. Since the energy transfer is depending on the distance, the overlap between the donor emission and the acceptor excitation spectra is crucial. In addition, the dipole moment of the fluorophores needs to be considered. In an ideal probe, the donor emission is nearly totally quenched, while due to the high FRET efficiency, the acceptor emission is significant. After cleavage, the two fluorophores separate to quasi indefinite distance. During cleavage, initial first order kinetics are

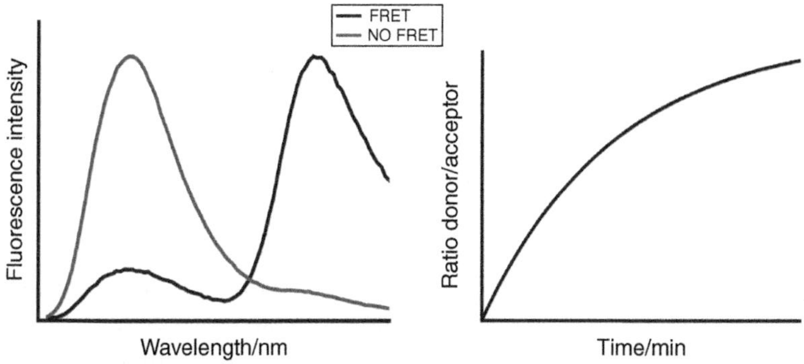

Fig. 6.12. Fluorescence spectra of an ideal proteinase probe before (black line) and after (gray line) enzymatic cleavage. Quantification of the hydrolysis rate is performed by measuring the change of ratio in donor/acceptor emission versus time.

observed as long as the reaction runs under substrate saturation conditions. As a final result, the donor fluorescence is totally unquenched and the sensitized acceptor emission is lost, leaving a minor directly excited acceptor signal at the donor excitation wavelength. This would generate huge ratio changes which translate into a high dynamic range of the probe and extremely large changes in the final readout. In most examples, the conditions are not as ideal. The most frequently encountered problems are direct acceptor excitation, a less than optimal FRET efficiency, and, depending on the probe size, direct fluorophore–fluorophore interaction (see Fig. 6.9).

In aqueous solutions, some pairs of fluorophores, including cyanines, fluorescein, and rhodamine derivatives, tend to come into close contact due to hydrophobic interactions and form ground state complexes that are in equilibrium with free fluorophore, as described in the previous section. Far from being beneficial for the transfer of energy, this proximity contributes to the quenching of both fluorophores. This type of static quenching leads to non- or weakly fluorescent probes where the emission of the donor is highly quenched. The main drawback of static quenching is that it reduces the sensitivity of the probe and sometimes the fluorescence of the donor is quenched to an extent that it almost eliminates the possibility of ratiometric measurements. The formation of ground-state complexes can be prevented, if physical contact of the two fluorophores is avoided. For peptidic sensors, this has been achieved with the use of conformationally restricted proline-rich peptides as linkers between donor and acceptor or by employing rigid linkers such as phenyl groups that decrease the flexibility [100, 101]. As an alternative, host molecules, such as cyclodextrins, forming inclusion complexes with coumarins, have been shown to prevent dye–dye interaction [95]. Although this might be useful for the design of some sensors, it is not possible to incorporate these features on a regular basis. Generally, it is beneficial to choose less hydrophobic dyes.

While the use of FRET probes in vitro is inflicted with a limited number of complications, the use in biological samples or in living cells needs much more careful considerations since factors such as enzyme specificity, cell toxicity, and spatio-temporal resolution usually play an important role.

6.3.2. Small probes based on conformational or structural changes

Changes in the conformation may bring together parts of a molecule that were initially spatially separated or vice versa. In fact, most sensors based on this design are genetically encoded molecules with two fluorescent proteins attached to the termini, or DNA- or RNA-based hybridization probes (molecular beacons) [102, 103]. These molecules have often the caveat that the emission ratio change is limited, thereby restricting the use of these sensors to high performance microscopy applications in living cells. Small molecule sensors might give a higher dynamic range because of the intrinsically higher change in fluorophore–fluorophore distance. As for large molecules, in small sensors ideally donor and acceptor dyes are attached to different positions of the molecule giving us information about the dynamic distance of the dyes before and after the conformational change, provided that the dye molecules can rotate freely and without a change in the average dipole moment. A strong form of a conformational change can be achieved by intramolecular cyclization [104]. This type of probe may also be used to determine the structure of small molecules that can adopt different conformations such as peptides [105–107].

6.3.3. Monitoring bond formation

To monitor bond formation a different strategy is needed. In this case, the fluorophores that form the FRET pair are attached to different molecules that are not interacting in the first place. The catalytic activity of the enzyme of interest attaches the two fluorophores sufficiently close to exhibit FRET. Ligases are enzymes that catalyze bond formation and their activity can indeed be monitored by this type of probe [96]. A big concern when using probes where the donor and the acceptor are attached to two different molecules is the tendency to obtain false positive and false negative results. False negatives might occur because the reactants have reduced affinity for the enzyme due to the attached fluorophores. A false positive might result from direct interaction of the dyes due to their hydrophobic nature without bond formation or by quenching produced by another source than the acceptor.

6.3.4. Sensing environmental changes

A fluorophore can undergo a change in their spectral properties as a result of pH variations or enzymatic activity. For example, fluorescein is such as fluorophore due to its two possible isoforms, lactone, and quinoid form. While the lactone form only absorbs in the UV and is not fluorescent, the quinoid form is excited at 490 nm and fluoresces. Only in this last form, there is the possibility for FRET. Therefore altering the pH or by enzymatically modifying fluorescein, it is possible to increase or reduce FRET [108, 109].

Many fluorophores are sensitive to changes in the hydrophobicity of the immediate environment. Therefore, bringing these fluorophores into a different environment may also produce a change in FRET, when a second fluorophore is affected by the emission change of the first. Fluorophores like Nile Red with changes of up to 100 nm when transferred from water to an aprotic organic solvent are principally suitable for such an approach [71]. Molecular rotors have the characteristic of having a quantum yield that depends on the viscosity. Such dyes are formed by an electron donor unit and an electron acceptor unit that can rotate relative to each other upon photoexcitation with a behavior that depends on the viscosity of the environment. These dyes have been included in FRET probes for viscosity studies [53].

6.3.5. Small molecule–protein interaction

For obvious reasons, it would be very attractive to monitor the binding of small molecules to proteins by FRET [110]. Two different approaches exist. One was developed by Blum et al., based on a quenched activity-based probe (qABP) for cathepsins [111]. This type of probe consists of a molecule flanked by a fluorophore and a quencher molecule that keeps the whole sensor non-fluorescent. It binds covalently to the target enzyme inhibiting its activity. Upon binding, the quencher is released and fluorescence is restored allowing localization of the target enzyme (see Fig. 6.13). The second approach requires the labeling of the target enzyme with a donor fluorophore and the synthesis of a substrate labeled with the acceptor. Upon substrate–protein interaction, sensitized emission, or FLIM can be used to monitor enzymatic activity and localization [99].

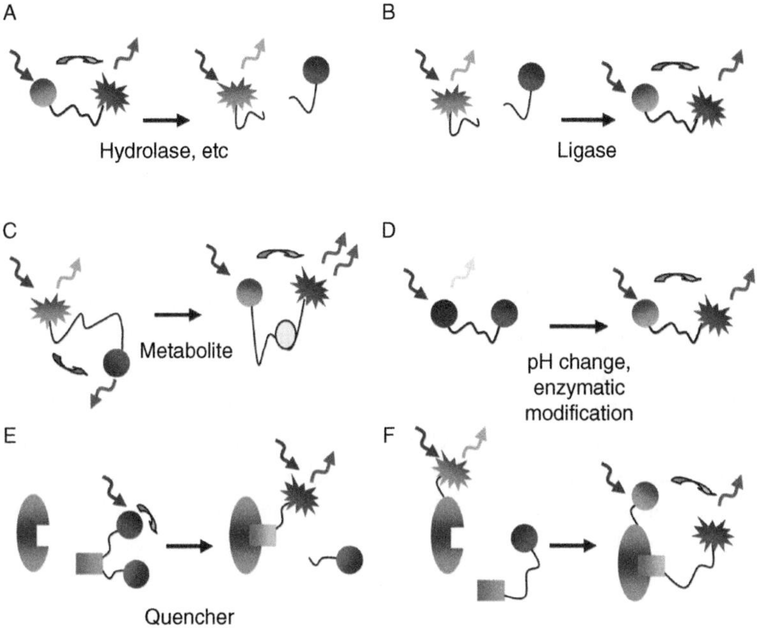

Fig. 6.13. Different designs of FRET sensors. (A) Substrates for hydrolytic enzymes. (B) Sensors for bond formation. (C) Sensors based on conformational or structural change. (D) Environmentally sensitive probes. (E) Quenched activity-based probe to monitor small molecule–enzyme interaction. (F) Small molecule–enzyme interaction using a labeled protein. (See Color Insert.)

6.4. Small molecule-based FRET sensors and their applications

A significant number of fluorogenic sensors are based purely on an increase or decrease of fluorescence. As discussed above, these probes, though often very useful, lack the internal reference that ratiometric sensors provide and require therefore difficult calibration. Especially hydrolytic enzymes are easily monitored by a fluorophore that is held in a non-fluorescent stage until the enzyme of interest removes the quencher. Certainly, it is beneficial if the quencher is a fluorophore itself and FRET can be monitored by measuring the emission ratio. In the following we will discuss various sensor applications including their relevant advantages and limitations.

6.4.1. Protease probes

Proteases are one of the largest families of enzymes and are involved in a multitude of vital processes. Due to their biological relevance and diversity, multiple fluorescent reporters monitoring their activity have been designed and successfully applied in vitro and in vivo [112–114]. Standard small molecule FRET probes for proteases consist of an amino acid sequence flanked by a FRET pair, consisting of two fluorophores or one fluorophore and a quencher molecule. Upon cleavage of the peptide sequence, the emission of the donor fluorophore is dequenched and the intensity increases whereas the emission of the acceptor decreases and vanishes more or less completely in those cases where the acceptor is fluorescent (see Fig. 6.11).

Following this strategy, Matayoshi et al. designed the first FRET probe to measure activity of the enzyme responsible for the correct maturation of the HIV-1 virus, HIV-1 PR, in vitro. The probe consisted of an octapeptide mimicking a naturally occurring substrate for HIV-1 PR and labeled with the fluorophore Edans [5-((2-aminoethyl)amino)naphthalene-1-sulfonic acid] and the quenching acceptor Dabcyl [4((4(dimethylamino)phenyl)azo) benzoic acid] [93]. Since then, a large number of peptide probes have been developed keeping the same pattern but improving the fluorescent characteristics with the use of better dyes and eliminating the artifacts by changing the quencher for a second fluorophore. These improvements have allowed the use of these small probes in cell studies and more recently in in vivo applications. Both extracellular and intracellular proteases have been studied with FRET probes. The measurement of intracellular activity requires the step of cell delivery. Some examples of proteinases monitored with FRET probes are: cathepsins [115], cysteine peptidases [116], antimicrobial proteases [117], serine proteases such as kallikrein 6 [118], and neutrophil elastase (NE) [119]. Within the protease family, matrix metalloproteinases (MMPs) have received especial attention due to their destructive role in several pathological diseases. There is an extensive review published compiling the fluorogenic substrates designed to assay MMPs [120]. Most of these sensors were created for high-throughput screening of protease inhibitors or for the creation of peptide libraries to identify protease substrates. Therefore they have been tested only in vitro.

A different strategy for measuring protease activity is based on the property of xanthene dyes to form H-type dimers (see Sect. 6.2.3) when they are in close proximity. These dimers are accompanied with a characteristic quenching of their fluorescence and, particularly for rhodamines, with a blue shift in the absorption spectrum [121, 122]. The probe D-NorFES-D designed to measure activity of elastase in HL-60 cells consists of an undecapeptide derivatized with one tetramethylrhodamine dye on each side. The sequence contains proline residues to create a bent structure and bring the two fluorophores in close proximity. Intact D-NorFES-D shows 90% of its fluorescence quenched plus a blue shift of the absorption spectrum. After addition of the serine protease elastase, an increase in the fluorescence and a bathochromic shift of the absorption spectrum is observed, resulting in an increase in the emission ratio [80].

The introduction of FRET pairs with improved properties allows the generation of probes suitable for the measurement of spatio-temporal activities in living cells and in vivo. One example is the imaging of caspase-3 activation in HeLa cells after inducing apoptosis. Previous fluorogenic substrates, such as DEVD-MCA, could not be used in living cells because the autofluorescence interfered with the low emission wavelength of the coumarin dye MCA [123]. With the incorporation of the FRET pair 6-carboxydichlorofluorescein and 5-carboxytetramethylrhodamine into another peptide sequence cleavable by caspase-3, Mizukami et al. measured caspase-3-like activity in HeLa cells stimulated by an inducer of apoptosis [124].

The same FRET pair was used to create a probe to monitor cathepsins in HeLa cells [94]. Since cathepsins are important in proteolysis within endosomes, it was in this case crucial to make the probe membrane-permeant. For this purpose a nona-arginine peptide was coupled to the N terminus to allow for endocytotic uptake.

The general interest in in vivo imaging and the development of fluorophores excitable at long wavelengths, stimulates the elaboration of near-infrared (NIR) probes that allow visualization of protease activity in whole organisms with noninvasive methods [125–127]. Until now, the main strategy to design NIR probes relies on the property of cyanine fluorophores to quench their fluorescence when they are close to each other. Based on this approach, it has been possible to measure tumor associated lysosomal cysteine/serine proteases [128]. Matrix metalloprotease-2 expression and the effect of inhibition of its activity after treatment with the MMP inhibitor prinomastat in tumors of nude mice is another example of NIR probe applications [129]. Recently, the use of a quenched activity-based probe (qABP) was reported by Blum et al. to image cathepsin activity (see Sect. 6.3.4) [111]. First generation probes contained the fluorescein/Dabcyl pair and were suitable for in vitro measurements. In order to obtain cell permeability, they incorporated the FRET pair BODIPY-TMR/QSY7 in a second generation probe. The major advantage of this approach is the feasibility for high-resolution microscopy studies of enzyme localization, outperforming in this aspect classical FRET probes that might show rapid diffusion inside cells. Specificity is controlled by the fact that only the probe that reacts with the target labeled enzyme is detected and activity of other enzymes is ignored (see Fig. 6.14).

6.4.2. Lipase probes

Like proteases, phospholipases (PLs) are enzymes that catalyze the hydrolysis of a covalent bond, in this particular case of phospholipids. For this reason, a similar approach for the design of FRET

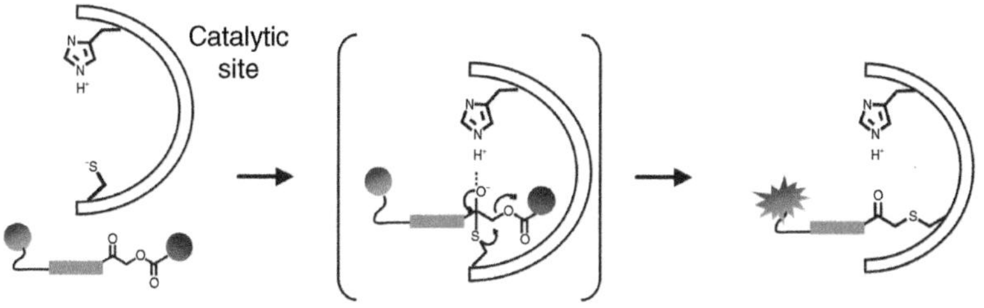

Fig. 6.14. Quenched activity based probe for the imaging of cathepsins. Upon covalent binding of the sensor to histidine and serine residues in the active site of the enzyme, the quencher is released and increased fluorescence indicates the now covalently labeled enzyme of interest.

sensors is used for proteases and lipases. PLs are located throughout living organisms and are active mostly on membrane surfaces where they play an important role in inflammation and signal transduction [130]. Among the different types of PLs, the activities of PLC and PLA_2 have been monitored with small molecule FRET probes. In a first approach, Burack et al. used the excimer formation of pyrene dyes to monitor the hydrolysis of phosphatidylcholine by PLA_2 [131]. Later, Hendrickson synthesized a quenched probe labeled with two identical BODIPY dyes flanking the cPLA2 substrate phosphatidylcholine. The change of pyrenes for BODIPY dyes provided the major advantage of allowing in vivo visualization of cPLA2 activity inside the gastrointestinal tract of zebrafish embryos with spatial resolution [132]. Some intramolecular quenched sensors and monolabeled fluorogenic probes have been synthesized to monitor PL activity and screen for inhibitors [133–135], but these lacked the property of being suitable to be used for ratiometric measurements which permit more accurate quantification. Another disadvantage was the deficiency of cell-permeability. Wichmann et al. synthesized the first truly ratiometric probe for measuring the activity of PLA_2 in vitro [89]. The probes were built using phosphatidylethanolamine (PE) and choline (PC) derivatives, which are known substrates of PLs, with NBD and Nile red attached to the ends of the fatty acids. The PE-based probe was termed

PENN. The specificity for PLA_2 over other PLs was ensured by modifiying the ester at the *sn*1 position to an ether preventing cleavage by PLA_1 or other PLs in this position (see Fig. 6.15). The ratio change of PENN was 30-fold, allowing detection of very small enzyme concentrations. Later, PENN was improved by preparing a

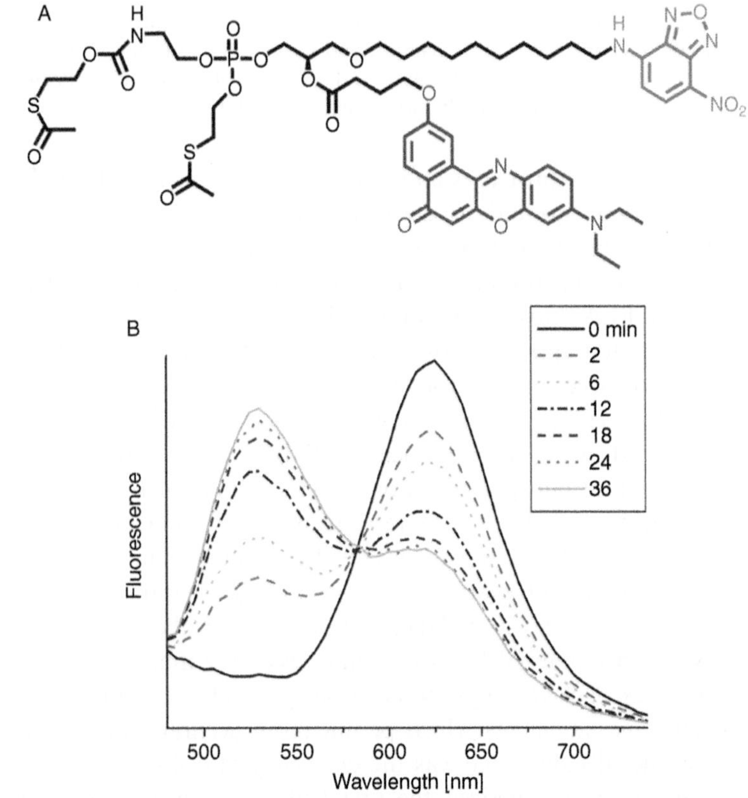

Fig. 6.15. (A) Structure of the PLA_2 probe PENN/SATE. (B) Fluorescence spectra change of the PLA_2 probe measured over time. The probe (1 μM) was added to a 3 mM solution of Triton X100 micelles and incubated with bee venom PLA_2. The excitation wavelength is at 440 nm. After 36 min the reaction was complete leading to a 30-fold increase in donor/acceptor ratio.

membrane-permeant derivative (PENN/SATE). Here, the charged groups were masked with the bioactivatable phosphate protecting group S-acetylthioethyl (SATE). With this probe it was possible to load cells and entire organisms non-disruptively. Consequently, it was possible to monitor the onset of PLA_2 activity during the development of a Medaka embryo in the entire tissue of the organism [136].

FRET probes have not only been generated to measure the phospholipase activity but to study its substrate specificity as well. Several substrates of PLA_2 with a variety of head groups and labeled with a BODIPY dye and a Dabcyl quencher were created by Rose et al. and tested against different PLAs in cells to determine substrate specificity and intracellular localization [137]. The specificity of PLA_2 isoforms towards the number of double bonds in the sn2 position was evaluated with a small series of PENN derivatives. It was demonstrated that the cytosolic type V PLA_2 preferred substrates with a single double bond [138].

6.4.3. β-Lactamase probes

The resistance of bacteria to β-lactame antibiotics is due to the presence of β-lactamases, enzymes that efficiently cleave penicillin and cephalosporin β-lactam rings of β-lactam antibiotics. Since this enzymatic activity is absent in mammalian cells, it serves well as a reporter gene when under the control of a promoter of interest. When a suitable fluorogenic probe was supplied to cells, the expression of the reporter gene would indicate promoter activity. In 1998, Roger Tsien and co-workers published a suitable ratiometric FRET probe to measure lactamase activity in cells: CCF2 [92]. It is based on a cephalosporin derivative equipped with 6-chloro-7-hydroxycoumarin and fluorescein. The fluorophores are attached to different positions of the molecule (see Fig. 6.16). To deliver the probe to the cytoplasm the sensor molecule was prepared as a membrane-permeant derivative (CCF2/AM) by masking the polar carboxylic

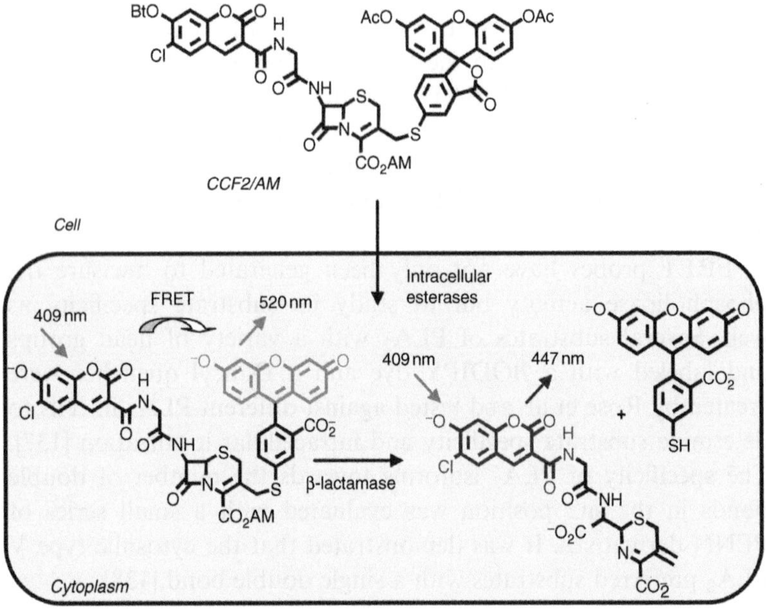

Fig. 6.16. The membrane-permeant probe CCF2/AM monitors lactamase activity in living cells after removal of the bioactivatable protecting groups by endogenous esterases.

acid of the molecule with an acetoxymethyl ester (AM ester) and other polar groups with butyryl and acetyl esters. This design allows for two important features: first, the sensor stays non-fluorescent until non-specific esterases cleave the protecting group in the cytoplasm allowing the monitoring of cellular uptake; and second, the masked probe is sufficiently unpolar to freely cross membranes. After release of the masking groups the sensor becomes negatively charged and stays trapped in the cytoplasm. Moreover, after β-lactam hydrolysis, the released mercaptofluorescein is rendered non-fluorescent avoiding interference with the measurement. The probe proved to be very sensitive and its large ratio change of 70-fold permitted to detect picomolar concentrations of β-lactamase.

Several attempts were made to get β-lactamase reporters more generally applicable to in vivo measurements [139]. CCF2, although useful for screening and studies in mammalian cells, is not helpful for applications in tissue or cells with thick walls, such as yeast. This is due to its high molecular weight, low aqueous solubility, and low excitation wavelength. Xing et al. tried to overcome the problem of cell autofluorescence synthesizing some cell-permeable near-infrared β-lactamase probes, CNIR 1–4 [140]. The core of the probe is the cephalosporin ring where the donor fluorophore Cy5 is attached via a glycyl linker and the acceptor quencher QSY21 is attached via a linker that includes a cysteine and the good leaving group amino thiophenol. The latter helps in the fragmentation of the molecule after hydrolysis. The fluorescence of Cy5 is totally quenched in the intact probe and an increase of 60 fluorescent units is observed after lactamase cleavage making the probe able to detect very low concentrations. In this case the main drawback is that the ratiometric measurements that render the experiments independent of cell size, probe concentration, as well as some other artifacts, are not possible. To improve solubility in aqueous media, a derivative with sulfonate groups on QSY21 was created. As detailed before, it is advisable to perpetrate the measurements in vivo without disrupting the cell wall. For this purpose the authors coupled acetylated D-glycosamine to the sensor, attached to the carboxylate of cysteine via a D-aminobutyric linker, mainly in order to avoid steric hindrance. D-Glycosamine is a sugar known to help carrying molecules into cells. With this strategy β-lactamase activity can be observed in stably transfected C6 glioma cells.

6.4.4. Kinase probes

Since kinases are not hydrolytic enzymes, a small molecule-based FRET probe does not seem to be a straight forward solution for this enzyme activity. Nevertheless, quite a number of fluorescent probes based on small substrate peptides have been prepared in the past. The most common approaches are based on intensity changes of a single fluorophore after phosphorylation of the peptide, often due to environmental changes [141, 142]. For a recent review see: [172, 143]. For energy transfer, either non-attached chromophores were employed or, in case of tyrosine kinase substrates, the tyrosine fluorescence was quenched by a nearby fluorophore attached to the substrate peptide. Phosphorylation then led to a dequenching effect [144]. These reporter molecules are only suitable for in vitro experiments. A true FRET sensor applicable to living cells was prepared adding 7-dimethylaminocoumarin and fluorescein diacetate to a PKA substrate sequence. Phosphorylation resulted in a loss of fluorescein acceptor emission [145].

6.4.5. Phosphatase probes

Phosphatase activity is easily monitored by attaching a phosphate to a fluorescein or coumarin derivative. These phosphate monoesters are non-fluorescent, but become strongly fluorescent upon hydrolytic cleavage by the enzyme of interest. This principle was readily extended to prepare FRET sensors. Enzymatically cleavable phosphates were used to prevent fluorescein from acting as an acceptor for a coumarin donor. Upon cleavage by phosphotyrosine phosphatase PTP1B or CD45, the fluorescein moiety became available for quenching the donor fluorescence and to exhibit FRET acceptor properties [101]. The use of this type of sensor in living cells was mentioned [108].

Very recently our group helped to provide a completely novel FRET method to look at PTP1B activity in living cells. This method is based on monitoring the formation of enzyme-substrate (ES) complexes, rather than the generation of the product from the enzymatic reaction [99]. The FRET pair was based on a genetically encoded fusion protein of PTP1B and EGFP or EYFP. A synthetic phosphorylated peptide equipped with a rhodamine served as a small molecule acceptor. After confirmation in vitro that the two molecules exhibited FRET and determination of K_d and K_m values, the labeled

PTP1B was expressed in cells and the substrate peptide was delivered either by microinjection or with the help of a penetratin peptide reversibly attached to a cysteine residue. In order to avoid premature dephosphorylation of the probe, the phosphate group was initially blocked by a photo-activatable protecting group. Photo-induced removal of the "cage" permitted the formation of the ES complex as was monitored by FLIM of the donor fluorescence. This technique has the advantage that only the overexpressed enzyme of interest is able to give a signal. Hence, its activity can be specifically measured against a huge background of other enzyme activities. The largest advantage, however, is the fact that the substrate may be re-phosphorylated by endogenous or expressed kinases. This was shown to lead to a steady-state of phosphorylated and unphosphorylated acceptor peptide levels. The amount of ES complex can be directly determined via the FRET population. When this amount was normalized over the overall enzyme concentration, it was possible to deduct the topology of PTP1B activity with spatial resolution throughout the cell. High amounts of ES complex reflect low enzyme activity while low FRET levels correspond to high enzyme activity. This method should principally be applicable for many enzyme activities (see Fig. 6.17).

6.4.6. Phosphodiesterase probes

Another important group of hydrolytic enzymes are phospho- and cyclophosphodiesterases. They catalyze the hydrolysis of phosphodiester bonds and many of the most relevant biological substrates are nucleic acids. Phospholipase C and D are also important examples. Initial attempts to measure phosphodiesterase activity placed a phosphodiester between a fluorophore and a quencher and the probe was tested in vitro [146]. This system was slightly modified by Caturla and used for the identification of catalysts with phosphodiesterase activity [147]. More recently, Nagano and co-workers used a coumarin donor and fluorescein as a FRET

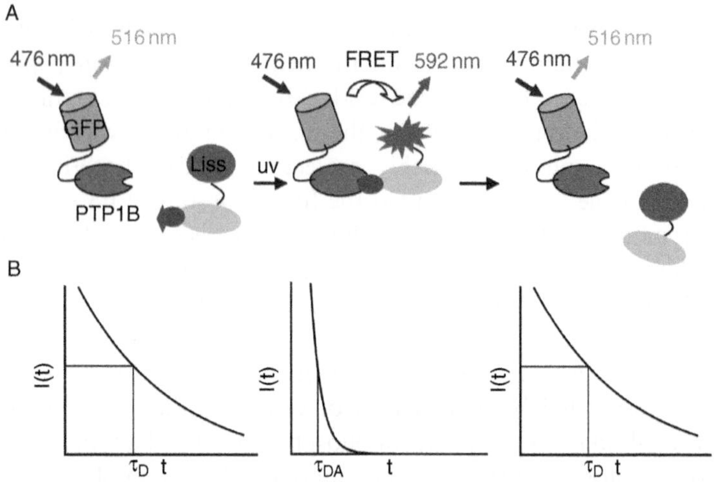

Fig. 6.17. PTP1B activity monitored by the formation of an enzyme–substrate complex. The probe is formed by the GFP-labeled enzyme PTP1B and the rhodamine-labeled peptide DADEYL (in orange). (A) The peptide DADEYL is synthesized with a caged phosphate on the tyrosine residue. Upon uncaging, the phosphate becomes available and PTP1B dephosphorylates DADEYL leading to a loss of enzyme–substrate complex. (B) During the three stages of the reaction, the GFP (donor) exhibits characteristic lifetime changes measured by FLIM. Only a FRET situation results in a short lifetime of the excited state (middle panel).

acceptor (see Fig. 6.18). In a significant synthetic effort, probe performance was optimized by varying the linker structure. It was shown that too short linkers resulted in direct contact of the fluorophores and hence inefficient acceptor fluorescence. However, sufficiently rigid linkers improved sensor performance significantly, provided that the affinity properties of the phosphodiesterase substrate were not compromised [108].

6.4.7. Probes for glycoside hydrolases and glycoside transferases

The same principle described above to monitor endo-peptidase activity was used by Matsuoka to monitor a ceramide glycanase

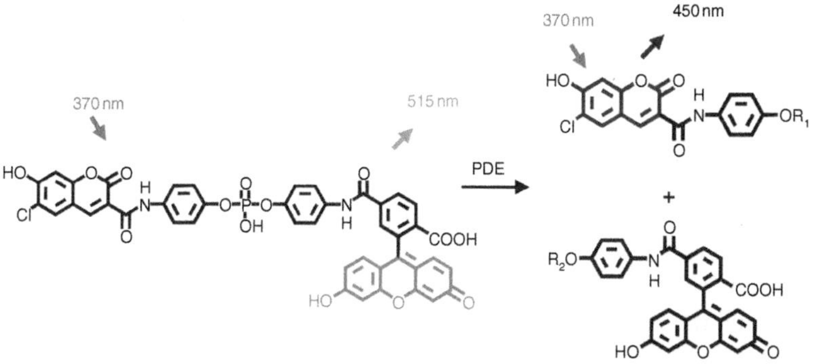

Fig. 6.18. Design and performance of the FRET probe CPF4 for measuring phosphodiesterase activity. R_1 and R_2 are phosphate and hydrogen or vice versa.

for the first time. This enzyme is an endo-carbohydrolase that cleaves the glycosidic bond between an oligosaccharide and ceramide. The bi-labeled probe NLD (Naphthyl-lactose-dansyl) was tested in vitro and the kinetic constants were determined [148]. Later, other FRET probes were synthesized to monitor glycoside hydrolases using EDANS as a donor fluorophore [149, 150]. Following the same principle, Cottaz et al. developed a probe able to discriminate between the activities of endo- and exo-glycoside hydrolases. The donor/acceptor pair is EDANS/dimethylaminophenylazophenyl (DAB). In order to couple the fluorophore EDANS synthetically to the reducing end of the tetrasaccharide, the derivatized α-glutamyl-EDANS had to be synthesized chemically [98].

Although many probes have been created with the purpose of monitoring hydrolytic enzymes, very few targeted ligases. The creation of such probes requires a different approach which is described in Sect. 6.3.3. One enzyme that has been monitored this way is sialyltransferase, a key player in the construction of glycoconjugates by catalyzing the transfer of one glycosyl group, from a sugar nucleotide to an oligosaccharide. Washiya et al. chemically synthesized a glycosyl donor (Nap-CMP-NANA) labeled with a naphthylmethyl group and a glycosyl acceptor (LacNAc-Dan) labeled with dansyl. The formation of the glycoside bond was followed by an increase of the dansyl emission and a decrease of the naphthyl emission and the kinetic constants of rat liver 2,6-sialyltransferase against these substrates was calculated in cuvette experiment [96].

An ingenious strategy developed for the assessment of β-galactosidase activity uses quinone methide chemistry. The probe has galactose bound to the hydroxyl group of phenol and two fluorophores attached to the *para*-position via two different linkers (see Fig. 6.19). Initially, the probe exhibits FRET. When β-galactosidase cleaves the glycosidic galactose bond, quinine is formed and the acceptor fluorophore acts as an excellent leaving group. In addition, a nucleophilic residue in the catalytic site of the enzyme reacts with the probe covalently. By doing so, the probe gets bound irreversibly to the enzyme and this is monitored by a loss of FRET [151].

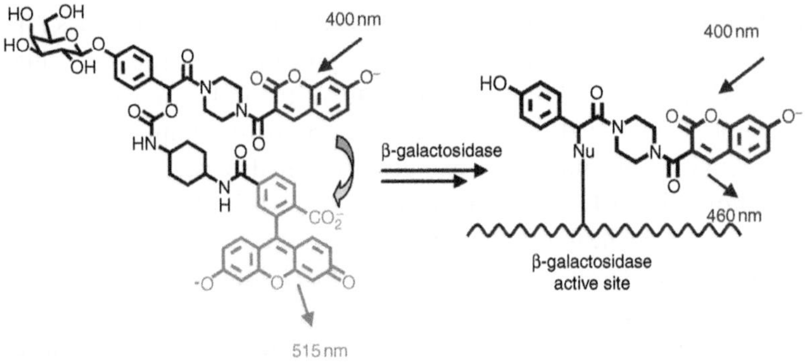

Fig. 6.19. Sensor CMFβ-gal to measure β-galactosidase activity in living cells.

6.4.8. *Detection of DNA modifying enzymes*

FRET probes based on DNA are strong potential tools to investigate cellular processes where nucleases and a set of enzymes acting on DNA play a role. Most probes of this class are not particularly small since they are usually formed by sequences longer than 10-mer but it is worth to include some examples in this chapter. One synthetic difference with respect to peptide-based probes is the introduction of fluorophores into the probe. In DNA-based probes this is generally achieved by direct incorporation of nucleosides modified with fluorescent dyes rather than by attaching the fluorophore after DNA synthesis [152, 153].

Within the group of enzymes acting on DNA are DNA ligases, enzymes that catalyze DNA recombination and nucleases that catalyze DNA cleavage. FRET probes to monitor all these enzymes have been developed. One FRET probe example for detection of DNA cleavage is the one designed to measure the endonucleolytic activity of HIV-1 integrase [154]. This enzyme cleaved viral DNA with high substrate specificity. A duplex formed by a 49- and 31-mer mimicking the substrate of the integrase was labeled with fluorescein and eosin, respectively. It produced initial FRET and is loss after cleavage. Other examples of DNA-based FRET sensors are reviewed in literature [155, 156].

6.4.9. Sensors for lipid trafficking and membrane fusion

An exploited use of FRET is found in the study of lipid trafficking and transport between membranes, induced either spontaneously or by transfer proteins [157]. Generally, two different strategies are used: self-quenching analysis and FRET assays (see Fig. 6.20). The self-quenching approach consists on the introduction of mono-labeled lipids, usually with the fluorophore pyrene, into a donor membrane. Due to lipid transport from the donor to an unlabeled acceptor membrane there is a decrease in concentration of the labeled lipid in the donor membrane and dequenching occurs producing an increase in fluorescence emission [158, 159]. FRET assays are based on the synthesis of two different classes of lipids each labeled with a different fluorophore forming the FRET pair. One lipid is transferable while the second is relatively immobile.

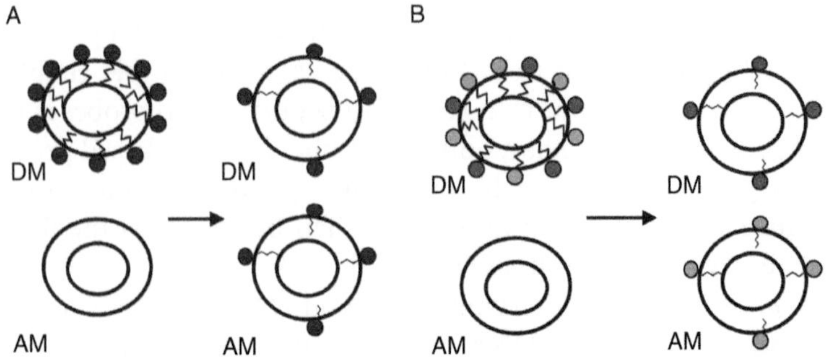

Fig. 6.20. Types of assays to visualize lipid trafficking in membranes. (A) Self-quenching method. Here the self-quenching is released upon transfer to unlabeled acceptor membranes that are usually in large excess. (B) FRET assays. Here the donor membrane contains a transferable lipid (green) that is quenched by FRET to a non transferable acceptor lipid (red). Upon transfer to an unlabeled acceptor membrane the green-labeled lipid becomes unquenched. DM = donor membrane, AM = acceptor membrane. (See Color Insert.)

In this case, a loss of FRET is observed when the transferable lipid moves to an acceptor membrane. The most common FRET pair used for these studies is the NBD/rhodamine pair [160, 161]. Some other FRET pairs have been used for the study of membrane sideness conservation [162] and intermembrane lipid transfer [163]. Distinguishably, a study of membrane activation by molecular selective recognition was performed using FRET [97].

6.4.10. Reporters of conformational change and structure determination

One of the most characteristic features of FRET is its sensitive dependency on the fluorophore distance. This is advantageously used to evaluate structures and conformational changes of peptides, glycopeptides, and proteins among other molecules [164–166]. The conformational change of the lipopeptide antibiotic daptomycin from an inactive linear form to a biological active cyclic form catalyzed by thioesterase domains has been evaluated using FRET [104]. Daptomycin has a peptide sequence of 13 amino acids. Originally, a tryptophan residue is located at the N-terminus and kynurenine at the C-terminus. These two fluorophores form a naturally occurring FRET pair with a spectral overlap between the emission of tryptophan ($\lambda = 330$ nm) and the excitation of kynurenine ($\lambda = 350$ nm). In the cyclic conformation, the two fluorophores are situated close to each other due to a lactone bond between the threonine at position 4 and the kynurenine residue at the C-terminus, facilitating energy transfer. In the linear form the two fluorophores are separated and FRET is less efficient. A family of cyclic and linear daptomycin derivatives peptides was synthesized on solid phase and the distance between the two fluorophores was evaluated by FRET. With these probes enzyme-catalyzed cyclization was studied in real-time.

The organization of lipids around the plasma membrane Ca^{2+}-transport ATPase of erythrocytes has been also determined by FRET. Taking advantage of the intrinsic fluorescence of the ATPase due to tryptophan residues and labeling different types of lipids with pyrene, it was demonstrated that the transporter is preferentially surrounded by negatively charged lipids such as phosphoinositides [167].

6.4.11. Sensors that detect small biological molecules

Apart from proteins and their action small molecules play an important role in the cell and their detection may be accomplished by using FRET probes. Although most of these sensors are protein-based, a few small molecule sensors have been introduced. A very recent example is a sensor for LPS (lipopolysaccharide), although the latter is not a particularly small molecule itself. The sensor is formed by an amino acid sequence derived from the binding domain of CD14 (a LPS-binding protein) and two fluorophores, fluorescein and tetramethylrhodamine, attached to the ends. The sensor relies on the ground state complex formed by these two xanthene fluorophores. When they are situated sufficiently close to each other their fluorescence is quenched. Upon LPS binding, the complex formation is disturbed and the emitted fluorescence recovers (see Fig. 6.21) [168].

In a similar fashion, steroids are molecules that have been investigated by disruption of FRET. The sensor is a double labeled peptide with cyclodextrin bound to one side chain. The latter keeps the fluorophores closely together by accommodating the coumarin into its cavity thereby ensuring efficient FRET. Steroids compete for the cavity of cyclodextrin and displace the coumarin reducing FRET efficiency. This model, although useful for in vitro applications, seems to be poorly selective for its application in biological samples [95].

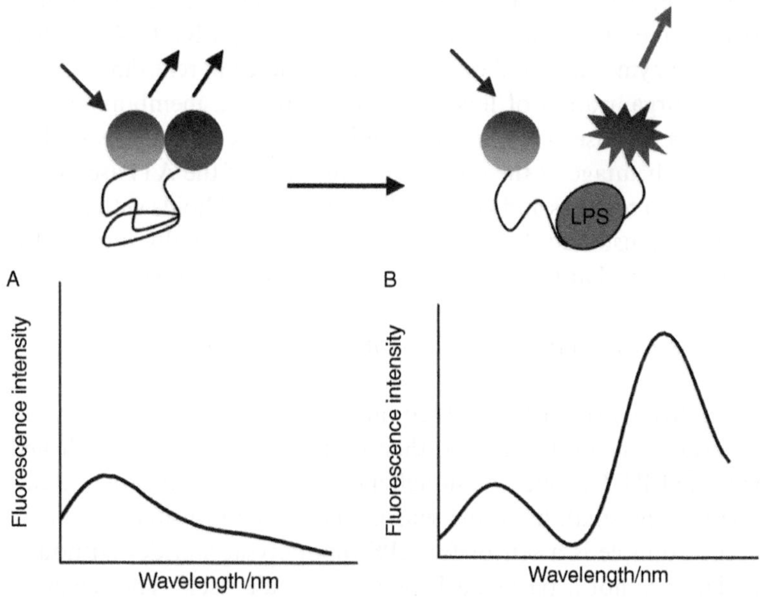

Fig. 6.21. Principle of detection of lipopolysaccharide (LPS) with the CD14-derived probe. It relies on the formation of a ground state complex between fluorescein and rhodamine in aqueous solution with quenching of donor and acceptor fluorescence. Spectrum A shows hypothetical fluorescence emission spectra of this complex. After LPS binding, the peptide sequence gets straightened prohibiting the close contact between the two fluorophores and leading to the recovery of red fluorescence (Spectra B).

Another recent sensor for small molecules is RPF1, a ratiometric reporter for hydrogen peroxide (H_2O_2). The donor/acceptor pair consists of coumarin and fluorescein but the last one is protected with boronate groups that maintain the fluorescein in the non-fluorescent lactone form. Upon reaction with H_2O_2 the boronate groups are released and FRET is observed allowing quantification of the metabolite [178].

6.4.12. Sensors for dynamics of biomolecular processes

The suitability of FRET for time-course studies permits the construction of probes appropriate for the monitoring of dynamics processes that occur in biological environments. One example is the disulfide reduction in endosomal compartments. With the Folate-FRET reporter, Yang et al. were able to visualize where and when disulfide bond reduction occurs in KB cells [91, 134]. The design of this sensor is based on folate, a member of the vitamin B family, with BODIPY FL conjugated through a lysine residue and rhodamine attached via a disulfide bridge. The folate moiety is recognized by the folate receptor situated on the surface of the KB cells and its internalization into endosomes followed by reduction of the disulfide bond was visualized (see Fig. 6.22).

Variations in viscosity have been also followed with FRET probes. An innovative FRET pair constituted by a viscosity sensitive dye (CMAM) and a viscosity-independent coumarin, which acts as internal reference, demonstrated a high sensitivity for detecting viscosity changes in vitro [52].

6.5. Perspectives and future directions

Although there is now a limited set of FRET sensors available, the gap between the demand in biology and the development of new reporters is not closing. Especially small molecule-based sensors are

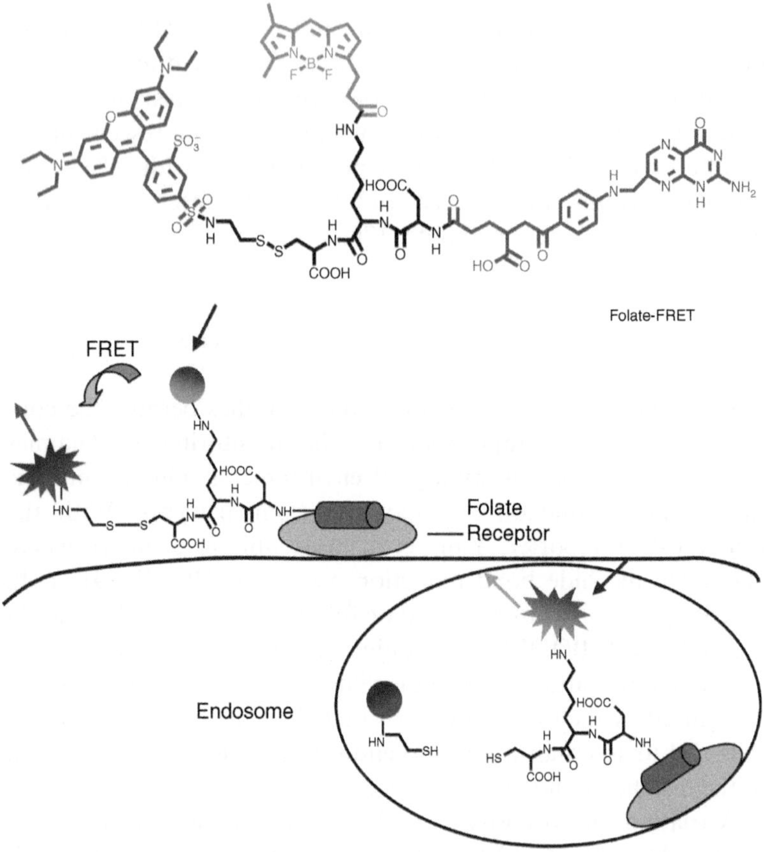

Fig. 6.22. Folate-FRET sensor structure and its application to measure disulfide bond reduction in endosomes. The molecule contains the folate moiety which is recognized by the folate receptor situated at the plasma membrane. This recognition leads to endocytosis and after some time to cleavage of the probe. The latter is monitored by a loss in FRET. (See Color Insert.)

still the exception, despite their obvious advantages such as large changes in the emission ratio or direct availability for application to live specimen. Both features make small molecule sensors a prime tool for screening in the pharmaceutical industry. However, the development and preparation is up to now very cumbersome. This could be helped by platform techniques based on solid phase synthesis which have the potential to generate larger arrays of sensor candidates. In addition, we will need new fluorophores with particular properties, for instance the lack of forming self-quenching aggregates. Finally, the use of quantum dots has become available for applications in living cells through bioconjugate chemistry (see Chap. 12). This offers the opportunity to significantly improve existing sensors and to develop new ones.

Acknowledgment

The authors gratefully acknowledge funding received for research projects on probe design and imaging from the EU (LSHG-CT-2003-503259), the Volkswagen Stiftung (I/78 989, I/81 597 and I/81 798), the Human Frontiers Science Programme, and the EMBL.

References

[1] Lakowicz, J. R. (2006). "Principles of fluorescence spectroscopy." Springer, New York.

[2] Rettig, W., Strehmel, B., Schrader, S. and Seifert, H. (1999). Applied fluorescence in chemistry, biology and medicine. Springer, Berlin.

[3] Wolf, D. E. (2007). Fundamentals of fluorescence and fluorescence microscopy. Methods Cell Biol. 63–91.

[4] Lichtman, J. W. and Conchello, J. A. (2005). Fluorescence microscopy. Nat. Methods *2*, 910–919.

[5] Jameson, D. M., Croney, J. C. and Moens, P. D. J. (2003). Fluorescence: Basic concepts, practical aspects, and some anecdotes. Biophotonics, Pt A *360*, 1–43.

[6] Karstens, T. and Kobs, K. (1980). Rhodamine-B and rhodamine-101 as reference substances for fluorescence quantum yield measurements. J. Phys. Chem. *84*, 1871–1872.

[7] Eggeling, C., Widengren, J., Rigler, R. and Seidel, C. A. M. (1998). Photobleaching of fluorescent dyes under conditions used for single-molecule detection: Evidence of two-step photolysis. Anal. Chem. *70*, 2651–2659.

[8] Mukherjee, S., Chattopadhyay, A., Samanta, A. and Soujanya, T. (1994). Dipole-moment change of Nbd group upon excitation Studied using solvatochromic and quantum-chemical approaches – Implications in membrane research. J. Phys. Chem. *98*, 2809–2812.
[9] Farinotti, R., Siard, P., Bourson, J., Kirkiacharian, S., Valeur, B. and Mahuzier, G. (1983). 4-Bromomethyl-6,7-dimethoxycoumarin as a fluorescent label for carboxylic-acids in chromatographic detection. J. Chromatogr. *269*, 81–90.
[10] Hermanson, G. T. (1996). Bioconjugate techniques.. Academic Press, San Diego.
[11] Meallier, P., Guittonneau, S., Emmelin, C. and Konstantinova, T. (1999). Photochemistry of fluorescein and eosin derivatives. Dyes and Pigments *40*, 95–98.
[12] Saletsky, A. M., Shekunov, V. A. and Yuzhakov, V. I. (1986). Solvatochromism and solvatofluorochromism of erythrosine. Teoreticheskaya I Eksperimentalnaya Khimiya *22*, 196–202.
[13] Haugland, R. P. (2005). A guide to fluorescent probes and labeling technologies. Invitrogen, San Diego.
[14] Fujita, S., Nakanishi, S. and Toru, T. (1998). Convenient preparation of fluorescein derivatives. Synth. Commun. *28*, 387–393.
[15] Nguyen, T. and Francis, M. B. (2003). Practical synthetic route to functionalized rhodamine dyes. Org. Lett. *5*, 3245–3248.
[16] Miura, T., Urano, Y., Tanaka, K., Nagano, T., Ohkubo, K. and Fukuzumi, S. (2003). Rational design principle for modulating fluorescence properties of fluorescein-based probes by photoinduced electron transfer. J. Am. Chem. Soc. *125*, 8666–8671.
[17] Ueno, T., Urano, Y., Setsukinai, K., Takakusa, H., Kojima, H., Kikuchi, K., Ohkubo, K., Fukuzumi, S. and Nagano, T. (2004). Rational principles for modulating fluorescence properties of fluorescein. J. Am. Chem. Soc. *126*, 14079–14085.
[18] Diehl, H. and Markuszewski, R. (1989). Studies on Fluorescein. 7. The fluorescence of fluorescein as a function of Ph. Talanta *36*, 416–418.
[19] Song, L. L., Hennink, E. J., Young, I. T. and Tanke, H. J. (1995). Photobleaching kinetics of fluorescein in quantitative fluorescence microscopy. Biophys. J. *68*, 2588–2600.
[20] Lundberg, P. and Langel, U. (2006). Uptake mechanisms of cell-penetrating peptides derived from the Alzheimer's disease associated gamma-secretase complex. Int. J. Pept. Res. Therap. *12*, 105–114.
[21] Grieshaber, S., Swanson, J. A. and Hackstadt, T. (2002). Determination of the physical environment within the Chlamydia trachomatis inclusion using ion-selective ratiometric probes. Cell. Microbiol. *4*, 273–283.

[22] Kubin, R. F. and Fletcher, A. N. (1982). Fluorescence quantum yield of some rhodamine dyes. J. Luminiscence *27*, 455–462.

[23] Snare, M. J., Treloar, F. E., Ghiggino, K. P. and Thistlethwaite, P. J. (1982). The photophysics of rhodamine B. J. Photochem. *18*, 335–346.

[24] Whitaker, J. E., Haugland, R. P., Ryan, D., Hewitt, P. C. and Prendergast, F. G. (1992). Fluorescent rhodol derivatives: versatile, photostable labels and tracers. Anal. Biochem. *207*, 267–279.

[25] Kulkarni, M. V., Kulkarni, G. M., Lin, C. H. and Sun, C. M. (2006). Recent advances in coumarins and 1-azacoumarins as versatile biodynamic agents. Curr. Med. Chem. *13*, 2795–2818.

[26] Pochet, L., Frederick, R. and Masereel, B. (2004). Coumarin and isocoumarin as serine protease inhibitors. Curr. Pharm. Des. *10*, 3781–3796.

[27] Schiedel, M. S., Briehn, C. A. and Bauerle, P. (2001). Single-compound libraries of organic materials: Parallel synthesis and screening of fluorescent dyes. Angew. Chem. Int. Ed. Engl. *40*, 4677–4680.

[28] Wheelock, C. E. (1958). The fluorescence of some coumarins. J. Am. Chem. Soc. *81*, 1348–1352.

[29] Murata, C., Masuda, T., Kamochi, Y., Todoroki, K., Yoshida, H., Nohta, H., Yamaguchi, M. and Takadate, A. (2005). Improvement of fluorescence characteristics of coumarins: Syntheses and fluorescence properties of 6-methoxycoumarin and benzocoumarin. Derivatives as novel fluorophores emitting in the longer wavelength region and their application to analytical reagents. Chem. Pharm. Bull. *53*, 750–758.

[30] Sherman, W. R. and Robins, E. (1968). Fluorescence of substituted 7-hydroxycoumarins. Anal. Chem. *40*, 803–805.

[31] Sivakumar, K., Xie, F., Cash, B. M., Long, S., Barnhill, H. N. and Wang, Q. (2004). A fluorogenic 1,3-dipolar cycloaddition reaction of 3-azidocoumarins and acetylenes. Org. Lett. *6*, 4603–4606.

[32] Muthuramu, K. and Ramamurthy, V. (1984). 7-Alkoxy coumarins as fluorescence probes for microenvironments. J. Photochem. *26*, 57–64.

[33] Satpati, A., Senthilkumar, S., Kumbhakar, M., Nath, S., Maity, D. K. and Pal, H. (2005). Investigations of the solvent polarity effect on the photophysical properties of coumarin-7 dye. Photochem. Photobiol. *81*, 270–278.

[34] Furuta, T., Wang, S. S. H., Dantzker, J. L., Dore, T. M., Bybee, W. J., Callaway, E. M., Denk, W. and Tsien, R. Y. (1999). Brominated 7-hydroxycoumarin-4-ylmethyls: Photolabile protecting groups with biologically useful cross-sections for two photon photolysis. Proc. Natl. Acad. Sci. USA. *96*, 1193–1200.

[35] Hagen, V., Dekowski, B., Nache, V., Schmidt, R., Geissler, D., Lorenz, D., Eichhorst, J., Keller, S., Kaneko, H.,, Benndorf, K. and

Wiesner, B. (2005). Coumarinylmethyl esters for ultrafast release of high concentrations of cyclic nucleotides upon one- and two-photon photolysis. Angew. Chem. Int. Ed. *44*, 7887–7891.

[36] Gagey, N., Neveu, P. and Jullien, L. (2007). Two-photon uncaging with the efficient 3,5-dibromo-2,4-dihydroxycinnamic caging group. Angew. Chem., Int. Ed. *46*, 2467–2469.

[37] Berthelot, T., Talbot, J. C., Lain, G., Deleris, G. and Latxague, L. (2005). Synthesis of N epsilon-(7-diethylaminocoumarin-3-carboxyl)- and N epsilon-(7-methoxycoumarin-3-carboxyl)-L-Fmoc lysine as tools for protease cleavage detection by fluorescence. J. Pept. Sci. *11*, 153–160.

[38] Rohand, T., Baruah, M., Qin, W. W., Boens, N. and Dehaen, W. (2006). Functionalisation of fluorescent BODIPY dyes by nucleophilic substitution. Chem. Commun. 266–268.

[39] Marushchak, D., Kalinin, S., Mikhalyov, I., Gretskaya, N. and Johansson, L. B. A. (2006). Pyrromethene dyes (BODIPY) can form ground state homo and hetero dimers: Photophysics and spectral properties. Spectrochim. Acta Part A-Mol. Biomol. Spectrosc. *65*, 113–122.

[40] Berlier, J. E., Rothe, A., Buller, G., Bradford, J., Gray, D. R., Filanoski, B. J., Telford, W. G., Yue, S., Liu, J. X. Cheung, C. Y. et al. (2003). Quantitative comparison of long-wavelength Alexa Fluor dyes to Cy dyes: Fluorescence of the dyes and their bioconjugates. J. Histochem. Cytochem. *51*, 1699–1712.

[41] Kassab, K. (2002). Photophysical and photosensitizing properties of selected cyanines. J. Photochem. Photobiol. B Biol. *68*, 15–22.

[42] Kirstein, S. and Daehne, S. (2006). J-aggregates of amphiphilic cyanine dyes: self-organization of artificial light harvesting complexes. Int. J. Photoenergy *5*, 1–21.

[43] Isacsson, J. and Westman, G. (2001). Solid-phase synthesis of asymmetric cyanine dyes. Tetrahedron Lett. *42*, 3207–3210.

[44] Silva, G. L., Ediz, V., Yaron, D. and Armitage, B. A. (2007). Experimental and computational investigation of unsymmetrical cyanine dyes: Understanding torsionally responsive fluorogenic dyes. J. Am. Chem. Soc. *129*, 5710–5718.

[45] Rye, H. S., Yue, S., Wemmer, D. E., Quesada, M. A., Haugland, R. P., Mathies, R. A. and Glazer, A. N. (1992). Stable fluorescent complexes of double-stranded DNA with bis-intercalating asymmetric cyanine dyes – Properties and applications. Nucleic Acids Res. *20*, 2803–2812.

[46] Thomas, L., Netzel, K. N. and Zhao, M. (1995). Base-content dependence of emission enhancements, quantum yields, and lifetimes for cyanine dyes bound to double-strand DNA: Photophysical properties of monomeric and bichromophoric DNA stains. J. Phys. Chem. *99*, 17936–17947.

[47] Liu, Z. J., Bushnell, W. R. and Brambl, R. (1987). Potentiometric cyanine dyes are sensitive probes for mitochondria in intact plant-cells – Kinetin enhances mitochondrial fluorescence. Plant Physiol. *84*, 1385–1390.

[48] Oseroff, A. R., Ohuoha, D., Ara, G., Whittaker, D., Murphy, G., Foley, J. and Cincotta, L. (1986). Intramitochondrial photosensitizers allow selective photochemotherapy of human squamous-cell carcinoma (Scc). J. Invest. Dermatol. *86*, 498.

[49] Schummer, U., Schiefer, H. G. and Gerhardt, U. (1979). Mycoplasma membrane-potentials determined by potential-sensitive fluorescent dyes. Curr. Microbiol. *2*, 191–194.

[50] Rotkiewicz, K., Grellmann, K. H. and Grabowski, Z. R. (1973). Reinterpretation of the anomalous fluorescense of p-n,n-dimethylamino-benzonitrile. Chem. Phys. Lett. *19*, 315–318.

[51] Haidekker, M. A. and Theodorakis, E. A. (2007). Molecular rotors – fluorescent biosensors for viscosity and flow. Org. Biomol. Chem. *5*, 1669–1678.

[52] Haidekker, M. A., Brady, T. P., Lichlyter, D. and Theodorakis, E. A. (2006). A ratiometric fluorescent viscosity sensor. J. Am. Chem. Soc. *128*, 398–399.

[53] Haidekker, M. A., Akers, W., Lichlyter, D., Brady, T. P. and Theodorakis, E. A. (2005). Sensing of flow and shear stress using fluorescent molecular rotors. Sensor Lett. *3*, 42–48.

[54] Birks, J. B. and Chirstophorou, L. G. (1970). Excimer fluorescence of aromatic compounds. Prog. React. Kinet. *5*, 181–272.

[55] Lehrer, S. S. (1997). Intramolecular pyrene excimer fluorescence: A probe of proximity and protein conformational change. Fluorescence Spectroscopy *278*, 286–295.

[56] Pal, R., Barenholz, Y. and Wagner, R. R. (1988). Pyrene Phospholipid as a biological fluorescent-probe for studying fusion of virus membrane with liposomes. Biochemistry (Mosc). *27*, 30–36.

[57] Sklar, L. A., Hudson, B. S. and Simoni, R. D. (1976a). Conjugated linear polyene fatty-acids as membrane probes. Fed. Proc. *35*, 1531–1531.

[58] Sklar, L. A., Hudson, B. S. and Simoni, R. D. (1976b). Conjugated polyene fatty-acids as fluorescent membrane probes – Model system studies. J. Supramol. Struct. *4*, 449–465.

[59] Sklar, L. A., Hudson, B. S. and Simoni, R. D. (1977). Conjugated polyene fatty-acids as fluorescent-probes – Synthetic phospholipid membrane studies. Biochemistry (Mosc). *16*, 819–828.

[60] Berlman, I. B. (1971). Handbook of fluorescence spectra of aromatic molecules. Academic Press, New York.

[61] Kuerschner, L., Ejsing, C. S., Ekroos, K., Shevchenko, A., Anderson, K. I. and Thiele, C. (2005). Polyene-lipids: A new tool to image lipids. Nat. Methods *2*, 39–45.

[62] Selvin, P. R. (1996). Lanthanide-based resonance energy transfer. IEEE Journal of selected topics in quantum electronics: Lasers Biol. *2*, 1077–1087.
[63] Terai, T., Kikuchi, K., Iwasawa, S., Kawabe, T., Hirata, Y., Urano, Y. and Nagano, T. (2006). Modulation of luminescence intensity of lanthanide complexes by photoinduced electron transfer and its application to a long-lived protease probe (vol. 128, pp. 6938, 2006). J. Am. Chem. Soc. *128*, 8699–8700.
[64] Karvinen, J., Laitala, V., Makinen, M. L., Mulari, O., Tamminen, J., Hermonen, J., Hurskainen, P. and Hemmila, I. (2004). Fluorescence quenching-based assays for hydrolyzing enzymes. Application of time-resolved fluorometry in assays for caspase, helicase, and phosphatase. Anal. Chem. *76*, 1429–1436.
[65] Selvin, P. R., Ha, T., Enderle, T., Ogletree, D. F., Chemla, D. S. and Weiss, S. (1996). Fluorescence resonance energy transfer between a single donor and a single acceptor molecule. Biophys. J. *70*, Wp302–Wp302.
[66] Hanaoka, K., Kikuchi, K., Kojima, H., Urano, Y. and Nagano, T. (2004). Development of a zinc ion-selective luminescent lanthanide chemosensor for biological applications. J. Am. Chem. Soc. *126*, 12470–12476.
[67] Yu, J. H., Parker, D., Pal, R., Poole, R. A. and Cann, M. J. (2006). A europium complex that selectively stains nucleoli of cells. J. Am. Chem. Soc. *128*, 2294–2299.
[68] Uchiyama, S., Santa, T., Okiyama, N., Azuma, K. and Imai, K. (2000). Semi-empirical PM3 calculations predict the fluorescence quantum yields (Phi) of 4-monosubstituted benzofurazan compounds. J. Chem. Soc. Perkin Trans. II *6*, 1199–1207.
[69] Greenspan, P., Mayer, E. P. and Fowler, S. D. (1985). Nile Red – a selective fluorescent stain for intracellular lipid droplets. J. Cell Biol. *100*, 965–973.
[70] Sackett, D. L. and Wolff, J. (1987). Nile red as a polarity-sensitive fluorescent-probe of hydrophobic protein surfaces. Anal. Biochem. *167*, 228–234.
[71] Brown, M., Edmonds, T., Miller, J. and Seare, N. (1993). Use of nile red as a long-wavelength fluorophore in dual-probe studies of ligand–protein interactions. J. Fluorescence *3*, 129–130.
[72] Black, S. L., Stanley, W. A., Fillpp, F. V., Bhairo, M., Verma, A., Wichmann, O., Sattler, M., Wilmanns, M. and Schultz, C. (2008). Probing lipid- and drug-binding domains with fluorescent dyes. Bioorg. Med. Chem. *16*, 1162–1173.
[73] Stanley, W. A., Versluis, K., Schultz, C., Heck, A. J. R. and Wilmanns, M. (2007). Investigation of the ligand spectrum of human

sterol carrier protein 2 using a direct mass spectrometry assay. Arch. Biochem. Biophys. *461*, 50–58.
[74] Jose, J. and Burgess, K. (2006). Syntheses and properties of water-soluble Nile Red derivatives. J. Org. Chem. *71*, 7835–7839.
[75] Stern, O. and Volmer, M. (1919). Über die Abklingungszeit der Fluoreszenz. Physikalische Zeitschrift *20*, 183–188.
[76] Lancet, D. and Pecht, I. (1977). Spectroscopic and immunochemical studies with nitrobenzoxadiazolealanine, a fluorescent dinitrophenyl analog. Biochemistry *16*, 5150–5157.
[77] Marme, N., Knemeyer, J. P., Sauer, M. and Wolfrum, J. (2003). Inter- and intramolecular fluorescence quenching of organic dyes by tryptophan. Bioconjug. Chem. *14*, 1133–1139.
[78] Foerster, T. (1948). Intermolecular energy migration and fluorescence. Ann. Phys. (Leipzig) *2*, 55–75.
[79] Valdesaguilera, O. and Neckers, D. C. (1989). Aggregation Phenomena in Xanthene Dyes. Acc. Chem. Res. *22*, 171–177.
[80] Packard, B. Z., Toptygin, D. D., Komoriya, A. and Brand, L. (1996). Profluorescent protease substrates: intramolecular dimers described by the exciton model. Proc. Natl. Acad. Sci. USA. *93*, 11640–11645.
[81] Packard, B. Z., Komoriya, A., Toptygin, D. D. and Brand, L. (1997). Structural characteristics of fluorophores that form intramolecular H-type dimers in a protease substrate. J. Phys. Chem. B *101*, 5070–5074.
[82] Green, N. M. (1964). Avidin. Quenching of fluorescence by dinitrophenyl groups. Biochem. J. *90*, 564–568.
[83] Jares-Erijman, E. A. and Jovin, T. M. (2003). FRET imaging. Nat. Biotechnol. *21*, 1387–1395.
[84] Vogel, S. S., Thaler, C. and Koushik, S. V. (2006). Fanciful FRET. Sci. STKE *331*, 1–8.
[85] Ha, T. (2001). Single-molecule fluorescence resonance energy transfer. Methods *25*, 78–86.
[86] Sapsford, K. E., Berti, L. and Medintz, I. L. (2006). Materials for fluorescence resonance energy transfer analysis: Beyond traditional donor–acceptor combinations. Angew. Chem. Int. Ed. *45*, 4562–4588.
[87] Sinev, M., Landsmann, P., Sineva, E., Ittah, V. and Haas, E. (2000). Design consideration and probes for fluorescence resonance energy transfer studies. Bioconjug. Chem. *11*, 352–362.
[88] Wu, P. G. and Brand, L. (1994). Resonance energy-transfer – Methods and applications. Anal. Biochem. *218*, 1–13.
[89] Wichmann, O. and Schultz, C. (2001). FRET probes to monitor phospholipase A(2) activity. Chem. Commun. 2500–2501.

[90] Bark, S. J. and Hahn, K. M. (2000). Fluorescent indicators of peptide cleavage in the trafficking compartments of living cells: Peptides site-specifically labeled with two dyes. Methods *20*, 429–435.
[91] Yang, J., Chen, H., Vlahov, I. R., Cheng, J. X. and Low, P. S. (2006a). Evaluation of disulfide reduction during receptor-mediated endocytosis by using FRET imaging. Proc. Natl. Acad. Sci. USA *103*, 13872–13877.
[92] Zlokarnik, G., Negulescu, P. A., Knapp, T. E., Mere, L., Burres, N., Feng, L., Whitney, M., Roemer, K. and Tsien, R. Y. (1998). Quantitation of transcription and clonal selection of single living cells with beta-lactamase as reporter. Science *279*, 84–88.
[93] Matayoshi, E. D., Wang, G. T., Krafft, G. A. and Erickson, J. (1990). Novel fluorogenic substrates for assaying retroviral proteases by resonance energy transfer. Science *247*, 954–958.
[94] Fischer, R., Bachle, D., Fotin-Mleczek, M., Jung, G., Kalbacher, H. and Brock, R. (2006). A targeted protease substrate for a quantitative determination of protease activities in the endolysosomal pathway. Chembiochem *7*, 1428–1434.
[95] Hossain, M. A., Mihara, H. and Ueno, A. (2003). Novel peptides bearing pyrene and coumarin units with or without beta-cyclodextrin in their side chains exhibit intramolecular fluorescence resonance energy transfer. J. Am. Chem. Soc. *125*, 11178–11179.
[96] Washiya, K., Furuike, T., Nakajima, F., Lee, Y. C. and Nishimura, S. I. (2000). Design of fluorogenic substrates for continuous assay of sialyltransferase by resonance energy transfer. Anal. Biochem. *283*, 39–48.
[97] Gong, Y., Luo, Y. M. and Bong, D. (2006). Membrane activation: Selective vesicle fusion via small molecule recognition. J. Am. Chem. Soc. *128*, 14430–14431.
[98] Cottaz, S., Brasme, B. and Driguez, H. (2000). A fluorescence-quenched chitopentaose for the study of endo-chitinases and chitobiosidases. Eur. J. Biochem. *267*, 5593–5600.
[99] Yudushkin, I. A., Schleifenbaum, A., Kinkhabwala, A., Neel, B. G., Schultz, C. and Bastiaens, P. I. H. (2007). Live-cell Imaging of enzyme-substrate interaction reveals spatial regulation of PTP1B. Science *315*, 115–119.
[100] Li, Y. L. and Glazer, A. N. (1999). Design, synthesis, and spectroscopic properties of peptide-bridged fluorescence energy-transfer cassettes. Bioconjug. Chem. *10*, 241–245.
[101] Takakusa, H., Kikuchi, K., Urano, Y., Sakamoto, S., Yamaguchi, K. and Nagano, T. (2002). Design and synthesis of an enzyme-cleavable sensor molecule for phosphodiesterase activity based on fluorescence resonance energy transfer. J. Am. Chem. Soc. *124*, 1653–1657.

[102] Schleifenbaum, A., Stier, G., Gasch, A., Sattler, M. and Schultz, C. (2004). Genetically encoded FRET probe for PKC activity based on pleckstrin. J. Am. Chem. Soc. *126*, 11786–11787.

[103] Zhang, J., Campbell, R. E., Ting, A. Y. and Tsien, R. Y. (2002). Creating new fluorescent probes for cell biology. Nat. Rev. Mol. Cell Biol. *3*, 906–918.

[104] Grunewald, J., Kopp, F., Mahlert, C., Linne, U., Sieber, S. A. and Marahiel, M. A. (2005). Fluorescence resonance energy transfer as a probe of peptide cyclization catalyzed by nonribosomal thioesterase domains. Chem. Biol. *12*, 873–881.

[105] Sahoo, H., Roccatano, D., Zacharias, M. and Nau, W. M. (2006). Distance distributions of short polypeptides recovered by fluorescence resonance energy transfer in the 10 A domain. J. Am. Chem. Soc. *128*, 8118–8119.

[106] Usui, K., Ojima, T., Takahashi, M., Nokihara, K. and Mihara, H. (2004a). Peptide arrays with designed secondary structures for protein characterization using fluorescent fingerprint patterns. Biopolymers *76*, 129–139.

[107] Usui, K., Takahashi, M., Nokihara, K. and Mihara, H. (2004b). Peptide arrays with designed alpha-helical structures for characterization of proteins from FRET fingerprint patterns. Mol. Divers. *8*, 209–218.

[108] Kikuchi, K. (2004). Recent advances in the design of small molecule-based FRET sensors for cell biology. Trends Anal. Chem. *23*, 407–415.

[109] Majeti, R. and Weiss, A. (2001). Regulatory mechanisms for receptor protein tyrosine phosphatases. Chem. Rev. *101*, 2441–2448.

[110] Johnsson, N. and Johnsson, K. (2003). A fusion of disciplines: chemical approaches to exploit fusion proteins for functional genomics. Chembiochem *4*, 803–810.

[111] Blum, G., Mullins, S. R., Keren, K., Fonovic, M., Jedeszko, C., Rice, M. J., Sloane, B. F. and Bogyo, M. (2005). Dynamic imaging of protease activity with fluorescently quenched activity-based probes. Nat. Chem. Biol. *1*, 203–209.

[112] Baruch, A., Jeffery, D. A. and Bogyo, M. (2004). Enzyme activity – it's all about image. Trends Cell Biol. *14*, 29–35.

[113] Thomas, D. A., Francis, P., Smith, C., Ratcliffe, S., Ede, N. J., Kay, C., Wayne, G., Martin, S. L., Moore, K. Amour, A. et al. (2006). A broad-spectrum fluorescence-based peptide library for the rapid identification of protease substrates. Proteomics *6*, 2112–2120.

[114] Turk, B. (2006). Targeting proteases: successes, failures and future prospects. Nat Rev Drug Discov *5*, 785–799.

[115] Puzer, L., Barros, N. M., Oliveira, V., Juliano, M. A., Lu, G., Hassanein, M., Juliano, L., Mason, R. W. and Carmona, A. K. (2005). Defining the substrate specificity of mouse cathepsin P. Arch. Biochem. Biophys. *435*, 190–196.

[116] Gouvea, I. E., Judice, W. A., Cezari, M. H., Juliano, M. A., Juhasz, T., Szeltner, Z., Polgar, L. and Juliano, L. (2006). Kosmotropic salt activation and substrate specificity of poliovirus protease 3C. Biochemistry *45*, 12083–12089.

[117] Warfield, R., Bardelang, P., Saunders, H., Chan, W. C., Penfold, C., James, R. and Thomas, N. R. (2006). Internally quenched peptides for the study of lysostaphin: An antimicrobial protease that kills Staphylococcus aureus. Org. Biomol. Chem. *4*, 3626–3638.

[118] Angelo, P. F., Lima, A. R., Alves, F. M., Blaber, S. I., Scarisbrick, I. A., Blaber, M., Juliano, L. and Juliano, M. A. (2006). Substrate specificity of human kallikrein 6: salt and glycosaminoglycan activation effects. J. Biol. Chem. *281*, 3116–3126.

[119] Korkmaz, B., Attucci, S., Moreau, T., Godat, E., Juliano, L. and Gauthier, F. (2004). Design and use of highly specific substrates of neutrophil elastase and proteinase 3. Am. J. Respir. Cell Mol. Biol. *30*, 801–807.

[120] Lombard, C., Saulnier, J. and Wallach, J. (2005). Assays of matrix metalloproteinases (MMPs) activities: a review. Biochimie *87*, 265–272.

[121] Arbeloa, F. L., Gonzalez, I. L., Ojeda, P. R. and Arbeloa, I. L. (1982). Aggregate formation of rhodamine-6g in aqueous-solution. J. Chem. Soc. Faraday Trans. Ii *78*, 989–994.

[122] Arbeloa, I. L. and Ojeda, P. R. (1982). Dimeric states of rhodamine-B. Chem. Phys. Lett. *87*, 556–560.

[123] Gurtu, V., Kain, S. R. and Zhang, G. (1997). Fluorometric and colorimetric detection of caspase activity associated with apoptosis. Anal. Biochem. *251*, 98–102.

[124] Mizukami, S., Kikuchi, K., Higuchi, T., Urano, Y., Mashima, T., Tsuruo, T. and Nagano, T. (1999). Imaging of caspase-3 activation in HeLa cells stimulated with etoposide using a novel fluorescent probe. FEBS Lett. *453*, 356–360.

[125] Jaffer, F. A., Tung, C. H., Gerszten, R. E. and Weissleder, R. (2002). In vivo imaging of thrombin activity in experimental thrombi with thrombin-sensitive near-infrared molecular probe. Arterioscler. Thromb. Vasc. Biol. *22*, 1929–1935.

[126] Tung, C. H., Gerszten, R. E., Jaffer, F. A. and Weissleder, R. (2002). A novel near-infrared fluorescence sensor for detection of thrombin activation in blood. Chembiochem *3*, 207–211.

[127] Tung, C. H., Ho, N. H., Zeng, Q., Tang, Y., Jaffer, F. A., Reed, G. L. and Weissleder, R. (2003). Novel factor XIII probes for blood coagulation imaging. Chembiochem *4*, 897–899.

[128] Weissleder, R., Tung, C. H., Mahmood, U. and Bogdanov, A., Jr. (1999). In vivo imaging of tumors with protease-activated near-infrared fluorescent probes. Nat. Biotechnol. *17*, 375–378.

[129] Bremer, C., Bredow, S., Mahmood, U., Weissleder, R. and Tung, C. H. (2001). Optical imaging of matrix metalloproteinase-2 activity in tumors: feasibility study in a mouse model. Radiology *221*, 523–529.
[130] Oestvang, J. and Johansen, B. (2006). PhospholipaseA(2): A key regulator of inflammatory signalling and a connector to fibrosis development in atherosclerosis. Biochim. Biophys. Acta – Mol. Cell Biol. Lipids *1761*, 1309–1316.
[131] Burack, W. R., Yuan, Q. and Biltonen, R. L. (1993). Role of lateral phase separation in the modulation of phospholipase A2 activity. Biochemistry *32*, 583–589.
[132] Farber, S. A., Olson, E. S., Clark, J. D. and Halpern, M. E. (1999). Characterization of Ca2+-dependent phospholipase A2 activity during zebrafish embryogenesis. J. Biol. Chem. *274*, 19338–19346.
[133] Hendrickson, H. S., Hendrickson, E. K., Johnson, I. D. and Farber, S. A. (1999). Intramolecularly quenched BODIPY-labeled phospholipid analogs in phospholipase A(2) and platelet-activating factor acetylhydrolase assays and in vivo fluorescence imaging. Anal. Biochem. *276*, 27–35.
[134] Yang, J., Chen, H., Vlahov, I. R., Cheng, J. X. and Low, P. S. (2006a). Evaluation of disulfide reduction during receptor-mediated endocytosis by using FRET imaging. Proc. Natl. Acad. Sci. USA. *103*, 13872–13877.
[135] Yang, Y. Z., Babiak, P. and Reymond, J. L. (2006b). Low background FRET-substrates for lipases and esterases suitable for high-throughput screening under basic (pH 11) conditions. Org. Biomol. Chem. *4*, 1746–1754.
[136] Zaikova, T. (2001). Synthesis of fluorogenic substrates for continuous assay of phosphatidylinositol-specific phospholipase C. Bioconjug. Chem. *12*, 307–313.
[137] Wichmann, O., Wittbrodt, J. and Schultz, C. (2006). A small-molecule FRET probe to monitor phospholipase A2 activity in cells and organisms. Angew. Chem. Int. Ed. *45*, 508–512.
[138] Rose, T. M. and Prestwich, G. D. (2006). Fluorogenic phospholipids as head group-selective reporters of phospholipase A activity. ACS Chem. Biol. *1*, 83–92.
[139] Wichmann, O., Gelb, M. H. and Schultz, C. (2007). Probing Phospholipase A(2) with Fluorescent Phospholipid Substrates. Chembiochem *8*, 1555–1569.
[140] Gao, W., Xing, B., Tsien, R. Y. and Rao, J. (2003). Novel fluorogenic substrates for imaging beta-lactamase gene expression. J. Am. Chem. Soc. *125*, 11146–11147.
[141] Xing, B., Khanamiryan, A. and Rao, J. (2005). Cell-permeable near-infrared fluorogenic substrates for imaging beta-lactamase activity. J. Am. Chem. Soc. *127*, 4158–4159.

[142] Shults, M. D. and Imperiali, B. (2003). Versatile fluorescence probes of protein kinase activity. J. Am. Chem. Soc. *125*, 14248–14249.

[143] Yeh, R. H., Yan, X. W., Cammer, M., Bresnick, A. R. and Lawrence, D. S. (2002). Real time visualization of protein kinase activity in living cells. J. Biol. Chem. *277*, 11527–11532.

[144] Sharma, S., Sharma, S., Singh, R. K. and Vaishampayan, A. (2008). Colonization behavior of bacterium Burkholderia cepacia inside the Oryza sativa roots visualized using green fluorescent protein reporter. World J. Microbiol. Biotechnol. *24*, 1169–1175.

[145] Wang, Q. Z., Dai, Z. H., Cahill, S. M., Blumenstein, M. and Lawrence, D. S. (2006). Light-regulated sampling of protein tyrosine kinase activity. J. Am. Chem. Soc. *128*, 14016–14017.

[146] Ohuchi, Y., Katayama, Y. and Maeda, M. (2001). Fluorescence-based sensing of protein kinase A activity using the dual fluorescent-labeled peptide. Anal. Sci. *17*(Suppl), i1465–i1467.

[147] Berkessel, A. and Riedl, R. (1997). Fluorescence reporters for phosphodiesterase activity. Angew. Chem., Int. Ed. Engl. *36*, 1481–1483.

[148] Caturla, F., Enjo, J., Bernabeu, M. C. and Le Serre, S. (2004). New fluorescent probes for testing combinatorial catalysts with phosphodiesterase and esterase activities. Tetrahedron *60*, 1903–1911.

[149] Matsuoka, K., Nishimura, S. I. and Lee, Y. C. (1995). A bi-fluorescence-labeled substrate for ceramide glycanase based on fluorescence energy-transfer. Carbohydr. Res. *276*, 31–42.

[150] Armand, S., Drouillard, S., Schulein, M., Henrissat, B. and Driguez, H. (1997). A bifunctionalized fluorogenic tetrasaccharide as a substrate to study cellulases. J. Biol. Chem. *272*, 2709–2713.

[151] Payre, N., Cottaz, S. and Driguez, H. (1995). Chemoenzymatic synthesis of a modified pentasaccharide as a specific substrate for a sensitive assay of alpha-amylase by fluorescence quenching. Angew. Chem. Int. Ed. Engl. *34*, 1239–1241.

[152] Komatsu, T., Kikuchi, K., Takakusa, H., Hanaoka, K., Ueno, T., Kamiya, M., Urano, Y. and Nagano, T. (2006). Design and synthesis of an enzyme activity-based labeling molecule with fluorescence spectral change. J. Am. Chem. Soc. *128*, 15946–15947.

[153] Amann, N. and Wagenknecht, H. A. (2002). Preparation of pyrenyl-modified nucleosides via Suzuki-Miyaura cross-coupling reactions. Synlett 687–691.

[154] Mayer, E., Valis, L., Huber, R., Amann, N. and Wagenknecht, H. A. (2003). Preparation of pyrene-modified purine and pyrimidine nucleosides via Suzuki-Miyaura cross-couplings and characterization of their fluorescent properties. Synthesis 2335–2340.

[155] Lee, S. P., Censullo, M. L., Kim, H. G., Knutson, J. R. and Han, M. K. (1995). Characterization of endonucleolytic activity of Hiv-1 integrase using a fluorogenic substrate. Anal. Biochem. *227*, 295–301.
[156] Didenko, V. V. (2001). DNA probes using fluorescence resonance energy transfer (FRET): Designs and applications. Biotechniques *31*, 1106–1116.
[157] Lee, S. P. and Han, M. K. (1997). Fluorescence assays for DNA cleavage. Fluoresc. Spectrosc. *278*, 343–363.
[158] Somerharju, P. (2002). Pyrene-labeled lipids as tools in membrane biophysics and cell biology. Chem. Phys. Lipids *116*, 57–74.
[159] Bisgaier, C. L., Minton, L. L., Essenburg, A. D., White, A. and Homan, R. (1993). Use of fluorescent cholesteryl ester microemulsions in cholesteryl ester transfer protein assays. J. Lipid Res. *34*, 1625–1634.
[160] Van Paridon, P. A., Gadella, J. T. W. J., Somerharju, P. J. and Wirtz, K. W. A. (1987). On the relationship between the dual specificity of the bovine brain phosphatidylinositol transfer protein and membrane phosphatidylinositol levels. Biochim. Biophys. Acta *903*, 68–77.
[161] Nichols, J. W. and Pagano, R. E. (1983). Resonance energy-transfer assay of protein-mediated lipid transfer between vesicles. J. Biol. Chem. *258*, 5368–5371.
[162] Schwarzmann, G., Wendeler, M. and Sandhoff, K. (2005). Synthesis of novel NBD-GM1 and NBD-GM2 for the transfer activity of GM2-activator protein by a FRET-based assay system. Glycobiology *15*, 1302–1311.
[163] Razinkov, V. I., HernandezJimenez, E. I., Mikhalyov, I. I., Cohen, F. S. and Molotkovsky, J. G. (1997). New fluorescent lysolipids: Preparation and selective labeling of inner liposome leaflet. Biochim. Biophys. Acta-Biomembranes *1329*, 149–158.
[164] Mattjus, P., Molotkovsky, J. G., Smaby, J. M. and Brown, R. E. (1999). A fluorescence resonance energy transfer approach for monitoring protein-mediated glycolipid transfer between vesicle membranes. Anal. Biochem. *268*, 297–304.
[165] Rice, K. G., Wu, P. G., Brand, L. and Lee, Y. C. (1991). Interterminal distance and flexibility of a triantennary glycopeptide as measured by resonance energy-transfer. Biochemistry *30*, 6646–6655.
[166] Saikumari, Y. K., Ravindra, G. and Balaram, P. (2006). Structure formation in short designed peptides probed by proteolytic cleavage. Protein Pept. Lett. *13*, 471–476.
[167] Young, S. H., Dong, W. J. and Jacobs, R. R. (2000). Observation of a partially opened triple-helix conformation in 1–>3-beta-glucan by fluorescence resonance energy transfer spectroscopy. J. Biol. Chem. *275*, 11874–11879.

[168] Verbist, J., Gadella, T. W. J., Raeymaekers, L., Wuytack, F., Wirtz, K. W. A. and Casteels, R. (1991). Phosphoinositide–protein Interactions of the plasma-membrane Ca2+-transport ATPase as revealed by fluorescence energy-transfer. Biochim. Biophys. Acta *1063*, 1–6.
[169] Voss, S., Fischer, R., Jung, G., Wiesmuller, K. H. and Brock, R. (2007). A fluorescence-based synthetic LPS sensor. J. Am. Chem. Soc. *129*, 554–561.
[170] White, B. R. and Holcombe, J. A. (2007). Fluorescent peptide sensor for the selective detection of Cu2+. Talanta *71*, 2015–2020.
[171] Okamura, Y. and Watanabe, Y. (2006). Detecting RNA/DNA hybridization using double-labeled donor probes with enhanced fluorescence resonance energy transfer signals. Methods Mol. Biol. *335*, 43–56.
[172] Blagoi, G., Rosenzweig, N. and Rosenzweig, Z. (2005). Design, synthesis, and application of particle-based fluorescence resonance energy transfer sensors for carbohydrates and glycoproteins. Anal. Chem. *77*, 393–399.
[173] Sharma, V., Wang, Q. and Lawrence, D. S. (2008). Peptide-based fluorescent sensors of protein kinase activity: Design and applications. Biochim Biophys. Acta. *1784*, 94–99.
[174] Drexhage, K. H., Erikson, G. R., Hawks, G. H. and Reynolds, G. A. (1975). Water-Soluble Coumarin Dyes for Flashlamp-Pumped Dye Lasers. Opt. Commun. *15*, 399–403.
[175] Seybold, P. G., Gouterman, M. and Callis, J. (1969). Calorimetric, photometric and lifetime determinations of fluorescence yields of fluorescein dyes. Photochem. Photobiol. *9*, 229–242.
[176] Smith, S. N. and Steer, R. P. (2001). The photophysics of Lissamine rhodamine-B sulphonyl chloride in aqueous solution: implications for fluorescent protein-dye conjugates. J. Photochem. Photobiol. A Chem. *139*, 151–156.
[177] Waggoner, A., DeBiasio, R., Conrad, P., Bright, G. R., Ernst, L., Ryan, K., Nederlof, M. and Taylor, D. (1989). Multiple spectral parameter imaging. Methods Cell. Biol. *30*, 449–478.
[178] Kenner, R. A. and Aboderin, A. A. (1971). A new fluorescent probe for protein and nucleoprotein conformation. Binding of 7-(p-Methoxybenzylamino) -4-nitrobenzoxadiazole to bovine trypsinogen and bacterial ribosomes. Biochemistry *10*, 4433–4440.
[179] Albers, A. E., Okreglak, V. S. and Chang, C. J. (2006). A FRET-based approach to ratiometric fluorescence detection of hydrogen peroxide. J. Am. Chem. Soc. *128*, 9640–9641.
[180] Sackett, D. L. and Wolff, J. (1987). Nile Red as a Polarity-Sensitive Fluorescent-Probe of Hydrophobic Protein Surfaces. Anal. Biochem. *167*, 228–234.

Laboratory Techniques in Biochemistry and Molecular Biology, Volume 33
FRET and FLIM Techniques
T. W. J. Gadella (Editor)

CHAPTER 7

FilterFRET: Quantitative imaging of sensitized emission

Kees Jalink[1] and Jacco van Rheenen[2]

[1]*Department of Cell Biology, The Netherlands Cancer Institute, Plesmanlaan 121, 1066 CX Amsterdam, The Netherlands*
[2]*Hubrecht Institute-KNAW and University Medical Center Utrecht, Uppsalalaan 8, 3584CT, Utrecht, The Netherlands*

Previous chapters in this volume were dedicated to advanced imaging techniques such as fluorescence lifetime imaging (FLIM) that require complicated, dedicated, and expensive equipment. Fluorescence resonance energy transfer (FRET) can also be assessed from simple fluorescence images taken with conventional wide-field or confocal microscopes readily available in most research institutes. To this goal, cells expressing donor- and acceptor-tagged proteins ("FRET cells") are imaged, along with control cells expressing either donor- or acceptor alone, under different spectral recording conditions (i.e., with different filter sets). Quantitative FRET images are then calculated by mathematically correcting for filter leak-through terms. Algorithms for "filterFRET" with wide-field microscopes have been reported by several groups but for confocal imaging, separate formalisms had to be developed.

In this chapter, we intend to give the reader an understanding of the possibilities and pitfalls of filterFRET. Following a brief historical overview describing early nonquantitative incarnations of FRET imaging, the theory that allows quantitative wide-field and

DOI: 10.1016/S0075-7535(08)00007-7

confocal FRET imaging will be treated. Special emphasis will be on calibration of the FRET setup and images. Next, major sources of error and noise that have hampered, until recently, application of these algorithms in a very quantitative way will be discussed. Finally, image enhancement strategies, their possible bias on the results and some useful presentation aids will be treated. Taken together, with proper attention for image recording conditions and enhancement strategies, imaging of sensitized emission is a quantitative, fast, photon-efficient, and easy way to determine FRET.

7.1. Introduction

7.1.1. Definition

The term filterFRET here refers to intensity-based methods for calculating fluorescence resonance energy transfer (FRET) from sets of images of the preparation collected at different excitation and/or emission wavelength. The term is not intended to imply that interference filters are actually present in the setup; very similar considerations apply when donor- and acceptor fluorophores are spectrally resolved by other means, such as monochromators or spectral detectors.

7.1.2. Sensitized emission

In previous chapters it was shown that FRET can be reliably detected by donor fluorescence lifetime imaging. Here, we will focus on what is perhaps the most intuitive and straightforward way to record FRET: imaging of sensitized emission (s.e., that is, the amount of acceptor emission that results from energy transferred by the donor through resonance) by filterFRET. While simple in principle, determinations of s.e. are complicated by overlap of excitation and emission spectra of the donors and acceptors, and by several imperfections of the recording optics, light sources and detectors.

7.1.3. The sensitized emission problem

To explain what problems complicate filterFRET, consider the model neuronal cell in Fig. 7.1A. It contains two independently expressed fluorophores: donor molecules at the membrane and in the nucleus, and acceptor molecules at the membrane and around the nucleus. Thus, FRET is only possible at the membrane and sensitized emission should be restricted to the membrane. However, if we collect an s.e. image (**S**) by exciting donors while collecting the fluorescence of acceptors (Fig. 7.1A) the intensity distribution differs from the predicted FRET distribution (compare the s.e. panel, bottom right, to the FRET panel). In particular, **S** also shows some signal from donors in the nucleus and from acceptors in the region around the nucleus. This is due to overlap of the excitation and emission spectra (Fig. 7.1B), which is particularly apparent when genetically encoded fluorescent proteins (FPs) are used as labels. First, overlap of the donor emission spectrum with the acceptor detection channel causes some emission of the donor to appear in the **S** image (leak-through; Fig 7.1C), and second, due to overlap of the acceptor excitation spectrum, some acceptors are directly (inappropriately) excited at the donor excitation wavelength (cross-excitation; Fig 7.1C) and this causes acceptor fluorescence independent from FRET. Thus, the first problem is:

FilterFRET Problem 1: Spectral overlap

S *is a mixture of sensitized emission, donor- and acceptor fluorescence.*

Even if we forget, for a moment, the overlap problem and assume that we obtained a "pure" sensitized emission image, interpretation of this image is still ambiguous. That is because first, the intensity of **S** varies linearly with the excitation intensity and with the detector sensitivity. The exact same preparation will, when measured on a different microscope, yield different s.e. intensities. In fact, as much as renewing the arc lamp would impede comparison of results obtained on the same microscope. Second, the interpretation

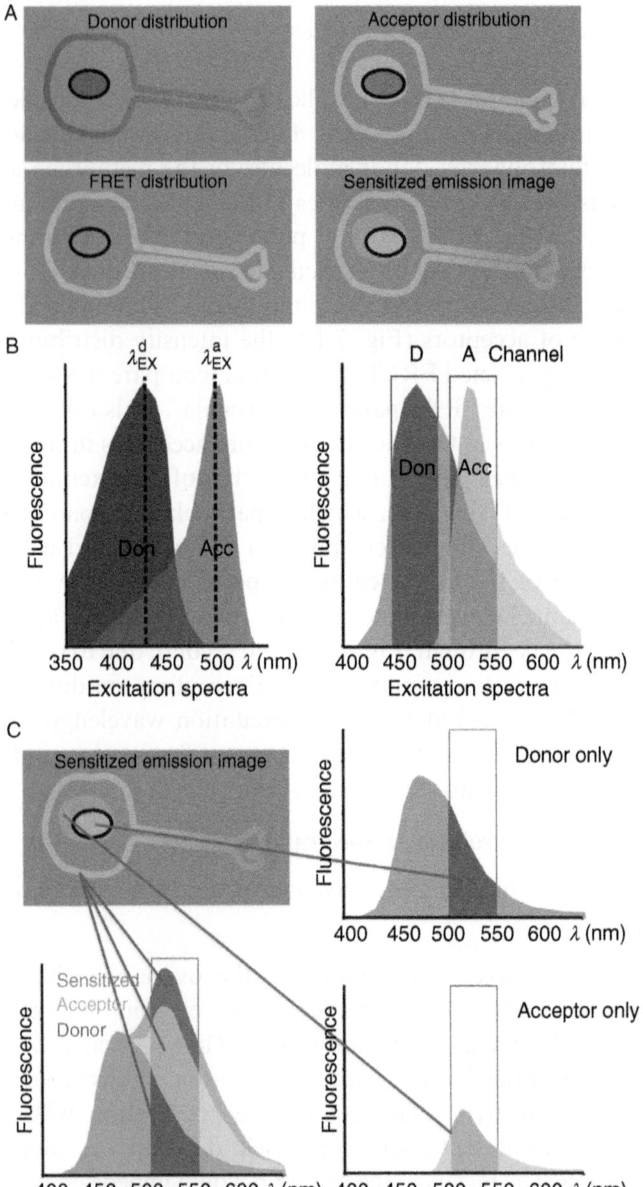

Fig. 7.1. (*Continued*)

of local differences in s.e. within a single image is also ambiguous. For example, the diminished s.e. in the axon of our model neuron (Fig. 7.1C) may either stem from less efficient interaction between donors and acceptors or from locally decreased abundance of donors and/or acceptors. As one often is interested in the distribution of the *degree* of protein interaction, the s.e. image must thus be normalized by relating it to the local concentration of the donor, the acceptor, or both. Thus:

FilterFRET Problem 2: Normalization

S *depends on FRET efficiency but also on fluorophore concentration, donor excitation, and detector sensitivity.*

To perform such normalization, images of the distribution of the donors and acceptors are needed. For the donor image (**D**), donor emission is collected while exciting at donor wavelength, and for the acceptor image (**A**), the acceptor fluorescence is imaged while exciting at acceptor wavelength. Unfortunately, these images also contain components of spectral overlap. In addition, **D** also contains a (negative) component of FRET because inevitably some donor emission will be lost due to resonance. This creates a recurrent problem: we cannot calculate s.e. because we do not know the

---

Fig. 7.1. *Detecting sensitized emission.* (A) Neuronal cell expressing donors (blue) at the membrane and in the nucleus, and acceptors (green) at the membrane and around the nucleus. Lower abundance of the donor in the axon is also depicted. FRET can only occur at the membrane of this cell (lower left panel). (B) Normalized excitation spectra of donors (CFP) and acceptors (YFP). Indicated are the donor- and acceptor excitation lines (left panel) and the two detection channels (bandpass filters). (C) Appearance of signals in the s.e. image. Whereas FRET is restricted to the membrane, due to leak-through of donor signal in the s.e. channel (e.g., in the nucleus; rightmost spectrum) and false excitation of acceptors (e.g., around the nucleus; lower right spectrum and figure B) additional signals are apparent. Note that leak-through and cross-excitation are not restricted to areas stained with either donors or acceptors alone (lower left panel). (See Color Insert.)

donor distribution, and the reason that we do not know this is that we cannot correct it for the unknown s.e. Similarly, the acceptor levels may not be directly derived from **A** if this image also contains some emission of donor molecules that have been cross-excited at the acceptor excitation wavelength and that bleed into the acceptor channel. It would appear that this latter term is always very small, but as we will see later on in this chapter its contribution may become significant when acceptors are present very sparsely and the sensitivity of the acceptor channel is increased to cope with that fact. Thus:

FilterFRET Problem 3: Reference donor- and acceptor images

Relative donor and acceptor levels cannot be determined directly from **D** *and* **A** *images.*

Once proper corrections have been applied to donor, acceptor, and s.e. images (i.e., Problems 1 and 3 have been tackled), we can proceed with Problem 2 and normalize the data. Here, a final important issue is raised: that of absolute quantification. Let us define here the apparent donor FRET efficiency E_D as the fraction of energy quanta absorbed by all donors (whether in complex with an acceptor or not) in a given pixel that is transferred to acceptors (note the difference with FRET efficiency E as defined in other chapters, which is the chance that excitation of the donor in a *donor–acceptor complex* leads to transfer to the acceptor). By definition, both E_D and E should be corrected, normalized, and quantitative measures for interactions. However, the quantitative E_D cannot be simply obtained by dividing corrected **S** and **D** images because **S** and **D** are not to the same scale. That is, even if **S** and **D** share the same excitation settings (which cancels out excitation changes as a source of variance), **D** and **S** are obtained with different "sensitivity" because filter settings, detector gain, and quantum yields of fluorophores are not the same. To be able to directly compare results obtained in different labs and with different setups, we thus have to find a scaling parameter that relates the sensitivity of the setup for donors to that for acceptors.

FilterFRET Problem 4: Getting it quantitative

Corrected s.e. and donor images are not to the same scale because they have been recorded under different conditions.

Luckily, a mathematical framework to solve these problems has been worked out by several groups [1–6] who showed that from just three acquired images **S**, **D**, and **A** quantitative FRET efficiency images can be calculated. This framework relies on calibrations taken from cells expressing either donors only or acceptors only and it allows direct comparison of results obtained around the world.

Before embarking on a detailed treatment of filterFRET, for completion we will briefly treat earlier nonquantitative FRET imaging methods that rely on calculating the ratio of **S** and either **D** or **A**. While not quantitative, ratio imaging is still widely in use because it is very simple and, depending on the biological application, it often gives enough information to provide answers.

7.2. Two-channel ratio imaging

7.2.1. Emission ratio

Emission ratio imaging is extremely popular due to its simplicity and speed. In essence, cells expressing donors and acceptors are illuminated at the donor wavelength and fluorescence intensity data are collected both at donor (**D**) and at acceptor (**S**) channels. Collected data may be either images, or, in case high acquisition speed is crucial and spatial information is not required, dual-channel photometer readings (see Textbox 1). **S** and **D** are not overlap-corrected and "FRET" is simply expressed as the ratio of intensities[1] as: ratio = **S/D**.

[1]Variations on the simple ratio are also encountered in literature, for example, S/(S + D).

Emission ratioing yields some form of normalization (provided that the FRET efficiency is small) and it nicely cancels out light source intensity fluctuations. It does not, however, provide sufficient data to calculate FRET quantitatively in most cases. Nevertheless, there are cases where quantitative FRET data are not needed to still be able to draw biological conclusions. For example, to study agonist-induced changes in FRET, emission ratio data from time-lapse series provide good information on the time course and localization of the induced FRET changes, and a reasonable impression about the magnitude. In addition, a better quantitative feel for the data can be obtained if endpoint calibrations are applicable, for example, if FRET can be experimentally maximized (see Textbox 1) [7–9].

Ratio imaging is particularly suited for single-polypeptide FRET sensors. In these constructs FRET changes are due to altered distance and/or orientation of the donor and acceptor, and since the fluorophores are tethered their stoichiometry is always fixed. Thus, the filterFRET problems are easier to address and, assuming full maturation of both FPs [4], it can in fact be shown that under these circumstances two images suffice to calculate FRET quantitatively (see Textbox 1 and Appendix 7.A.6).

7.2.2. Excitation ratio

In principle, similar information can be obtained from excitation ratioing where acceptor images are acquired at both donor (**S**) and acceptor (**A**) excitation wavelength, and FRET is apparent from ratio = **S/A**.

Because of the double exposure, the preparation suffers from increased bleaching and photodamage. Furthermore, split-imaging on charge-coupled-device (CCD) systems (see Textbox 1) is not an option. Nevertheless, excitation ratioing may be an economic choice for laboratories that have an old Fura-imaging setup. These microscopes often allow very fast excitation switching

Textbox 1. **Ratio imaging**

Emission ratio imaging involves collection of **S** and **D** images from the preparation. On wide-field microscopes, **D** and **S** can be sequentially acquired by emission filter switching, for example, using a filter wheel. This requires two consecutive exposures, causing unnecessary photobleaching and raising the risk of errors due to cell movements (Fig. 7.T1). Therefore it is better to collect the images simultaneously, for example, using a commercially available image-splitting device (Fig. 7.T1B) that projects the channels on two halves of the same CCD camera chip. Note, however, that these devices require precise calibration to ensure perfect co-registration of the images. On point-scanning confocal microscopes, simultaneous acquisition of **D** and **S** is also possible and the two images will usually overlap quite well.

Ratio imaging nicely cancels out some of the main complications in the interpretation of wide-field images in that it normalizes fluorescence intensity differences caused by for example, cell height (Fig. 7.T1) as well as possible slow drift in excitation intensity. Light sources invariably are much less stable than detectors. Incidentally, for these reasons emission ratio imaging has been applied for over 3 decades by the Ca^{2+} imaging community.

In Fig. 7.T1C, a special case of emission ratioing is shown. Rather than forming an image, the objective is used to project the emission on a beamsplitter/dual-photometer assembly that simply records the total emission in **S** and **D** channels. Pooling of all the emission photons allows for dimming of the excitation intensity by several orders of magnitude, effectively eliminating photodamage. Whereas spatial resolution is given up, this setup is ideally suited for fast kinetic experiments because it can easily be tuned for sub-ms temporal resolution.

(*Continued*)

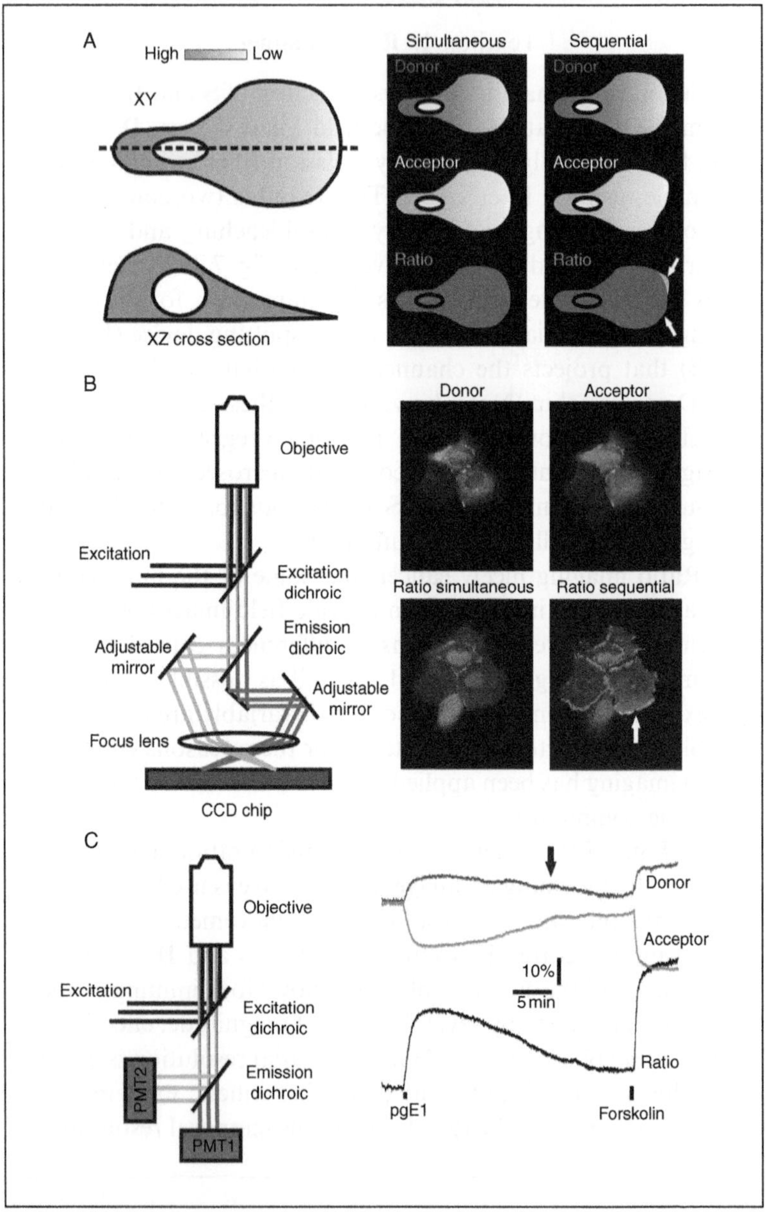
A
High
Low
XY
XZ cross section
Simultaneous
Sequential
Donor
Acceptor
Ratio
B
Objective
Excitation
Excitation dichroic
Emission dichroic
Adjustable mirror
Adjustable mirror
Focus lens
CCD chip
Donor
Acceptor
Ratio simultaneous
Ratio sequential
C
Objective
Excitation
Excitation dichroic
Emission dichroic
PMT2
PMT1
Donor
Acceptor
10%
5 min
Ratio
pgE1
Forskolin

The traces in Fig. 7.T1C show a typical result and also illustrate the use of FRET endpoint calibrations by manipulating the preparation to a state of maximal FRET.

In general, ratio imaging is not quantitative nor is it, strictly spoken, normalized because the acquired data do not permit Problems 1–4 (see Sect. 7.1.1) to be properly addressed. One important exception is the case where donors and acceptors are present at a fixed stoichiometry. Examples of that are the popular single-polypeptide FRET sensors. In this case, the normalization problem (2) is inherently solved and the overlap- and reference-image problems (1 and 3) simplify considerably. It can be shown [1 and Appendix 7.A.6] that in that case FRET efficiency (*E*) can be calculated from **D** and **S** images.

Fig. 7.T1. *Emission ratio imaging of the cAMP FRET sensor CFP-Epac-YFP* [8]. (A) Ratio imaging largely corrects intensity differences that are due to cell morphology. However, if channels are collected consecutively, any shape changes of the cells cause errors (arrows) in the ratio image that may easily be mistaken for FRET differences. (B). Simultaneous image collection. **D** and **S** are projected side by side on a CCD chip using a commercially available image splitting device fitted to a widefield epifluorescence microscope. The ratio image (lower left photomicrograph) is calculated using Image J; no attempt was made to correct for bleed-through. The lower right ratio image shows errors due to cell movement (images taken 5 s apart) (C) Emission ratioing using a beam splitter and dual photometer setup. Traces represent **D** (blue), **S** (green), and the ratio (black). Since the Epac sensor looses FRET upon binding to cAMP, in this case the ratio was calculated as **D/S** to have upward ratio changes correspond with increased [cAMP]. Note that excitation fluctuations (arrow) disappear in the ratio. All data courtesy of B. Ponsioen. (See Color Insert.)

and may be adapted for FRET by a mere filter change. Furthermore, unlike the emission ratio, the ratio **S/A** changes linearly with FRET.

7.3. Three-channel measurements: FilterFRET

In this section, FRET will be calculated from sets of three separately acquired images that are chosen to optimally detect s.e. (**S**), donor emission (**D**), and acceptor emission (**A**). Because both the excitation and emission spectra of donor and acceptor overlap extensively, **S**, **D,** and **A** likely also contain leak-through components that have to be subtracted (Fig. 7.2). Just how much leak-through needs to be subtracted depends on calibration values that must be calculated from measurements obtained with special reference samples. Note that in essence our treatment follows and builds on the work of several authors [1–3, 5, 6, 10–12]. Most of the published correction schemes have been worked out for images acquired with a wide-field fluorescence microscope equipped with digital camera [1, 2, 5, 6, 10]; however, a variety of mathematical terminologies has been used. For confocal images, we showed that correction is distinctly more complex because the sensitivities of the detection channels are varied independently [3]. We here present a generalized mathematical framework, with an attempt to arrive at a compromise in terminology.

7.3.1. Sensitized emission

In the following treatment it is assumed that detector gain and offset are correctly adjusted, and that autofluorescence of cells is either negligible or properly subtracted. In addition, it is assumed that the images are shading-corrected; see Sect. 7.4.4. For details on shading techniques, the reader is referred to Nagy et al. and Tomazevic et al. [10, 13]. Provided that independent measurements on the magnitude of cross talk terms can be made, straightforward corrections can be carried out from three acquired images:

- Donor excitation with donor emission, **D**
- Donor excitation with acceptor emission, **S**
- Acceptor excitation with acceptor emission, **A**

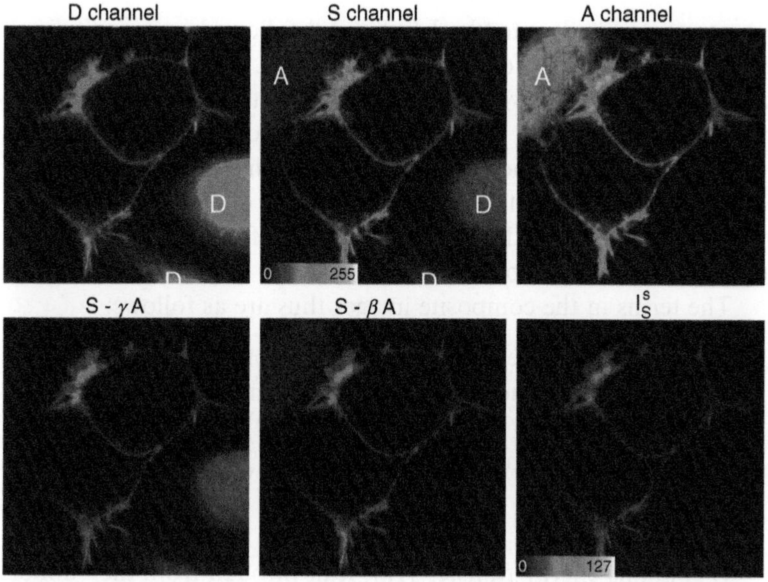

Fig. 7.2. *Sensitized emission calculated from confocal images.* Cells expressing CFP- and YFP-tagged Pleckstrin homology (PH) domains were seeded together with control cells expressing either CFP (marked "D") or YFP (marked "B"). Top row shows raw input files and illustrates donor leak-through (middle panel, "D") and cross-excitation (middle panel, "A"). In the bottom row, **S** images are corrected for cross-excited YFP (left), for CFP leak-through (middle) or according to Eq. (7.8) (right panel). The contrast of the I_S^s panel is stretched twofold as indicated. All images in this chapter are collected with Leica TCS SP2 or SP5 confocal microscopes except for Fig. 7.T1, which was acquired with a Leica ASMDW wide-field epifluorescence microscope equipped with dual-view adapter (Optical Insights). Image acquisition and specimen refocusing were automated from within a custom-made Visual Basic (v6.0) program by calling commands from the Leica macro tool package. ROIs were manually assigned to cells expressing only CFP or YFP and from these, correction factors were measured and calculated. Using these factors, sensitized emission was calculated as outlined in the text. (See Color Insert.)

The acquired images are composite images that consist of fluorescence stemming from different molecular species: donors, acceptors, or FRET pairs (Figs. 7.1 and 7.2). These fluorescent components are denoted by I (intensity) followed by a capitalized subscript ($_D$, $_A$, or $_S$, for respectively donors, acceptors, or donor/acceptor FRET pairs) to indicate the particular population of molecules responsible for emission of I, and a lower-case superscript (^{d}, ^{a}, or ^{s}) that indicates the detection channel (or filter cube). For example, I_D^d denotes the intensity of the donors as detected in the donor channel and reads as "Intensity of donors in the donor channel," etc. (see Table 7.1).

The terms in the composite images thus are as follows:

– **D** is the sum of the remaining donor fluorescence in the donor channel (I_{D-S}^d), and of leak-through components of sensitized emission back into the donor channel (I_S^d) and of cross-excited acceptors back into the donor channel (I_A^d).

$$\mathbf{D} = I_{D-S}^d + I_S^d + I_A^d \tag{7.1}$$

– **S** contains energy transfer (I_S^s), leak-through from the "donor minus FRET" population (I_{D-S}^s) and emission from cross-excited acceptors (I_A^s).

$$\mathbf{S} = I_{D-S}^s + I_S^s + I_A^s \tag{7.2}$$

– Finally, **A** contains acceptor fluorescence (I_A^a) and two usually very minor leak-through components: that of the (partially quenched) donor population inappropriately excited at acceptor wavelength and leaking into the acceptor channel (I_{D-S}^a), and the small amount of sensitized emission that stems from FRET after inappropriate excitation of donors at acceptor wavelength (I_S^a).

$$\mathbf{A} = I_{D-S}^a + I_S^a + I_A^a \tag{7.3}$$

In the majority of cases, the two cross-terms in this equation can be ignored (i.e., are $\ll 0.01$) and Eq. (7.3) simplifies to:

$$\mathbf{A} = I_A^a \tag{7.3'}$$

TABLE 7.1
Glossary of terms

Symbol	Excitation	Emission	Fluorophore	Indicates
D	Don	Don		raw donor image collected at λ_{ex}^{d} with donor emission filter (donor channel)
S	Don	s.e.		raw sensitized emission image collected at λ_{ex}^{d} with the sensitized emission filter (s.e. channel)
A	Acc	Acc		raw acceptor image collected at λ_{ex}^{a} with the acceptor emission filter (acceptor channel)
I_D^d	Don	Don	Don	unquenched donor signal in the donor channel
I_S^d	Don	Don	s.e.	(acceptor) s.e. leaking through in the donor channel
I_A^d	Don	Don	Acc	cross-excited (at λ_{ex}^{d}) acceptor emission leaking through in the donor channel
I_D^s	Don(S)[1]	s.e.[2]	Don	leak-through of unquenched donor excited at λ_{ex}^{d} in s.e. channel
I_S^s	Don(S)	s.e.	s.e.	(acceptor) s.e. signal detected in the s.e. channel
I_A^s	Don(S)	s.e.	Acc	emission of acceptors cross-excited at λ_{ex}^{d} in the s.e. channel
I_D^a	Acc	Acc	Don	cross-excited (at λ_{ex}^{a}) signal of unquenched donors leaking through in the acceptor channel
I_S^a	Acc	Acc	s.e.	(acceptor) s.e. in acceptor channel at λ_{ex}^{a}; this signal derives from the small population of cross-excited donors that leads to FRET

(*Continued*)

TABLE 7.1 (*Continued*)

Symbol	Excitation	Emission	Fluorophore	Indicates
I_A^{a}	Acc	Acc	Acc	(directly excited) acceptor signal in the acceptor channel
I_{D-S}^{d}	Don	Don	Don[3]	(partly) quenched donor signal in the donor channel
I_{D-S}^{s}	Don(S)	s.e.	Don[3]	leak-through of the (partly) quenched donor signal in the s.e. channel
I_{D-S}^{a}	Acc	Acc	Don[3]	leak-through of the cross-excited, (partly) quenched donor signal in the s.e. channel

The fluorescent components are denoted by I (intensity) followed by a capitalized subscript ($_{D}$, $_{A}$, or $_{S}$, for respectively Donors, Acceptors, or s.e.) to indicate the particular population of molecules responsible for emission and a lower-case superscript (^{d}, ^{a}, or ^{s}) that indicates the detection channel (or filter cube). For example, I_D^{d} denotes the intensity of the donors as detected in the donor channel and reads as "Intensity of donors in the donor channel," etc. Notes: (1) The excitation in the s.e. channel is generally set up to be equal to that in the donor channel. In case a separate filter cube is used, slight differences may occur, which is denoted by Don(S). See the text and appendix for further details. (2) The s.e. emission filter is usually the same as the acceptor emission filter in confocal determinations. We here designate a different filter to accommodate those wide-field/digital camera experiments that employ different filters for A and S. (3) Here the notation D–S indicates the residual (quenched) donor fluorescence in the presence of the acceptor. In the other chapters this is indicated as DA. Hence: $I_{D-S}^{d} = I_{DA}^{d}$; $I_{D-S}^{s} = I_{DA}^{s}$; and $I_{D-S}^{a} = I_{DA}^{a}$.

For those cases where cross-terms cannot be ignored, we derive an expression in Sect. 7.A.5.

Now, note that each of the leak-through terms in Eqs. (7.1) and (7.2) is just a fixed fraction of the intensity in its "own" channel:

$$I_A^{d} = \alpha I_A^{a} \ (= \alpha A) \tag{7.4}$$

$$I_{D-S}^{s} = \beta I_{D-S}^{d} \tag{7.5}$$

$$I_A^{\mathrm{s}} = \gamma I_A^{\mathrm{a}} \; (= \gamma A) \tag{7.6}$$

$$I_S^{\mathrm{d}} = \delta I_S^{\mathrm{s}} \tag{7.7}$$

where α is the ratio of pure acceptor fluorescence detected using donor/acceptor filters, β is the leak-through of pure donor fluorescence in the acceptor (s.e.) channel, δ that for leak-through of sensitized emission back into the donor channel, and γ relates acceptor fluorescence excited at donor wavelength and detected in the s.e. channel to the acceptor fluorescence excited at acceptor wavelength and detected in the acceptor channel. Further, it can be easily shown that $\alpha = \gamma\delta$ (see Textbox 2 and Appendix).

We can thus rewrite Eqs. (7.1) and (7.2) to:

$$\mathbf{D} = I_{D-S}^{\mathrm{d}} + \delta I_S^{\mathrm{s}} + \gamma\delta \mathbf{A} \tag{7.1B}$$

$$\mathbf{S} = \beta I_{D-S}^{\mathrm{d}} + I_S^{\mathrm{s}} + \gamma \mathbf{A} \tag{7.2B}$$

which rearranges to [3]:

$$I_S^{\mathrm{s}} = \frac{\mathbf{S} - \beta\mathbf{D} - \gamma(1 - \beta\delta)\mathbf{A}}{1 - \beta\delta} \tag{7.8}$$

Sensitized emission $\left(I_S^{\mathrm{s}}\right)$, as defined in Eq. (7.8), reliably measures the *relative* amount of energy transfer occurring in each pixel (Fig. 7.2, lower right panel). I_S^{s} is corrected for spectral overlap (i.e., Problem 1 has been taken care of); however, unlike E, it is not a normalized measure for interaction nor is it quantitative in absolute terms. It depends on the specific biological question which of the two yields the most relevant information.

7.3.2. *FRET efficiency*

Having obtained the s.e. image I_S^{s}, which provides a spatial map of molecular interactions in the cell, the next steps are normalization and absolute quantification of the interactions. Normalization can

Textbox 2. **The leak-through parameters**

The correction factors α, β, γ, and δ must be determined independently. From Eqs. (7.4) to (7.7), it is clear that estimates for α, γ, and δ can be obtained by imaging a sample with only acceptor molecules and calculating:

$\alpha = \mathbf{D}/\mathbf{A}$ leak-through of cross-excited acceptors back into the donor channel
$\gamma = \mathbf{S}/\mathbf{A}$ cross excitation of acceptors
$\delta = \mathbf{D}/\mathbf{S}$ leak-through of s.e. back into the donor channel

Similarly, β is estimated from a sample with only donor molecules:

$\beta = \mathbf{S}/\mathbf{D}$ leak-through of donors into the s.e. channel

Note that parameters β and δ depend on signal amplifications in the utilized detectors and on the elements in the optical path (optical filter, spectral detection bands) only, while α and γ are additionally influenced by relative excitation intensity. This is usually a fixed constant in wide-field microscopy but in confocal imaging laser line intensities are adjusted independently. Furthermore, note that the α factor equals δ multiplied by γ (see Appendix for further detail).

be carried out through division of the s.e. image by a pure donor image to arrive at the "*apparent* FRET efficiency" (not to be confused with the quantitative E_D):

$$^{\mathrm{app}}\mathrm{FRET}_D = I_S^{\mathrm{s}}/I_D^{\mathrm{d}} \tag{7.9}$$

where the denominator represents the total donor fluorescence as it would appear in the absence of FRET. That is, the pure donor image in the denominator has to be corrected for leak-through components and for loss of donor emission due to FRET (Problem 3) [1, 3, 5, 6].

Sometimes normalization to the acceptor image is encountered:

$$^{\mathrm{app}}\mathrm{FRET}_A = I^{\mathrm{s}}_S / I^{\mathrm{a}}_A \tag{7.10}$$

This is easier because the acceptor image is readily available (Eq. (7.3′)) without further corrections. $^{\mathrm{app}}\mathrm{FRET}_A$ and $^{\mathrm{app}}\mathrm{FRET}_D$ are not quantitative in that nominator and denominator images are acquired in different channels. Thus, their relative magnitude depends on filter settings, detector gain, on donor and acceptor quantum yield and, in the case of $^{\mathrm{app}}\mathrm{FRET}_A$, also on relative excitation intensities. While neither $^{\mathrm{app}}\mathrm{FRET}_A$ nor $^{\mathrm{app}}\mathrm{FRET}_D$ solve the problem of quantification, both are frequently encountered in the literature and both do allow quantitative comparison of interactions *within the same image*.

In order to obtain the desired quantitative measure of FRET (Fig. 7.3), an additional correction factor must scale the nominator to the denominator in Eq. (7.9) [1–3, 6]. In other words, we must relate the FRET-induced sensitized emission in the **S** channel to the loss of donor emission in the **D** channel as in:

$$E_D = \frac{\text{Loss in donor emission due to FRET}}{\text{Total donor emission in the absence of FRET}} \tag{7.11}$$

or in our terminology:

$$E_D = \frac{I^{\mathrm{d}}_D - I^{\mathrm{d}}_{D-S}}{I^{\mathrm{d}}_D} = 1 - \frac{I^{\mathrm{d}}_{D-S}}{I^{\mathrm{d}}_D} \left(= 1 - \frac{I_{DA}}{I_D} \right) \tag{7.11A}$$

Note that the "*Loss in donor emission due to FRET*" (Eq. (7.11)) is just a constant times the "*sensitized emission*" (Eq. (7.8)) for given acquisition settings, or $I^{\mathrm{d}}_{\mathrm{Loss}} = \phi I^{\mathrm{s}}_S$. Thus (noting that both $I^{\mathrm{d}}_{\mathrm{Loss}}$ and ϕ have negative values):

$$I^{\mathrm{d}}_{D-S} = I^{\mathrm{d}}_D + I^{\mathrm{d}}_{\mathrm{Loss}} = I^{\mathrm{d}}_D + \phi I^{\mathrm{s}}_S \tag{7.12}$$

Now it is straightforward to solve for I^{d}_D and I^{d}_{D-S}. Substituting Eq. (7.12) in Eq. (7.2B), **S** and **D** become:

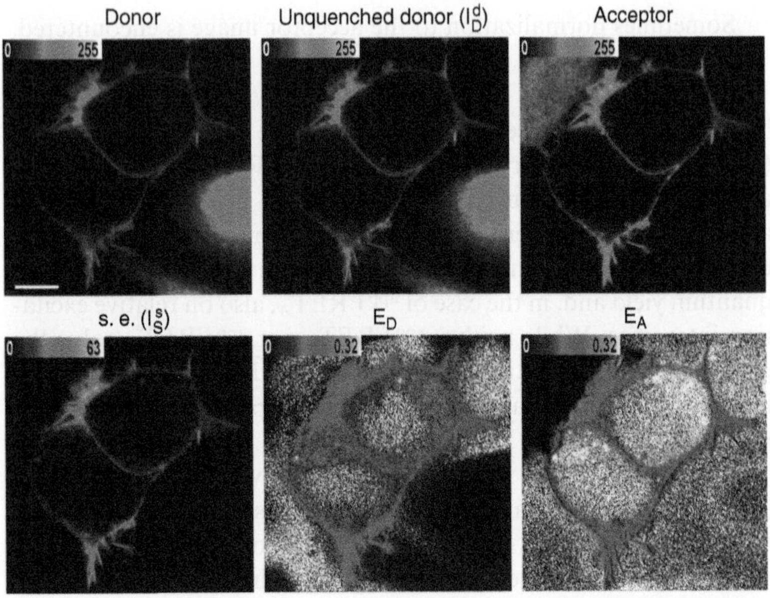

Fig. 7.3. Fret efficiency. The unquenched donor image (top row, middle panel), as calculated according to Eq. (7.13), and the acceptor image (top right panel) are used to normalize the s.e. image. The resulting images E_D and E_A (Eqs. (7.10) and (7.11), respectively) are quantitative, as detailed in the text. Unfiltered, raw data are shown. Scale bar is 12 μm. (See Color Insert.)

$$\mathbf{D} = I_D^{\mathrm{d}} + (\phi + \delta)I_S^{\mathrm{s}} + \gamma\delta\mathbf{A} \tag{7.1C}$$

$$\mathbf{S} = \beta I_D^d + (\beta\phi + 1)I_S^s + \gamma\mathbf{A} \tag{7.2C}$$

Combining Eqs. (7.8) and (7.1C):

$$I_D^{\mathrm{d}} = \frac{\beta\phi + 1}{1 - \beta\delta}\mathbf{D} - \frac{\phi + \delta}{1 - \beta\delta}S + \gamma\phi\mathbf{A} \tag{7.13}$$

defining

$$\zeta = \frac{\beta(\phi + \delta)}{1 - \beta\delta} \tag{7.14}$$

then:

$$I_D^{\mathrm{d}} = (\zeta + 1)\mathbf{D} - \frac{\zeta}{\beta}\mathbf{S} + \gamma\left(\delta - \frac{\zeta}{\beta} + \delta\zeta\right)\mathbf{A} \tag{7.13B}$$

Note that this equation is identical to the expression for unquenched donor fluorescence of van Rheenen et al. (Eq. (A17)).

And for I_{D-S}^{d} we can derive the following expression after combining Eqs. (7.1B) and (7.2B):

$$I_{D-S}^{\mathrm{d}} = \frac{\mathbf{D} - \delta\mathbf{S}}{1 - \beta\delta} \tag{7.15}$$

Thus, we now have the results to express E_D as:

$$E_D = 1 - \frac{I_{D-S}^{\mathrm{d}}}{I_D^{\mathrm{d}}} = 1 - \frac{\mathbf{D} - \delta\mathbf{S}}{(\beta\phi + 1)\mathbf{D} - (\delta + \phi)\mathbf{S} + \phi(1 - \alpha\beta)\mathbf{A}} \tag{7.16}$$

7.3.3. Making it quantitative

One final step is needed to wrap things up: the factor ϕ (which relates the s.e. in **S** to loss of donor fluorescence in **D**) that was introduced to solve for I_D^{d} must be determined. Note that for a given combination of filter settings and fluorophores ϕ is a constant, independent from expression levels and excitation intensity. For the popular cyan fluorescent protein (CFP)/yellow fluorescent protein (YFP) FRET pair, we have mostly used a very intuitive approach that employs Yellow Cameleon, the well-known single-polypeptide intracellular Ca^{2+} sensor. This construct shows significant FRET change upon raising intracellular Ca^{2+} concentration with, for example, ionomycin [7]. Recording **D** and **S** before and after ionomycin-induced Ca^{2+} saturation of Yellow Cameleon gives paired observations for Eqs. (7.1C) and (7.2C) that differ only in FRET efficiency. The increase in **S** is:

$$\frac{\begin{array}{l}\mathbf{S}(\text{post}) = \beta I_D^{\text{d}} + (\beta\phi + 1)I_S^{\text{s}}(\text{post}) + \gamma\mathbf{A} \\ \mathbf{S}(\text{pre}) \;\; = \beta I_D^{\text{d}} + (\beta\phi + 1)I_S^{\text{s}}(\text{pre}) \;\; + \gamma\mathbf{A}\end{array}}{\mathbf{S}(\text{post}) - \mathbf{S}(\text{pre}) = (\beta\phi + 1)\left(I_S^{\text{s}}(\text{post}) - I_S^{\text{s}}(\text{pre})\right)} \text{subtract} \quad (7.17)$$

And for the change in D:

$$\frac{\begin{array}{l}\mathbf{D}(\text{pre}) \;\; = I_D^{\text{d}} + (\delta + \phi)I_S^{\text{s}}(\text{pre}) \;\; + \gamma\delta\mathbf{A} \\ \mathbf{D}(\text{post}) = I_D^{\text{d}} + (\delta + \phi)I_S^{\text{s}}(\text{post}) + \gamma\delta\mathbf{A}\end{array}}{\mathbf{D}(\text{pre}) - \mathbf{D}(\text{post}) = -(\delta + \phi)\left(I_S^{\text{s}}(\text{post}) - I_S^{\text{s}}(\text{pre})\right)} \text{subtract} \quad (7.18)$$

Dividing Eq. (7.17) by Eq. (7.18) gives:

$$\frac{\mathbf{S}(\text{post}) - \mathbf{S}(\text{pre})}{\mathbf{D}(\text{pre}) - \mathbf{D}(\text{post})} = -\frac{\beta\phi + 1}{\phi + \delta} \equiv G \quad (7.19)$$

or[2]

$$\phi = -\frac{1 + \delta G}{\beta + G} \quad (7.20)$$

Note that G as derived here relates the FRET-induced sensitized emission in the **S** channel to the loss of donor emission in the **D** channel and that it is identical to the correction factor γ/ξ [2] or G [6, 14]. Note however, that if the correction factors β or δ change, G and ϕ change as well. In contrast, our correction factor ζ [3] is a constant that depends only on fluorophore properties and filter settings, and therefore it does not change with excitation intensity or detector gain. This is a clear advantage for confocal filterFRET. ζ (Eq. (7.14)) and G (Eq. (7.19)) are related as:

$$G = -\beta\left(1 + \frac{1}{\zeta}\right) \quad (7.21)$$

[2]Note that we have slightly rearranged the math from van Rheenen et al, BJ 2004 to adopt the correction factor G that is used in several publications.

and

$$\zeta = -\frac{\beta}{\beta + G} \tag{7.22}$$

Of course any of the many cytosolic constructs that can be forced to change FRET are useful for calibration, as long as the fluorophores are the same as those used in the experiments. We have for example also successfully used caspase-cleavable GFP-mRFP chimera (unpublished results).

Several other approaches to solve the quantitation problem have been proposed. Hoppe et al. [2] determined γ/ξ by calibrating it against constructs with known FRET efficiency. We and others [3, 6] have used data from a cell before and after acceptor photobleaching to relate the FRET-induced sensitized emission in the **S** channel to the loss of donor emission in the **D** channel by factors termed ζ or G, respectively. For the CFP/YFP pair this works very well on confocal microscopes with a 514-nm Argon ion laser line, but on wide-field systems, selective acceptor photobleaching reportedly causes problems [14]. Finally, G can also be determined by comparison of several constructs that differ in FRET efficiency, a bit analogous to the Yellow Cameleon calibration described above [10, 14].

7.3.4. Stoichiometry

The FRET efficiency E_D as determined above is the fraction of energy quanta absorbed by all donor molecules that is transferred to acceptors. For a given pixel, E_D effectively reflects both the efficiency with which paired donor–acceptors transfer energy (E) and the fraction of molecules in that pixel that pair up (f_D). This means, for example, that a pixel with $E_D = 0.2$ may result from 100% of donors having $E = 0.2$, or from 20% of donors having $E = 1$, or anything in between. The FRET efficiency E of a donor/acceptor pair (termed characteristic FRET efficiency, E_c in some literature [2, 3]) is most often unknown.

However, if a good estimate of E can be made, the fraction of donors in complex can be readily calculated as

$$f_D = \frac{E_D}{E} \tag{7.23}$$

The cases where reliable determination of E is possible are those where good assumptions can be made based on known fluorophore dipole alignment and distance and those where the donors and acceptors can be induced to quantitatively engage in interactions (for example, by determining FRET in an (in vitro) preparation of a 1:1 donor–acceptor mix, or by inducing maximal interaction in a single-polypeptide FRET sensor such as Yellow Cameleon or CFP-Epac-YFP, see Figure in Textbox 1). In vivo, usually many uncertainties exist that prevent determination of E. For example, distance and orientation of donors and acceptors may be variable and FRET may also occur between molecules that just happen to come within resonance range (sometimes called spurious- or collision FRET). This latter effect is very small for molecules in solution, but it dramatically increases when the donors and acceptors are concentrated in cell organelles. For example, even donor- and acceptor-tagged molecules that are distributed randomly (i.e., not clustered) at the membrane will yield significant FRET despite moderate expression levels (for analysis, see Appendix of van Rheenen et al. [15]). Thus, E_D reflects both intra- and inter FRET-pair resonance, and consequently f_D would be overestimated (which incidentally also holds true for the determination of E or f_D using FLIM). Usually I_S^{s}, E_D, and E_A can be reliably determined, but in the vast majority of FilterFRET or FLIM experiments E and f_D will be unknown.

Unlike donor-based FRET methods like FLIM, filterFRET also yields spatial information on the acceptor population. This means that in addition to querying donor-FRET (by solving for E_D or I_D^{d}), we can also assess the relationship between sensitized emission and the acceptor population. At 1:1 stoichiometry obviously E_D should equal the acceptor-normalized efficiency E_A. In other cases, E_A deviates from E but sometimes can yield biologically more relevant information than E_D or E. For example, dislocation of 50% of the donors from an organelle decorated with donor–acceptor pairs into the cytosol might leave 50% of acceptors unpaired at the organelle. E_D would in this case report that the remaining 50% donors still interact just as efficiently, whereas E_A clearly reveals the lowered acceptor occupation by dropping with 50%.

What is the donor/acceptor ratio in a given cell? Again, this ratio cannot be directly derived because it concerns two quantities that stem from fluorophores with different properties (absorption coefficient, quantum yield, spectra) and that emit into two channels differing in gain, filters, and excitation intensity. Thus, the (overlap corrected) intensity of acceptors in channel **A** will be a factor k times that of donors in D, at equimolar concentrations,[3] or:

$$kI_A^{\mathrm{a}} = I_D^{\mathrm{d}} \tag{7.24}$$

For a first approximation, k can be simply calculated by dividing I_D^{d} by I_A^{a} for a donor–acceptor fusion construct, because both quantities are corrected for overlap and FRET. Note however that this requires ϕ to be known (see Sect. 7.3.3). For the ratio of donor to acceptor concentration, we simply find:

$$\frac{[\mathrm{donor}]}{[\mathrm{acceptor}]} = k\frac{I_D^{\mathrm{d}}}{I_A^{\mathrm{a}}} \tag{7.25}$$

Or, in the case of confocal acquisition[4]

$$\frac{[\mathrm{donor}]}{[\mathrm{acceptor}]} = \gamma\sigma\frac{I_D^{\mathrm{d}}}{I_A^{\mathrm{a}}} \tag{7.25B}$$

---

[3]Termed k in [1]; α in [10]; R in [7], $\kappa\gamma$ in [17].

[4]For confocal imaging with tunable gains, k is not constant. Rather, we can distinguish a fixed part (which relates the efficiency of donor excitation at donor wavelength to that of acceptor excitation at acceptor wavelength) and a part that depends on relative excitation intensities and gains. The former was termed κ in van Rheenen et al. [3] but to keep κ in line with the terminology used in this volume that factor will here be renamed to σ, such that $k = \gamma\sigma$. Also, see Appendix.

Of actual interest to the biologist is usually not the quantification of donor, acceptor, and donor–acceptor, but rather to estimate the concentrations of the interacting proteins and the extent of their interactions, regardless of their labeling state. It is obvious that if the interacting proteins are incompletely labeled (for example, due to the presence of an endogenous population of untagged proteins, or due to imperfect maturation of FP labels) FRET recordings will significantly underestimate the amount of interactions between the proteins. Formalisms to cope with incomplete labeling have been put forward by several groups [2, 4, 16].[5] Inasmuch as such formalisms rely on calibration using donor–acceptor tandem constructs, it is important to note that we observed that speed of maturation of a given FP may dramatically vary from construct to construct [17].

7.4. Optimizing image acquisition

As the I_S^s, E_D, and E_A images are calculated from the raw input images, it is extremely important that **D, S,** and **A** are of the best possible quality. In addition, care must be taken that correction factors are derived from reference images taken at exactly the same

[5]Wlodarczyk and coworkers consider that labeling may be incomplete for 2 two reasons: first, not all proteins may become labeled (either in the chemical crosslinking process, or in the case of FP-labeling, due to the presence of endogenous proteins), and second, existing labels may be non-fluorescent, for example due to poor maturation of FPs, or due to photobleaching. Be p_d and p_a the probability that a given molecule of type a and d receive a functional label, respectively, then the concentration of fluorescent donors equals p_d times the total concentration of d plus $(1-p_d)$ times the concentration of complexes, or, in their terminology,

$[D] = p_d \cdot [d] + (1-p_a)[da]$. Analogously,
$[A] = p_a \cdot [a] + (1-p_d)[da]$ and
$[AD] = p_d p_a [da]$.

Provided that proper estimates of p_a and p_d are present, quantitative estimates of interactions can be readily achieved for certain types of experiments.

imaging conditions. In this part, possible pitfalls will be discussed and strategies to improve image quality will be lined out.

7.4.1. Wide-field versus confocal FilterFRET: A comparison

FRET imaging differs enough between wide-field fluorescence microscopes and confocal microscopes to warrant a comparison of the two techniques in this chapter. Confocal imaging offers significant advantages over wide-field imaging because it produces crisp optical sections of the preparation. Furthermore, point-scanning confocals offer greater freedom in image acquisition by allowing free choice of zoom and resolution and independent tuning of channel sensitivities through adjustment of the voltage of the photomultiplier tubes (PMTs). Confocal imaging is also more easily combined with acceptor photo bleaching and with fluorescence recovery after photobleaching (FRAP) experiments. On the other hand, wide-field fluorescence setups offer the freedom to filter-select whatever excitation wavelength desired, and the CCD detectors are more sensitive than PMTs. CCD imaging also is less harsh for the cells than laser point-scanning, although careful tuning of the confocal excitation regime can remedy that for a large part.

These differences have important consequences for filterFRET imaging. The major complication posed by confocal acquisition is that relative sensitivities for **D**, **S**, and **A** are tunable. This is true even if identical filter and pinhole settings are used from experiment to experiment since in general the user wants to fine-tune the excitation line intensities and to control individual PMT gain (high voltage) and offset settings for the channels. With CCD acquisition, weaker fluorescent cells are imaged by increasing the integration time, which causes both I_D^d, I_S^s, and I_A^a as well as the leak-through terms to increase linearly. As a result, when leak-through factors for particular fluorophores and a particular filter set have been quantified once, alterations in integration time can be easily compensated for. In the confocal case, however, unless laser intensity and PMT settings are kept fixed, this is not possible, due to the nonlinear dependence of gain and offset on the PMT high voltage. This necessitates that β, γ, and δ are determined after each change in setting. The advantage—on the other hand—is that the added flexibility allows simultaneous optimized acquisition of the often weak FRET signals without compromising acquisition time.

Because on CCD setups excitation for **D**, **S**, and **A** images is usually filter-selected from a single white light source the relative intensity of excitation is approximately fixed. Confocal microscopes use separate laser lines, often from distinct lasers, that can (and for optimal imaging should) be independently adjusted. Thus, on CCD setups γ (Eq. (7.6)) is constant for a given set of filters whereas on the confocal, it varies from image to image (also, see Sect. 7.4.2).

A final distinction is that on confocal microscopes **S** and **A** images are commonly acquired with the exact same emission filter settings whereas for CCD microscopes they typically involve physically separate- and therefore slightly different—filter cubes.[6] This simplifies the calculation of leak-through terms [3]. In Appendix of this chapter, we rather generalized the treatment of filterFRET by not making assumptions on the filter settings for **S** and **A**.

These differences add up to one major distinction: on wide-field imaging setups, it suffices to calibrate the setup *just once* for a given set of filters and fluorophores, and then use it for weeks or months without bothering about it. In contrast, for confocal filterFRET imaging, calibrations must be made every time a gain setting or laser line is adjusted, and preferably, for every image.

7.4.2. Temporal errors: Laser intensity fluctuation

On a variety of confocal microscopes, we and others [3, 18] observed considerable drift as well as oscillations (on a time scale of

[6]Unless a dual-excitation filter cube is used, in combination with a excitation switcher.

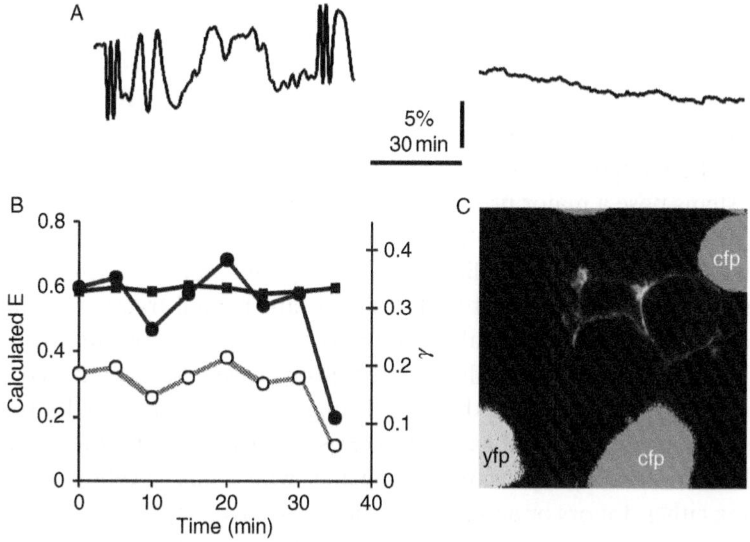

Fig. 7.4. *Correcting excitation fluctuations by inclusion of reference cells.* (A) Fluctuations in the intensity of a 514-nm argon ion laser line (left) and a mercury arc lamp (right), measured every 20 s for a 3 h time period. (B) Calculated E_D (solid circles; left axis) is seen to fluctuate significantly in time-lapse experiments. After 30 min a large intensity fluctuation in acceptor excitation was simulated by manually diminishing laser power with 60%. The open circles depict the correction factor γ, calculated according to Eq. (7.6) from cells expressing acceptors only. Calculating E_D with the online-updated γ-factor (solid squares) abolished the effects of excitation fluctuations. (C) Preparation containing FRET cells (gray) and CFP- and YFP reference cells (blue and yellow), as recognized by automated segmentation based on the ratio of intensity of donors and acceptors.

one to a few minutes) in excitation line intensity (Fig. 7.4A). Changes of several percent are common, while worst-case variations of up to 20% were found in poorly aligned systems. Importantly, individual laser lines fluctuate independently, even when derived from the same laser. Excitation stability is extremely important because it influences γ. While intensity variations may also occur in arc lamps on wide-field fluorescence microscopes, these changes are often much smaller (compare Fig. 7.4 A left and right panels). Furthermore, slow arc lamp intensity variations affect **D**, **S,** and **A** to the same degree if images are gathered in rapid succession, and thus have no effect on the apparent FRET image (Eqs. (7.9) and (7.10)).

The independent variations in laser line intensity on confocal systems pose a major problem for time-lapse FRET measurements. The supplier of our TCS-SP2 confocal installed a stabilization loop that improved the stability considerably, but not completely. In particular when expected FRET signals are a small fraction of the total fluorescence, the realized stability of ~3% will prevent acquisition of meaningful results. We therefore implemented online correction by recalculating the leak-through factors α, β, γ, and δ for each image [3], as well as G or ϕ (Eqs. (7.4)–(7.19)). To this goal, the cells under study are plated together with a mix of cells expressing either donors or acceptors on the same cover slip (Figs. 7.2 and 7.3). In an image taken at low zoom factor, regions of interest (ROIs) are assigned to single donor- or acceptor transfected cells (Fig. 7.4B and C). Then correction factors are determined from these ROIs as detailed in Eqs. (7.4)–(7.7). E_D and s.e. images are thus calculated using correction factors taken simultaneously with (or just before, in case one wants to zoom in) the FRET cell. This procedure completely removes the effect of laser fluctuations, resulting in superior registration of FRET during acquisition of time-lapse series.

As an alternative, changes in relative intensity of the laser lines may be directly recorded using for example, reflection images or a transmission detector, and γ may be adjusted accordingly. In our experience, this works significantly less reliable.

7.4.3. Co-registration of the input images

Obviously, it is of the utmost importance that the three input channels spatially overlap tightly, both in lateral and in axial direction. Co-registration (i.e., the precise, pixel-by-pixel correspondence of

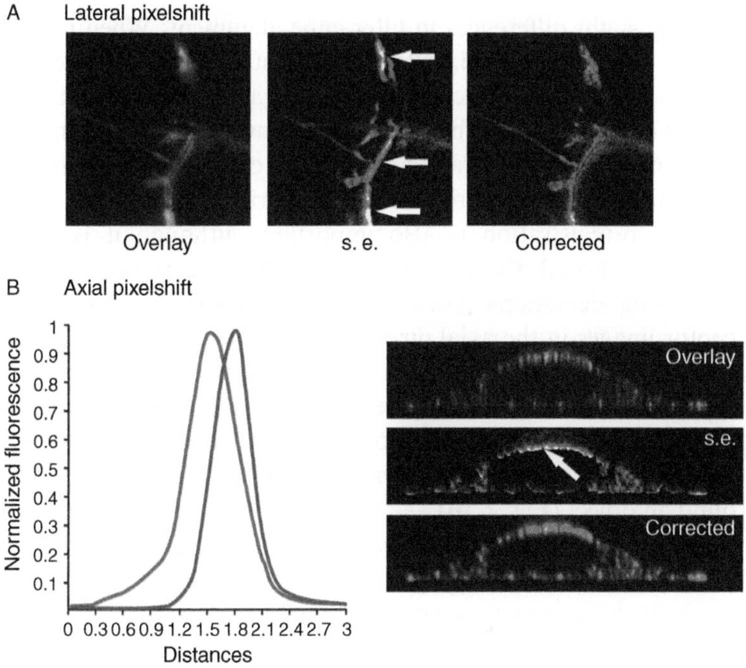

Fig. 7.5. *Effects of poor co-registration on calculated FRET images.* (A) Typical artifacts due to improper alignment (left) of raw input images caused by switching between unmatched filter cubes. The consistent appearance of high FRET values at the right side of bright structures (middle) is a sure indication to check image alignment. Right panel, proper alignment of the images corrects FRET artifacts. (B) Left panel, profiles of fluorescence intensities in a confocal X/Z image of the green emission (525 nm) of a 0.17-microm bead was registered using a HCX PL APO CS 63× objective upon 430-nm (blue line) and 514-nm (red line) excitation. Both scans use detection at 525 nm, demonstrating the extent of axial offset. Right panels, confocal images were acquired from a cell expressing CFP- and YFP-tagged membrane anchors. Top image, green–red overlay illustrates axial offset. Erroneous values (middle image) in the calculated I_S^s (s.e.) image are effectively corrected by using the refocusing macro routine (lower image). Shown are extreme examples. (See Color Insert.)

D, S, and **A**) has to be checked meticulously by the experimenter, using for example, color overlay images (Fig. 7.5A). Pixel-shift deviations are common on CCD imaging setups where they are caused by slight differences in filter cube alignment. When image-splitting devices are used, extensive adjustment for optimal co-registration is always necessary. In contrast, lateral overlay of confocal channels should be excellent for a well-maintained instrument. If needed, co-registration of channels can be easily optimized postacquisition by software pixel-shift algorithms.

Axial co-registration is also important, although it is often completely ignored. Compared with wide-field microscopy, possible focusing deviations (deviations due to offset of donor- and acceptor images in the axial direction) are emphasized by the confocals inherent optical sectioning. When the input images are effectively taken from slightly different planes in the cell, erroneous results occur during calculation of the sensitized emission that are often apparent as margins of unexpected high or low FRET values around an object (Fig. 7.5B).

Two main sources for this type of deviation exist: chromatic aberrations within the objective and other optics, and, for confocals, slight differences in the collimation of the laser beams. Chromatic aberrations are due to the wavelength dependency of the refractive index of optical glasses, which causes axial misregistration of images taken at different wavelengths [19]. Depending on the objective used, chromatic aberrations may be several micrometers (worst case). Chromatically corrected objectives are available, but it should be stressed that these are optimized only for a limited part of the spectrum, typically the mid-visible range. Therefore, significant chromatic aberration may still be present outside this range. For example, using a good, standard corrected 63×, 1.32 NA oil immersion objective (HCX PL APO CS, #506180, Leica), we noticed focusing deviations of about 400 nm (Fig. 7.5B) between the 430 and 514 nm laser lines used to excite the CFP/YFP FRET pair. Use of a UV-corrected 63× objective (HCX PL APO lbd.BL, #506192, Leica) significantly, but not completely, remedied this chromatic aberration. Chromatic focusing deviations are not limited to violet wavelengths because significant deviations exist for dye pairs excited throughout the visible spectrum (Table 7.1 in [3]). In addition, chromatic aberrations vary with lens types, and even for different objectives of the same type (L. Oomen and K.J., unpublished; [18]). Axial focusing errors also exist in CCD images but here they usually go unnoticed because of the poor axial resolution of wide-field fluorescence microscopes.

A more generic approach to overcome focusing deviations can be implemented if the setup is equipped for fast fine-focusing. First, **D** and **S** images are recorded at donor excitation. Then, before taking the **A** image at acceptor excitation, the preparation is refocused to minimize chromatic aberration. Because for a given combination of objective and excitation lines the focus deviation is constant, the correction distance needs to be determined only once. We used XZ-scanning of fixed cells or fluorescent beads for this goal. Applying this focus correction in an automated acquisition routine (macro), **D**, **S**, and **A** images are collected from the same focal plane in the biological sample (Fig. 7.5B).

7.4.4. Shading

Lateral intensity errors may be present over the entire image and occur on CCD and confocal systems alike. For CCD systems, a standard correction algorithm exists: corrections are carried out by normalizing pixel intensities using a reference image, a procedure called shading correction [13]. On the confocal system with independent excitation lines, these deviations are often more pronounced because spatial excitation intensities vary independently (L. Oomen, L. Brocks and K.J., unpublished; [18]). For example, when measuring excitation inhomogeneities for the 430 and 514 nm CFP/YFP lines by imaging a solution of the FRET calcium sensor Yellow Cameleon [7], we observed very significant deviations from unity flatness (Fig. 7.6). The 430 nm excitation intensity dropped by as much as 50% at the image corners, while 514 nm deviated by about 15% (data not shown). Importantly, significant differences (up to 20%) also occurred in the center of the images. Deviations of

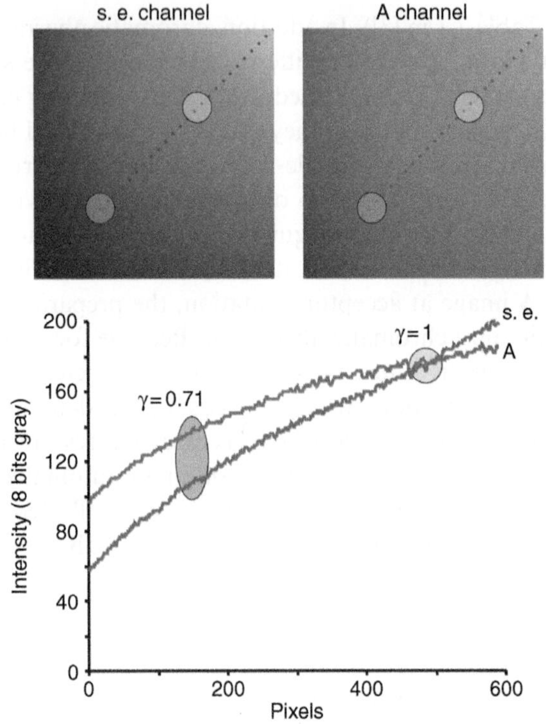

Fig. 7.6. *Shading errors: Lateral image intensity errors.* Shown are parts of reference images (1024 × 1024) that were acquired by averaging eight confocal images of a solution of Yellow Cameleon at 430-nm excitation (upper left panel) and at 514-nm excitation (upper right panel), both detected at 525–570 nm. Note that due to significant differences in shading at these two excitation wavelengths, the γ value (calculated according to Eq. (7.6)) may vary by ~25% in either direction, causing significant errors in calculated FRET. Images were collected with a 63× oil immersion objective without zooming. Note that similar, albeit smaller, differences were observed when 458 and 514 nm laser lines (both derived from the same argon ion laser) were compared.

this magnitude are not uncommon in confocal systems [18], and they are often diminished by increasing the zoom factor.

Shading correction is simply carried out by measuring the fluorescence of solutions of dyes that are spectrally similar to the donor and acceptor. The fluorescence of these reference images is then normalized to a standard, and all **D, S,** and **A** images are divided by their cognate reference standard. Correction factors and FRET images are only calculated after applying the shading correction. In our experience, shading correction is crucial to obtain good FRET images.

7.5. Postacquisition improvements and analysis

When optimal input pictures are obtained whilst observing all the above corrections and precautions, the raw calculated FRET images nonetheless often are quite disappointing and complicated to interpret. See for example the I_S^s, E_D, and E_A images in Fig. 7.3. To blame are noise in the FRET images and the way our eyes handle that.

7.5.1. Noise in FRET images

Even in the nominal absence of laser fluctuations or other image-degrading aberrations, the number of photons that hit the detector during the data collection period of the image (i.e., the exposure time for a CCD image or the pixel dwell time for a confocal image) will contain considerable noise. The photon count x follows a Poisson distribution (Fig. 7.7A) with mean value μ as

$$p(x) = \frac{\mu^x e^{-\mu}}{x!} \tag{7.26}$$

It can be shown that the standard deviation (SD) of this distribution is also just $\sqrt{\mu}$. In other words, if one would repeatedly measure the same pixel that on average collects 100 photons during a single dwell time (a normal value for a rather bright confocal image!!) one would record less than $100–2\times\sqrt{100} = 80$ photons or more than $100 + 2\times\sqrt{100} = 120$ photons just by coincidence in ~5% of the measurements. This uncertainty is expressed as the

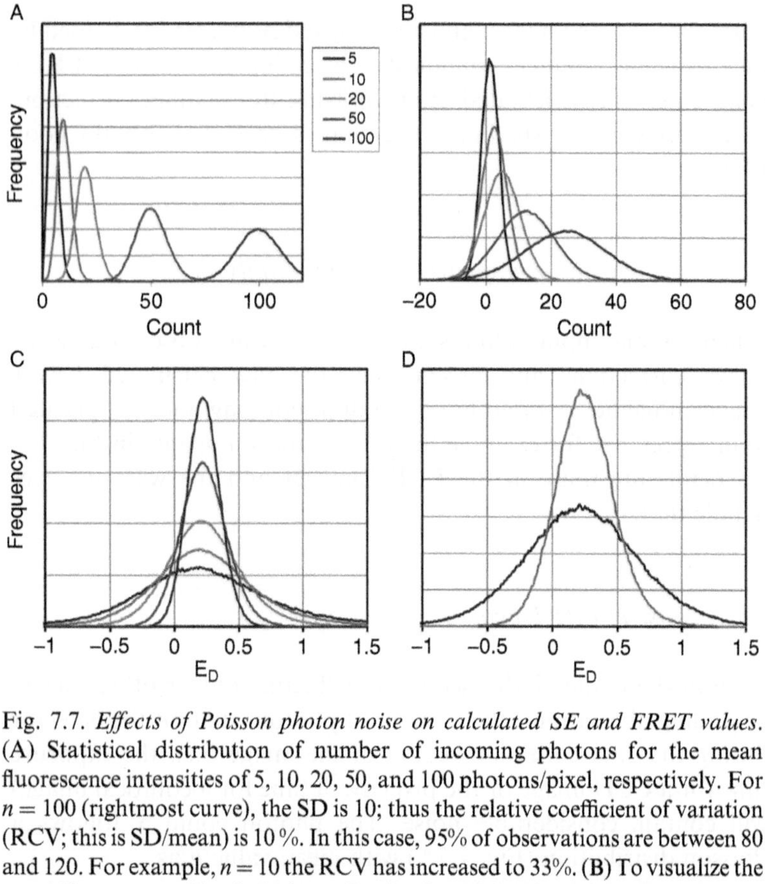

Fig. 7.7. *Effects of Poisson photon noise on calculated SE and FRET values.* (A) Statistical distribution of number of incoming photons for the mean fluorescence intensities of 5, 10, 20, 50, and 100 photons/pixel, respectively. For $n = 100$ (rightmost curve), the SD is 10; thus the relative coefficient of variation (RCV; this is SD/mean) is 10 %. In this case, 95% of observations are between 80 and 120. For example, $n = 10$ the RCV has increased to 33%. **(B)** To visualize the spread in s.e. caused by the Poisson distribution of pixel intensities that averaged 100 photons for each **A**, **D**, and **S** (right-most curve), s.e. was calculated repeatedly using a Monte Carlo simulation approach. Realistic correction factors were used ($\alpha = 0.0023, \beta = 0.59, \gamma = 0.15, \delta = 0.0015$) that determine 25% FRET efficiency. Note that spread in s.e. based on a population of pixels with RCV = 10 % amounts to RCV = ~60 % for these particular settings! Other curves: for photon counts decreasing as in (A), the uncertainty further grows and an increasing fraction of calculated s.e. values are actually below zero. (C) Spread in E_D values for photon counts as in (A). Note that whereas the value of the mean remains the same, the spread (RCV) increases to several hundred percent. (D) Spread depends not only on photon counts but also on values of the correction

relative coefficient of variation (RCV, defined as the SD divided by the mean value). In this example, RCV = 10%. The situation becomes considerably worse for dimmer pixels (Fig. 7.7A). Thus, in dim sections of the image the degrading influence of noise is worst (Fig. 7.3; see also Fig. 7.8).

As the s.e. values are calculated from individual images, each subject to noise, the errors multiply. This leads to a dramatic spreading of calculated values (Figs. 7.7B and 7.8). Thus, calculated E_D pixel values less than 0 or larger than 1 may be quite common, depending on the intensities of individual images in that pixel and the values of the correction factors. Clearly, such outliers cannot be simply rejected as this would introduce systematic errors in the calculated FRET results. For E_D, s.e. is divided by I_D^d which further increases the spread (Fig. 7.7C). In such pictures, single-pixel RCVs in the order of 100% are not uncommon. It is obvious that the RCV strongly depends on the values of the correction factors α, β, γ, and δ (Fig. 7.7D) and on the magnitude of the FRET efficiency.

Clearly, a major route towards better pictures is to maximize the photon count. On confocal microscopes this can be accomplished in different ways. More photons are collected when increased laser power is used (which incidentally also allows using lower PMT voltages which reduces detector noise). Furthermore, the pinhole can be opened, the dwell time can be increased by lowering the scan speed and acquired images can be averaged. However, these measures come at the expense of increased fluorophore bleaching, prolonged imaging time, and degraded resolution. As this is usually not desired, we will next cover alternative procedures to clear up the images in an—as much as possible—unbiased manner.

In performing the operations described in the next sections, it is absolutely necessary to use imaging software that can handle

---

factors. For the green trace, mean photon counts for **A**, **D**, and **S** were 30 each. The factors were $\beta = 0.645$ and $\gamma = 0.11$. For the lower trace, counts for **A**, **D**, and **S** were 100, 100, and 30, and correction factor values were $\beta = 0.81$ and $\gamma = 0.4$. Despite higher photon counts, the spread in this example is significantly larger.

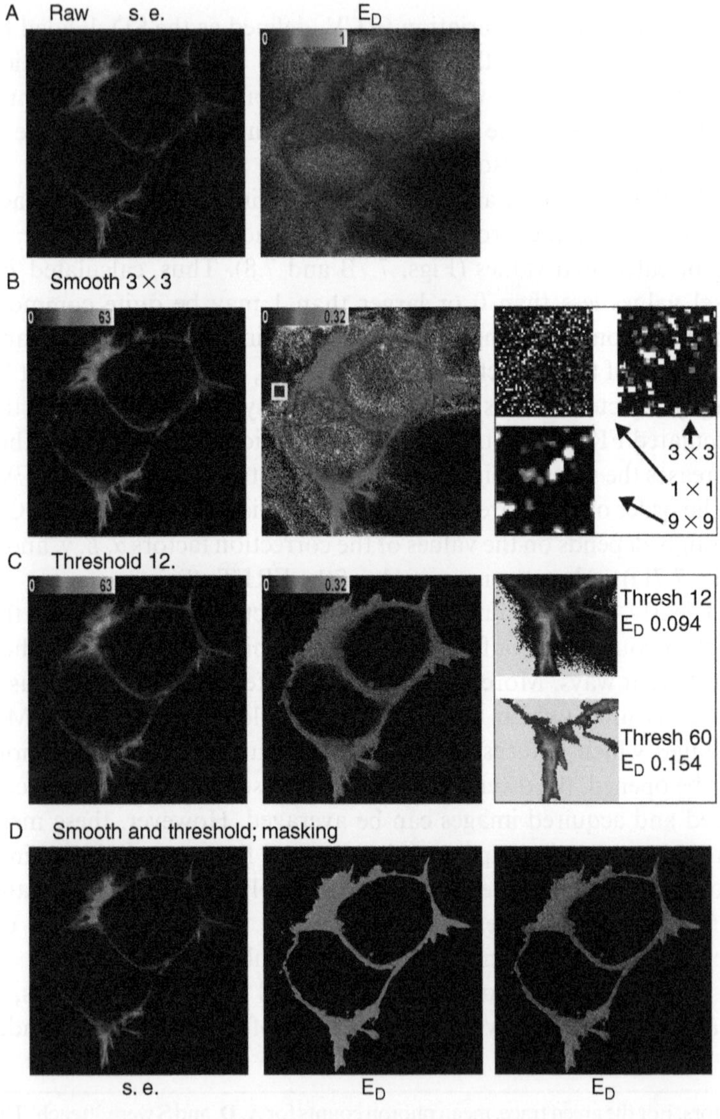

Fig. 7.8. *Postacquisition improvements.* (A) Unaltered "raw" 1024 × 1024 confocal s.e. and E_D images. Note the appearance of excessive noise in E_D at low-signal locations. (B) Lateral averaging (smoothing) using a 3 × 3 kernel

floating point (Negative and broken numbers) calculations. The freeware package Image J (Rasband, W., NIH, Bethesda, Md; *http://rsb.info.nih.gov/ij*) for example performs the described image analysis steps excellently, if 8- or 12-bit images are converted to 32-bit images, and it allows automation of frequently occurring analysis sequences.

7.5.2. Smoothing/filtering

Lateral averaging of the pixel values of the raw images ("smoothing") is the easiest way to reduce the noise effects due to photon statistics significantly, but of course it also reduces the resolution of the picture. A simple mathematical averaging with a 3×3 kernel applied to the background-subtracted, shading-corrected input images already reduces photon noise by ~3-fold (Fig. 7.8A and B). On the other hand, smoothing also spreads out the effect of incidental very high noise pixels that are present in just one of the images (Fig. 7.8B, right detail panels). The effect of this is that it replaces a single outlier pixel in the FRET image with an island of (albeit less pronounced) outliers. To our eye, the effect is the same. Smoothing works better on s.e. images than on the FRET efficiency image. As an alternative, smoothing may be applied to the calculated images as well. It depends on the raw input images which approach gives the best results.

---

cleans up the s.e. image but is less effective on the E_D image. Note the difference in Look-Up Table. Right panel: detail of the middle panel (yellow box) showing how smoothing with the indicated kernel sizes influences E_D. (C) Application of lower threshold on the input images rejects any pixels for which either donor or acceptor intensity is below 12 gray levels. Right panel: the detail taken from the lower portion of the middle panel demonstrates that thresholding significantly influences calculated E_D values. Yellow pixels are excluded from the calculations. (D) A combination of smoothing and thresholding with settings as in B and C cleans up the s.e. and E_D images. Right panel, a mask, derived from the smoothed and thresholded E_D image, is applied to the unfiltered E_D data to preserve fine details in the FRET image. (See Color Insert.)

7.5.3. *Thresholding*

As the effects of photon noise are most confusing in very dim regions in the image, significant improvements may be expected from simply setting a low-end threshold for each of the input images (Fig. 7.8C). Commonly, separate thresholds have to be set for each of the images, depending on background level and brightness of the image. Indeed thresholding clears up a lot of the noise in background areas, but the highest outliers will still exceed the threshold, and conversely any accidental low-value pixels in the regions that are of interest will be removed. In addition, if the FRET efficiency depends on the expression level of the donor and acceptor constructs [15] thresholding is guaranteed to bias the results. Independent estimation of the magnitude of this bias is necessary. This can be simply performed by comparing calculated FRET values from images differing only in threshold setting (Fig. 7.8C, right detail panels).

Setting an upper threshold may also be necessary. This is evident when saturated pixels are present in the image, a situation that can not always be avoided because expression of some GFP-tagged constructs sometimes causes appearance of extremely bright aggregates in the cytosol that outshine the structures of interest. To retain full dynamic range in the ROI, it is best to allow image saturation in the aggregates while rejecting them from the analysis using an upper threshold. Note that the upper threshold has to be set well below the highest gray level to prevent systematic bias, in particular when the input images are acquired with averaging.

7.5.4. *Masking*

A masking strategy may be used to dismiss false high FRET values in dim and therefore noisy image regions while simultaneously preventing loss of resolution in the brighter regions (Fig. 7.8D, right panel). In the apparent FRET image, resonance can be distinguished from incidental noise pixels by smoothing the image with a spatial filter. Isolated noise pixels become averaged out, while consecutive adjacent pixels with positive FRET remain visible. Setting thresholds for each image to just above background intensity generates a mask that contains only regions of true FRET. The thus obtained binary mask may be further edited by for example, erosion, dilation, or floodfill if necessary. The mask is subsequently applied to the original, unfiltered FRET image resulting in near-complete rejection of noise pixels (Fig. 7.8D). Masking tends to be a very powerful way to clean up the images, but since it involves thresholding one has to verify that it does not bias the results in experiments where dimmer pixels are likely to contain less FRET.

7.5.5. Unbiased cleaning up: Mixing FRET efficiency with image intensity information

Completely unbiased visualization of FRET results with strong rejection of noise in dim regions is possible by combining the FRET efficiency picture with the original image intensity (Fig. 7.9A). Here, a pseudocolor RGB FRET image is made from the unprocessed (i.e., without any thresholding or masking) FRET image by overlaying it with a pseudocolor table, also called Look-Up Table or LUT. The intensity of the RGB image is then modulated with the intensity information present in the original input images (see Fig. 7.9). Intensity information may be derived from **D**, **A**, I_S^s or a combination thereof (e.g., the maximum value of **D** or **A**). Thus, background noise pixels that yield high FRET values are still retained in the image, but they will be displayed very dim, just as they appear to the eye.[7] Prior to the modulation step the intensity

[7]Many image analysis packages are not capable to perform this operation correctly in a single step. One approach is to unravel the pseudocolor FRET image into individual R, G, and B channel images. Each of these images is then multiplied with the normalized intensity image, and the final image is regenerated by combining the channels to an RGB image.

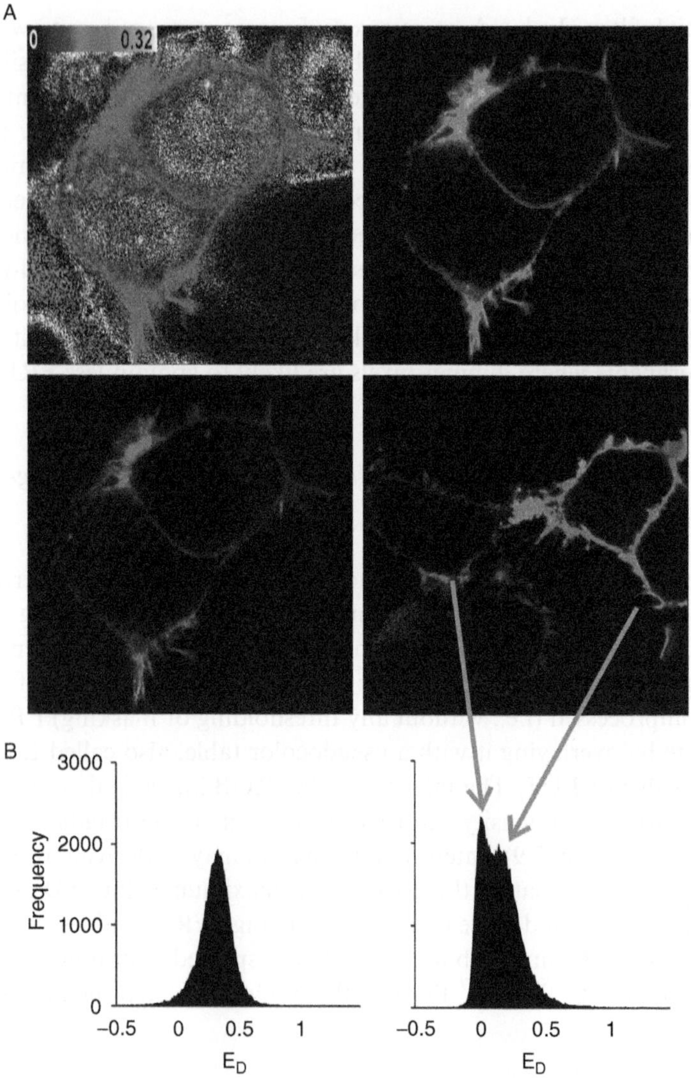

Fig. 7.9. *Further FRET efficiency analysis.* (A) Unbiased display of FRET efficiency. The E_D image (upper left panel) is modulated with an intensity picture (in this case, s.e., upper right panel) to yield the lower left image. See text for further details. Lower right panel, example with several cells

image may optionally be enhanced by smoothing, brightness/contrast adjustment, and masking.

7.5.6. *FRET efficiency histograms*

A final postanalysis step that helps interpreting the data is to present frequency histograms of pixel FRET values in selected ROI (Fig. 7.8B). As outlined before, correctly acquired and calculated FRET results are expected to display a Gaussian distribution, in first approximation. Inspection of the distribution of FRET values is a quick way to identify possible deviations from this rule. For example, failing to apply proper shading correction may lead to significant broadening of the distribution. In addition, multimodal or very skewed distributions may draw attention to specific problems with the input images or, alternatively, to interesting cell-to-cell variability.

7.6. *Discussion*

In this chapter, it was shown that filterFRET is an easy, intuitive and quantitative alternative to record sensitized emission and FRET efficiency. The major advantages of filterFRET over donor-based FRET detection methods (FLIM) are that it can be carried out with standard wide-field or confocal fluorescence microscopes that are available in most laboratories, and that it yields additional data on the acceptor population. FilterFRET is also fast, requiring just two confocal scans (if need be on a line-by-line basis) which minimizes the risk of artifacts due to, for example, organelle movement in living cells, and acquisition can be optimized for each channel independently. However, quantitative

---

displaying different E_D. (B) Frequency histograms of E_D in the pictures in the lower panels in A. Note that whereas the distribution on the left hand displays merely stochastic noise (compare Fig. 7.7C), the rightmost histogram reveals the heterogeneous FRET efficiency in the cells of the corresponding image. (See Color Insert.)

filterFRET requires significant attention for corrections and calibration, whereas FLIM-based FRET techniques are inherently quantitative from first physical principles.

The corrections and calibration of filterFRET differ significantly for CCD microscopes and confocal microscopes. This is because in confocal experiments, channel sensitivities are adjusted at will by the experimenter, and because relative excitation intensities show intended–as well as unintended variations (adjustments and drift, respectively). Confocal filterFRET therefore requires frequent, if not in-line, recalibration; however, if properly streamlined this should not take more than 15 min a day. It also slightly complicates the mathematical framework, as compared to CCD imaging filterFRET. We aimed to arrive at a comprehensive theory that is equally applicable to both imaging modes. We also proposed mathematical jargon that is a compromise between the widely differing terminologies used in the various publications on this topic.

What degree of precision is to be expected from filterFRET? Obviously, precision is governed by the quality of the input pictures, in particular by the lateral and axial co-registration, shading correction, and by photon Poisson noise. For typical confocal images, averaging around a 100 pictures is necessary to arrive at 1% uncertainty in each pixel value of the input pictures, and the FRET calculation will further degrade that figure. Clearly, excessive averaging is not realistic for live-cell imaging. In addition, FPs perform poorly by displaying ill-characterized behavior such as bleaching, maturation, excitation-induced dark states, and pH dependence. Thus, going for one percent variance in pixel FRET values almost certainly stretches the data. However, when multiple pixels can be pooled (average of a ROI), this type of precision may be obtained even without image averaging. Of course, these considerations also hold true for FLIM. From a noise-point of view, simple 2-channel ratio determinations (Sect. 7.22) are preferred. We also want to emphasize again that quantitative FRET efficiency images are not necessarily the "holy grail" for cell biologists. For example, if in the soma of the neuron in Fig. 7.1, a kinase-FRET construct would be 100 times more abundant but only half as active as in the axon, E_D images would suggest half the kinase activity in the soma, whereas sensitized emission images would correctly report a 50-fold enhanced kinase activity.

The area of filterFRET is evolving actively. Recent developments include its combination with other imaging modes including total internal reflection microscopy (TIRF, see also Chap. 9) and FRAP [20]. Furthermore, several laboratories are testing novel FP donor/acceptor combinations to minimize spectral overlap and issues related to FP maturation. In our hands, especially poor red FP (mRFP1, mCherry, and Tomato) maturation interferes with reliable FRET imaging because of the green emission of the immature proteins [17]. In addition, reliable photo-switchable acceptors [21] are an obvious idea with potential. Likewise, correction algorithms keep evolving. For example, Zal and Gasgoigne [6] included correction for photo bleaching in their treatment, and Elangovan et al. [22] introduced concentration-dependency in the leak-through factors. The applicability of such corrections depends very much on imaging conditions and equipment, and we therefore choose not to incorporate them in this chapter. With all these developments, filterFRET has emerged as a full-grown, adaptable, efficient, and fun way to study molecular interactions in vivo.

7.A. Appendix

7.A.1. Factorization

In this appendix, we will assume that a cell expressing donor- and acceptor molecules is excited at appropriate wavelength λ_{ex}^{d} and λ_{ex}^{a} to image FRET. As detailed in the main text, three images are collected that allow independent estimates of cross talk magnitude to perform correction of leak-through:

- **D**, excitation and emission at donor wavelength
- **S**, excitation at donor wavelength and emission at acceptor wavelength
- **A**, excitation and emission at acceptor wavelength

Furthermore, we assume the more general, but also more complex case in which confocal detection is used. This allows for example flexibility in setting independent detector gains for **D**, **S,** and **A**. When this flexibility is not required the expressions simplify considerably.

Before proceeding, an important note must be made. In literature, two different but fully equivalent approaches have been taken in s.e. The first approach considers a cell that contains (unknown) numbers of donors and acceptors N_D and N_A. When energy transfer takes place (be it from collisional encounters or because a stable population of FRET pairs exist with FRET efficiency E) this diminishes the *effective* number of emitting donors with N_S [3]; that is, the FRET efficiency for this population is unity. Thus, the residual donor emission results from $(N_D - N_S)$ unquenched donor molecules, and the N_S population emits *only* sensitized emission. This approach is intuitive in case no assumptions are being made on the presence of a stable population of FRET pairs or on the magnitude of E in a donor–acceptor complex.

A second approach also considers three populations: free (unquenched) donors N_D, free acceptors N_A, and a population engaged in FRET pairs N_S that transfer energy with characteristic efficiency E (between 0 and 1). However, in this case, the N_S population emits *both* donor fluorescence (quenched by a fraction $(1 - E)$) *and* sensitized emission (proportional to EN_S). To keep in line with the treatment and terminology in other chapters in this volume, this latter approach will be followed here. Note, however, that in other chapters the population of FRET pairs is indicated by the subscript $_{DA}$ whereas we stick to the notation N_S to indicate that this quantity is based on photons emitted from sensitized emission (**S** image) and to keep the close synonymy with the former approach. Thus, our I_{D-S} equals I_{DA} and $I_S + I_A$ equals I_{AD}. Both ways yield essentially identical results.

Bearing this in mind, the acquired images are composite images that consist of multiple terms (see Fig. 7.1; for symbols, see Appendix Table 7.A1) as follows:

TABLE 7.A1
Factorization of symbols

Symbol	Factorization
I_D^d	$N_D \ell^d \varepsilon_D^d Q_D F_D^d g^d$
I_A^d	$N_A \ell^d \varepsilon_A^d Q_A F_A^d g^d$
I_S^d	$N_S E \ell^d \varepsilon_D^d Q_A F_A^d g^d$
I_{D-S}^d	$(N_D - N_S)\ell^d \varepsilon_D^d Q_D F_D^d g^d + N_S(1-E)\ell^d \varepsilon_D^d Q_D F_D^d g^d$
I_{D-S}^s	$(N_D - N_S)\ell^s \varepsilon_D^d Q_D F_D^s g^s + N_S(1-E)\ell^s \varepsilon_D^d Q_D F_D^s g^s$
I_S^s	$N_S E \ell^s \varepsilon_D^d Q_A F_A^s g^s$
I_A^s	$N_A \ell^s \varepsilon_A^d Q_A F_A^s g^s$
I_A^a	$N_A \ell^a \varepsilon_A^a Q_A F_A^a g^a$

The fluorescent components are denoted by I (intensity) followed by a capitalized subscript ($_D$, $_A$, or $_S$, for respectively Donors, Acceptors, or Donor/Acceptor FRET pairs) to indicate the particular population of molecules responsible for emission of$I$and a lower-case superscript (^{d,} ^{a,} or ^{s}) that indicates the detection channel (or filter cube). For example, I_D^d denotes the intensity of the donors as detected in the donor channel and reads as "Intensity of donors in the donor channel," etc. Similarly, properties of molecules (number of molecules, N; quantum yield, Q) are specified with capitalized subscript and properties of channels (laser intensity, ℓ; gain, g) are specified with lowercase superscript. Factors that depend on both molecular species and on detection channel (excitation efficiency, ε; fraction of the emission spectrum detected in a channel, F) are indexed with both. Note that for all factorized symbols it is assumed that we work in the linear (excitation-fluorescence) regime with negligible donor or acceptor saturation or triplet states. In case such conditions are not met, the FRET estimation will not be correct. See Chap. 12 (FRET calculator) for more details.

– **D** is the output gray scale value after amplification[8] in the donor channel (g^d) of the sum of the fraction $\left(F_D^d\right)$ of donor fluorescence in the donor channel and the fraction $\left(F_A^d\right)$ of acceptor fluorescence in the donor channel. The fluorescence of donors depends on the number of donor molecules (N_D) diminished by

[8]The factor g may account for integration time and electron multiplication in CCD imaging, or for the PMT gain in confocal imaging.

those donors that lose their excited state energy due to FRET (EN_S), the molar extinction coefficient of the donor ($\varepsilon_D^{\mathrm{d}}$), the laser intensity ($\ell^{\mathrm{d}}$) at $\lambda_{\mathrm{ex}}^{\mathrm{d}}$, and the donor quantum yield (Q_D). The fluorescence of acceptors depends on their quantum yield (Q_A), and on the sum of the number of acceptor molecules (N_A) cross-excited at $\lambda_{\mathrm{ex}}^{\mathrm{d}}$ ($\ell^{\mathrm{d}}\varepsilon_A^{\mathrm{d}}$) and those excited by FRET ($N_S E$).

$$\begin{aligned}\mathbf{D} = (N_D - \mathrm{EN_S})\ell^{\mathrm{d}}\varepsilon_D^{\mathrm{d}}Q_D F_D^{\mathrm{d}} g^{\mathrm{d}} + N_A \ell^{\mathrm{d}}\varepsilon_A^{\mathrm{d}} Q_A F_A^{\mathrm{d}} g^{\mathrm{d}} \\ + EN_S \ell^{\mathrm{d}}\varepsilon_D^{\mathrm{d}} Q_A F_A^{\mathrm{d}} g^{\mathrm{d}}\end{aligned} \qquad (7.\mathrm{A1})$$

- **S** is the output gray value after the s.e. channel detector scaling (g^{s}) of the sum of the fractions of donor fluorescence in the s.e. channel (F_D^{s}) and of acceptor fluorescence in the s.e. channel (F_A^{s}). The donor fluorescence depends on Q_D, the excitation efficiency at $\lambda_{\mathrm{ex}}^{\mathrm{d}}$ (that is, $\ell^{\mathrm{s}}\varepsilon_D^{\mathrm{d}}$), the number of donors (N_D), and the population of donors that lose their energy by FRET ($\mathrm{EN_S}$). The fluorescence of acceptors depends on Q_A, the amount of acceptor molecules (N_A) excited with $\lambda_{\mathrm{ex}}^{\mathrm{d}}$ ($\ell^{\mathrm{s}}\varepsilon_A^{\mathrm{s}}$) and on the amount of acceptor molecules excited by FRET ($\mathrm{EN_S}$, which is linear to $\ell^{\mathrm{s}}\varepsilon_D^{\mathrm{s}}$).

$$\begin{aligned}\mathbf{S} = (N_D - \mathrm{EN_S})\ell^{\mathrm{s}}\varepsilon_D^{\mathrm{s}}Q_D F_D^{\mathrm{s}} g^{\mathrm{s}} + N_A \ell^{\mathrm{s}}\varepsilon_A^{\mathrm{s}} Q_A F_A^{\mathrm{s}} g^{\mathrm{s}} \\ + \mathrm{EN_S} \ell^{\mathrm{s}}\varepsilon_D^{\mathrm{s}} Q_A F_A^{\mathrm{s}} g^{\mathrm{s}}\end{aligned} \qquad (7.\mathrm{A2})$$

- Finally, **A** is the output gray value after the acceptor channel[9] scaling (g^{a}) of the fraction of acceptor fluorescence in the acceptor channel (F_A^{a}), which depends on the acceptor quantum yield (Q_A) and on the amount of acceptors N_A excited at $\lambda_{\mathrm{ex}}^{\mathrm{a}}$ ($\ell^{\mathrm{a}}\varepsilon_A^{\mathrm{a}}$), of (usually very minor) contributions of donor fluorescence cross-excited at $\lambda_{\mathrm{ex}}^{\mathrm{a}}$ and leaking into the acceptor channel ($(N_D - \mathrm{EN_S})\ell^{\mathrm{a}}\varepsilon_D^{\mathrm{a}}Q_D F_D^{\mathrm{a}} g^{\mathrm{a}}$) and of sensitized emission resulting from cross-excitation at $\lambda_{\mathrm{ex}}^{\mathrm{a}}$ ($\mathrm{EN}_S \ell^{\mathrm{a}}\varepsilon_D^{\mathrm{a}} Q_A F_A^{\mathrm{a}} g^{\mathrm{a}}$). However, as

[9]Note that in confocal imaging, the PMT will generally be the same physical detector as that of the s.e. channel, but operated at a different gain setting.

$\ell^{\mathrm{a}}\varepsilon_D^{\mathrm{a}}$ is very small, the latter two (leakthrough and FRET) terms in most cases essentially are zero.

$$\mathbf{A} = (N_D - \mathrm{EN_S})\ell^{\mathrm{a}}\varepsilon_D^{\mathrm{a}}Q_DF_D^{\mathrm{a}}g^{\mathrm{a}} + N_A\ell^{\mathrm{a}}\varepsilon_A^{\mathrm{a}}Q_AF_A^{\mathrm{a}}g^{\mathrm{a}} + \mathrm{EN_S}\ell^{\mathrm{a}}\varepsilon_D^{\mathrm{a}}Q_AF_A^{\mathrm{a}}g^{\mathrm{a}} \tag{7.A3}$$

As mentioned before, we will assume that a cell expressing donor- and acceptor molecules is excited at appropriate wavelength $\lambda_{\mathrm{ex}}^{\mathrm{d}}$ and $\lambda_{\mathrm{ex}}^{\mathrm{a}}$ in all three channels, so that the donor and acceptor extinction coefficients in **D** and **S** are the same:

$$\varepsilon_D^{\mathrm{s}} = \varepsilon_D^{\mathrm{d}} \text{ and } \varepsilon_A^{\mathrm{s}} = \varepsilon_A^{\mathrm{d}} \tag{7.A4}$$

Note that this is always the case for wide-field and confocal determination where **D** and **S** are collected simultaneously using the same excitation filters or lasers. In case three separate filters are used, care should be taken to match the filters so as to fulfill Eq. (7.A4).

By taking (7.A4) in account, (7.A1)–(7.A3) can be simplified as:

$$\begin{aligned}\mathbf{D} &= pN_D + q\mathrm{EN_S} + rN_A\\ \mathbf{S} &= tN_D + u\mathrm{EN_S} + vN_A,\\ \mathbf{A} &= xN_D + y\mathrm{EN_S} + zN_A\end{aligned}$$

$$\begin{array}{lll} p = \ell^{\mathrm{d}}\varepsilon_D^{\mathrm{d}}Q_DF_D^{\mathrm{d}}g^{\mathrm{d}} & q = -p + \ell^{\mathrm{d}}\varepsilon_D^{\mathrm{d}}Q_AF_A^{\mathrm{d}}g^{\mathrm{d}} & r = \ell^{\mathrm{d}}\varepsilon_A^{\mathrm{d}}Q_AF_A^{\mathrm{d}}g^{\mathrm{d}}\\ t = \ell^{\mathrm{s}}\varepsilon_D^{\mathrm{d}}Q_DF_D^{\mathrm{s}}g^{\mathrm{s}} & u = -t + \ell^{\mathrm{s}}\varepsilon_D^{\mathrm{d}}Q_AF_A^{\mathrm{s}}g^{\mathrm{s}} & v = \ell^{\mathrm{s}}\varepsilon_A^{\mathrm{d}}Q_AF_A^{\mathrm{s}}g^{\mathrm{s}}\\ x = \ell^{\mathrm{a}}\varepsilon_D^{\mathrm{a}}Q_DF_D^{\mathrm{a}}g^{\mathrm{a}} & y = -x + \ell^{\mathrm{a}}\varepsilon_D^{\mathrm{a}}Q_AF_A^{\mathrm{a}}g^{\mathrm{a}} & z = \ell^{\mathrm{a}}\varepsilon_A^{\mathrm{a}}Q_AF_A^{\mathrm{a}}g^{\mathrm{a}}\end{array} \tag{7.A1a–7.A3a}$$

Based on the three basic images, **D**, **A**, and **S**, expressions can be derived for the intensity of donors, acceptors, and sensitized emission in their own channel (i.e., the leak-through- and FRET-corrected quantities I_D^{d}, I_A^{a}, and I_S^{s}, respectively). Subsequently, to obtain FRET efficiency, I_S^{s} will be scaled to I_D^{d} analogous to the treatment in the main text.

7.A.2. Sensitized emission

Below, we work out the situation for $x = y = 0$, since in almost all cases, these factors can be neglected. For image sets in which these factors cannot be neglected a corrected image **A*** must first be determined from **A** by subtracting xN_D and $y\mathrm{EN_S}$ (7.A.5). In this case, **A*** should be used in all equations below. If $x = y = 0$ then

$$A = zN_A = I_A^{\mathrm{a}} \tag{7.A5}$$

and

$$N_D = \frac{\mathbf{D}}{p} - \frac{\mathrm{EN}_s q}{p} - \frac{\mathbf{A}r}{zp} \tag{7.A6}$$

and Eqs. (7.A5) and (7.A6) are substituted into the equation for **S** (Eq. 7.A2a)), yielding,

$$u\mathrm{EN_s} = S - \left(\frac{\mathbf{D}}{p} - \frac{\mathrm{EN}_s q}{p} - \frac{\mathbf{A}r}{zp}\right)t - \mathbf{A}v/z$$

hence:

$$\mathrm{EN_s} = \frac{\mathbf{S} - \mathbf{D}t/p - \mathbf{A}(v/2 - rt/pz)}{u - qt/p}$$

or

$$\mathrm{EN_S} = \frac{S - D\frac{\ell^{\mathrm{s}} F_D^{\mathrm{s}} g^{\mathrm{s}}}{\ell^{\mathrm{d}} F_D^{\mathrm{d}} g^{\mathrm{d}}} - A\left(\frac{\ell^{\mathrm{s}} \varepsilon_A^{\mathrm{d}} F_A^{\mathrm{s}} g^{\mathrm{s}}}{\ell^{\mathrm{a}} \varepsilon_A^{\mathrm{a}} F_A^{\mathrm{a}} g^{\mathrm{a}}} - \frac{\ell^{\mathrm{d}} \varepsilon_A^{\mathrm{d}} F_A^{\mathrm{d}} g^{\mathrm{d}}}{\ell^{\mathrm{a}} \varepsilon_A^{\mathrm{a}} F_A^{\mathrm{a}} g^{\mathrm{a}}} \frac{\ell^{\mathrm{s}} F_D^{\mathrm{s}} g^{\mathrm{s}}}{\ell^{\mathrm{d}} F_D^{\mathrm{d}} g^{\mathrm{d}}}\right)}{\ell^{\mathrm{s}} \varepsilon_D^{\mathrm{d}} Q_A F_A^{\mathrm{s}} g^{\mathrm{s}} \left(1 - \frac{\ell^{\mathrm{s}} F_D^{\mathrm{s}} g^{\mathrm{s}} \ell^{\mathrm{d}} F_A^{\mathrm{d}} g^{\mathrm{d}}}{\ell^{\mathrm{d}} F_D^{\mathrm{d}} g^{\mathrm{d}} \ell^{\mathrm{s}} F_A^{\mathrm{s}} g^{\mathrm{s}}}\right)} \tag{7.A7}$$

Relating back to Eq. (7.8) from the main text, the sensitized emission gray scale image I_S^{s} is composed of the emission from $\mathrm{EN_S}$, which depends on the acceptor quantum yield Q_A, scaled by factors for sensitized emission channel gain (g^{s}), fraction of acceptor fluorescence in the sensitized emission channel F_A^{s}, and donor excitation efficiency $\ell^{\mathrm{s}}\varepsilon_D^{\mathrm{d}}$;

$$I_S^{\mathrm{s}} = \mathrm{EN_S}\ell^{\mathrm{s}}\varepsilon_D^{\mathrm{d}}Q_A F_A^{\mathrm{s}} g^{\mathrm{s}} = \frac{S - D\frac{\ell^{\mathrm{s}} F_D^{\mathrm{s}} g^{\mathrm{s}}}{\ell^d F_D^d g^d} - A\left(\frac{\ell^{\mathrm{s}}\varepsilon_A^{\mathrm{d}} F_A^{\mathrm{s}} g^{\mathrm{s}}}{\ell^a \varepsilon_A^a F_A^a g^a} - \frac{\ell^{\mathrm{d}}\varepsilon_A^{\mathrm{d}} F_A^{\mathrm{d}} g^{\mathrm{d}}}{\ell^{\mathrm{a}}\varepsilon_A^{\mathrm{a}} F_A^{\mathrm{a}} g^{\mathrm{a}}}\frac{\ell^{\mathrm{s}} F_D^{\mathrm{s}} g^{\mathrm{s}}}{\ell^{\mathrm{d}} F_D^{\mathrm{d}} g^{\mathrm{d}}}\right)}{\left(1 - \frac{\ell^{\mathrm{s}} F_D^{\mathrm{s}} g^{\mathrm{s}}}{\ell^{\mathrm{d}} F_D^{\mathrm{d}} g^{\mathrm{d}}}\frac{\ell^{\mathrm{d}} F_A^{\mathrm{d}} g^{\mathrm{d}}}{\ell^{\mathrm{s}} F_A^{\mathrm{s}} g^{\mathrm{s}}}\right)} \tag{7.A8}$$

In Eq. (7.A8), the constants α, β, γ, and δ (see main text and Table 7.A2) are identified as detailed in Eqs. (7.A9)–(7.A12). Values for α, γ, and δ can be deduced from imaging of a sample with only acceptor molecules:

$$\frac{^{\mathrm{Acc}}D}{^{\mathrm{Acc}}A} = \frac{r}{z} = \frac{N_A \ell^d \varepsilon_A^d Q_A F_A^d g^d}{N_A \ell^a \varepsilon_A^a Q_A F_A^a g^a} = \frac{\ell^d \varepsilon_a^d F_A^d g^d}{\ell^a \varepsilon_A^a F_A^a g^a} = \alpha \tag{7.A9}$$

$$\frac{^{\mathrm{Acc}}D}{^{\mathrm{Acc}}S} = \frac{r}{v} = \frac{N_A \ell^d \varepsilon_A^d Q_A F_A^d g^d}{N_A \ell^s \varepsilon_A^d Q_A F_A^s g^s} = \frac{\ell^d F_A^d g^d}{\ell^s F_A^s g^s} = \delta \tag{7.A10}$$

$$\frac{^{\mathrm{Acc}}S}{^{\mathrm{Acc}}A} = \frac{N_A \ell^s \varepsilon_A^d Q_A F_A^s g^s}{N_A \ell^a \varepsilon_A^a Q_A F_A^a g^a} = \frac{\ell^s \varepsilon_A^d F_A^s g^s}{\ell^s \varepsilon_A^a F_A^a g^a} = \gamma \tag{7.A11}$$

Where again (compare Textbox 2), $\alpha = \gamma\delta$. Similarly, β is calculated from a sample with only donor molecules:

$$\frac{^{\mathrm{Acc}}S}{^{\mathrm{Acc}}D} = \frac{t}{p} = \frac{N_D \ell^s \varepsilon_D^d Q_D F_D^s g^s}{N_D \ell^d \varepsilon_D^d Q_D F_D^d g^d} = \frac{\ell^s F_D^s g^s}{\ell^d F_D^d g^d} = \beta \tag{7.A12}$$

Note that when **S** and **D** are collected simultaneously (typically for confocal imaging) β and δ are independent of relative laser line intensities.

Analogous to Eq. (7.8) (text) we can thus rewrite Eq. (7.A8) as:

$$I_S^{\mathrm{s}} = \frac{S - \beta D - \gamma A(1 - \beta\delta)}{1 - \beta\delta} \tag{7.A13}$$

Eqs. (7.A8) and (7.A13) are valid not only in case there is a single FRET population N_{S} with characteristic FRET efficiency E, but also when different FRET populations each with different characteristic FRET efficiency E_i are present $\left(N_{\mathrm{S},i}\right)N_{\mathrm{S},i}$. One can simply verify this by substituting N_S by $\sum_i N_{S,i}$ and $N_S E$ by $\sum_i N_{S,i} E_i$ in Appendix Table 7.A1 and Eqs. (7.A1)–(7.A7).

7.A.3. The unquenched donor image and E_D

To derive an expression for I_D^{d} (the unquenched donor image), **D** is corrected for leak-through as well as for signal lost to FRET. For this latter correction, the factor ϕ that relates the lost signal I_S^{d} in **D**, that is $\left(\mathrm{EN}_S \ell^{\mathrm{d}} \varepsilon_D^{\mathrm{d}} Q_D F_D^{\mathrm{d}} g^{\mathrm{d}}\right)$, to the gain I_S^{s} in **S**, or ($\mathrm{EN}_S \ell^{\mathrm{s}} \varepsilon_D^{\mathrm{d}} Q_A F_A^{\mathrm{s}} g^{\mathrm{s}}$, Eq. (7.A13)) is:

$$\phi = -\frac{\mathrm{EN}_S \ell^{\mathrm{d}} \varepsilon_D^{\mathrm{d}} Q_D F_D^{\mathrm{d}} g^{\mathrm{d}}}{\mathrm{EN}_S \ell^{\mathrm{s}} \varepsilon_D^{\mathrm{d}} Q_A F_A^{\mathrm{s}} g^{\mathrm{s}}} = \frac{-p}{u+t} = -\frac{\ell^{\mathrm{d}} Q_D F_D^{\mathrm{d}} g^{\mathrm{d}}}{\ell^{\mathrm{s}} Q_A F_A^{\mathrm{s}} g^{\mathrm{s}}} \qquad (7.\mathrm{A}14)$$

or:

$$\mathrm{EN}_S \ell^{\mathrm{d}} \varepsilon_D^{\mathrm{d}} Q_D F_D^{\mathrm{d}} g^{\mathrm{d}} = -\phi \mathrm{EN}_S \ell^{\mathrm{s}} \varepsilon_D^{\mathrm{d}} Q_A F_A^{\mathrm{s}} g^{\mathrm{s}} \qquad (7.\mathrm{A}15)$$

In order to solve I_D^{d}, Eq. (7.A1) is rearranged with information from Eqs. (7.A5), (7.A9), (7.A10), and (7.A14):

$$I_D^{\mathrm{d}} = N_D \ell^{\mathrm{d}} \varepsilon_D^{\mathrm{d}} Q_D F_D^{\mathrm{d}} g^{\mathrm{d}} = \mathbf{D} - (\delta + \phi) \mathrm{EN}_S \ell^{\mathrm{s}} \varepsilon_D^{\mathrm{d}} Q_A F_A^{\mathrm{s}} g^{\mathrm{s}} - \alpha \mathbf{A} \qquad (7.\mathrm{A}16)$$

Combined with Eqs. (7.A8) and (7.A13) this rearranges Eq. (7.A16), which is identical to Eq. (7.13) in the main text.

$$I_D^{\mathrm{d}} = \mathbf{D} - (\delta + \phi) I_S^{\mathrm{s}} - \alpha \mathbf{A} = \frac{\beta\phi + 1}{1 - \beta\delta} \mathbf{D} - \frac{\phi + \delta}{1 - \beta\delta} \mathbf{S} + \gamma\phi \mathbf{A} \qquad (7.\mathrm{A}17)$$

Also ζ (Eq. (7.14) in main text) can be redefined after substitution of Eqs. (7.A9)–(7.A12) according to Eq. (7.A18):

$$\zeta = \frac{\beta(\delta + \phi)}{(1 - \beta\delta)} = \frac{\frac{\mathrm{tr}}{pv} - \frac{t}{u+t}}{1 - \frac{\mathrm{tr}}{pv}} = \frac{\frac{F_D^{\mathrm{s}}}{F_A^{\mathrm{s}}} \left(\frac{F_A^{\mathrm{d}}}{F_D^{\mathrm{d}}} - \frac{Q_D}{Q_A} \right)}{1 - \frac{F_D^{\mathrm{s}}}{F_A^{\mathrm{s}}} \frac{F_A^{\mathrm{d}}}{F_D^{\mathrm{d}}}} \qquad (7.\mathrm{A}18)$$

From Eq. (7.A18) it is clear that ζ only depends on filter throughput and the ratio of quantum yields for donor and acceptor and hence is independent on laser or detector gain settings. For a given combination of confocal filter settings and fluorophores ζ is therefore a constant (for our confocal settings, using CFP and YFP, $\zeta = -0.248$). Therefore, using ζ is distinctly advantageous during confocal imaging where excitation intensities and channel sensitivities are varied independently, whereas G (see main text) varies with detector settings. Now, Eq. (7.A17) can be rewritten:

$$I_D^{\mathrm{d}} = N_D \ell^{\mathrm{d}} \varepsilon_D^{\mathrm{d}} Q_D F_{DS}^{\mathrm{d}} g^{\mathrm{d}} = (1+\zeta)\mathbf{D} - \frac{\zeta}{\beta}\mathbf{S} - \gamma\left(\delta - \frac{\zeta}{\beta} + \delta\zeta\right)\mathbf{A} \tag{7.A19}$$

By definition the E_D is calculated from the loss in donor signal, which can be defined from the symbols in Appendix Table 7.A1, and from the experimental images **S**, **D**, and **A** using Eqs. (7.12), (7.15), and (7.A16):

$$E_D = \frac{N_S}{N_D} E \equiv 1 - \frac{I_{D-S}^{\mathrm{d}}}{I_D^{\mathrm{d}}} = \frac{\beta\phi\mathbf{D} - \phi\mathbf{S} + \gamma\phi(1-\beta\delta)\mathbf{A}}{(\beta\phi+1)\mathbf{D} - (\delta+\phi)\mathbf{S} + \gamma\phi(1-\beta\delta)\mathbf{A}}$$

$$= 1 - \frac{\mathbf{D}}{(1+\zeta)D - \frac{\zeta}{\beta}\mathbf{S} - \gamma\left(\delta - \frac{\zeta}{\beta} + \delta\zeta\right)\mathbf{A}} \tag{7.A20}$$

ζ can be reliably determined, for example by acquiring the **D**, **S**, and **A** images before and after complete acceptor photobleaching. Since postbleach **D** is equal to $N_D \ell^{\mathrm{d}} \varepsilon_D^{\mathrm{d}} Q_D F_{DS}^{\mathrm{d}} g^{\mathrm{d}}$ $(= I_D^{\mathrm{d}})$ (see Eqs. (7.A1) and (7.A19)), ζ is found from:

$$\zeta = \frac{{}^{postbleach}\mathbf{D} - {}^{postbleach}\mathbf{D} + {}^{postbleach}\mathbf{A}\alpha}{{}^{postbleach}\mathbf{D} - {}^{postbleach}\mathbf{S}\frac{1}{\beta} + {}^{postbleach}\mathbf{A}(\frac{\gamma}{\beta} - \alpha)} \tag{7.A21}$$

Table 7.A2
Definition of correction factors and constants

Factor	Equations	Factorization	Calculate from	Prep	Comments
α	(7.4), (7.A9)	$\frac{\ell^{d}\varepsilon_A^{d}F_A^{d}g^{d}}{\ell^{a}\varepsilon_A^{a}F_A^{a}g^{a}}$	$\frac{\mathbf{D}}{\mathbf{A}}$	acc.	Leak-through of cross-excited acc. into **D**
β	(7.5), (7.A12)	$\frac{\ell^{s}F_D^{s}g^{s}}{\ell^{d}F_D^{d}g^{d}}$	$\frac{\mathbf{S}}{\mathbf{D}}$	donor	Leak-through of donors into **S**
γ	(7.6), (7.A11)	$\frac{\ell^{s}\varepsilon_A^{d}F_A^{s}g^{s}}{\ell^{a}\varepsilon_A^{a}F_A^{a}g^{a}}$	$\frac{\mathbf{S}}{\mathbf{A}}$	acc.	Cross excitation of acceptors
δ	(7.7), (7.A10)	$\frac{\ell^{d}F_A^{d}g^{d}}{\ell^{s}F_A^{s}g^{s}}$	$\frac{\mathbf{D}}{\mathbf{S}}$	acc.	Leak-through of s.e. into **D**
ϕ	(7.12), (7.A14)	$-\frac{\ell^{d}Q_D F_D^{d}g^{d}}{\ell^{s}Q_A F_A^{s}g^{s}}$	$-\frac{1+\delta G}{\beta+G}$	sensor[1]	Relates the loss in **D** due to FRET (I_S^{d}) to the gain in **S** due to FRET (I_S^{s}); negative
ρ	(7.A36)	$\frac{\ell^{a}\varepsilon_D^{a}g^{a}}{\ell^{s}\varepsilon_D^{d}g^{s}}$	$\frac{\mathbf{A}}{\mathbf{S}}$	donor	Relates signal from cross-excited donors in **S** to that in **A** (provided that emission filters are identical)
σ	(7.A26)	$\frac{\varepsilon_D^{d}}{\varepsilon_A^{d}}$	$-1/\phi\upsilon$	zero-FRET construct	Relates extinction coefficient of donors at λ_{ex}^{d} to that of acceptors at λ_{ex}^{d}
υ	(7.A28), (7.A32)	$\frac{\ell^{s}\varepsilon_A^{d}Q_A F_A^{s}g^{s}}{\ell^{d}\varepsilon_D^{d}Q_D F_D^{d}g^{d}}$	$-1/\phi\sigma$ or $\frac{\beta\mathbf{D}_0-\mathbf{S}_0}{\delta\mathbf{S}_0-\mathbf{D}_0}$	zero-FRET construct	Relates the visibility of acceptors in the **S** channel to the visibility of the same number of donors in the **D** channel
ζ	(7.14), (7.22), (7.A18), (7.A21)	$\frac{\frac{F_D^{s}}{F_A^{s}}\left(\frac{F_A^{d}}{F_D^{d}}-\frac{Q_D}{Q_A}\right)}{1-\frac{F_D^{s}F_A^{d}}{F_A^{s}F_D^{d}}}$	$\frac{\beta(\delta+\phi)}{(1-\beta\delta)}$ *or:* $-\beta/(\beta+G)$	sensor	Used to relate the loss in **D** due to FRET to the gain in **S** due to FRET; independent from excitation intensity and gain and therefore constant for given filters and fluorophores; negative
G	(7.19), (7.A24)	$\frac{\frac{Q_D F_D^{s}g^{s}}{Q_A F_A^{s}g^{s}}-1}{\frac{\ell^{d}F_A^{d}g^{d}}{\ell^{s}F_A^{s}g^{s}}-\frac{\ell^{d}Q_D F_D^{d}g^{d}}{\ell^{s}Q_A F_A^{s}g^{s}}}$	$-\frac{\Delta S}{\Delta D}$ or: $-\frac{\beta\phi+1}{\phi+\delta}$	sensor	Alternative constant that relates the loss in **D** due to FRET to the gain in **S** due to FRET used in literature; note that G depends on changes in β and δ; positive

[1]Sensor: any single-polypeptide construct (containing the fluorophores to be used in the experiment) that can be induced to change FRET significantly and homogeneously; e.g., Yellow Cameleon in combination with ionomycin.

Alternatively, ζ can be related to G (see main text, Eq. (7.22)):

$$\zeta = -\frac{\beta}{\beta + G} \tag{7.A22}$$

Or conversely, G can be expressed in terms of β, δ, and ϕ by combining Eqs. (7.A22) and (7.A18);

$$\zeta = -\frac{\beta}{\beta + G} = \frac{\beta(\delta + \phi)}{(1 - \beta\delta)} \tag{7.A23}$$

The constant G is isolated:

$$G = -\frac{\beta\phi + 1}{\phi + \delta} = -\frac{\frac{u}{u+t}}{-\frac{p}{u+t} + \frac{r}{v}} = \frac{\frac{Q_D F_D^{\mathrm{s}} g^{\mathrm{s}}}{Q_A F_A^{\mathrm{s}} g^{\mathrm{s}}} - 1}{\frac{\ell^{\mathrm{d}} F_A^{\mathrm{d}} g^{\mathrm{d}}}{\ell^{\mathrm{s}} F_A^{\mathrm{s}} g^{\mathrm{s}}} - \frac{\ell^{\mathrm{d}} Q_D \mathrm{F}_D^{\mathrm{d}} g^{\mathrm{d}}}{\ell^{\mathrm{s}} Q_A F_A^{\mathrm{s}} g^{\mathrm{s}}}} \tag{7.A24}$$

from which dependency of G on detector gain is evident.

7.A.4. Direct acceptor excitation and E_A estimation

Apart from the unquenched donor image providing a FRET estimate E_D, another FRET estimate can be deduced directly from the pure sensitized emission $\left(I_S^{\mathrm{s}}\right)$ and direct acceptor excitation $\left(I_A^{\mathrm{a}}\right)$ components. From Appendix Table 7.A1, the following ratio can be defined[10]:

$$E_A \equiv \frac{N_{\mathrm{S}}}{N_A} E = E \frac{N_{\mathrm{S}} \ell^{\mathrm{s}} \varepsilon_D^{\mathrm{d}} Q_A F_A^{\mathrm{s}} g^{\mathrm{s}}}{N_A \ell^{\mathrm{a}} \varepsilon_A^{\mathrm{a}} Q_A F_A^{\mathrm{a}} g^{\mathrm{a}}} \frac{1}{\gamma\sigma} = \frac{I_S^{\mathrm{s}}}{I_A^{\mathrm{a}}} \frac{1}{\gamma\sigma} \left(= \frac{I_S^{\mathrm{s}}}{I_A^{\mathrm{s}}} \frac{1}{\sigma} \right) \tag{7.A25}$$

where

$$\sigma = \frac{\varepsilon_D^{\mathrm{d}}}{\varepsilon_A^{\mathrm{d}}} \tag{7.A26}$$

[10]Note the similarity of equation A25 between brackets and Eq.equation (16) of Chap. 1.

Combining Eqs. (7.A25) with (7.A5) and (7.A13) yields

$$E_A = \frac{I_S^s}{\gamma\sigma A} = \frac{\mathbf{S} - \beta\mathbf{D} - \gamma\mathbf{A}(1 - \beta\delta)}{\gamma\sigma(1 - \beta\delta)\mathbf{A}} \tag{7.A27}$$

Hence the quantity of E_A can be simply calculated from the corrected sensitized emission image and the acceptor only image provided the ratio of the molar extinction coefficients of the donor and acceptor at the donor excitation wavelength is known (σ). This quantity can be determined from absorption spectra of purified labeled components or can be experimentally determined as follows. First, let us define a factor υ that relates the signal of N acceptors in the **S** channel to the signal of the same number of donors in the **D** channel:

$$\upsilon = \frac{v}{p} = \frac{N\ell^s\varepsilon_A^d Q_A F_A^s g^s}{N\ell^d\varepsilon_D^d Q_D F_D^d g^d} = \frac{1}{-\phi\sigma} \tag{7.A28}$$

One easy way to determine υ is by using a donor–acceptor fusion chimera that displays no detectable FRET. In this case, $N = N_D = N_A$ and $N_S = 0$, and Eqs. (7.A1a) and (7.A2a) simplify to:

$$\mathbf{D}_0 = pN + rN \tag{7.A29}$$

$$\mathbf{S}_0 = tN + vN \tag{7.A30}$$

Combining Eqs. (7.A29) and (7.A30) with (7.A28), (7.A10), and (7.A12) it follows that:

$$\frac{\mathbf{S}_0}{\mathbf{D}_0} = \frac{t + v}{p + r} = \frac{\frac{t}{p} + \frac{v}{p}}{1 + \frac{r}{v}\frac{v}{p}} = \frac{\beta + \upsilon}{1 + \delta\upsilon} \tag{7.A31}$$

Here υ can be isolated to yield:

$$\upsilon = \frac{\beta\mathbf{D}_0 - \mathbf{S}_0}{\delta\mathbf{S}_0 - \mathbf{D}_0} \tag{7.A32}$$

7.A.5. The corrected acceptor image

For the acceptor image $\left(I_A^{\mathrm{a}}\right)$ it suffices in almost all cases to simply use the **A** image, because the donor excitation at acceptor wavelength is essentially zero. However, in special cases (e.g., when FRETting between spectrally similar FPs such as CFP and GFP) $\varepsilon_D^{\mathrm{a}}$ may be larger and it may become necessary to correct for the leak-through terms. To accommodate such cases, we here derive **A***, a leak-through-corrected version that than must replace **A** in all calculations.[11] First, copy the expressions for **A** and **S** (Eqs. (7.A2a), (7.A3a), using (7.A4)), substituting (see note 3)F_A^{s} by $F_A^{\mathrm{a}'}$ and F_D^{s} by $F_D^{\mathrm{a}'}$:

$$\begin{cases} \mathbf{S} = t'N_D + u'N_S E + vN_{\mathrm{a}} \\ \mathbf{A} = xN_D + yN_S E + zN_{\mathrm{a}} \end{cases} \tag{7.A33}$$

with

$$\begin{aligned} t' &= \ell^{\mathrm{s}} \varepsilon_D^{\mathrm{d}} Q_D F_D^{\mathrm{a}} g^{\mathrm{s}}, \\ u' &= -t' + \ell^{\mathrm{s}} \varepsilon_D^{\mathrm{d}} Q_A F_A^{\mathrm{a}} g^{\mathrm{s}}, \\ \rho &= \frac{x}{t'} = \frac{y}{u'} = \frac{\ell^{\mathrm{a}} \varepsilon_D^{\mathrm{a}} g^{\mathrm{a}}}{\ell^{\mathrm{s}} \varepsilon_D^{\mathrm{d}} g^{\mathrm{s}}} \end{aligned} \tag{7.A34}$$

From combining Eqs. (7.A10), (7.A33), and (7.A34) it can be shown that:

$$\mathbf{A}^* = zN_A = \frac{\mathbf{A} - \rho \mathbf{S}}{1 - \rho\gamma} \tag{7.A35}$$

Note that this equals Eq. (7.8a) of [1].

[11] This is straightforward in case the **A** and **S** filters are identical (i.e., $F_A^{\mathrm{a}} = F_A^{\mathrm{s}}$ and $F_D^{\mathrm{a}} = F_D^{\mathrm{s}}$). With confocal FRET this is commonly the case; with CCD imaging, it requires matching the filters. Without this assumption, an analogous result can be obtained, although derivation is significantly more complicated.

ρ can be determined experimentally in cells labeled with only donor molecules since:

$$\rho = \frac{^{\text{Don}}\mathbf{D}}{^{\text{Don}}\mathbf{S}} = \frac{x}{t'} = \frac{N_D \ell^a \varepsilon_D^a F_D^a Q_D g^a}{N_D \ell^s \varepsilon_D^d F_D^a Q_D g^s} = \frac{\ell^a \varepsilon_D^a g^a}{\ell^s \varepsilon_D^d g^s} \tag{7.A36}$$

7.A.6. Imaging FRET in tethered constructs using two channels

In this appendix, FRET in tethered constructs (where thus donor: acceptor stoichiometry is exactly 1) is calculated from just two images. Provided that independent estimates of cross talk magnitude are available, and the excitation power in both images is the same, full correction of leak through is possible from images:

- **D**, excitation and emission at donor wavelength
- **S**, excitation at donor wavelength and emission at acceptor wavelength. Approach and terminology are as detailed above, except that $N_D = N_A = N_S = N$ and $\ell^d = \ell^s = \ell$

Hence Eqs. (7.A1a) and (7.A2a) can be rewritten as:

$$\begin{cases} \mathbf{D} = pN + q\text{NE} + rN \\ \mathbf{S} = tN + u\text{NE} + vN \end{cases} \tag{7.A37}$$

This set of equations can be solved for N and NE yielding:

$$\begin{cases} N = \dfrac{q\mathbf{S} - u\mathbf{D}}{(t+v)q - (p+r)u} \\ \text{NE} = \dfrac{(t+v)\mathbf{D} - (p+r)\mathbf{S}}{(t+v)q - (p+r)u} \end{cases} \tag{7.A38}$$

Hence:

$$E = E_D = E_A = \frac{\text{NE}}{N} = \frac{(t+v)\mathbf{D} - (p+r)\mathbf{S}}{q\mathbf{S} - u\mathbf{D}} = \frac{\left(\frac{t}{p} + \frac{v}{p}\right)\mathbf{D} - \left(1 + \frac{r}{p}\right)\mathbf{S}}{\frac{q}{p}S - \frac{u}{p}\mathbf{D}} \tag{7.A39}$$

From Eqs. (7.A1)–(7.3A), and (7.A26) it follows that:

$$q = -p + \frac{\varepsilon_D^{\mathrm{d}}}{\varepsilon_A^{\mathrm{d}}} r = -p + \sigma r \qquad (7.A40)$$

$$u = -t + \frac{\varepsilon_D^{\mathrm{s}}}{\varepsilon_A^{\mathrm{s}}} v = -t + \sigma v \qquad (7.A41)$$

By referring to Eqs. (7.A10), (7.A12), (7.A14), and (7.A28) it follows that:

$$\delta = \frac{r}{v};\, \beta = \frac{t}{p};\, \phi = \frac{-p}{u+t};\, \upsilon = \frac{v}{p};\, \text{and } \upsilon = \frac{-1}{\sigma\phi} \qquad (7.A42)$$

Combining Eqs. (7.A40)–(7.A42) yields:

$$\frac{r}{p} = \frac{r}{v}\frac{v}{p} = \delta\upsilon \qquad (7.A43)$$

$$\frac{q}{p} = -1 + \sigma\frac{r}{p} = -1 + \sigma\delta\upsilon \qquad (7.A44)$$

$$\frac{u}{p} = -\frac{t}{p} + \sigma\frac{v}{p} = -\beta + \sigma\upsilon \qquad (7.A45)$$

Substituting Eqs. (7.A42)–(7.A45) into Eq. (7.A39) yields:

$$E = \frac{(\beta + \upsilon)\mathbf{D} - (1 + \delta\upsilon)\mathbf{S}}{(\delta\sigma\upsilon - 1)\mathbf{S} - (\sigma\upsilon - \beta)\mathbf{D}} = \phi\frac{(\beta + \upsilon)\mathbf{D} - (1 + \delta\upsilon)\mathbf{S}}{(\delta + \phi)\mathbf{S} - (1 + \beta\phi)\mathbf{D}} \qquad (7.A46)$$

A calibration procedure (using only **D** and **S** images) to yield β, δ, ϕ, and υ has been described in Eqs. (7.A12), (7.A10), (7.19)–(7.20), and (7.A32), respectively.

Acknowledgments

We are indepted to Drs L.Oomen and L. Brocks for sharing unpublished data; to Drs L.Oomen, G. vd. Krogt and M. Langeslag for critical reading and comments and to Dr G. vd. Krogt, B. Ponsioen and W. Zwart for preparation of samples. Financially supported by NWO, the Netherlands Cancer Society and by the Josephine Nefkens Stichting.

References

[1] Gordon, G. W., Berry, G., Liang, X. H., Levine, B. and Herman, B. (1998). Quantitative fluorescence resonance energy transfer measurements using fluorescence microscopy. Biophys. J. *74*, 2702–13.

[2] Hoppe, A., Christensen, K. and Swanson, J. A. (2002). Fluorescence resonance energy transfer-based stoichiometry in living cells. Biophys. J. *83*, 3652–64.

[3] van Rheenen, J., Langeslag, M. and Jalink, K. (2004). Correcting confocal acquisition to optimize imaging of fluorescence resonance energy transfer by sensitized emission. Biophys. J. *86*, 2517–29.

[4] Wlodarczyk, J., Woehler, A., Kobe, F., Ponimaskin, E., Zeug, A. and Neher, E. (2008). Analysis of FRET signals in the presence of free donors and acceptors. Biophys. J. *94*, 986–1000.

[5] Xia, Z. and Liu, Y. (2001). Reliable and global measurement of fluorescence resonance energy transfer using fluorescence microscopes. Biophys. J. *81*, 2395–402.

[6] Zal, T. and Gascoigne, N. R. (2004). Photobleaching-corrected FRET efficiency imaging of live cells. Biophys. J. *86*, 3923–39.

[7] Miyawaki, A., Llopis, J., Heim, R., McCaffery, J. M., Adams, J. A., Ikura, M. and Tsien, R. Y. (1997). Fluorescent indicators for Ca^{2+} based on green fluorescent proteins and calmodulin. Nature *388*, 882–7.

[8] Ponsioen, B., Zhao, J., Riedl, J., Zwartkruis, F., van der Krogt, G., Zaccolo, M., Moolenaar, W. H., Bos, J. L. and Jalink, K. (2004). Detecting cAMP-induced Epac activation by fluorescence resonance energy transfer: Epac as a novel cAMP indicator. EMBO Rep. *5*, 1176–80.

[9] van der Wal, J., Habets, R., Varnai, P., Balla, T. and Jalink, K. (2001). Monitoring agonist-induced phospholipase C activation in live cells by fluorescence resonance energy transfer. J. Biol. Chem. *276*, 15337–44.

[10] Nagy, P., Vamosi, G., Bodnar, A., Lockett, S. J. and Szollosi, J. (1998). Intensity-based energy transfer measurements in digital imaging microscopy. Eur. Biophys. J. *27*, 377–89.

[11] Tron, L., Szollosi, J., Damjanovich, S., Helliwell, S. H., Arndt-Jovin, D. J. and Jovin, T. M. (1984). Flow cytometric measurement of fluorescence resonance energy transfer on cell surfaces. Quantitative

evaluation of the transfer efficiency on a cell-by-cell basis. Biophys. J. *45*, 939–46.
[12] Wouters, F. S., Verveer, P. J. and Bastiaens, P. I. (2001). Imaging biochemistry inside cells. Trends Cell Biol. *11*, 203–11.
[13] Tomazevic, D., Likar, B. and Pernus, F. (2002). Comparative evaluation of retrospective shading correction methods. J. Microsc. *208*, 212–23.
[14] Chen, H., , Puhl, H. L.III, Koushik, S. V., Vogel, S. S. and Ikeda, S. R. (2006). Measurement of FRET efficiency and ratio of donor to acceptor concentration in living cells. Biophys. J. *91*, L39–L41.
[15] van Rheenen, J., Achame, E. M., Janssen, H., Calafat, J. and Jalink, K. (2005). PIP2 signaling in lipid domains: A critical re-evaluation. EMBO J. *24*, 1664–73.
[16] Clegg, R. M. (1992). Fluorescence resonance energy transfer and nucleic acids. Methods Enzymol. *211*, 353–88.
[17] van der Krogt, G. N., Ogink, J., Ponsioen, B. and Jalink, K. (2008). A comparison of donor-acceptor pairs for genetically encoded FRET sensors: Application to the Epac cAMP sensor as an example. PLoS ONE. *3*, e1916.
[18] Zucker, R. M. and Price, O. (2001). Evaluation of confocal microscopy system performance. Cytometry *44*, 273–94.
[19] Cogswell, C. J. and Larkin, K. G. (1995). The specimen illumination path and its effect on image quality. In: "Handbook of Biological Confocal Microscopy"(Pawley, J. B., ed.). Plenum press, New York, pp. 127–37.
[20] van Royen, M. E., Cunha, S. M., Brink, M. C., Mattern, K. A., Nigg, A. L., Dubbink, H. J., Verschure, P. J., Trapman, J. and Houtsmuller, A. B. (2007). Compartmentalization of androgen receptor protein-protein interactions in living cells. J. Cell Biol. *177*, 63–72.
[21] Demarco, I. A., Periasamy, A., Booker, C. F. and Day, R. N. (2006). Monitoring dynamic protein interactions with photoquenching FRET. Nat. Methods *3*, 519–24.
[22] Elangovan, M., Wallrabe, H., Chen, Y., Day, R. N., Barroso, M. and Periasamy, A. (2003). Characterization of one- and two-photon excitation fluorescence resonance energy transfer microscopy. Methods *29*, 58–73.

Laboratory Techniques in Biochemistry and Molecular Biology, Volume 33
FRET and FLIM Techniques
T. W. J. Gadella (Editor)

CHAPTER 8

Spectral imaging and its use in the measurement of Förster resonance energy transfer in living cells

Steven S. Vogel,[1] Paul S. Blank,[2] Srinagesh V. Koushik,[1] and Christopher Thaler[1]

[1]National Institute on Alcohol Abuse and Alcoholism, National Institutes of Health, 5625 Fishers Lane, Rockville, Maryland 20892
[2]National Institute of Child Health and Human Development, National Institutes of Health, 10 Center Drive Bldg. 10 Bethesda, Maryland 20892

Förster resonance energy transfer (FRET) imaging is a form of microscopy that allows the visualization of interaction between two fluorophores on a 1–10 nm scale. FRET imaging is based on measuring subtle changes in fluorescence that arises from nonradiative energy transfer from a "donor" fluorophore to an "acceptor." Interest in applying FRET to image molecular interactions inside living cells has been growing, but has been hampered by technical problems encountered when accurate and precise measurements of fluorescence is necessary. Ironically, a requirement for FRET to occur, spectral overlap of donor emission with acceptor absorption makes it technically difficult to accurately and efficiently measure the fluorescent signals required to quantify FRET. Spectral imaging is a relatively new form of fluorescence light microscopy where

DOI: 10.1016/S0075-7535(08)00008-9

a fluorescent emission spectrum is recorded at each location in an image. Many of the technical problems encountered when acquiring FRET images can be eliminated by analyzing the data encoded in spectral images with an image analysis algorithm called linear unmixing. In this chapter, we will cover the theory of linear unmixing of spectral images, and describe how it can be used to acquire accurate FRET measurements.

8.1. Introduction

Förster resonance energy transfer (FRET) is a physical phenomena in which photon energy absorbed by a fluorophore is transferred by nonradiative dipole–dipole coupling to a nearby chromophore [1]. While this arcane phenomenon was first observed in the 1920s, there has been a reemergence of interest in FRET driven in part by the need for a microscope based assay for monitoring protein–protein interactions inside living cells. Commercial interest in FRET has also grown, driven by the conviction that FRET can be used effectively as the basis for developing new biosensors. Spectral imaging microscopy [2–4] is a relatively new form of multidimensional fluorescence microscopy that can potentially eliminate several of the obstacles one encounters in FRET imaging [5–7]. In spectral imaging, each picture element or pixel maps to a specific Cartesian coordinate within a sample and encodes the complex spectrum emitted from the population of fluorophores present at each specific location. Compared with more conventional forms of fluorescence microscopy in which the emission intensity is detected through a filter, spectral imaging holds the promise of potentially revealing information about the abundance and identity of the fluorescent species present. Vis-à-vis FRET, the emission spectrum of a donor should be attenuated and the emission spectrum of acceptors should be potentiated as energy transfer increases. Moreover, spectral imaging has the potential to detect these FRET related spectral changes in a photon efficient manner, which is critical for effective live-cell imaging. One goal of this chapter is to describe how spectral imaging can be used to detect and measure FRET and to convey the strengths and weaknesses of the approach. FRET is not a rare event in biology as it is fundamental to the process of photosynthesis [8]. FRET can also occur, often unintentionally, upon the introduction of fluorophores into a biological milieu, particularly into crowded environments such as in membranes [9]. Fluorophore tagged membrane proteins, that are not commonly thought to specifically interact, often undergo non-specific FRET by virtue of their close proximity [10, 11]. Thus, it should be clear that the apparent abundance of fluorophores discerned by both conventional light microscopy through emission filters, as well as spectral imaging will be erroneous if energy transfer is occurring [12]. We call this the FRET problem. Accordingly, a second goal of this chapter will be to convey an appreciation of this fundamental problem in quantitative fluorescence microscopy, and to outline how spectral imaging can be used to measure the true abundance of fluorophores, even when FRET is occurring. Finally, we wish to state that this chapter is not intended to be a review of the literature addressing spectral imaging or how it has been used to measure FRET, rather, our goal is to convey a more intuitive appreciation of the spectral FRET method, the strengths and weaknesses of the approach, as well as to identify some of the current technical limitations that we are hopeful will soon be overcome, perhaps by some of our readers.

8.2. Understanding spectral imaging

As mentioned above, spectral imaging microscopy is a form of multidimensional fluorescent microscopy where a fluorescent emission spectrum is acquired at each coordinate location in the sample. This mode of imaging has been implemented for wide field, confocal, and two-photon laser scanning microscopy, and several excellent reviews on spectral imaging microscopy have been published [2, 4, 13–18]. To illustrate and contrast spectral imaging with conventional filter-based imaging, we imaged three glass capillaries (Fig. 8.1). The first capillary (from top to bottom) contained Cerulean (10 μM) [19], a cyan spectral variant of green-fluorescent protein (GFP); the second contained Venus (10 μM) [20], a yellow spectral variant of GFP; and the third capillary contained a mixture of Cerulean and Venus at unknown concentrations. The monomeric variant of Cerulean and Venus were used to avoid nonspecific interactions between these proteins [21]. Two-photon microscopy with 900 nm infrared excitation light was used to acquire a spectral image of the three capillaries (Fig. 8.1). In panel A, we see 32 individual images of the capillaries. In the implementation of spectral imaging used to acquire these images (a Zeiss 510 NLO/META), the fluorescence emitted from a sample is spectrally dispersed by a diffraction grating and projected onto a 32-node photomultiplier tube array. Each of the images depicted in panel A was acquired from a single node of this photomultiplier array and represents a ~10–11 nm spectral slice of the emission spectrum. The center wavelength is indicated in yellow in the upper left corner of each image. You can see that none of the three capillaries contained samples that emit below 446 or above 596 nm. Between 457 and 510 nm, the top and bottom capillaries are emitting but not the center. Between 521 and 596 nm, all three capillaries emit. The emission wavelength and intensity information encoded in the stack of images depicted in Fig. 8.1A can be distilled down into a color-coded 'spectral image' shown in panel B. This image depicts how the sample would appear if observed by the human eye. The color at each pixel is based on the weighted mean of the emitted photon wavelengths, and the intensity is based on the number of photons detected at that pixel location. In panel C, we show the emission spectra for regions of interest (ROI) centered over the Cerulean capillary (blue solid line), Venus capillary (yellow solid line), the mixture capillary (green dashed line), and from an ROI centered over a region that did not have a capillary (background, red dotted line).

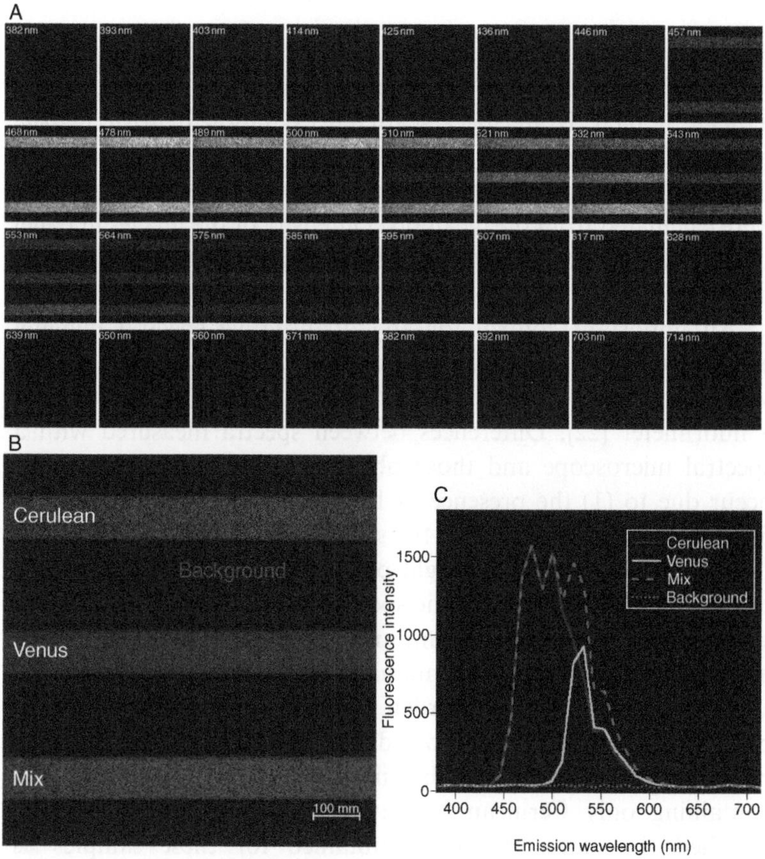

Fig. 8.1. Spectral Imaging. Panel A shows 32 individual images that comprise a spectral image of three capillaries containing from top to bottom 10 μM Cerulean, 10 μM Venus, and a mixture of Cerulean and Venus of unknown stoichiometry acquired on a Zeiss 510 META/NLO laser scanning confocal microscope with 900 nm two-photon excitation. Each image was acquired from a single node of a 32 node photomultiplier array, and represents sequential ~10 nm portion of the emission spectrum that are detected on each node after being dispersed by a diffraction grating. The emission wavelength measured by each node is indicated in yellow. Panel B shows a color-coded representation of the images shown in panel A. The color (wavelength) is the intensity weighted average of the 32 individual emission wavelengths, and the brightness is proportional to the total number of photons detected at each pixel. Each capillary is labeled as is a background region. Size bar is 100 μm. Panel C shows emission spectra calculated measured from ROI centered over the Cerulean capillary, Venus capillary, mixture capillary or from a background region. (See Color Insert.)

It is useful to monitor the background spectrum because it can indicate if there is any back-scattered excitation light or higher-order scattering reaching the detector [12], as well as an indicator of the presence of any autofluorescence in a sample. These spectra were calculated from the mean pixel intensity of the images in panel 1A for the specific ROI regions indicated, and plotted as a function of wavelength.

An important control for any quantitative spectral imaging experiment is to compare emission spectra of pure fluorophore samples obtained on the microscope with spectra obtained using a fluorimeter [22]. Differences between spectra measured with a spectral microscope and those obtained using a fluorimeter can occur due to (1) the presence of filters or dichroic beam splitters in the emission light path, (2) the spectral throughput of the objective and other optics in the light-path, (3) wavelength aliasing due to the limited bandwidth of the spectral detector, (4) misalignment of the dispersed emission beam and the spectral detector, and as mentioned previously, (5) contamination by backscattered excitation light including higher-order scattering. The impact of these potential artifacts must be considered when quantitative spectral imaging is desired. Because the emission spectra of the capillaries containing only Cerulean or Venus shown in panel C closely matched the emission spectra obtained for these samples as measured on conventional nonimaging fluorimeters, these artifacts could be eliminated from further consideration.

One of the major advantages of spectral imaging can be appreciated by noting that the emission spectrum of Cerulean in Fig. 8.1C completely overlaps the emission spectrum of Venus. At the excitation wavelengths used here, conventional filter-based imaging can never isolate Venus emission from Cerulean emission. To quantify the abundance of Venus in the presence of Cerulean using filter-based technology requires the existence and use of excitation wavelengths that excite Venus without exciting Cerulean. As we shall see shortly, this is not a requirement for quantifying Venus and Cerulean using spectral imaging.

Next, we will explore how spectral images change as a function of excitation wavelength. In Fig. 8.2A, we see spectral images of the same three capillaries depicted in Fig. 8.1 but now obtained with 820, 900, 920, and 940 nm two-photon excitation. With 820 nm excitation Cerulean is readily excited while Venus is poorly excited. In contrast, with 940 nm excitation, Cerulean is poorly excited while Venus is excited well. With 900 and 920 nm excitation, intermediate excitation behavior is observed. The Cerulean capillary emitted blue fluorescence at all excitation wavelengths. Similarly, the middle Venus capillary appeared green at all excitation wavelengths.

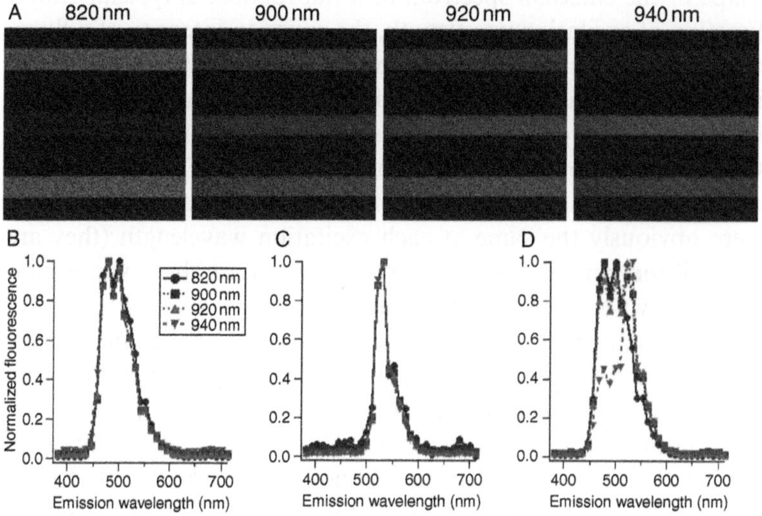

Fig. 8.2. Spectral Images can change with excitation wavelength. Spectral images of the same three capillaries depicted in Fig. 8.1 were imaged at four different excitation wavelengths: 820, 900, 920, and 940 nm (A). Normalized emission spectra (at all four excitation wavelengths) from ROI's centered over the Cerulean capillary (B), the Venus capillary (C), and the mixture capillary (D) are co-plotted. Note that the color and normalized emission spectra do not change for a sample containing a single fluorophore. In contrast the color and normalized spectrum changed as a function of excitation wavelength in the capillary containing both fluorophores. (See Color Insert.)

In contrast, the color of the Cerulean–Venus mixture capillary (bottom) appeared blue at 820 nm, blue-green at 900 and 920 nm, and green at 940 nm. The normalized emission spectra of the Cerulean (Fig. 8.2 B) and Venus (Fig. 8.2 C) capillaries did not change as a function of excitation wavelength, yielding the expected characteristic spectrum of these fluorophores. In stark contrast, the normalized emission spectra of the Cerulean–Venus mixture capillary was different at each excitation wavelength used (Fig. 8.2 D) with subtle differences between the emission spectra observed with 820, 900, and 920 nm excitation, and a dramatic difference with 940 nm excitation. These differences can be understood if one considers that while the shape of the emission spectrum of a fluorophore is typically not a function of excitation wavelength, the emission intensity of a fluorophore is a function of excitation wavelength [23]. The emission intensity is also a function of the fluorophore's abundance and the intensity of the excitation light source [23]. With regard to the Cerulean and Venus mixture capillary shown in Fig. 8.2, the concentrations of the two fluorophores and their relative abundance were obviously the same at each excitation wavelength (they are after all the same sample). In this demonstration, however, excitation energy was not the same at each excitation wavelength (though the excitation energy for the mixture capillary was the same for the Cerulean and Venus capillaries), and it is known that Cerulean and Venus have significantly different absorption spectra [19, 20] and two-photon absorption cross sections [12]. At a Venus concentration far less than 1–3 mM, there should not be any appreciable FRET between Cerulean "donors" and Venus "acceptors" as a result of molecular crowding [23]. Thus, for the mixture capillary, the complex spectrum observed should be a linear sum of the Cerulean and Venus emission spectrum (a function of excitation light intensity and their absorption coefficients at each excitation wavelength) weighted by their respective abundance. A mathematical formalism, called *linear unmixing*, based on this idea of the abundance-weighted summation of individual fluorophore spectra, can be used to measure the concentration of individual fluorophores in a mixed population, but only if energy transfer between different fluorophores is not occurring [12, 24].

8.2.1. *Linear unmixing and its limitations*

As mentioned previously, the complex emission spectrum $F^i(\lambda)$ of samples containing multiple fluorophores is assumed to be the linear sum of individual component spectra $F_1(\lambda)$, $F_2(\lambda)$, $F_3(\lambda)$, weighted by their abundance x_1, x_2, x_3. Let $F_1^i(\lambda)$ and $F_2^i(\lambda)$ be the reference emission spectra of pure samples of fluorophore (e.g., Cerulean and Venus). The term *reference emission spectra* is used because these spectra describe the emission at excitation wavelength λ_{ex}^i of a defined concentration of fluorophore (e.g.,10 μM) acquired using the same excitation light intensity as was used to acquire an emission spectra of an unknown sample mixture. Under these conditions, the shape and magnitude of the fluorophore mixture spectra will be:

$$F^i(\lambda) = x_1 F_1^i(\lambda) + x_2 F_2^i(\lambda) + x_3 F_3^i(\lambda) + \ldots \qquad (8.1)$$

If reference emission spectra of a set of pure fluorophores are available, and if an emission spectrum of an unknown mixture of any combination of these fluorophores is acquired under the same conditions, this equation can be used to determine the abundance of the different fluorophores in the mixture. The use of this equation to determine the abundance of the fluorophores present is called linear unmixing. To illustrate the basis of linear unmixing, we will first use this equation to analyze the emission spectra of the mix capillary containing an unknown mixture of Cerulean and Venus depicted in Fig. 8.1. The unmixing approach we describe will utilize reasonable guesses for the values of x_1 (representing the abundance of Cerulean) and x_2 (representing the abundance of

Venus) in the linear unmixing equation, and then use these guesses to generate spectra to compare with actual data. In Fig. 8.3A, we see a 10 μM reference emission spectrum of Cerulean and Venus, as well as a spectrum of our unknown mixture. All three spectra were acquired at the same excitation wavelength (900 nm) at the same laser intensity. Thus, all of the requirements for linear unmixing outlined above have been met. In Fig. 8.3 B, the linear unmixing equation is used to model complex emission spectra of mixtures containing 10 μM Venus and either 14 μM Cerulean (blue dashed trace), 10 μM Cerulean (white dotted trace), or 6 μM Cerulean (red dashed trace). These models can be compared with the actual emission spectra from the mixture capillary (green circles). The model represented by the white trace is the closest match to the data set, particularly below 500 nm. Next, we explore models where we hold the Cerulean concentration at 10 μM and vary the Venus concentration (Fig. 8.3 B). In the blue trace, Venus is held at 10 μM; in the white trace, it has been reduced to 6 μM; and in the red trace, it has been reduced further to 2 μM. We can now see that while all three models are reasonable matches to the data, the blue model overestimates emissions between 500 and 600 nm and the red model underestimates emissions from this same spectral region. In contrast, the white model (10 μM Cerulean, 6 μM Venus) matched the experimental data well.

In this example of linear unmixing, the value x_2 was first held constant while the value of x_1 was varied. Next, the value of x_1 was held constant and the value of x_2 was varied. This illustration was used to provide an intuitive example of how linear unmixing finds values for the abundance of each fluorophore that are consistent with a given complex emission spectrum. In practice, linear unmixing software utilizes curve fitting algorithms [25] to rapidly find values of x_1, x_2,... which can generate spectra that best match the experimental data. These calculations are repeated for every pixel in an image. Ultimately, separate images are created representing the abundance of each fluorophore present. An example set of images generated by linear unmixing of the data set presented in

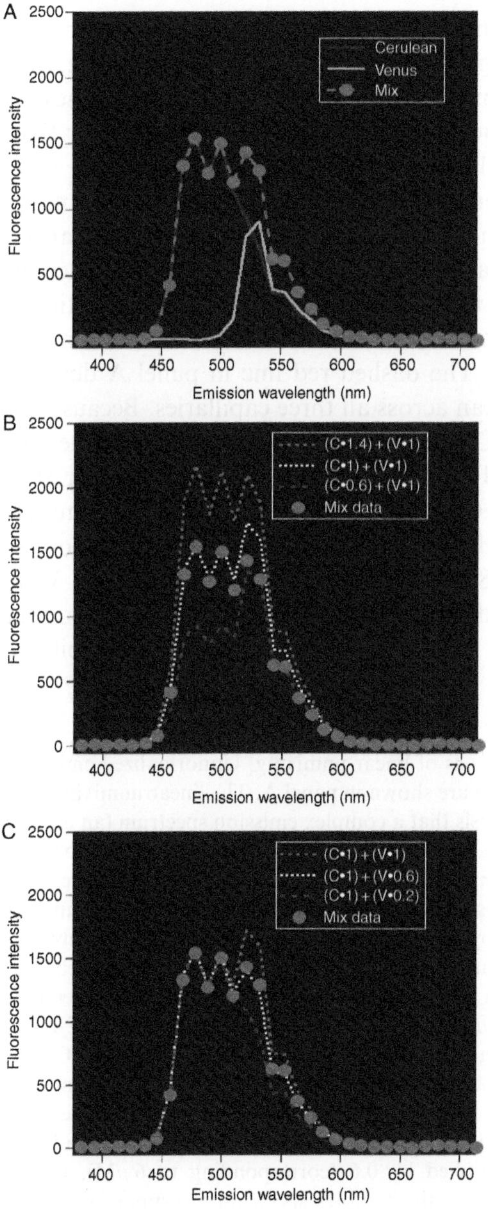
A
Cerulean
Venus
Mix
Fluorescence intensity
Emission wavelength (nm)
B
(C•1.4) + (V•1)
(C•1) + (V•1)
(C•0.6) + (V•1)
Mix data
Fluorescence intensity
Emission wavelength (nm)
C
(C•1) + (V•1)
(C•1) + (V•0.6)
(C•1) + (V•0.2)
Mix data
Fluorescence intensity
Emission wavelength (nm)

Fig. 8.1 is shown in Fig. 8.4. In panel A, we show the Cerulean channel image (left) generated by linear unmixing, the Venus channel image (middle), and an overlay image of these two channels (right). In the Cerulean channel image, it is clear that the top and bottom capillaries each contain Cerulean at the same concentration. The Venus channel image revealed that the middle and lower capillaries each contain Venus, but the concentration of Venus in the middle capillary was higher than in the lower capillary. A more quantitative view of this data can be seen by plotting the intensities (calibrated to concentration) of a line scan across these images (Fig. 8.4 B). The dashed red line in panel A depicts the position of the line scan across all three capillaries. Because known concentrations of Cerulean and Venus were used in the top and middle capillaries (10 μM), the intensities measured for these capillaries (blue for the Cerulean channel, yellow for the Venus channel) could be calibrated to that concentration of fluorophore (dotted black line in Fig. 8.4 B). In essence, the Cerulean and Venus capillaries can be considered as calibration controls for interpreting the signals obtained by linear unmixing of the mixture capillary. In the line

Fig. 8.3. The basis of linear unmixing. Unnormalized emission spectra of the three capillaries are shown in panel A. The linear unmixing algorithm is based on the hypothesis that a complex emission spectrum (an emission spectrum of a sample containing 2 or more fluorophores) can be modeled as a weighted sum of the emission spectra of the individual fluorophores present. Thus, the Mix spectrum should be the sum of the Cerulean and Venus spectra after each is multiplied by an abundance factor. In panel B the abundance factor for Venus is held at a value of 1, while the value of the Cerulean abundance factor is varied from 0.6 to 1.4. Because the Cerulean and Venus capillaries each contained 10 μM of fluorophore, an abundance range of 0.6–1.4 corresponds to a concentration range of 6–14 μM. In panel C the Cerulean abundance factor is held at a value of 1 (10 μM) while the abundance factor for Venus was altered from 0.2 to 1 (2–10 μM). Note that when the Cerulean spectrum was multiplied by 1 (corresponding to 10 μM) and added to the Venus spectrum multiplied by 0.6 (corresponding to 6 μM), the linear unmixing model matched the complex spectrum measured for the mix capillary.

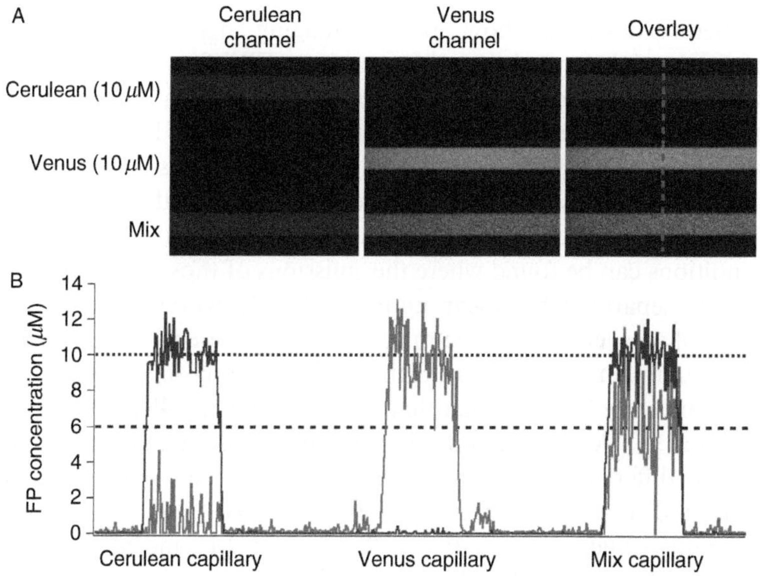

Fig. 8.4. Linear unmixing with curve fitting algorithms. Linear unmixing image processing software utilize least-square curve fitting routines to fit a spectra from each pixel of a spectral image to the linear unmixing equation and predict values for the abundance factors for each fluorophore. These abundance factors for Cerulean and Venus are then multiplied by the concentration of the individual reference spectrum samples (10 μM) to produce a Cerulean Channel image (blue), a Venus Channel image (yellow), and an overlay image (panel A). Now it can be seen that the top and bottom capillaries have the same concentration of Cerulean. The middle and bottom capillaries both have Venus, but at different concentrations. The dashed red line indicates the location of a line scan across the two image channels. Line scan plots for each channel are useful for measuring the actual concentration of fluorophores observed in a sample, and are plotted in panel B. (See Color Insert.)

scan trace corresponding to the position of the mixture capillary (the far right), we see a Cerulean signal (blue trace) of 10 μM (dotted line), and a Venus signal (yellow trace) of 6 μM (dashed line). These values are consistent with the values generated by spectral analysis in Fig. 8.3.

Before describing how spectral imaging in conjunction with linear unmixing can be used to measure FRET, we should first consider what limitations exist for successful quantitative linear unmixing of spectral images. First and foremost, linear mixing will fail if the emission spectra of the fluorophores present in a sample are nearly indistinguishable. As a rule of thumb, if two fluorophores have emission maxima separated by at least 10 nm, conditions can be found where the emissions of those fluorophores can be separated by linear unmixing [2]. However, even if the fluorophores emission spectra are clearly different, quantitative linear unmixing can still fail if the emission from one fluorophore is significantly brighter than that from the others [24]. An obvious question is "At what point will quantitative linear unmixing fail due to differences in fluorophore intensity?" This question was addressed in a study using mixtures of purified CFP and YFP at different defined molar ratios (1:9, 1:1, and 9:1) and at different emission intensity ratios [24]. Spectral images of these mixtures were acquired and used to determine how well the CFP:YFP ratio predicted by linear unmixing matched the specific samples. It was found that linear unmixing could accurately predict the correct CFP:YFP ratio for all three mixtures. Linear unmixing yielded the correct molar ratios even under conditions where one fluorophore was 90 times brighter than the other. It must be pointed out, however, that the variance in the fluorophore ratio measurement observed under these extreme differences in intensity was quite large. Because of signal-to-noise limitations in spectral microscopes in conjunction with limitations in the dynamic range of data acquisition instrumentation (typically 12 bits), it is difficult to use linear unmixing to accurately separate signals whose intensities are greater than an order of magnitude apart. Essentially, if the dynamic range of a spectral detector is set to capture the peak emission of the brightest fluorophore (without clipping the signal), that dynamic range will be poorly matched to accurately measure the emission of a much dimmer fluorophore. Under these conditions, linear unmixing will yield accurate measurements for the abundance of the bright fluorophore, but the accuracy of the dim fluorophores abundance may be compromised. This problem is compounded by the fact that it is often difficult to control the abundance of fluorophores in a biological experiment, and even more difficult to control their relative abundance in living cells. The relative brightness of the light emitted from the fluorophores present in a mixed sample is a function of their absorption spectra (and thus the excitation wavelength), their quantum yields, as well as their relative abundance. Thus, one solution to this problem is to empirically select an excitation wavelength for a given sample that yields a complex emission spectrum whose shape is significantly different from the emission spectrum of the individual fluorophores alone. To illustrate this approach, linear unmixing was applied to the four spectral images depicted in Fig. 8.2. The same set of capillaries was imaged at four different excitation wavelengths. At 820 nm, Cerulean was excited well but Venus was barely excited. At 900 and 920 nm, both were excited well, and at 940 nm Venus was excited more efficiently than Cerulean. In Fig. 8.5A, we replot the emission spectra of the Cerulean capillary (blue) and the Mix capillary (red) at each excitation wavelength, but here each spectrum is normalized to the intensity of its Cerulean emission peak at 478 nm. Because the Cerulean spectrum is overlaying the Mix spectrum in these graphs, the visible part of the Mix spectrum (red) represents the portion of the Mix spectrum resulting from Venus emission. With 820 nm excitation, the Mix emission spectrum is almost identical to the Cerulean spectrum (note the token red signal between 500 and 600 nm). In contrast, the Mix spectra resulting from 900, 920, and 940 nm excitation were all significantly different than the Cerulean alone spectrum (blue), as well as from a Venus alone spectrum (data not shown). Furthermore, at these three excitation wavelengths the fractional contribution of Cerulean and Venus to the Mix spectra emissions was all within an order of magnitude of each other. Thus, we would predict that the 900, 920, and 940 spectral images (but not the 820 nm spectral image) will yield accurate estimates of the abundance of Cerulean and

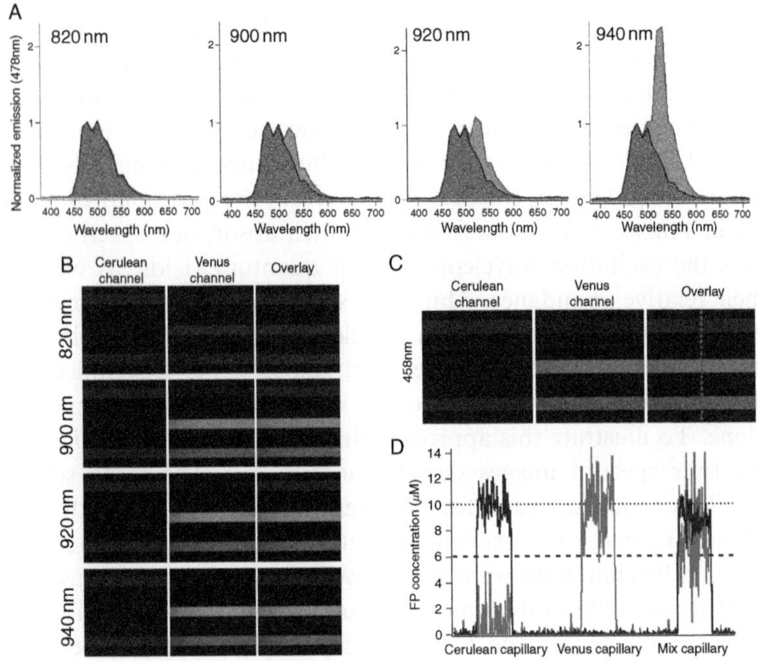

Fig. 8.5. The effects of excitation wavelength on linear unmixing. In panel A we see the Cerulean normalized emission spectra of the mixture capillary (red) overlaid with the normalized emission spectrum of Cerulean alone (blue) at the four different two-photon excitation wavelengths used in Fig. 8.2. Note that with 820 nm excitation the mix spectrum was not significantly different than the Cerulean spectrum, while with 900, 920, and 940 nm excitation they were different. When the four spectral images depicted in Fig. 8.2 were processed by linear unmixing they produced the four unmixing channel-sets depicted in panel B. The unmixed images with 900, 920, and 940 nm excitation were all identical, while with 820 nm excitation erroneous results were obtained. In panel C the results of linear unmixing of a spectral image of the same samples, obtained with one-photon excitation at 458 nm is shown. These images looked identical to the images obtained with 900, 920, and 940 nm two-photon excitation, and a line scan across the one-photon image confirmed this conclusion (compare panel D with Fig. 8.4B). (See Color Insert.)

Venus with linear unmixing analysis. To test this prediction, the linear unmixing algorithm was applied to these data sets (Fig. 8.5 B). As expected, the Cerulean and Venus unmixing channels generated by linear unmixing of the 900, 920, and 940 nm excitation spectral images (as well as the overlay channel) were virtually identical. In contrast, the unmixing of the 820 nm spectral image dramatically underestimated the abundance of Venus in the Mix capillary. Clearly, judicious selection of excitation wavelength is important for quantitative linear unmixing of spectral images, and the guidelines mentioned earlier can help in the selection of those wavelengths.

8.2.2. Single- versus multiphoton spectral imaging

Can conventional one-photon excitation be used to acquire spectral images for quantitative linear unmixing, and if so, are there any advantages or disadvantages of two-photon excitation over one-photon? Fig. 8.5C depicts the Cerulean, Venus, and overlay channels generated by linear unmixing of a spectral image of our three capillaries acquired with one-photon excitation using the 458 nm laser line of an argon laser. These images were nearly identical to the linear unmixing images generated for the same sample excited with 900, 920, and 940 nm two-photon excitation (Fig. 8.5B). Similarly, a line scan across this unmixed one-photon image (red dotted line in Fig. 8.5C) was equivalent to a line scan across a linear unmixed two-photon image (compare Fig. 8.5D with Fig. 8.4B). Thus, we conclude that quantitative linear unmixing can work well with spectral images acquired with either one-photon or two-photon excitation. There are differences between two-photon and one-photon microscopy that may impact on their suitability for quantitative spectral imaging microscopy. First, the spectral band involved in two-photon excitation (typically from 700 to 1000 nm) is well separated from the spectral region where most commonly used fluorophores emit (400–600 nm). Thus, with two-photon excitation, the whole emission spectrum of a sample can be easily collected. In contrast, because the wavelength for one-photon excitation often overlaps with a samples emission spectrum a portion of the emission spectrum typically is lost. Even when one-photon excitation is left shifted from emission, the emission filters and dichroic beam splitters involved in preventing one-photon excitation from reaching the spectral detector can attenuate the emission signal over a portion of the emission spectrum. The ability to collect a full emission spectrum potentially translates into a more photon-efficient microscope, and this is key for successful live cell imaging. Second, because the lasers used in most two-photon microscopes are tunable, it is relatively easy and straightforward to select an excitation wavelength that is optimized for the specific fluorophores present in a sample as well as to their stoichiometry in the sample. In contrast, one-photon laser scanning confocal microscopes typically use lasers with a limited number of fixed laser lines and are thus poorly suited for optimizing the excitation wavelength for spectral imaging. Recently, however, new white light laser sources have become available which can support one-photon excitation over a broad range of wavelengths. Wide-field conventional fluorescence microscopes typically use broad wavelength light sources in conjunction with excitation filters. Although the excitation wavelength can be optimized for a particular sample, this would require having a large set of different excitation filters on hand that can be rapidly changed to optimize excitation. This is typically not available. One alternative is to replace excitation filters with a tunable programmable excitation filter. This could potentially allow easy and convenient matching of one-photon excitation to a particular sample. Third, the depletion of fluorophores by bleaching with two-photon excitation is typically much slower than with one-photon excitation because with two-photon excitation there is no fluorophore excitation occurring above and below the image plane [26]. If different fluorophores are bleached at different rates in a sample, emission spectra will change as a function of acquisition time. This will greatly complicate the quantitative analysis of spectral images. Alternatively, for some fluorophores, multiphoton excitation can have different selection rules than one-photon excitation. Thus, it may be possible with two-photon excitation to directly populate a fluorophores triplet state. If this happens, the bleach rate, at least in the image plane, can be much faster [27]. Finally, two-photon absorption cross-sections are in general broader than that of one-photon absorption spectra. Thus, with two-photon excitation it is relatively easy to find wavelengths that excite multiple fluorophores simultaneously.

8.2.3. Baseline correction and emission wavelength selection

We have demonstrated how the selection of the excitation wavelength for acquiring spectral images can influence how accurate linear unmixing can be in predicting the abundance of fluorophores present in a mixed sample. For this reason, there is a great advantage in having a spectral microscope with a large, preferably continuous selection of excitation wavelengths so that the relative intensities of the fluorophores present can be modulated to optimize the signal to noise ratio obtainable for all fluorophores present. Two-photon microscopy is well suited for this task. We will now address two other factors that can influence how quantitative the results of linear unmixing can be: (1) baseline corrections of reference spectra, and (2) the wavelength range and resolution of the spectral detector used. The Cerulean, Venus, and Mix capillary spectra shown in Fig. 8.3A all have baseline intensities of zero (seen below 425 and above 625 nm). In practice, the baseline of spectra obtained using a spectral microscope will have a positive offset from zero as a result of dark noise and instrumentation noise that is added to the true signal. If these offsets are not corrected before applying the linear unmixing algorithm, significant errors can be introduced [24]. Baseline correction can be implemented in hardware by properly configuring the gain and offset of the spectral detector. Alternatively, Baseline corrections can also be implemented with postprocessing by measuring the offset for the spectra and subtracting it from the signals. A third approach for Baseline correction is only partially effective. In this approach, in addition to using reference spectra for each fluorophore present in a sample, the spectrum of the background signal (e.g., see red trace in Fig. 8.1 C) can also be used by the linear unmixing algorithm. This will produce, in addition to image channels for each fluorophore, an additional image channel of the background which will not be used in any subsequent overlay images. While this approach can effectively remove the offset from the sample image, it does not remove offset from the individual fluorophore unit reference spectra. These offsets must be corrected before the linear unmixing algorithm is applied. For the best photon efficiency, the wavelength range of a spectral detector used for linear unmixing should cover the full range of a samples complex emission spectrum. For the sample depicted in Fig. 8.1, this range would be from 450 to 600 nm. To effectively implement the Baseline corrections described earlier, however, this range should be extended a bit (i.e., from 400 to 650 nm), so that the true baseline can be defined. In the example depicted in Fig. 8.1, the spectral detector used captured a spectral image comprised of 32 separate images (Fig. 8.1 A). Theoretically, linear unmixing of a sample containing n fluorophores will require a spectral image composed of at least n spectral points (or subimages) [28], and those images must cover the portions of the spectrum where the individual fluorophores emission spectra differ. Thus, theoretically, Cerulean and Venus emissions from the mixture capillary could be accurately separated using the linear unmixing algorithm if applied to a spectral image composed of only two spectral points; one recording from 450 to 500 nm, and the other covering 500–600 nm. Although the theory describing the influence of detector wavelength range, and resolution on linear unmixing has been described [28], experimental confirmation would be a welcome addition to the literature.

8.3. *What is the spectral signature of FRET?*

Finally, an underlying assumption of quantitative linear unmixing is the absence of mechanisms that can selectively alter the emission spectra of the individual fluorophores present in a sample. Because FRET shifts energy from donor to acceptor fluorophores, FRET activity will decrease the magnitude of a donors emission spectrum and increase the magnitude of the acceptors emission spectrum. In practice, this means that quantitative linear unmixing of a sample with a FRET efficiency greater than zero will fail to yield accurate estimates of the abundance of donors and acceptors present. This has been confirmed experimentally [12]. This is both good and bad news. The bad news is that studies that have used quantitative linear unmixing of spectral images without regard to the possibilities that the fluorophores present in their sample might be transferring energy by FRET may be based on erroneous unmixing data. The good news is that because linear unmixing will yield different results in the presence or absence of FRET, this behavior can be used to measure FRET from spectral data sets.

It has been shown that the linear unmixing equation used to estimate the abundance of fluorophores in a sample will fail if FRET is occurring between the fluorophores in a sample. Devising a strategy for using spectral imaging to measure FRET requires a quantitative understanding of how the shape and magnitude of a complex emission spectrum changes with FRET. A *FRET adjusted spectral equation* which accurately describes the shape and magnitude of the complex emission spectrum ($F^i(\lambda)$) from a population containing two fluorophores that might be transferring energy by FRET is:

$$F^i(\lambda) = \boldsymbol{d}(1 - E_D)F^i_{D,\mathrm{ref}}(\lambda) + \boldsymbol{a}F^i_{A,\mathrm{ref}}(\lambda) + \boldsymbol{d}E_D\frac{Q_A}{Q_D}k^iF^i_{A,\mathrm{ref}}(\lambda) \tag{8.2}$$

where d and a represents the donor and acceptor fluorophores concentrations, $F^i_{D,\mathrm{ref}}(\lambda)$ and $F^i_{A,\mathrm{ref}}(\lambda)$ are the reference emission spectra of the donor and acceptor at the same concentration as measured on the spectral microscope being used. It is important to note that these spectra are a function of the excitation wavelength, λ^i_{ex}. E_D is the *apparent* FRET efficiency of the sample, that is the fraction of donor excitation events that results in energy transfer to an acceptor for all of the donors present at a specific pixel in an image. Q_D and Q_A are the quantum efficiency of the donor or acceptor, and k^i is a transfer factor (see the Appendix) [12]. This equation can be understood intuitively by realizing that the complex emission spectrum of a population of donor and acceptor fluorophores (that may be transferring energy by FRET) is comprised of the sum of three parts. The first part describes the fluorescent emission from directly excited donor fluorophores in a sample:

$$\boldsymbol{d}(1 - E_D)F^i_{D,\mathrm{ref}}(\lambda) \tag{8.3}$$

It is the reference emission spectrum of the donor, multiplied by its abundance, but attenuated by the fraction of donors that are not transferring energy by FRET. The second part of the equation describes emission from directly excited acceptors:

$$\boldsymbol{a}F^i_{A,\mathrm{ref}}(\lambda) \tag{8.4}$$

It is simply the reference emission spectrum of the acceptor, multiplied by the abundance of acceptor. Finally, the last part of the equation represents the energy transferred from directly excited donors to acceptors by FRET, and then emitted as acceptor fluorescence:

$$\boldsymbol{d}E_D\frac{Q_A}{Q_D}k^iF^i_{A,\mathrm{ref}}(\lambda) \tag{8.5}$$

It is the emission spectrum of the acceptor multiplied by the abundance of the *donor* (attenuated by the fraction of donor emissions that actually result in FRET), then multiplied by a factor

$Q_A/Q_D \cdot k^i$ that equates how the emission intensity increase of the acceptor corresponds to the attenuation of the donors emission. This factor is equal to the extinction coefficient ratio of the donor and acceptor at the used excitation wavelength, λ^i_{ex}, see appendix. If FRET is not occurring in a sample (i.e., $E_D = 0$), the whole equation reduces to:

$$F^i(\lambda) = \boldsymbol{d}F^i_{D,\text{ref}}(\lambda) + \boldsymbol{a}F^i_{A,\text{ref}}(\lambda) \tag{8.6}$$

This is the same equation as the standard linear unmixing equation for two fluorophores.

Standard linear unmixing of a spectral image of a sample composed of two fluorophores yields a measure of the concentration of each fluorophore present for each pixel. If FRET is occurring, linear unmixing will produce an *apparent* donor concentration (d_{apparent}) that underestimates the *true* donor concentration (d) by a factor of 1-E_D:

$$\boldsymbol{d}_{\text{apparent}} = \boldsymbol{d}(1 - E_D) \tag{8.7}$$

Linear unmixing will also produce an apparent acceptor concentration (a_{apparent}) that will over-estimate the true abundance of the acceptor (a):

$$\boldsymbol{a}_{\text{apparent}} = \boldsymbol{a} + \boldsymbol{d}E_D \frac{Q_A}{Q_D} k^i F^i_{A,\text{ref}}(\lambda) \tag{8.8}$$

A prediction based on these two equations is that the value of d_{apparent} will remain constant for spectral images taken at different excitation wavelengths, while values for a_{apparent} can change with different excitation wavelength (because a_{apparent} is a function of excitation wavelength) but only when FRET is occurring. This prediction has been experimentally verified [12], and is in essence the spectral signature of FRET. This behavior can also be used as a simple test to determine if FRET is or is not occurring in a sample. Spectral images are taken at two excitation wavelengths that yield good signal to noise ratios for both donors and acceptors.

Linear unmixing is applied to each spectral image to produce four measurements at each pixel, $\boldsymbol{d}^1_{\text{apparent}}$ and $\boldsymbol{a}^1_{\text{apparent}}$ at λ^1_{ex}; and $\boldsymbol{d}^2_{\text{apparent}}$ and $\boldsymbol{a}^2_{\text{apparent}}$ at λ^2_{ex}. If the ratio of $(\boldsymbol{d}^1_{\text{apparent}} + \boldsymbol{a}^1_{\text{apparent}})/(\boldsymbol{d}^2_{\text{apparent}} + \boldsymbol{a}^2_{\text{apparent}})$ equals 1 then FRET is not occurring.

8.3.1. Strategies for measuring FRET using spectral imaging

Strategies for using spectral imaging to measure FRET begin with the realization that the FRET adjusted spectral equation described above has *three* independent variables, d the abundance of donor, a the abundance of acceptor, and E_D the FRET efficiency. In contrast, at a single excitation wavelength, quantitative linear unmixing produces only *two* observables, d_{apparent}, the apparent donor concentration in a sample and a_{apparent}, the apparent acceptor concentration. Thus, with three unknown variables and two observables there can and will be multiple combinations of values for d, a, and E_D that can produce the same complex emission spectrum observed in a sample [7, 12]. Strategies for using spectral imaging to measure d, a, and E_D therefore must either reduce the number of independent variables in the FRET adjusted spectral equation, or produce at least one additional observable to supplement the two values measured by spectral unmixing.

One way to reduce the number of independent variables in the FRET-adjusted spectral equation is to use samples with a fixed donor-to-acceptor ratio. Under these conditions, the values of d and a are no longer independent, but rather the concentration of d is now a function of a and vice-versa. This approach is typical for the situation of FRET-based biosensor constructs. These sensors normally are designed to have a donor fluorophore attached to an acceptor by a domain whose structure is altered either as a result of a biological activity (such as proteolysis or phosphorylation), or by its interaction with a specific ligand with which it has high affinity. In general, FRET based biosensors have a stoichiometry of one donor to one acceptor. FRET activity for biosensors is typically reported as a "FRET-ratio" index value rather than as a FRET efficiency. This is unfortunate because FRET indices are often nonlinear [11, 29]. For this example, where a single donor is linked to a single acceptor, the FRET-adjusted spectral equation will have only two variables, x, the concentration of the donor and acceptor ($x = d = a$), and E_D, the FRET efficiency. Linear unmixing will yield:

$$\boldsymbol{d}_{\text{apparent}} = \boldsymbol{x}(1 - E_D) \tag{8.9a}$$

$$\boldsymbol{a}_{\text{apparent}} = \boldsymbol{x}\left(1 + E_D \frac{Q_A}{Q_D} k^i\right) \tag{8.9b}$$

In this situation because these equations have only two variables E_D and x, and two observables d_{apparent} and a_{apparent} this problem is determined, and these two simultaneous equations can be solved for E_D and x:

$$E_D = \frac{\boldsymbol{a}_{\text{apparent}} - \boldsymbol{d}_{\text{apparent}}}{\boldsymbol{a}_{\text{apparent}} + \boldsymbol{d}_{\text{apparent}} \frac{Q_A}{Q_D} k^i} \tag{10a}$$

$$x = \frac{\boldsymbol{a}_{\text{apparent}} + \boldsymbol{d}_{\text{apparent}} \frac{Q_A}{Q_D} k^i}{1 + \frac{Q_A}{Q_D} k^i} \tag{10b}$$

Using these equations, the values of E_D and x can be calculated from the measured values of d_{apparent} and a_{apparent}. The advantage of this approach is that it is simple and only requires a microscope capable of producing spectral images, as well as software that can perform quantitative linear unmixing to produce d_{apparent} and a_{apparent} images, and then process those images according to the equations listed earlier to yield x and E_D images. The main limitation of this approach is that it will only work for samples with fixed donor-to-acceptor stoichiometry such as FRET-based biosensors.

In many biological applications of FRET, the donor to acceptor ratio of a specimen is not fixed, and unknown. Even with samples that do have a fixed donor to acceptor ratio such as in a biosensor, it is possible that cellular activities such as proteolysis might alter this ratio in unknown ways. Under these circumstances, an occult change in donor to acceptor ratio might be misinterpreted as a change in the biological activity that the biosensor was designed to monitor. Obviously, knowing the real donor to acceptor ratio in a sample as well as the actual concentrations of these fluorophores can be very useful for avoiding these types of errors, as well as for interpreting the meaning of FRET. For example, if the donor concentration in a FRET experiment is known to be much greater than the acceptor concentration, a low FRET efficiency would be expected (because most donors will not have even a single acceptor to interact with) even if the molecule that the donor fluorophore is attached to does have a high affinity for the acceptor tagged component. In contrast, a low FRET efficiency when the donor concentration is known to be much lower than the acceptor concentration might indicate that the molecule with an attached donor does not interact with the acceptor tagged molecule (though other reasons for having a low FRET efficiency must also be considered). Donor and acceptor stoichiometry and concentrations can be obtained by acquiring a spectral image in conjunction with FRET efficiency measurements obtained by another imaging method. Typically, FRET efficiency can be measured by monitoring the sensitized emission before and after acceptor bleaching [30–34], or by monitoring the fluorescence lifetime of the donor in the presence or absence of acceptors [34–38]. Regardless of the auxiliary methods used to measure the FRET efficiency hybrid approaches reduce the number of independent variables in the FRET adjusted spectral equation from three (d, a, and E_D) to two (d and a), by independently finding the value of the FRET efficiency (E_D) at each location in an image. The benefit of this approach is that under these circumstances, the true donor (d) and acceptor (a) concentrations can be calculated at each pixel in a spectral image using the independently measured FRET efficiency (E_D), in conjunction with the d_{apparent} and a_{apparent} values produced by linear unmixing using the following equations:

$$\boldsymbol{d} = \frac{\boldsymbol{d}_{\text{apparent}}}{1 - E_D} \tag{8.11a}$$

$$\boldsymbol{a} = \boldsymbol{a}_{\text{apparent}} - \frac{\boldsymbol{d}_{\text{apparent}}}{1 - E_D} E_D \frac{Q_A}{Q_D} k^i \tag{8.11b}$$

One criticism of this approach, however, is that in addition to requiring the specialized hardware for obtaining spectral images, additional instrumentation is often required to measure the FRET efficiency. Furthermore, the limitations specific to the FRET method used in conjunction with spectral imaging will also apply to this hybrid approach.

As mentioned previously, strategies for using spectral imaging to measure FRET (as well as the concentrations of donors and acceptors) must either reduce the number of variables in the FRET adjusted spectral equation, or increase the observables measured using linear unmixing. All of the methods mentioned so far work by reducing the number of variables in the spectral equation. Next, a method will be described whose approach is to measure an additional observable by acquiring two spectral images that will each be analyzed by linear unmixing. At each excitation wavelength: (λ_{ex}^1 and λ_{ex}^2), a spectral image of the sample is acquired. Reference spectra of samples containing known concentrations of donor and acceptor are also acquired at the same excitation wavelengths and excitation intensities. Quantitative linear unmixing of the first spectral image obtained at λ_{ex}^1 using the references spectra also acquired at λ_{ex}^1 will yield two observables:

$$\boldsymbol{d}^1_{\text{apparent}} = \boldsymbol{d}(1 - E_D) \tag{8.12a}$$

$$\boldsymbol{a}^1_{\text{apparent}} = \boldsymbol{a} + \boldsymbol{d}E_D \frac{Q_A}{Q_D} k^1 \tag{8.12b}$$

Linear unmixing of the spectral image acquired at λ_{ex}^2 will also yield two observables:

$$\boldsymbol{d}_{\text{apparent}}^2 = \boldsymbol{d}(1 - E_D) \tag{8.13a}$$

$$\boldsymbol{a}_{\text{apparent}}^2 = \boldsymbol{a} + dE_D \frac{Q_A}{Q_D} k^2 \tag{8.13b}$$

Note that because the value of $\boldsymbol{d}_{\text{apparent}}$ is not a function of excitation wavelength, the same values for $\boldsymbol{d}_{\text{apparent}}$ should be observed at both excitation wavelengths (λ_{ex}^1 and λ_{ex}^2). In contrast, the values measured for $\boldsymbol{a}_{\text{apparent}}^1$ and $\boldsymbol{a}_{\text{apparent}}^2$ will be different if $k^1 \neq k^2$. Thus, acquiring spectral images of a sample at two different excitation wavelengths can allow three observables to be observed with linear unmixing; $\boldsymbol{d}_{\text{apparent}}$, $\boldsymbol{a}_{\text{apparent}}^1$, and $\boldsymbol{a}_{\text{apparent}}^2$. Because the FRET adjusted spectral equation has three unknowns (d, a, and E_D) and spectral imaging at two excitation wavelengths produces three observables this problem is determined, and the three simultaneous equations describing these observables can be solved for E_D, d, and a:

$$E_d = \frac{\Delta \boldsymbol{a}}{\Delta \boldsymbol{a} + \boldsymbol{d}_{\text{apparent}} \Omega \Delta k} \tag{8.14a}$$

$$\boldsymbol{d} = \frac{\Delta \boldsymbol{a} + \boldsymbol{d}_{\text{apparent}} \Omega \Delta k}{\Omega \Delta k} \tag{8.14b}$$

$$\boldsymbol{a} = \frac{k^2 \boldsymbol{a}_{\text{apparent}}^1 - k^1 \boldsymbol{a}_{\text{apparent}}^2}{\Delta k} \tag{8.14c}$$

Where the following substitutions have been made:

$$\Delta \boldsymbol{a} = \boldsymbol{a}_{\text{apparent}}^2 - \boldsymbol{a}_{\text{apparent}}^1 \tag{8.14d}$$

$$\Omega = \frac{Q_A}{Q_D} \tag{8.14e}$$

$$\Delta k = k^2 - k^1 \tag{8.14f}$$

Using these equations the values of E_D, d, and a can be calculated from the measured values of $\boldsymbol{d}_{\text{apparent}}$, $\boldsymbol{a}^1_{\text{apparent}}$, and $\boldsymbol{a}^2_{\text{apparent}}$. The advantage of this approach is that it only requires a microscope capable of producing spectral images as well as software for performing quantitative linear unmixing. Furthermore, not only does it not require knowledge of the donor–acceptor stoichiometry of a sample, *it will actually yield this information.* Another advantage of this approach is that the FRET measurements are based on changes in both the donors' fluorescence signal as well as the acceptors'. Most other FRET methods are based on monitoring changes in either the donors' fluorescence or the acceptors' and are thus susceptible to artifacts caused by nonspecific quenching or dequenching. The major limitation of this approach is that it requires acquiring two spectral images at two different excitation wavelengths. Thus, a light source with a choice of several excitation wavelengths is required. Furthermore, if cell components move between the acquisition of the first and second spectral images, motion artifacts will introduce errors in the calculations. This is true for any analysis whose calculations involve more than one image. Finally, any variance in the tuning, bandwidth, or power of the light source used in this approach will also necessitate obtaining reference spectra at the exact same settings. This typically means measuring reference spectra before tuning to the second wavelength.

8.3.2. Measuring FRET from spectral images: sRET

Many of the strategies for measuring FRET from spectral images that were mentioned above have been implemented to study FRET. We will now cover sRET [12], a specific implementation that uses the last approach where FRET is measured from a pair of spectral images collected at different excitation wavelengths. Recently, the sRET approach has been extended to explicitly consider paired and unpaired fluorophores, the impact of incomplete labeling (or for fluorescent proteins fractional maturation), and the implementation of a calibration procedure that does not require purified fluorophores [39]. While it is important to realize that all of the approaches mentioned in the previous section are valid and may have specific advantages for particular biological problems, we have chosen to highlight the sRET approach because: (1) it can measure FRET efficiencies from samples with unknown donor to acceptor stoichiometry, (2) in addition to measuring the FRET efficiency it also measures the abundance of donors and acceptors, (3) it does not require the destruction of the sample (i.e., with sensitized emission by acceptor bleaching), (4) It measures FRET solely based on spectral images of the donor and acceptor, and (5) sRET has been shown to yield the same FRET efficiencies as obtained by fluorescence lifetime imaging (FLIM–FRET) and by a variant of the three-cube method (E-FRET) [40].

Spectral images of cells transfected with DNA encoding either a fluorescent protein construct that has a low FRET efficiency (CTV) or a high FRET efficiency (C5V) [12] were acquired with two photon excitation at 890 and 940 nm. Spectral images of capillaries containing either 7.8 μM Cerulean or Venus were also acquired at these excitation wavelengths to serve as reference spectra for linear unmixing of these spectral images, as well as to measure k^1 at $\lambda_{ex}^1 = 890$nm and k^2 at $\lambda_{ex}^2 = 940$nm for the Cerulean–Venus pair. Values measured for k^i will be specific for a particular fluorophore pair, and for the microscope used to measure spectra. It can also change as a function of the intensity, wavelength, and bandwidth of the light source used, and is therefore best measured along with the sample. These values, as well as the quantum efficiencies of Cerulean ($Q_D = 0.62$) and Venus $Q_A = 0.57$) will be needed to convert the linear unmixed images of our transfected cells into a FRET-efficiency image, a Cerulean-concentration image, and a Venus-concentration image.

Linear unmixing of each pair of spectral images for a given sample will produce an apparent Cerulean-image at $\lambda_{ex}^1 = 890$ nm ($\boldsymbol{d}_{\text{apparent}}^1$), an apparent Venus Image at $\lambda_{ex}^1 = 890$ nm($\boldsymbol{a}_{\text{apparent}}^1$), an apparent Cerulean-image at $\lambda_{ex}^2 = 940$ nm ($\boldsymbol{d}_{\text{apparent}}^2$), and an apparent Venus-image at $\lambda_{ex}^2 = 940\,nm$ ($\boldsymbol{a}_{apparent}^2$). As mentioned above, $\boldsymbol{d}_{apparent}^1$ and $\boldsymbol{d}_{apparent}^2$ should be indistinguishable, and therefore an average of these two images are used for further processing ($\boldsymbol{d}_{apparent}$). A significant difference between these two images is indicative of quenching and/or bleaching in the sample. Additionally, sample motion during the period between acquiring the two spectral images can also be responsible for this type of artifact. Regardless, if a ratio of the pixel intensities of these two donor images are significantly different than 1 they should not be used for further processing. In contrast, $\boldsymbol{a}_{apparent}^1$ and $\boldsymbol{a}_{apparent}^2$ should be different if the FRET efficiency of the sample is greater than zero. Next, these three *apparent* images that were produced by linear unmixing are processed (for each pixel) using Eq. (8.14) explained above.

Image processing with these equations produces a donor-image (d), an acceptor-image (a), and a FRET efficiency image (E_D). Examples of images produced by this process can be observed for cells transfected with either CTV, a construct that has a low FRET efficiency (Fig. 8.6), or for cells transfected with C5V, a construct that has a high FRET efficiency (Figure 8.7). In both the figures, panel A shows the donor abundance (Cerulean concentration). Panel B shows the acceptor abundance (Venus concentration), and panel D shows the color-coded FRET-efficiency image where pink to red indicates increasing amounts of FRET and white indicates low FRET efficiencies. Panel C shows a color-coded ratio image formed by dividing the image in panel B by the image in panel A. This ratio image is useful for confirming that the proteins expressed and imaged have the same donor:acceptor stoichiometry as encoded in the construct and transfected into the cell. It is also important to realize that in live cell experiments, particularly when a FRET pair is composed of spectral variants of GFP, that differential expression, maturation, folding, as well as post-translation modifications (such as proteolysis) can all modify the observed acceptor/donor ratio of an expressed construct. Both the CTV and C5V constructs encode a single Cerulean molecule

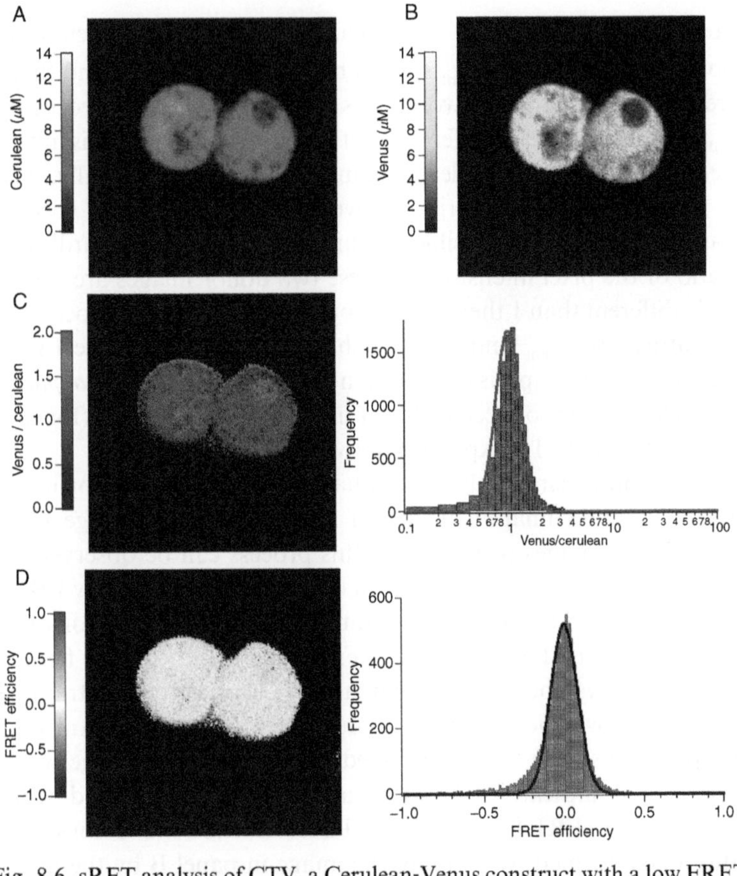

Fig. 8.6. sRET analysis of CTV, a Cerulean-Venus construct with a low FRET efficiency. sRET analysis is based on linear unmixing of two spectral images obtained at two different excitation wavelengths. Spectral images of cells expressing the CTV construct were acquired with 890 and 940 nm excitation. These spectral images, and their matching reference spectra were processed using the sRET algorithm to produce (A) a Cerulean concentration image, (B) a Venus concentration Image, (C) a Venus/Cerulean ratio image, and (D) a FRET-efficiency image. The graphs in panels C and D show frequency histograms of the pixel values for the corresponding images. The red trace in panel C is a log-normal fit to the Venus/Cerulean histogram, while the black trace in panel D is a Gaussian fit to the FRET-efficiency histogram. (See Color Insert.)

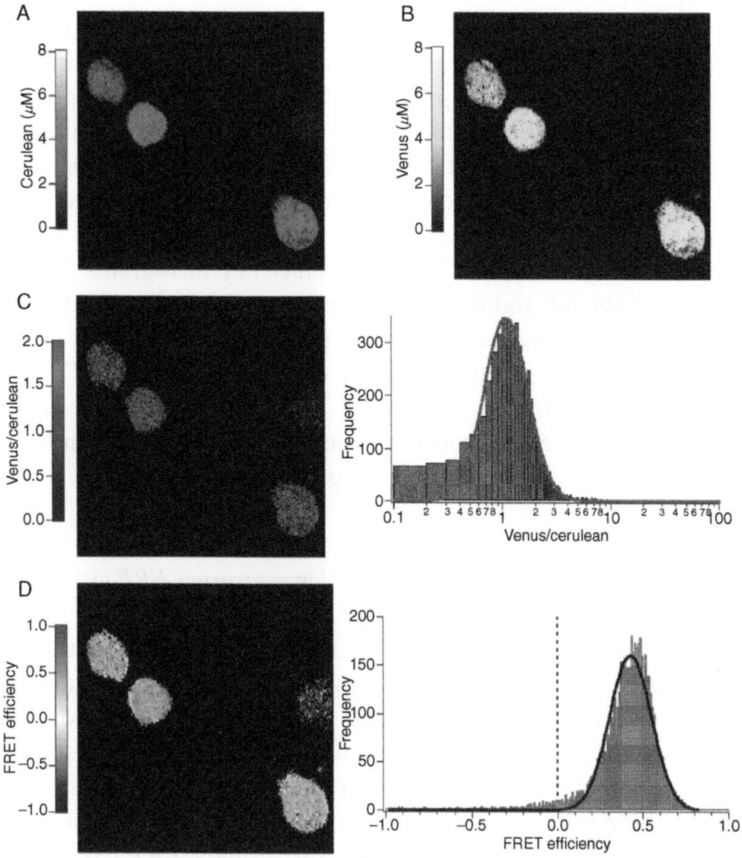

Fig. 8.7. sRET analysis of C5V, a Cerulean-Venus construct with a high FRET efficiency. Spectral images of cells expressing C5V, and their matching reference spectra were processed using the sRET algorithm to produce (A) a Cerulean concentration image, (B) a Venus concentration Image, (C) a Venus/Cerulean ratio image, and (D) a FRET-efficiency image. The graphs in panels C and D show frequency histograms of the pixel values for the corresponding images. (See Color Insert.)

concatenated to a single Venus molecule. Thus, they should have a Venus/Cerulean ratio of 1 in these images (red). To the right of panels C and D in Figs. 8.6 and 8.7 are frequency histograms of the pixel values measured for the Venus/Cerulean ratio image (panel C) or for the FRET-efficiency images (panel D). The Venus/Cerulean ratio-image histograms (in panels C of Figs. 8.6 and 8.7) are plotted on a log-scale (grey bars) and are fit to a log-normal distribution (red trace). We can see that for both CTV and C5V, these distributions peak near a value of 1 confirming that on average each Cerulean expressed in these cells is attached to a single Venus molecule and vice-versa. The FRET-efficiency histograms (in panels D of Figs. 8.6 and 8.7) are plotted on a linear-scale (red bars) and are fit to a Gaussian distribution (black trace). The peak of the FRET-efficiency distribution for the CTV image (Fig. 8.6 D) was near 0 indicating little if any FRET. In contrast, the peak of the distribution for the C5V construct was between 0.4 and 0.5 indicating an average FRET efficiency of approximately 45%.

8.3.3. Testing and validating methods for measuring FRET

What are the relative merits of different methods of measuring FRET, and is the spectral approach to FRET measurement appropriate for a specific project? First and foremost, a method for measuring FRET must produce accurate results. Only then should other factors, such as the precision of the measurements, equipment costs, photon efficiency, and speed of data acquisition be considered when selecting a FRET method. Surprisingly, FRET reference standards (i.e., compounds with known FRET efficiencies) have only become available over the past year [40]. Without reference standards it was difficult to compare one type of FRET measurement with another or even to determine if a specific implementation of FRET was in fact accurate [11]. To correct this problem, our laboratory produced three genetically encoded FRET standards with the following *'known'* FRET efficiencies: C5V ($E = 43 \pm 2\%$), C17V ($E = 38 \pm 3\%$), and C32V ($31 \pm 2\%$). These genetic constructs are based on two spectral variants of green fluorescent protein, Cerulean and Venus. The Förster radius (R_0) for energy transfer from Cerulean to Venus is 5.4 nm [41]. In these constructs, the donor was separated from the acceptor by amino acid linkers of different lengths so that they would have different FRET efficiencies. To date, we have provided these DNA encoding FRET standards to over 75 different labs. All of the constructs mentioned above have one donor and one acceptor. We have also produced other related constructs that have two donors and one acceptor (CVC) and one donor and two acceptors (VCV) [12]. These are particularly useful for evaluating how well spectral FRET methods can measure the relative abundance of donors and acceptors. Finally, we have also produced a construct (CTV) that has a very low FRET efficiency. The CTV construct encodes a single Cerulean donor attached to a single Venus acceptor. When this construct is expressed it assembles into trimers composed of three donors (in close proximity) and three acceptors (also in close proximity). The donors, however, are thought to be separated from the acceptors by at least 8 nm, and therefore this construct has a very low FRET efficiency. The CTV construct is useful as a negative FRET control, and can also be used to evaluate if different FRET methods are susceptible to errors caused by homo-FRET occurring between donors and/or acceptors. In Table 8.1 the FRET-efficiency values measured for these constructs by a spectral FRET method (sRET), as well as by fluorescence lifetime imaging (FLIM–FRET) [43, 44], and by a variant of the three-cube method (E-FRET) [42, 45] are shown. As can be seen, the spectral method produced FRET efficiencies that were similar to the other two methods. Furthermore, the sRET method also successfully determined the C5V, CVC, and VCV acceptor donor stoichiometries, demonstrating that the spectral FRET method can successfully and accurately measure FRET efficiency over a range of 3–69% with varying donor to acceptor ratios.

Another important lesson from Table 8.1 is that all three of the methods tested yielded virtually the same FRET efficiencies for the same samples. The sRET method as implemented used two-photon excitation on a Zeiss 510 META/NLO microscope, as did the FLIM–FRET method, but FLIM–FRET used auxiliary time

TABLE 8.1

FRET efficiencies and acceptor to donor ratios of FRET standards. The FRET efficiencies of six genetic constructs expressed in cell culture were evaluated by three different methods: sRET, FLIM-FRET, and E-FRET. The acceptor to donor ratio (V/C) was also measured for each construct by the sRET method. Key: C5V, Cerulean-5 amino acid linker-Venus; C17V, Cerulean-17 amino acid linker-Venus; C32V, Cerulean-32 amino acid linker-Venus; CVC, Venus flanked on each side by a Cerulean; VCV, Cerulean flanked on each side by a Venus; CTV, Cerulean-Traf2 protein domain-Venus

	FRET efficiency by Method				V/C
Construct	sRET	FLIM–FRET	E-FRET	V/C of constructs	measured by sRET
C5V	41 ± 9, $n = 62$[a]	44 ± 2, $n = 10$[a]	45 ± 4, $n = 10$[a]	1	1.0 ± 0.3, $n = 12$[b]
C17V	35 ± 9, $n = 91$[a]	39 ± 2, $n = 10$[a]	40 ± 4, $n = 18$[a]	1	n.d.
C32V	30 ± 8, $n = 81$[a]	33 ± 4, $n = 10$[a]	31 ± 2, $n = 16$[a]	1	n.d.
CVC	41 ± 5, $n = 6$[b]	41 ± 3, $n = 20$[b]	40 ± 1, $n = 28$[c]	0.5	0.5 ± 0.3, $n = 6$[b]
VCV	70 ± 6, $n = 6$[b]	65 ± 3, $n = 20$[b]	69 ± 1, $n = 21$[c]	2	2.1 ± 1.0, $n = 6$[b]
CTV	2 ± 7, $n = 12$[b]	6 ± 3, $n = 30$[b]	3 ± 1, $n = 13$[c]	1	0.9 ± 0.4, $n = 12$[b]

[a]From Ref. [40].
[b]From Ref. [12].
[c]From Ref. [42].

correlated single photon counting hardware, detectors, and software [46]. In contrast, the E-FRET method used one photon excitation on a standard automated fluorescence microscope, using a CCD detector and custom written software. For reasons that we do not fully understand, the standard deviations observed with the sRET measurements were at least two times greater than those observed with FLIM–FRET or with E-FRET. We suspect that this is not intrinsic to the spectral approach, but arises from some aspect of the sRET hardware/software implementation. Even though the standard deviations observed for FLIM–FRET and E-FRET were lower than that of sRET, all three methods were capable of differentiating a 5% change in FRET efficiency [40].

Another difference in these FRET methods is the cost of the microscopes. The two-photon microscope and its mode-locked laser used for sRET and FLIM–FRET cost approximately an order of magnitude more than the E-FRET system. Clearly, if cost is a limiting factor then the E-FRET approach is superior.

Theoretically, spectral imaging should be among the most photon efficient methods for measuring FRET because every photon emitted by either the donor or acceptor (if detected) can be used in the FRET calculation. With FLIM–FRET only emissions from the donor are typically used, and any part of the donors emission spectrum that overlaps with the acceptors emission spectrum is also not typically used. Clearly, FLIM–FRET, as implemented, is not very photon efficient. E-FRET is typically more photon-efficient than FLIM–FRET because emissions from both donors and acceptors are used. However, the use of emission filters to isolate donor emissions from acceptor emissions prevents many photons from being detected. Complicating this analysis of photon-efficiency is the fact that three separate excitation periods must occur to acquire the images required for E-FRET analysis. FLIM–FRET requires only one excitation period but it can last tens of minutes, and sRET requires two excitation periods, one for each excitation wavelength. Finally, not all photon detectors have the same quantum efficiency. Photomultipliers used for FLIM–FRET (and in confocal microscopes) typically have quantum efficiencies for detecting photons in the range of 10–40% [46]. A recent study compared the efficiencies of the spectral detector used in a Zeiss META confocal with nonspectral detectors and concluded that the spectral detectors were fivefold less efficient in detecting photons [47]. State of the art CCD cameras that can be used for E-FRET imaging, can have quantum efficiencies as high as 90%. Because of all of these different factors, it is often difficult to predict the photon efficiency of different FRET methods. Nonetheless, an empirical estimate of their relative photon efficiencies can be derived from the time each method requires to acquire a FRET-efficiency image. A single FRET image acquired on a time-domain

FLIM–FRET system (Becker and Hickl SPC 830 with a Hamamatsu R3809 detector mounted on a Zeiss 510 META/NLO) can take between 5 to 20 min to acquire. On a sRET system (Zeiss 510 META/NLO) a single FRET image (i.e., two spectral images) takes approximately 100 s to acquire, but tuning the laser (Coherent Chameleon) to the second excitation wavelength adds an additional minute or two. Acquiring a FRET-efficiency image on an E-FRET system typically requires only a few seconds. Thus, even though E-FRET does not produce thin optical sections as does sRET and FLIM–FRET (with a two-photon pulsed laser), E-FRET's low cost and rapid acquisition time affords it a great advantage, particularly for time-lapse studies. It is also important to realize that empirical comparisons like these only contrast specific implementations of these FRET methods. For example, a comparison of FRET methods similar to our own reached noticeably different conclusions [48]. With future improvements in the quantum efficiencies of spectral detectors, improvements in their signal to noise ratio, as well as technology to rapidly and reproducibly tune lasers [49], we predict that spectral imaging will ultimately become the most photon efficient, and rapid method for accurately measuring FRET while simultaneously measuring the abundance of donors and acceptors.

Acknowledgments

This work was supported by the intramural program of the National Institutes of Health, National Institute on Alcohol Abuse and Alcoholism, Bethesda, MD 20892.

Appendix

The detected fluorescence emission spectrum $F^i(\lambda)$ at excitation wavelength λ^i_{ex} is composed of three components. These three components are (analogous to the notation in Chapter 7) the spectrum originating from (partly) quenched donors $I^i_{D-S}(\lambda)$ (corresponding to I_{DA} in the general notation, referring to residual donor fluorescence in the presence of the acceptor), the spectrum originating from sensitized emission ($I^i_S(\lambda)$) and the spectrum due to direct acceptor excitation $I^i_A(\lambda)$.

$$F^i(\lambda) = I^i_{D-S}(\lambda) + I^i_S(\lambda) + I^i_A(\lambda) \qquad (8.A1)$$

The spectra are a product of the number of molecules (N), the laser intensity ℓ^i (at λ^i_{ex}), the (excitation wavelength dependent) extinction coefficient ε^i, the quantum yield Q, the corrected emission spectra $F(\lambda)$, and the instrument response or gain $g(\lambda)$ that is typically wavelength dependent. If we have N_s molecules that show FRET with an efficiency E, and N_D (total) donor molecules and N_A (total) acceptor molecules, Eq. (8.A1) can be rewritten:

$$\begin{aligned} F^i(\lambda) = \ell^i g(\lambda)\{&(N_D - N_S)\varepsilon^i_D Q_D F_D(\lambda) + N_S[\varepsilon^i_D Q_D(1-E)F_D(\lambda) \\ &+ \varepsilon^i_D E Q_A F_A(\lambda)] + N_A \varepsilon^i_A Q_A F_A(\lambda)\} \end{aligned} \qquad (8.A2)$$

If we define $E_D \equiv \frac{N_S}{N_D} E = f_D E$ where f_D is the fraction of Donor molecules involved in FRET (8.A2) can be rewritten as:

$$\begin{aligned} F^i(\lambda) = \ell^i g(\lambda)\ &\{N_D(1-E_D)\varepsilon^i_D Q_D F_D(\lambda) \\ &+ \left(N_A + N_D \frac{\varepsilon^i_D}{\varepsilon^i_A}\right)\varepsilon^i_A Q_A F_A(\lambda)\} \end{aligned} \qquad (8.A3)$$

If we consider two reference samples of pure donor and acceptor with a known concentration, two reference spectra can be recorded according to Eq. (8.A4):

$$\begin{aligned} F^i_{D,\mathrm{ref}}(\lambda) &= \ell^i g(\lambda)\varepsilon^i_D Q_D F_D(\lambda) N_{D,\mathrm{ref}} \\ F^i_{A,\mathrm{ref}}(\lambda) &= \ell^i g(\lambda)\varepsilon^i_A Q_A F_A(\lambda) N_{A,\mathrm{ref}} \end{aligned} \qquad (8.A4)$$

Then we can perform the following integration:

$$\begin{aligned}\int_\lambda \frac{F^i_{D,\text{ref}}(\lambda)}{F_D(\lambda)} &= \ell^i \varepsilon^i_D Q_D N_{D,\text{ref}} \int_\lambda g(\lambda) \\ \int_\lambda \frac{F^i_{A,\text{ref}}(\lambda)}{F_A(\lambda)} &= \ell^i \varepsilon^i_A Q_A N_{A,\text{ref}} \int_\lambda g(\lambda)\end{aligned} \tag{8.A5}$$

From which it is apparent that:

$$\frac{\varepsilon^i_D}{\varepsilon^i_A} = k^i \frac{N_{A,\text{ref}}}{N_{D,\text{ref}}} \frac{Q_A}{Q_D} \text{ with } k^i = \int_\lambda \frac{F^i_{D,\text{ref}}(\lambda)}{F_D(\lambda)} \bigg/ \int_\lambda \frac{F^i_{A,\text{ref}}(\lambda)}{F_A(\lambda)} \tag{8.A6}$$

Note that for the calculation of k^i, two reference samples of known concentrations are required in addition to calibrated unit area spectra of donor and acceptor fluorophores. In case the instrument response curve is not wavelength dependent ($g(\lambda) = g$) then unit reference spectra are not required because in this case:

$$k^i = \int_\lambda F^i_{D,\text{ref}}(\lambda) \bigg/ \int_\lambda F^i_{A,\text{ref}}(\lambda) \tag{8.A6b}$$

With Eqs. (8.A4–8.A6), (8.A4) can be reformulated according to:

$$\begin{aligned}F^i(\lambda) =& \frac{N_D}{N_{D,\text{ref}}}(1 - E_D) F^i_{D,\text{ref}}(\lambda) \\ &+ \frac{N_A}{N_{A,\text{ref}}} F^i_{A,\text{ref}}(\lambda) + \frac{N_D}{N_{D,\text{ref}}} E_D \frac{Q_A}{Q_D} k^i F^i_{A,\text{ref}}(\lambda)\end{aligned} \tag{8.A7}$$

If we define the relative donor and acceptor concentration as:

$$\boldsymbol{d} = \frac{N_D}{N_{D,\text{ref}}} \text{ and } \boldsymbol{a} = \frac{N_A}{N_{A,\text{ref}}} \tag{8.A8}$$

Then Eq. (8.A8) rewrites as:

$$F^i(\lambda) = \boldsymbol{d}(1 - E_D)F^i_{D,\text{ref}}(\lambda) + \boldsymbol{a}F^i_{A,\text{ref}}(\lambda) + \boldsymbol{d}E_D\frac{Q_A}{Q_D}k^iF^i_{A,\text{ref}}(\lambda) \tag{8.A9}$$

which is identical to Eq. (8.2).

Note that if from a separate experiment the molar extinction coefficients of the donor and acceptor at both excitation wavelengths are known, the determination of k^i is not required since Eq. (8.A9) rewrites in:

$$F^i(\lambda) = \boldsymbol{d}(1 - E_D)F^i_{D,\text{ref}}(\lambda) + \boldsymbol{a}F^i_{A,\text{ref}}(\lambda) + \boldsymbol{d}E_D\frac{\varepsilon^i_D}{\varepsilon^i_A}\frac{N_{D,\text{ref}}}{N_{A,\text{ref}}}F^i_{A,\text{ref}}(\lambda) \tag{8.A9a}$$

$$= \boldsymbol{d}(1 - E_D)F^i_{D,\text{ref}}(\lambda) + \boldsymbol{a}\left(1 + \frac{\varepsilon^i_D}{\varepsilon^i_A}E_A\right)F^i_{A,\text{ref}}(\lambda) \text{ with} \quad E_A \equiv \frac{N_S}{N_A}E = f_AE \tag{8.A9b}$$

However, in case of multiphoton excitation, the determination of $\frac{\varepsilon^i_D}{\varepsilon^i_A}$ will be difficult.

References

[1] Förster, T. (1948). Intermolecular energy migration and fluorescence. Ann. Phys. *2*, 55–75.

[2] Dickinson, M. E., Bearman, G., Tille, S., Lansford, R. and Fraser, S. E. (2001). Multi-spectral imaging and linear unmixing add a whole new dimension to laser scanning fluorescence microscopy. Biotechniques *31*, 1272–8.

[3] Haraguchi, T., Shimi, T., Koujin, T., Hashiguchi, N. and Hiraoka, Y. (2002). Spectral imaging fluorescence microscopy. Genes Cells *7*, 881–7.

[4] Zimmermann, T., Rietdorf, J. and Pepperkok, R. (2003). Spectral imaging and its applications in live cell microscopy. FEBS Lett. *546*, 87–92.

[5] Harpur, A. G., Wouters, F. S. and Bastiaens, P. I. (2001). Imaging FRET between spectrally similar GFP molecules in single cells. Nat. Biotechnol. *19*, 167–9.

[6] LaMorte, V. J., Zoumi, A. and Tromberg, B. J. (2003). Spectroscopic approach for monitoring two-photon excited fluorescence resonance energy transfer from homodimers at the subcellular level. J. Biomed. Opt. *8*, 357–61.

[7] Neher, R. A. and Neher, E. (2004b). Applying spectral fingerprinting to the analysis of FRET images. Microsc. Res. Technol. *64*, 185–95.

[8] Koepke, J., Hu, X., Muenke, C., Schulten, K. and Michel, H. (1996). The crystal structure of the light-harvesting complex II (B800–850) from *Rhodospirillum molischianum*. Structure *4*, 581–97.

[9] Fung, B. K. and Stryer, L. (1978). Surface density determination in membranes by fluorescence energy transfer. Biochemistry *17*, 5241–8.

[10] James, J. R., Oliveira, M. I., Carmo, A. M., Iaboni, A. and Davis, S. J. (2006). A rigorous experimental framework for detecting protein oligomerization using bioluminescence resonance energy transfer. Nat. Methods *3*, 1001–6.

[11] Vogel, S. S., Thaler, C. and Koushik, S. V. (2006). Fanciful FRET. Sci. STKE 2006, re2.

[12] Thaler, C., Koushik, S. V., Blank, P. S. and Vogel, S. S. (2005). Quantitative multiphoton spectral imaging and its use for measuring resonance energy transfer. Biophys. J. *89*, 2736–49.

[13] Ecker, R. C., de Martin, R., Steiner, G. E. and Schmid, J. A. (2004). Application of spectral imaging microscopy in cytomics and fluorescence resonance energy transfer (FRET) analysis. Cytometry A *59*, 172–81.

[14] Harris, A. T. (2006). Spectral mapping tools from the earth sciences applied to spectral microscopy data. Cytometry A *69*, 872–9.

[15] Larson, J. M. (2006). The Nikon C1si combines high spectral resolution, high sensitivity, and high acquisition speed. Cytometry A *69*, 825–34.

[16] Levenson, R. M. (2006). Spectral imaging perspective on cytomics. Cytometry A *69*, 592–600.

[17] Zimmermann, T. (2005). Spectral imaging and linear unmixing in light microscopy. Adv. Biochem. Eng. Biotechnol. *95*, 245–65.

[18] Zucker, R. M., Rigby, P., Clements, I., Salmon, W. and Chua, M. (2007). Reliability of confocal microscopy spectral imaging systems: Use of multispectral beads. Cytometry A *71*, 174–89.

[19] Rizzo, M. A., Springer, G. H., Granada, B. and Piston, D. W. (2004). An improved cyan fluorescent protein variant useful for FRET. Nat. Biotechnol. *22*, 445–9.

[20] Nagai, T., Ibata, K., Park, E. S., Kubota, M., Mikoshiba, K. and Miyawaki, A. (2002). A variant of yellow fluorescent protein with fast and efficient maturation for cell-biological applications. Nat. Biotechnol. *20*, 87–90.
[21] Zacharias, D. A., Violin, J. D., Newton, A. C. and Tsien, R. Y. (2002). Partitioning of lipid-modified monomeric GFPs into membrane micro-domains of live cells. Science *296*, 913–6.
[22] McNamara, G., Gupta, A., Reynaert, J., Coates, T. D. and Boswell, C. (2006). Spectral imaging microscopy web sites and data. Cytometry A *69*, 863–71.
[23] Lakowicz, J. R. (1999). Principles of Fluorescence Spectroscopy. Kluwer/Plenum, New York.
[24] Thaler, C. and Vogel, S. S. (2006). Quantitative linear unmixing of CFP and YFP from spectral images acquired with two-photon excitation. Cytometry A *69*, 904–11.
[25] Press, W. H., Flannery, B. P., Teukolsky, S. A. and Vetterling, W. T. (1986). Numerical Recipes. Cambridge University Press, Cambridge.
[26] Denk, W., Strickler, J. H. and Webb, W. W. (1990). Two-photon laser scanning fluorescence microscopy. Science *248*, 73–6.
[27] Patterson, G. H. and Piston, D. W. (2000). Photobleaching in two-photon excitation microscopy. Biophys. J. *78*, 2159–62.
[28] Neher, R. and Neher, E. (2004a). Optimizing imaging parameters for the separation of multiple labels in a fluorescence image. J. Microsc. *213*, 46–62.
[29] Berney, C. and Danuser, G. (2003). FRET or no FRET: A quantitative comparison. Biophys. J. *84*, 3992–4010.
[30] Gu, Y., Di, W. L., Kelsell, D. P. and Zicha, D. (2004). Quantitative fluorescence resonance energy transfer (FRET) measurement with acceptor photobleaching and spectral unmixing. J. Microsc. *215*, 162–73.
[31] Hanley, Q. S., Murray, P. I. and Forde, T. S. (2006). Microspectroscopic fluorescence analysis with prism-based imaging spectrometers: Review and current studies. Cytometry A *69*, 759–66.
[32] Kenworthy, A. K. (2001). Imaging protein–protein interactions using fluorescence resonance energy transfer microscopy. Methods *24*, 289–96.
[33] Raicu, V., Jansma, D. B., Miller, R. J. and Friesen, J. D. (2005). Protein interaction quantified *in vivo* by spectrally resolved fluorescence resonance energy transfer. Biochem. J. *385*, 265–77.
[34] Vermeer, J. E., Van Munster, E. B., Vischer, N. O. and Gadella, T. W. Jr. (2004). Probing plasma membrane microdomains in cowpea protoplasts using lipidated GFP-fusion proteins and multimode FRET microscopy. J. Microsc. *214*, 190–200.

[35] Bastiaens, P. I. and Squire, A. (1999). Fluorescence lifetime imaging microscopy: Spatial resolution of biochemical processes in the cell. Trends Cell Biol. *9*, 48–52.
[36] Gadella, T. W. J., Jovin, T. M. and Clegg, R. M. (1993). Fluorescence lifetime imaging microscopy (FLIM): Spacial resolution of microstructures on the nanosecond time scale. Biophys. Chem. *48*, 221–39.
[37] van Kuppeveld, F. J., Melchers, W. J., Willems, P. H. and Gadella, T. W., Jr. (2002). Homomultimerization of the coxsackievirus 2B protein in living cells visualized by fluorescence resonance energy transfer microscopy. J. Virol. *76*, 9446–56.
[38] Wang, X. F., Periasamy, A. and Herman, B. (1992). Fluorescence lifetime imaging microscopy (FLIM): Instrumentation and applications. Crit. Rev. Anal. Chem. *23*, 369–95.
[39] Wlodarczyk, J., Woehler, A., Kobe, F., Ponimaskin, E., Zeug, A. and Neher, E. (2008). Analysis of FRET signals in the presence of free donors and acceptors. Biophys. J. *94*, 986–1000.
[40] Koushik, S. V., Chen, H., Thaler, C., Puhl, H. L., III and Vogel, S. S. (2006). Cerulean, Venus, and VenusY67C FRET reference standards. Biophys. J. *91*, L99–L101.
[41] Rizzo, M. A., Springer, G., Segawa, K., Zipfel, W. R. and Piston, D. W. (2006). Optimization of pairings and detection conditions for measurment of FRET between cyan and yellow fluorescent proteins. Microsc. Microanal. *12*, 238–54.
[42] Chen, H., Puhl, H. L., III, Koushik, S. V., Vogel, S. S. and Ikeda, S. R. (2006). Measurement of FRET efficiency and ratio of donor to acceptor concentration in living cells. Biophys. J. *91*, L39–41.
[43] Gadella, T. W., Jr. and Jovin, T. M. (1995). Oligomerization of epidermal growth factor receptors on A431 cells studied by time-resolved fluorescence imaging microscopy. A stereochemical model for tyrosine kinase receptor activation. J. Cell Biol. *129*, 1543–58.
[44] Wallrabe, H. and Periasamy, A. (2005). Imaging protein molecules using FRET and FLIM microscopy. Curr. Opin. Biotechnol. *16*, 19–27.
[45] Zal, T. and Gascoigne, N. R. (2004). Photobleaching-corrected FRET efficiency imaging of live cells. Biophys. J. *86*, 3923–39.
[46] Becker, W. (2005). Advanced Time-Correlated Single Photon Counting Techniques. Springer, Berlin.
[47] Cho, E. H. and Lockett, S. J. (2006). Calibration and standardization of the emission light path of confocal microscopes. J. Microsc. *223*, 15–25.
[48] Pelet, S., Previte, M. J. and So, P. T. (2006). Comparing the quantification of Forster resonance energy transfer measurement accuracies based on intensity, spectral, and lifetime imaging. J. Biomed. Opt. *11*, 34017.
[49] Muller, B. K., Zaychikov, E., Brauchle, C. and Lamb, D. C. (2005). Pulsed interleaved excitation. Biophys. J. *89*, 3508–22.

Laboratory Techniques in Biochemistry and Molecular Biology, Volume 33
FRET and FLIM Techniques
T. W. J. Gadella (Editor)

CHAPTER 9

Total internal reflection fluorescence lifetime imaging microscopy

T. W. J. Gadella Jr.

Section of Molecular Cytology and Centre for Advanced Microscopy, Swammerdam Institute for Life Sciences, University of Amsterdam, Kruislaan 316, 1098 SM Amsterdam, The Netherlands

Since the first application of total internal reflection fluorescence (TIRF) microscopy in 1981, the technique has been used extensively for the study of events in or near plasma membranes of (living) cells adhered to glass surfaces. The evanescent wave produced by TIRF only penetrates approximately 100 nm away from the glass surface, thereby selectively exciting fluorophores in this region enabling very high contrast imaging, inducing no out-of-focus bleaching, and generating depth discrimination superior to confocal imaging.

The upgrade of a frequency-domain fluorescence lifetime imaging microscope (FLIM) to a prismless objective-based total internal reflection-FLIM (TIR-FLIM) system is described. By off-axis coupling of the intensity-modulated laser from a fiber and using a high numerical aperture oil objective, TIR-FLIM can be readily achieved. The usefulness of the technique is demonstrated by a fluorescence resonance energy transfer study of Annexin A4 relocation and two-dimensional crystal formation near the plasma membrane of cultured mammalian cells. Possible future applications and comparison to other techniques are discussed.

DOI: 10.1016/S0075-7535(08)00009-0

9.1. Introduction: Total internal reflection theory

Total internal reflection fluorescence (TIRF) microscopy was introduced by Axelrod in the 1980s [1]. The advantage of TIRF is that it provides a means to selectively excite fluorophores in an aqueous or cellular environment very near a glass surface (within ~100 nm) without inducing fluorescence from regions further away from the surface. The principle of the technique is that a surface separating a medium with high and low refractive index will totally reflect the incoming light if the incoming light is coming from a shallow angle through the higher refractive index medium. According to the Snellius law of refraction the incident angle θ_1 and output angle θ_2 are directly related to the respective refractive indices of the media n_1 and n_2 according to Eq. (9.1):

$$n_1 \sin\theta_1 = n_2 \sin\theta_2 \tag{9.1}$$

In Fig. 9.1, the situation is explained for a light beam traveling through a high refractive index medium (i.e., glass or immersion

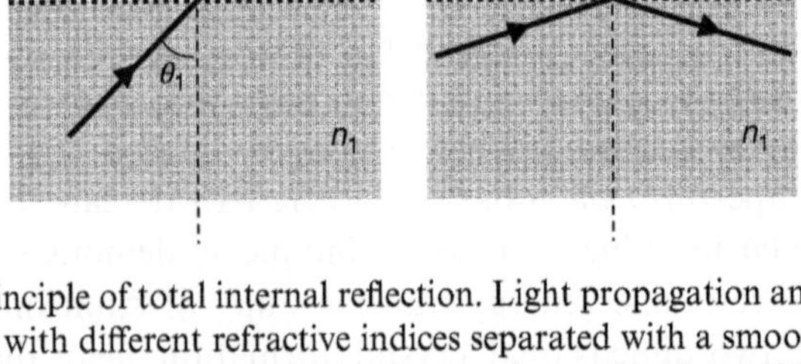

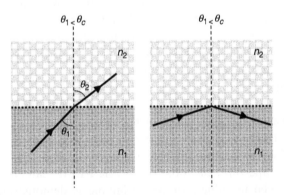

Fig. 9.1. Principle of total internal reflection. Light propagation and refraction in a system with different refractive indices separated with a smooth surface is shown. Left: The incident light is entering from the high refractive index medium under an angle θ_1 which is less than the critical angle θ_c. Right: total internal reflection because the incident light angle θ_1 is larger than θ_c.

$n_1 = 151$) and low refractive index medium (i.e., water $n_2 = 1.33$). At a certain incident critical angle θ_c for which $\theta_1 = \sin^{-1}(n_2/n_1)$, the output angle θ_2 will be higher than 90° according to Eq. (9.1) and hence the light is reflected back into the high refractive index medium. For air/glass interfaces the critical angle $\theta_c \approx 70°$. Very importantly, at the surface of total reflection, a so-called evanescent wave is produced capable of exciting fluorophores localized at the interface. This evanescent wave can be considered as an electromagnetic field of the reflected light that is penetrating into the low refractive index medium. However, the strength of this evanescent field is dropping exponentially with the distance z from the surface into the low refractive medium according to:

$$I(z) = I_0 \exp(-z/d) \tag{9.2}$$

The decay constant d is typically in the order of 80 nm. This distance depends on the angle of incidence and the wavelength λ according to:

$$d = \frac{\lambda}{4\pi\sqrt{n_1^2 \sin^2\theta_1 - n_2^2}} \tag{9.3}$$

Considering Eq. (9.3), the penetration depth can be "tuned" by varying the incident angle. At more shallow angles (higher θ_1) the penetration depth decreases.

The major advantage of TIRF is that fluorophores outside the evanescent wave (typically more than 200 nm away from the surface) are not excited. Hence, TIRF has an intrinsic sectioning capability. Of interest is that the section capability (z-resolution) is far better than for confocal microscopy systems, which typically have a z-resolution of about 1 μm. In addition and in contrast to confocal microscopy, TIRF does not cause out-of-focus bleaching because only the molecules at the surface will sense the evanescent wave. However, in comparison with confocal microscopy, a clear limitation of TIRF is that only one z-plane can be imaged: the molecules immediately adjacent to the surface. As a consequence, for biological applications, TIRF microscopy is limited to studies of cell surfaces: events at the plasma membrane including endocytosis, exocytosis, and membrane dynamics. Because of the high sectioning capability of TIRF, images usually have much more contrast and much lower background fluorescence than the corresponding confocal sections.

9.2. TIRF microscopy history and applications

TIRF was first coupled to microscopy by Daniel Axelrod in early 1980s [1]. However, the phenomenon of inducing fluorescence at a surface using evanescent waves was known and applied since 1965 [2]. Applications of TIRF microscopy have been reviewed extensively, for example see [3, 4]. Because of its high signal-to-noise characteristics TIRF-microscopy is increasingly popular for single molecule studies (for a review see [5]). The most popular application of TIRF microscopy is the study of membrane trafficking near/at the plasma membrane, initially mostly for the study of synaptic vesicle-fusion in excitable cells. Increasingly popular are now membrane trafficking (endocytosis/exocytosis) studies in nonexcitable cells, for reviews see Ref. [6, 7]. Technically, all these studies are time-lapse TIRF imaging studies of dynamics at membrane interphases. With the ability to perform dual emission TIRF microscopy, recently several TIR-fluorescence resonance energy transfer (FRET) studies have appeared. For instance, the dynamics of G-protein to receptor coupling has been studied using GFP-tagged G protein and receptor subunits [8, 9]. Also receptor–receptor complexes, heterologous channel complexes, and bacterial cytoskeletal protein complexes have been identified with TIRFM [10–12], respectively. Recently, even single molecule FRET studies have been described for monitoring actin dynamics [13] or complexin-SNARE complexes [14]. This clearly demonstrates the wide applicability of TIRF microscopy for a variety of important cell biological applications in general and the possibility to combine it with FRET contrast in particular. In most of the FRET-TIRF microscopy applications described above, the FRET contrast was obtained by ratio imaging of donor and acceptor fluorescence intensities upon donor excitation. This method produces useful FRET contrast, but is not very quantitative because the calculated ratio values will depend very much on filter throughput and (wavelength-dependent) detector sensitivity. Also differential (unintended) donor and acceptor bleaching, or unbalanced expression ratio's can seriously interfere with the FRET calculations. For agonist-induced fast ratio changes these limitations are not very serious, but for characterizing prolonged or very slowly changing interactions the ratio-method can prove to be particularly troublesome. A more quantitative and also simple method for determining FRET is acceptor photobleaching [15]. Acceptor photobleaching FRET studies have been performed using TIRF [11]. However, acceptor photobleaching is intrinsically destructive, cannot be repeated on the same specimen, is highly sensitive to motion before and after acceptor photobleaching, and is highly sensitive to moderate (unintended) donor photobleaching. In this respect, the intensity-independent and mostly bleach-insensitive FRET detection method of fluorescence lifetime imaging microscopy (FLIM) is of particular interest. As discussed elsewhere in detail (Chapters 2–4), FRET–FLIM does provide quantitative numbers and global analysis applied to FLIM data can be used to calculate populations of molecules in a certain physical state. Hence, given the enormous recent revival of both TIRF microscopy and FRET microscopy, the combination of TIRF microscopy and FLIM is of particular interest. Recently, in the lab of Dr. Schneckenburger a time-domain TIR-FLIM system was assembled. They published TIRF-lifetime maps of Rhodamine 123 [16], and in addition, a FRET TIR-FLIM study was published demonstrating FRET between Laurdan and DiI in membrane microdomains [17]. In this chapter, the combination of TIRF microscopy and FLIM is described and discussed.

9.3. Combining TIRF with FLIM

Since TIRF produces an evanescent wave of typically 80 nm depth and several tens of microns width, detection of TIRF-induced fluorescence requires a camera-based (imaging) detector. Hence, implementing TIRF on scanning FLIM systems or multiphoton FLIM systems is generally not possible. To combine it with FLIM, a nanosecond-gated or high-frequency-modulated imaging detector is required in addition to a pulsed or modulated laser source. In this chapter, the implementation with of TIRF into a frequency-domain wide-field FLIM system is described.

9.3.1. Optical designs

There are multiple optical designs possible for the generation of the evanescent wave. Generally, they are divided into prism and objective-based methods. For an excellent review of the different optical designs refer to [3]. In the prism method, a laser is introduced into a prism and the object (e.g., cells) are placed on top of the prism or onto a cover glass that is placed on top of the prism with immersion oil between the prism and the cover glass. This method has the advantage of producing large evanescent fields, but also has some practical disadvantages of hazardous high-intensity laser light exiting the prism and of manipulating the prism and cells relative to the objective for microscopic observation of the induced fluorescence. Especially, when manipulation of cells is required (e.g., microinjection or addition of stimuli), the prism implementation can be impractical because of limited workspace between the microscope objective and the cells. The other type of implementation is the objective method. A very high numerical aperture (NA) objective is used (typically NA $\geq$ 1.45). For a usual microscopy optical layout, the excitation light is centered at the back focal plane of the objective. For TIRF however, the excitation light enters the objective at the back focal plane but offset with respect to the center. Because of the high NA of the objective, the excitation light will exit the front end of the objective at a very shallow angle. This will produce an evanescent wave at a glass–water interface. Hence, for objective-type TIRF systems, no special sample holders are needed: normal immersion oil and a cover glass on which cells are attached can be used. Nowadays, all major microscopy manufacturers offer special microscope objectives suitable for TIRF. Nikon offers two CFI Plan Apochromat 60× and 100× NA 1.45 oil objectives with correction rings for cover glass thickness or temperature. Olympus offers several objectives including a new UAPO Apochromat 150 × O NA 1.45 oil objective and a PLAPON 60 × O NA 1.45 oil objective both with correction collars for cover glass thickness. In addition, they offer a PLAPO 100 × NA 1.45 oil objective without correction collar and a OHR 100 × NA 1.65 special immersion objective. This latter world record high NA objective requires special (but volatile and expensive) immersion oil with a high refractive index (1.78) for proper operation. Zeiss offers an α-Plan Fluar 100 × NA 1.45 oil objective without correction collar, and Leica offers a HCX Plan Apo 100 × NA 1.46 oil objective with correction collar. Most of the companies offer special microscope attachments to optimally adjust laser sources for TIRF. In addition, Nikon also offers optics to accommodate white light arc sources for TIRF. By combining this with a modulated light emitting diode as light source, Lambert Instruments offers a commercial frequency-domain objective (prismless) FLIM-TIRF setup (see http://www.lambert-instruments.com).

9.3.2. Upgrading wide-field FLIM to a TIR-FLIM microscope

Described below is how we upgraded our frequency-domain wide-field FLIM system in Amsterdam to incorporate TIRF. A TIRF upgrade to an existing wide-field FLIM setup is marginal both with respect to cost and time. Furthermore, after the upgrade, the system can still be used as a regular wide-field FLIM system.

The upgrade of an existing TIRF setup to TIRF-FLIM is much more serious, both with respect to costs and with respect to investments in software and high-frequency or nanosecond gating detection systems and control software.

Most wide-field FLIM systems use laser sources for excitation. These lasers are coupled into the microscope using a beam expander and lens or by a fiber optic and a lens. The simplest upgrade for TIRF is the use of a fiber optic system [3]. For a full description of our FLIM setup, refer to [18]. Briefly, we use three lasers as excitation sources a helium–cadmium laser (442 nm, 125 mW, Melles-Griot, USA), and two argon-ion lasers (488 and 514 nm, 150 mW, Melles-Griot). The laser beams are combined using dichroic mirrors (Q497LP and Q455LP, Chroma, USA). Lasers are modulated using a standing wave acoustooptical modulator (AOM) (FSML-37.5-5-500, Brimrose, USA) with a resonance peak at 75 MHz, of which only the zero-order light is selected, using an iris stop. The modulated light beam is coupled into a multimode fiber using a 10× microscope objective (Zeiss, Germany). We use a 5 m long 200 μm internal diameter multimode fiber (FT-200-UMT, Thorlabs, USA) with a NA of 0.39. This is coupled into the microscope (Zeiss Axiovert 200 M) with a 50 mm focal length lens attached to the side port of the microscope. Detection is provided by a modulated image intensifier (II-18M, Lambert Instruments, The Netherlands) coupled to a slow-scan cooled CCD camera (Coolsnap HQ, Roper Scientific, USA) using a relay lens incorporated in the intensifier. Both the AOM and the image intensifier were coupled to HF generators and software for control, acquisition, processing, and analysis of the data was written in C++, using Matlab 6.1 (The Mathworks, USA) and the image processing library DIPlib (Pattern Recognition Group, TU Delft, The Netherlands, http://www.ph.tn.tudelft.nl/DIPlib/) (see Fig. 9.2).

To implement TIRF, we needed two modifications: firstly we incorporated a Zeiss α-Plan Fluar 100 × NA 1.45 oil objective, and secondly, we modified the configuration of the output of the multimode fiber to incorporate off axis coupling into the microscope. By using a micromanipulator moving the output of the fiber off axis

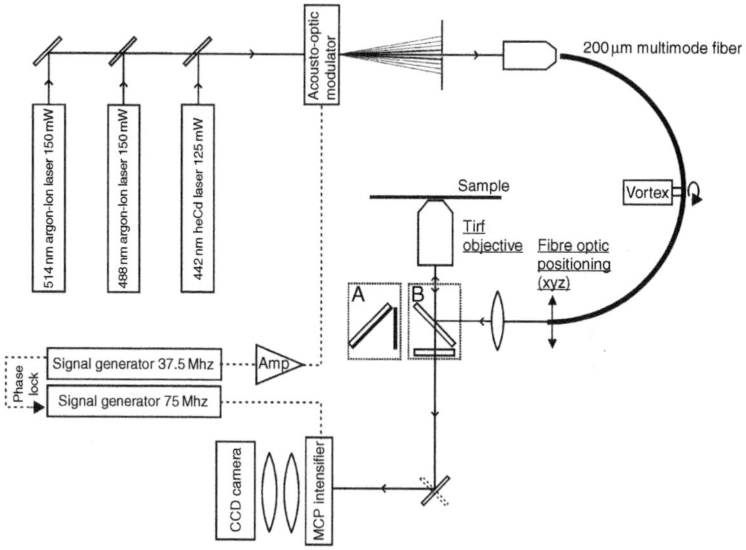

Fig. 9.2. Schematic representation of the frequency domain TIR-FLIM setup. The essential components required for upgrading a wide-field frequency domain system to incorporate TIRF are underlined. Two filter cubes are shown: (A) for TIR-FLIM imaging incorporating a dichroic mirror and an emission bandpass filter, and (B) the special reference filter cube with an neutral density filter (OD 3) for attenuating the excitation light and a 90° tilted reflective mirror instead of a dichroic mirror for passing on the excitation light immediately to the modulated detector. For further explanation see text Sections 9.3.2 and 9.5.2.

with respect to the lens at the side port of the microscope, one can toggle between TIRF and regular wide field illumination (see Fig. 9.3). In case total internal reflection (TIR) is obtained, the same micromanipulator can be used to adjust the incident angle of the laser light as to change the depth of the evanescent wave (see Eq. (9.2)). Using this setup one can switch between TIRF and wide-field FLIM within 10 s. Yet for obtaining a good TIRF field, some practical considerations are important. Since much more detailed literature is available describing practical tips for optical alignment of excitation sources for TIRF, refer to these for further information [3].

Fig. 9.3. Optical alignment of the fiber-optic output with respect to the microscope axis (black line). A close up is shown of the side-port of the Axiovert 200 microscope and fiber-optic coupling of a modulated 514 nm laser source. Left: the fiber output (coming from the right) is aligned onto the microscope axis enabling wide-field excitation. Right: the fiber output is aligned slightly off axis, but sufficient to induce TIRF. The scale of the picture can be inferred from the optical table M6 screw mounts separated by 1 inch. (See Color Insert.)

However, below I describe some simple troubleshooting experience of the TIRF-FLIM system in Amsterdam.

9.3.3. Troubleshooting optical alignment in TIR-FLIM

Crucial for obtaining a good TIRF field is that the laser source enters the objective from a well-defined angle at the back focal plane. For a microscope with infinity-corrected optics, it is simple to check the optics of laser coupling by removing the objective and placing some piece of paper (e.g., lens paper) on top of the objective turret of the microscope. In order to prevent hazards due to high intensity and focused laser light, wearing protective glasses and/or insertion of a high optical density filter in the microscope is recommended. Most important is to check whether the laser light is focused in one (as small as possible) spot. This should not be done by changing the field diaphragm, but by changing the distance of the fiber output with respect to the lens attached at the side port of the microscope. Also care should be taken that the axis of the fiber and lens is exactly parallel to the optical axis of the microscope. Once in this way a perfect tiny laser spot is obtained at the center of the back focal plane of the objective, one can move the output of the fiber so that it is off axis (but still parallel) with respect to side port of the microscope (see Fig. 9.3). As a result, the laser spot should move away from the axis of the objective. If the objective is attached, one can first see the laser beam exit the objective parallel to the objective axis, but if the laser spot is moved off axis, the laser beam will exit the objective at increasingly higher angles with respect to the objective axis until TIR is obtained. By observing cells, one can further fine-tune the incident angle to achieve a good evanescent field. Usually the optimal angle is a compromise between desirable z-resolution and signal-to-noise. If the laser spot is not very small, the laser light will move through the objective at variable angles. As a result, in addition to the TIRF field also residual wide-field excitation will be obtained. Another point that needs to be addressed is the prevention of laser speckles in the TIRF field. Special optics can be used to scramble laser speckles before or after the laser beam is coupled into the fiber [19]. We, simply, mechanically shake the fiber using a vortex. If shaken at a higher frequency than the exposure time of the CCD camera taking FLIM stacks, this results in averaging the laser speckles producing a homogeneous TIRF field.

9.4. Results and discussion

To demonstrate an application of TIRF-FLIM, a FRET study of annexin A4 translocation and self-aggregation near the plasma membrane is shown in Fig. 9.4. This is a particularly useful application of TIRF-FLIM, since TIRF provides the spatial contrast of detecting only molecules immediately adjacent to the plasma membrane and the lifetime contrast reports on the aggregation state of annexin A4. Annexins are calcium-dependent lipid-binding domains with a different type of lipid binding domain compared to the common C2 domains (e.g., found in protein kinase C). Annexins consist of an N-terminal domain and a core domain binding calcium and phospholipids. The core domain is conserved in the

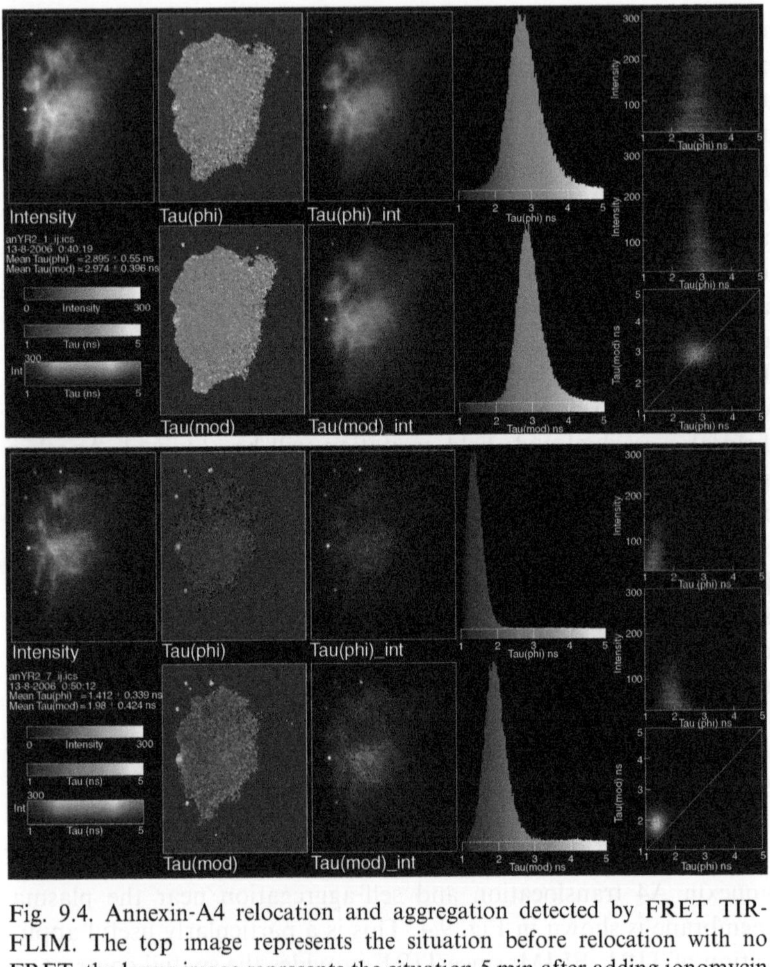

Fig. 9.4. Annexin-A4 relocation and aggregation detected by FRET TIR-FLIM. The top image represents the situation before relocation with no FRET, the lower image represents the situation 5 min after adding ionomycin inducing relocation and aggregation of annexin A4-EYFP and annexin A4-mCherry reflected by a substantial drop of the annexin A4-EYFP fluorescence lifetime. The bright fluorescent dots are fluorescent latex beads (0.1 μm diameter) that were incorporated into the medium for efficient focusing the plasma membrane area before translocation. Only beads adhered to the glass are excited in TIRF. The images represent from left to right: the DC fluorescence intensity image, the phase lifetime image, the pseudocolored intensity-phase lifetime image and the phase lifetime distribution histogram.

annexin family whereas the N-terminus is variable, presumably reflecting differential regulatory roles. Annexin A4, is self-associating in two-dimensional arrays on a membrane surface in case of a prolonged calcium increase [20–22]. For a more detailed discussion on annexin A4, refer to a recent excellent FRET study on annexin A4 relocation and complex formation by Piljic and Schultz and references cited therein [23].

For measuring annexin A4 aggregation, donor and acceptor-labeled Annexin A4 constructs were cotransfected into HeLa cells: annexin A4 fused to EYFP and annexin A4 fused to mCherry [24]. The YFP fluorescence lifetime can be imaged without interference from mCherry fluorescence by exciting EYFP using the 514 nm Argon laser line, and selecting the fluorescence using a 525 dichroic and 545/30 nm bandpass filter. In case of annexin self-aggregation, the EYFP fluorescence lifetime should decrease.

In Fig. 9.4, the results are shown of an experiment using a double (annexinA4-EYFP+ annexinA4-mCherry) transfected cell immediately after addition of the calcium ionophore ionomycin (top) and 5 min after application of ionomycin (bottom). Clearly visible is the diffuse localization of annexin A4 before relocation and the more structured localization (into membrane ruffles and filopodia) after relocation. More importantly, the EYFP fluorescence lifetime is quenched from 2.9 ns before translocation to

---

Below are from left to right the modulation lifetime image, the pseudocolored intensity-modulation lifetime image, and the modulation lifetime distribution histogram. The color bars depicting fluorescence intensity (arbitrary units), fluorescence lifetime (between 1 and 5 ns), and pseudocolored intensity-lifetimes are shown in the lower left corner of the image. At the right side of the images three dotplots or two-dimensional histograms are shown depicting from top to bottom the correlation between the fluorescence intensity with the phase lifetime; the fluorescence intensity with the modulation lifetime; and the correlation between phase and modulation lifetime. The composite layout was generated using a standard macro (Lifetime6) in image J using an input image stack of three images: the calculated DC intensity image, the phase, and the modulation fluorescence lifetime images. (See Color Insert.)

approximately 1.5 ns after translocation demonstrating significant FRET after translocation. These results show that translocation is accompanied by a strong decrease of the intermolecular distance between Annexin molecules near the plasma membrane reflecting 2D crystal formation.

The above study represents the first TIR-FRET study using frequency-domain FLIM. Because of the enormous revival of FRET–FLIM boosted by the availability of many visible fluorescent proteins yielding many (potential) FRET couples (see Chapter 5) and the many applications of TIRF microscopy, I foresee a bright future for TIR-FLIM. Given the ease of and low costs associated with upgrading wide-field FLIM with TIR optics, it is remarkable, that TIR-FLIM systems are not much more widespread. TIR-FLIM is possible on both time-domain [16] and frequency domain setups, and given the availability of a first commercial TIR-FLIM system by Lambert Instruments (Leutingewolde, The Netherlands) using a cheap LED light source and commercial arc-lamp coupling device, I expect a strong increase in applications for the technique. It is important to realize that TIR-FRET–FLIM is a far-field nanoscale imaging technique: a z-resolution in the order of 100 nm, is combined with 3–10 nm molecular FRET contrast at approximately 200 nm spatial resolution. The big advantage of FLIM over other FRET detection techniques is that the lifetime contrast is insensitive to moderate bleaching, to illumination efficiencies, and that FLIM allows quantitative estimation of molecular populations using global analysis techniques (see Chapter 20).

9.5. Methods

9.5.1. Cell culturing and transfection

HeLa cells were transfected using 1 μl lipofectamine (Invitrogen, Breda, The Netherlands), 0.5 μg plasmid DNA and 50 μl Opti-MEM per 35 mm dish holding a 24 mm Ø #1 coverslip. Samples were imaged 1 day after transfection. Coverslips with cells were mounted in an Attofluor cell chamber (Invitrogen, Breda, The Netherlands) and submerged in microscopy medium (20 mM HEPES (pH = 7.4), 137 mM NaCl, 5.4 mM KCl, 1.8 mM $CaCl_2$, 0.8 mM $MgCl_2$, and 20 mM glucose). Translocation of Annexin A4 was induced by addition of 1 μM ionomycin (Sigma-Aldrich, Zwijndrecht, The Netherlands) (final concentration). All measurements were done at room temperature.

9.5.2. TIR-FLIM

Fluorescence lifetime imaging was performed using the wide-field frequency-domain approach on a home-build instrument [18] adapted for TIRF microscopy (see above). The modulation frequency was set to 75.1 MHz. Twelve phase images with an exposure time of 50–100 ms were acquired in a random recording order to minimize artifacts due to photobleaching [25]. An argon-ion laser was used for excitation at 514 nm, passed onto the sample by a 525 nm dichroic and emission light was filtered by a 545/30 nm emission filter. A Zeiss α-Plan Fluar 100 $\times$ NA 1.45 oil objective was used for TIRF. Each FLIM measurement is calibrated by a reference measurement of the reflected laser light using a modified filter cube [18] for correcting the phase and modulation drift of the excitation light. The reference is calibrated by averaging three to five FLIM measurements of a freshly prepared 1 mg/ml solution of erythrosine B (EB) (cat # 198269, Sigma-Aldrich, Zwijndrecht, The Netherlands) in H_2O, which has a known short fluorescence lifetime of 0.086 ns ([25, 26]). This extra calibration corrects for path-length differences and possible optics-related reflections that are different between the FLIM and reference measurements. For TIRF measurements, the calibration cube obtained-reference phase and modulation proved to be accurate. However for the final EB calibration, it was necessary to move the laser fiber back to illuminate the center of the back focal plane of the objective producing wide-field excitation as we found that the lifetime of EB measured by TIRF was significantly higher than 0.086 ns. We hypothesize that with TIRF a small amount of EB adhered to the glass surface is selectively imaged, which apparently has a higher quantum yield than EB in water solution. In agreement with this assumption is the known higher lifetime of EB in more apolar solvents such as methanol (≈0.2 ns, [27]). For this reason, for the final EB-calibration measurements we focus the laser deeply inside a volume of EB as to minimize the contribution of surface-adsorbed EB molecules. Using this calibration scheme, no significant difference is detected between YFP fluorescence lifetimes observed in wide-field or with TIRF. From the phase sequence an intensity (DC) image and the phase and modulation lifetime images are calculated using Matlab macros. From these data false-color lifetime maps and 1D and 2D histograms are generated by an ImageJ (http://rsb.info.nih.gov/ij/) macro. For the presentation of lifetime maps (but not the DC fluorescence intensity image), a 3×3 smooth filter is applied to the raw data.

Acknowledgments

I am grateful to Erik van Munster, who implemented TIR-FLIM in our laboratory. Annexin-A4-EYFP and mCherry constructs in clontech N1 vectors were obtained from Dr. Carsten Schultz and Alen Piljic′ (EMBL, Heidelberg). Iam grateful to Merel Adjobo Hermans for transfecting HeLa cells with Annexin constructs and performing the translocation experiments. This work was supported by the EU integrated project on "Molecular Imaging" LSHG-CT-2003-503259.

References

[1] Axelrod, D. (1981). Cell-substrate contacts illuminated by total internal reflection fluorescence. J. Cell Biol. *89*, 141–5.

[2] Hirschfeld, T. (1965). Total reflection fluorescence. Can. Spectrosc. *10*, 128.

[3] Axelrod, D. (2001). Total internal reflection fluorescence microscopy in cell biology. Traffic *2*, 764–74.
[4] Axelrod, D., Thompson, N. L. and Burghardt, T. P. (1983). Total internal inflection fluorescent microscopy. J. Microsc. *129*, 19–28.
[5] Wazawa, T. and Ueda, M. (2005). Total internal reflection fluorescence microscopy in single molecule nanobioscience. Adv. Biochem. Eng. Biotechnol. *95*, 77–106.
[6] Beaumont, V. (2003). Visualizing membrane trafficking using total internal reflection fluorescence microscopy. Biochem. Soc. Trans. *31*, 819–23.
[7] Toomre, D. and Manstein, D. J. (2001). Lighting up the cell surface with evanescent wave microscopy. Trends Cell Biol. *11*, 298–303.
[8] Fowler, C. E., Aryal, P., Suen, K. F. and Slesinger, P. A. (2006). Evidence for association of GABAB receptors with Kir3 channels and RGS4 proteins. J. Physiol. *580*, 51–5.
[9] Riven, I., Iwanir, S. and Reuveny, E. (2006). GIRK channel activation involves a local rearrangement of a preformed G protein channel complex. Neuron *51*, 561–73.
[10] Garner, E. C., Campbell, C. S. and Mullins, R. D. (2004). Dynamic instability in a DNA-segregating prokaryotic actin homolog. Science *306*, 1021–5.
[11] Khakh, B. S., Fisher, J. A., Nashmi, R., Bowser, D. N. and Lester, H. A. (2005). An angstrom scale interaction between plasma membrane ATP-gated P2X2 and alpha4beta2 nicotinic channels measured with fluorescence resonance energy transfer and total internal reflection fluorescence microscopy. J. Neurosci. *25*, 6911–20.
[12] Tateyama, M., Abe, H., Nakata, H., Saito, O. and Kubo, Y. (2004). Ligand-induced rearrangement of the dimeric metabotropic glutamate receptor 1alpha. Nat. Struct. Mol. Biol. *11*, 637–42.
[13] Kozuka, J., Yokota, H., Arai, Y., Ishii, Y. and Yanagida, T. (2006). Dynamic polymorphism of single actin molecules in the actin filament. Nat. Chem. Biol. *2*, 83–6.
[14] Bowen, M. E., Weninger, K., Ernst, J., Chu, S. and Brunger, A. T. (2005). Single-molecule studies of synaptotagmin and complexin binding to the SNARE complex. Biophys. J. *89*, 690–702.
[15] Bastiaens, P., I. H., Majoul, I., Verveer, P. J., Soling, H. and Jovin, T. (1996). Imaging the intracellular trafficking and state of the AB5 quaternary structure of cholera toxin. EMBO J. *15*, 4246–53.
[16] Schneckenburger, H., Stock, K., Lyttek, M., Strauss, W. S. and Sailer, R. (2004a). Fluorescence lifetime imaging (FLIM) of rhodamine 123 in living cells. Photochem. Photobiol. Sci. *3*, 127–31.

[17] Schneckenburger, H., Wagner, M., Kretzschmar, M., Strauss, W. S. and Sailer, R. (2004b). Laser-assisted fluorescence microscopy for measuring cell membrane dynamics. Photochem. Photobiol. Sci. *3*, 817–22.
[18] Van Munster, E. B. and Gadella, T. W. J., Jr. (2004). phiFLIM: A new method to avoid aliasing in frequency-domain fluorescence lifetime imaging microscopy. J. Microsc. *213*, 29–38.
[19] Mattheyses, A. L., Shaw, K. and Axelrod, D. (2006). Effective elimination of laser interference fringing in fluorescence microscopy by spinning azimuthal incidence angle. Microsc. Res. Tech. *69*, 642–7.
[20] Kaetzel, M. A., Mo, Y. D., Mealy, T. R., Campos, B., Bergsma-Schutter, W., Brisson, A., Dedman, J. R. and Seaton, B. A. (2001). Phosphorylation mutants elucidate the mechanism of annexin IV-mediated membrane aggregation. Biochemistry *40*, 4192–9.
[21] Newman, R. H., Leonard, K. and Crumpton, M. J. (1991). 2D crystal forms of annexin IV on lipid monolayers. FEBS Lett. *279*, 21–4.
[22] Zanotti, G., Malpeli, G., Gliubich, F., Folli, C., Stoppini, M., Olivi, L., Savoia, A. and Berni, R. (1998). Structure of the trigonal crystal form of bovine annexin IV. Biochem. J. *329(Pt 1)*, 101–6.
[23] Piljic, A. and Schultz, C. (2006). Annexin A4 self-association modulates general membrane protein mobility in living cells. Mol. Biol. Cell *17*, 3318–28.
[24] Shaner, N. C., Campbell, R. E., Steinbach, P. A., Giepmans, B. N., Palmer, A. E. and Tsien, R. Y. (2004). Improved monomeric red, orange and yellow fluorescent proteins derived from Discosoma sp. red fluorescent protein. Nat. Biotechnol. *22*, 1567–72.
[25] van Munster, E. B. and Gadella, T. W. J., Jr. (2004). Suppression of photobleaching-induced artifacts in frequency-domain FLIM by permutation of the recording order. Cytometry A *58*, 185–94.
[26] Lakowicz, J. R. (1999). Principles of Fluorescence Spectroscopy. Kluwer Academic, New York.
[27] Bastiaens, P. I. H., van Hoek, A., Benen, J. A., Brochon, J. C. and Visser, A. J. W. G. (1992). Conformational dynamics and intersubunit energy transfer in wild-type and mutant lipoamide dehydrogenase from Azotobacter vinelandii. A multidimensional time-resolved polarized fluorescence study. Biophys. J. *63*, 839–53.

Laboratory Techniques in Biochemistry and Molecular Biology, Volume 33
FRET and FLIM Techniques
T. W. J. Gadella (Editor)

CHAPTER 10

FRET and FLIM applications in plants

Riyaz A. Bhat

Department of Biology, Indiana University, Bloomington, Indiana 47405

Recent advances in the use of fluorescence microscopy-based methods to visualize proteins and characterize protein–protein interactions in vivo with ever increasing quantitative and spatio-temporal resolution have revolutionized our understanding of cellular functions. Förster resonance energy transfer (FRET) and fluorescence lifetime imaging microscopy (FLIM) have emerged as invaluable tools to study and characterize dynamic protein complexes. Despite a rather slow start, a growing community of plant scientists also embraced these novel tools to study and understand in planta gene functions. In this chapter, I give an overview of FRET–FLIM techniques with particular emphasis on their applications in plant research. The potential use of new donor–acceptor FRET–FLIM pairs and ways of circumventing the problem of chlorophyll autofluorescence as a likely cause of false FRET signals are also discussed. Finally the most commonly used protocols for expressing fusion proteins transiently or stably in plant cells are described.

DOI: 10.1016/S0075-7535(08)00010-7

10.1. Introduction

The availability of whole genome sequences of many eukaryotic and prokaryotic species has made it possible to predict the entire set of gene products in an organism. Since proteins seldom function alone and are often subunits of complexes, determining the spatial and temporal dynamics of protein–protein associations within these complexes in a cell is of fundamental importance to understanding biological processes [1]. Over the past few decades genetic and biochemical approaches to studying gene functions have yielded valuable information. However, these approaches provide limited insight into protein behavior and organization in their native environments inside living cells. Recent advances in live cell imaging combined with state-of-the-art microscopy equipment and fluorescence techniques have heralded a new dawn for cell biology [2]. In light of these latest technological advances, fluorescence-based cell biology has undergone a renaissance in the last decade [3] and has become indispensable to understand in vivo protein functions in diverse eukaryotes. Among other commonly used fluorescence-based methods like bioluminescence resonance energy transfer (BRET [4]) and bimolecular fluorescence complementation (BiFC [5]), Förster resonance energy transfer also referred to as fluorescence resonance energy transfer (FRET) has emerged as a powerful and convenient tool to study in vivo protein dynamics and associations. This chapter will focus on the use and application of FRET imaging to understand in planta gene functions.

10.2. Multicolor and FRET imaging in plants

The finding that the gene encoding green fluorescent protein (GFP) from the jellyfish *Aequorea victorea* could be fused in-frame with a gene encoding an endogenous protein with subsequent expression in an appropriate cellular background enabled the use of GFP as a molecular marker for gene expression [6–9]. Furthermore, since GFP can be targeted to most subcellular sites, these fusion protein chimeras have been variously used to monitor protein dynamics and fate in different subcellular compartments, making it feasible to examine biochemical processes in real time [10]. The next milestone in live cell imaging was the identification and engineering of mutant versions of the wild-type GFP that resulted in fluorophores with distinct absorbance (abs) and emission (em) spectra [9]. The spectrally distinct variants of GFP (abs/em = 488/509 nm; see Section 10.3) made it possible to perform dual color imaging and thus follow the dynamics and fate of two proteins in a single colocalization experiment [11]. Identification of red fluorescent protein (RFP; abs/em = 558/583 nm) from the coral *Discosoma striata* [12] along with its monomeric forms (mRFP; abs/em = 556/586 nm; [13]) and different spectral variants (see Section 10.3 for details) has made it possible to discriminate between three or more different fluorescent fusion proteins in a single experiment [14, 15]; for a more extensive coverage on the subject see Chapter 5.

Fluorescence imaging with genetically encoded fluorophores is widely being used to study the dynamics and turnover of diverse proteins involved in different plant signal transduction pathways [16]. However, engineering of fluorescent marker proteins to determine subcellular protein localizations and associations in planta can be quite challenging since plant cells contain a number of autofluorescent compounds (e.g., lignin, chlorophyll, phenols, etc.,) whose emission spectra interfere with that of the most commonly used green or red fluorophores and their spectral variants [17]. For example, lignin fluorescence in roots, vascular tissues, and cell walls of aerial plant parts interferes with imaging at wavelengths between 490 and 620 nm while the chlorophyll autofluorescence in leaves and stems is most problematic between 630 and 770 nm. Thus imaging of GFP and its closest spectral variants such as CFP and YFP is most likely to be problematic in roots, whereas RFPs may be hard to discriminate in chloroplast containing aerial plant tissues [17]. These problems have only recently been effectively alleviated with the development and use of high-end microscopes that can differentiate between closely emitting fluorophores and autofluorescence using spectroscopic methods (see also Chapters 4 and 8).

Besides multicolor imaging, perhaps the most exciting use of GFP and/or RFP and their spectral variants has been the development of FRET-based assays to probe bimolecular interactions in living cells with spatiotemporal resolution and sensitivity [18]; for definition and principle of FRET refer to Chapter 1. During the last 2 decades, these assays have been increasingly used to detect protein–protein interactions or protein conformational changes in diverse organisms ranging from yeast [19] to animals [20, 21] and plants (reviewed in [22]). However, despite its widespread popularity in yeast and animal systems, the use of FRET in plant sciences is still relatively limited (Fig. 10.1).

This relatively slow start is partly due to general unavailability of the high-end instruments and technical expertize required for in planta FRET measurements and secondly the largely genetics- and biochemistry-based approaches that are more prevalent among plant scientists. Despite these impediments, recent years have seen the publication of many high-profile papers from the plant sciences utilizing the potential of FRET to understand regulation of development, physiology, transcription, as well as disease signaling in plants [18, 23–30]. A new array of available fluorophores with improved biophysical properties is expected to further increase the popularity of FRET-based methods in studying dynamic protein trafficking and associations in living plant cells.

10.3. Ideal donor–acceptor pairs for in planta FRET applications

In living cells, FRET occurs when putative interacting proteins (or different domains within a single protein) equipped with suitable donor and acceptor fluorophores physically interact, thus bringing

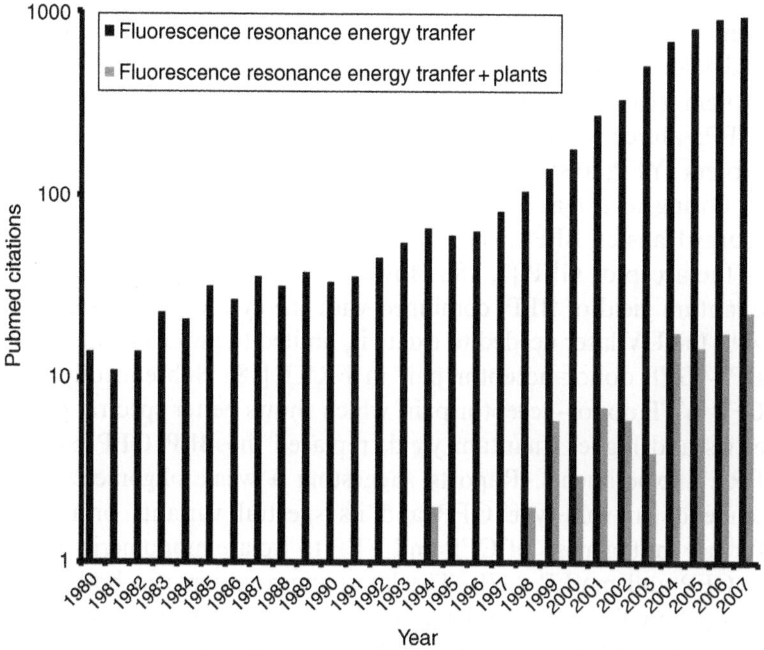

Fig. 10.1. The number of Pubmed Medline citations containing the keyword "Fluorescence resonance energy transfer" or "Fluorescence resonance energy transfer + Plants" anywhere in the text over the past 27 years. Although the technique has become very popular generally, its potential has yet to be fully exploited in plant research.

the donor and the acceptor within the favorable proximity for energy transfer [24]. Antibodies coupled to fluorescent donor–acceptor dyes like cy3-cy5, Alexa488-Alexa555, Alexa488-Cy3, Alexa546-Alexa594, Texas Red-Cy5, FITC-TRITC have been variously used to demonstrate FRET in several living systems [31–35] (see also Chapter 6). However, this is a rather indirect approach to measuring molecular associations. In living cells, it is highly desirable to have fluorescent labels directly attached to the proteins under investigation [36]. The identification and characterization of genetically encoded fluorophores such as GFP and RFP and their mutant derivatives have paved the way for using them as donor–acceptor FRET pairs in biological systems. The well-known and widely used mutants of wild-type GFP are blue-fluorescent protein (BFP; abs/em = 382/446 nm), cyan-fluorescent protein (CFP; abs/em = 434, 452/476, 505 nm), and yellow-fluorescent protein (YFP; abs/em = 516/529 nm; [9, 18]).

Owing to a reasonable spectral overlap necessary for efficient energy transfer, BFP was initially heralded as an ideal FRET donor to the acceptor GFP [7, 18]. However, the low photostability and quantum yield of BFP, combined with the cytotoxicity associated with the UV laser needed to excite it, limited the widespread use of BFP–GFP donor–acceptor pair in FRET [18]. Subsequently, the CFP–YFP donor–acceptor pair, which shows better spectral properties and higher quantum yield, replaced the BFP–GFP pair in FRET experiments. Reports suggesting a weak oligomerization tendency in wild-type GFP and its spectral variants prompted further mutagenesis of CFP and YFP to isolate monomeric forms (mCFP and mYFP; [37]). The latest improvements in both cyan and yellow fluorophores have resulted in a battery of even better fluorophores for both cyan (SCFP3a-brightest CFP available so far, cerulean, and CyPet), as well as yellow (Venus, SYFP2, mCitrine, and YPet) variants suitable for FRET studies [15, 38–40] (see also Chapter 5). With the availability of these improved variants, the widely used enhanced CFP and YFP versions have become obsolete [15] and thus should no longer be used.

mRFP (abs/em = 556/586 nm) and the recently described novel monomeric fluorescent proteins with longer red-shifted wavelengths such as mOrange (abs/em = 548/562 nm), mOrange2 (abs/em = 549/565 nm), TagRFP (abs/em = 555/584 nm), mStrawberry (abs/em = 574/596 nm), mCherry (abs/em = 587/610 nm), mKate (abs/em = 588/635 nm), and mPlum (abs/em = 590/649 nm) [14, 15, 41–43] have opened up the possibility of using these as acceptors with monoexponential donor fluorophores such as mGFP (eGFP with an A206K mutation) and T-Sapphire [15, 44]. However, the autofluorescence arising from phenols, lignin, and chlorophyll can limit the choice of fluorophores suitable for in planta FRET imaging. Because of their faster folding, superior solubility, and brightness, SCFP3A, and SYFP2 can be recommended as an ideal in planta FRET pair. Also CyPet and YPet can be used as the donor–acceptor pair, since they have been reported to be sevenfold more efficient in FRET than their progenitors [39, 45]. However, both CyPet and YPet are weak dimers and can be made monomeric by introducing an A206K mutation [15] to avoid false homodimerization in FRET experiments. In addition, it is possible to use mGFP, monomeric T-Sapphire (with an A206K mutation), or SYFP2 as donors with the above-mentioned red-shifted monomers as acceptors [46, 47]. mOrange and TagRFP may be good choices since both are bright red-shifted monomers [15] and their emission spectrum does not overlap with that of chlorophyll autofluorescence [41, 47]. An application of T-Sapphire and mOrange as a donor–acceptor FRET pair in intact plant cells is described in Section 10.4. A disadvantage of mOrange compared with other red-shifted monomers is that it is more prone to photobleaching, which makes it unsuitable for imaging for extended periods of time [15]. Recently described mOrange2 may have solved this problem [42]. This upgrade to mOrange is reportedly 25 times more photostable than the parent. However, the use of mOrange2 may be limited in situations where acceptor photobleaching (see Section 10.4) is the sole method to be used for determining FRET, since rapid photobleaching of the acceptor mOrange2 (a crucial prerequisite for acceptor photobleaching) may not be achieved fast enough. Although mCherry, mKate, and mPlum are fairly photostable, their use may be limited in aerial plant cells because of interference with chlorophyll pigments. In such tissues, it is quite critical to subtract the background fluorescence either by using very narrow bandpass emission filters, which compromises sensitivity, or by resolving the individual spectra. For FRET experiments in cells devoid of chlorophyll pigments such as roots or tobacco BY2 cells, donors such as mGFP or T-Sapphire may be used with any of the red-shifted acceptors, such as mOrange, TagRFP, mStrawberry, or mCherry, without complications.

10.4. Quantifying FRET—Methods and Pitfalls

Because FRET results in a decrease in donor fluorescence intensity and excited state lifetime with a corresponding increase in the acceptor fluorescence (sensitized emission), various methods for measuring FRET have been based on assessing one of the above photophysical consequences [22]. The most commonly employed methods for measuring FRET in living systems (described elsewhere in more detail in this volume) are:

1. Sensitized emission FRET (also called filter FRET or channel FRET)—a method that assesses the changes in donor and acceptor fluorescence intensity upon successful energy transfer [26, 48–50] (see Chapter 7);
2. Fluorescence spectral imaging microscopy (FSPIM)—a method that uses a spectroscopic approach to quantify changes in the acceptor intensity at donor excitation wavelengths [24, 27, 51, 52] (see Chapter 8);
3. Donor fluorescence recovery after acceptor photobleaching (also called acceptor photobleaching or acceptor depletion FRET) [22, 23, 26, 28, 30, 48, 53–56] (see Chapters 1 and 7); and
4. Fluorescence lifetime imaging microscopy (FLIM) that measures the donor excited state lifetime in the presence and absence of an acceptor [23, 25, 28, 47, 52, 53, 57, 58] (see Chapters 2–4).

The calculation of protein proximity and hence association on the basis of sensitized emission or FSPIM requires correction for direct acceptor excitation and donor bleed through using several mathematical models and instrument correction factors [22, 59–61], which is difficult to control [22] (see also Chapters 7 and 8). A high detected acceptor to donor signal ratio in these techniques may also reflect other phenomena than FRET. For instance, this ratio is dependent on cellular expression levels and subcellular localizations, which are difficult to control. Additionally, for the widely used

CFP–YFP donor–acceptor pair, YFP is several times brighter than CFP [62]. Lastly, for studying dynamic protein associations in plants, the presence of chlorophyll pigments in leaf and stem cells is an additional limitation. These pigments directly absorb the fluorescence, which decreases blue fluorescence intensity for BFP and CFP donors that can be erroneously interpreted as reduced donor fluorescence quantum yield caused by FRET [18]. If sensitized emission or FSPIM is the only available method for quantifying FRET, then it is very important to restrict measurements to chlorophyll free areas within the cells.

Fig. 10.2 depicts the use of FSPIM to study the interaction between two putative maize transcriptional coactivators—a histone acetyltransferase GCN5 (general control nonderepressible-5) and an adaptor protein ADA2 (alteration/deficiency in activation)—in the nuclei of plant protoplasts. The interaction between the *Zm*GCN5 and *Zm*ADA2 results in quenching of the donor (CFP) fluorescence while the acceptor (YFP) fluorescence gets sensitized, resulting in a strongly increased fluorescence emission at around 530 nm as compared with 480 nm. Subsequent acceptor photobleaching and FLIM analysis confirmed the sensitized acceptor emission seen in the above studied interaction [53].

Acceptor photobleaching is a widely used alternative method for measuring FRET. The rationale behind this approach is that energy transfer between the donor and the acceptor gets disrupted upon bleaching of the acceptor leading to an enhanced (dequenched) donor fluorescence emission [66]. To quantify energy transfer by this method, donor emission is compared before and after acceptor photobleaching, and the difference between the two reflects the amount of energy transfer [22, 56]. In the absence of energy transfer, donor fluorescence usually remains unaltered or may even (unintentionally) decrease following acceptor photobleaching, thus an increase in donor fluorescence is considered a reliable indicator of successful FRET [56, 66]. Additionally, dequenched donor fluorescence cannot be attributed to acceptor bleed through since following acceptor photobleaching, the acceptor is

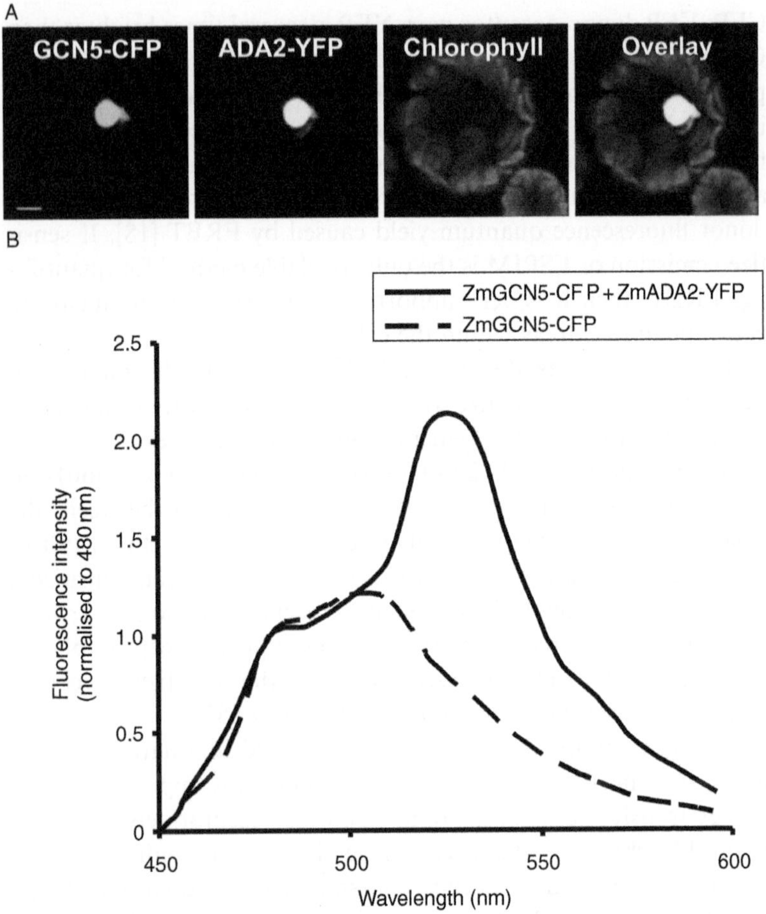

Fig. 10.2. FSPIM analysis of the interaction between maize transcriptional coactivators—GCN5 and ADA2—fused to CFP and YFP. GCN5 is a histone acetyltransferase that, in conjunction with adaptor protein ADA2, modulates transcription in diverse eukaryotes by affecting the acetylation status of the core histones in nucleosomes [63]. CFP- and YFP-tagged proteins, expressed in protoplasts, were excited by the 458 nm and the 514 nm laser lines sequentially. CFP fluorescence was selectively detected by an HFT 458 dichroic mirror and BP 470–500 band pass emission filter while YFP fluorescence was selectively detected by using an HFT 514 dichroic mirror and

no longer available [22]. Two variations of acceptor photobleaching—intensity and spectral-based—were used to confirm the in planta interaction between the plant-specific integral membrane protein MLO (mildew resistance locus "o") and the cytosolic calcium sensor protein calmodulin (CaM) in barley leaf epidermal cells (Fig. 10.3). Following bleaching of MLO–YFP, there is a sharp increase in the donor CFP–CaM, indicative of FRET and hence interaction between MLO and CaM [23].

A critical prerequisite for acceptor photobleaching is that the acceptor should be readily photobleachable to achieve nearly 95–100% acceptor bleaching in a relatively short period of time. While

---

BP 535–590 band pass emission filter. In both cases the chlorophyll autofluorescence was filtered out and detected in another channel using a LP650 long pass filter. A 25 × Plan–Neofluar water immersion objective lens (N.A 0.8) was used for scanning protoplasts. In order to avoid crosstalk between CFP and YFP, images were acquired in the multichannel tracking mode and analyzed with Zeiss LSM510 Meta software. (A) Colocalization of *Zm*GCN5-CFP and *Zm*ADA2-YFP in cowpea protoplasts imaged by a confocal laser-scanning microscope. Nuclear localized *Zm*GCN5-CFP and *Zm*ADA2-YFP along with chlorophyll autofluorescence and an overlay image of all the three is shown. (B) Nuclei of cowpea protoplasts expressing either *Zm*GCN5-CFP alone, or in combination with *Zm*ADA2-YFP, were subjected to FSPIM analysis as described [27, 64, 65]. For FSPIM, a 435 nm line from a 100 W Hg Arc lamp light source was selected by inserting a 435DF10 bandpass filter from Omega (Brattleboro, VT) in the excitation light path. Subsequently the excitation light was reflected onto the sample by an Omega 430DCLP dichroic mirror. Residual excitation light was rejected using a Schott (Mainz, Germany) GG455 longpass filter. Images were acquired using a 20 × Plan Neofluar objective. Spectra were obtained using a spectrograph (Chromex 250; Chromex, Albuquerque) with 150 grooves/mm grating set at a central wavelength of 500 nm. Protoplasts expressing CFP and YFP fusion proteins were aligned across the entrance slit of the spectrograph (set at 200 μm width corresponding to 10-μm width in the object plane). Acquisition time was 2–3 s. Data collection and background subtraction was performed as described [65]. Representative spectra normalized to the detected intensity at 480 nm from the nuclei of protoplasts expressing *Zm*GCN5-CFP + *Zm*ADA2-YFP (solid line) or *Zm*GCN5-CFP alone (dashed line) are shown. (See Color Insert.)

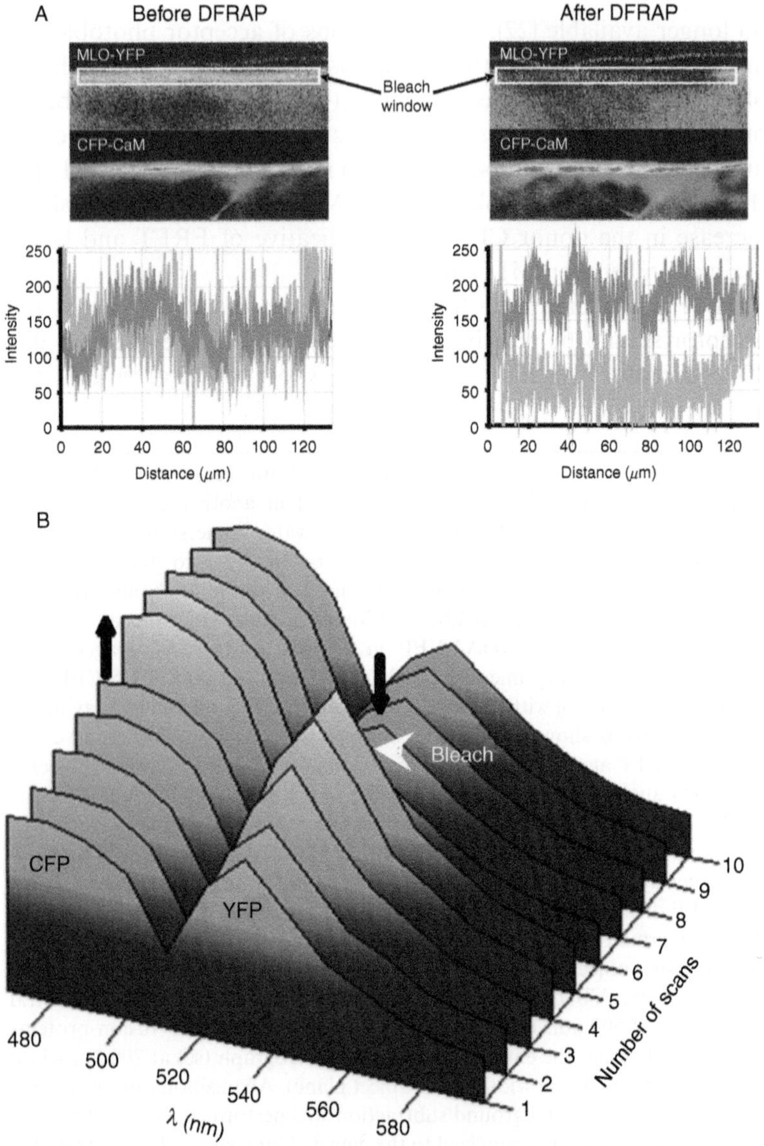

Fig. 10.3. Acceptor photobleaching analysis of interaction between barley MLO and calmodulin. Barley MLO is a plant-specific integral membrane protein that associates with the cytosolic calcium sensor protein Calmodulin

both YFP and RFP acceptors have been used in acceptor photobleaching experiments, it is relatively easier to photobleach the former [26, 56, 62]. From the red-shifted monomers, mOrange is easier to use as an acceptor in acceptor photobleaching experiments than any other fluorophore. Another consideration is that the acceptor bleach wavelength should not bleach the donor. For the CFP–YFP donor–acceptor pair, this is usually accomplished by using the 514 nm argon line for YFP photobleaching, but the 543 nm HeNe line can be used for this purpose as well [36]. Compared with sensitized emission and FSPIM techniques, acceptor photobleaching is quite robust [68]. However, if not used with caution, the acceptor photobleaching technique has pitfalls of its

---

(CaM) and functions as a modulator of defense against the common powdery mildew pathogen [67]. Wild-type MLO fused to YFP was coexpressed with CaM fused to CFP in barley leaf epidermal cells [23]. 458 and 514 nm argon laser lines were used to excite CFP and YFP respectively. The experimental conditions and filter settings were the same as in Figure 10.2 except that a 40 × Plan–Neofluar water immersible objective lens (N.A 0.75) was used for scanning the leaf epidermal cells. Acceptor photobleaching was performed as described [22, 53, 56]. (A) Intensity-based acceptor photobleaching on MLO–CaM interaction. Following a rapid photobleaching of the acceptor YFP (white rectangular bleach window in the top panels), the donor CFP intensity before and after acceptor photobleaching was compared. The photochemical destruction of the acceptor YFP abolishes any energy transfer from CFP–CaM to MLO–YFP resulting in a rapid increase in the CFP–CaM emission intensity (compare line plot intensity profile panels in bottom panels; light gray–MLO–YEP; dark grey–CFP–CaM) indicating a FRET and hence an association between MLO and CaM (see also [23]) (B) Spectral acceptor photobleaching analysis on MLO–CaM interaction. A 458 nm laser argon line was used to co-excite both CFP–CaM and MLO–YFP and fluorescence spectra were recorded in a lambda mode from 470 to 600 nm using a Zeiss LSM510 Meta confocal laser-scanning microscope. The rapid photobleaching pulse was given at the 5th scan (white arrow head). With the bleaching of the YFP (530 nm, black downward arrow), the CFP emission spectrum at 480 nm sharply increases (upwards black arrow), indicative of FRET and hence the association between MLO and CaM. The dichroic mirror used to reflect 514 nm laser line causes the dip in the emission spectra. (See Color Insert.)

own. First, this method is applicable either in fixed cells or in living cells with proteins that are largely immobile (e.g., membrane associated proteins). In the case of highly mobile cytosolic proteins, the recovery of the donor fluorescence, following acceptor photobleaching, may just reflect the diffusion of the donor into the photobleached areas of the cell, resulting in false enhanced FRET values. This process can be avoided either by photobleaching the whole cell (problematic for relatively large plant cells), or by selecting very small areas of the cell for acceptor photobleaching combined with a very intense but relatively short photobleaching pulse (Fig. 10.3; [22]). Second, acceptor photobleaching irreversibly damages the acceptor, thus severely compromising the ability to perform time course experiments. Finally, the intense excitation required for photobleaching may cause damage to the living cells [36].

When a fluorophore is excited, it stays in the excited state for a very short period of time before emitting a photon and returning back to its ground state. The average time that the fluorophore spends in the excited state is in the order of pico- to nanoseconds and is termed the fluorescence lifetime. Fluorescence lifetime is a sensitive measure of the fluorophore environment and is influenced by many environmental variables such as pH, ion concentrations, oxygen saturation by fluorescence quenching and FRET [36, 69, 70] (see Chapters 1–4). FRET between a suitable donor–acceptor pair results in a decrease of the donor fluorescence lifetime that can be measured using the technique of fluorescence lifetime imaging (FLIM) microscopy. The efficiency of energy transfer (E) based on the fluorescence lifetime (τ) is calculated as:

$$E = 1 - \tau_{DA}/\tau_D \tag{10.1}$$

where τ_{DA} is donor fluorescence lifetime in the presence of acceptor while τ_D is the donor fluorescence lifetime in the absence of acceptor [18].

Two different approaches for measuring fluorescence lifetimes are commonly employed to study FRET-related phenomenon: the frequency-domain FLIM (see Chapter 2) and the time-domain

FLIM (see Chapter 3). FRET measurements by FLIM are superior to both sensitized emission and acceptor photobleaching, since fluorescence lifetimes are independent of local chromophore concentrations or interference by spectral cross talk, direct absorption of donor fluorescence by chlorophyll pigments, or moderate photobleaching [18, 22, 24]. Additionally, if multiple lifetimes can be resolved, FLIM is able to differentiate subpopulations with different amounts of energy transfer [71] (see also Chapter 2) and thus provides a quantitative interaction map of a cell with a single measurement [22]. Fig. 10.4 illustrates the time-domain FLIM measurement of energy transfer from T-Sapphire as a donor to red-shifted mOrange as an acceptor in *Nicotiana benthamiana* leaf epidermal cells.

However, FLIM measurements by both frequency and time domain using commercially available software suffers from variation in the measured lifetime from region to region in cells [36], which can be confusing. In the light of the potential pitfalls associated with each of the above-mentioned FRET techniques, potential protein associations should ideally be tested independently by a combination of two or more FRET-based methods in addition to biochemical techniques such as co-immunoprecipitations.

10.5. Protoplasts versus intact tissues: Which cell-type to FRET

One of the first reports describing in planta FRET between two putative interacting proteins was the study of Mas and colleagues [26] who used a combination of sensitized emission and acceptor photobleaching techniques to study the interaction between the *Arabidopsis* photoreceptors phytochrome B (phyB) and cryptochrome 2 (cry2) involved in the control of flowering time, hypocotyl elongation and circadian rhythm. PHYB–GFP and CRY2–RFP were seen to colocalize and interact in the light-induced nuclear speckles in tobacco protoplasts. This was the first study that used

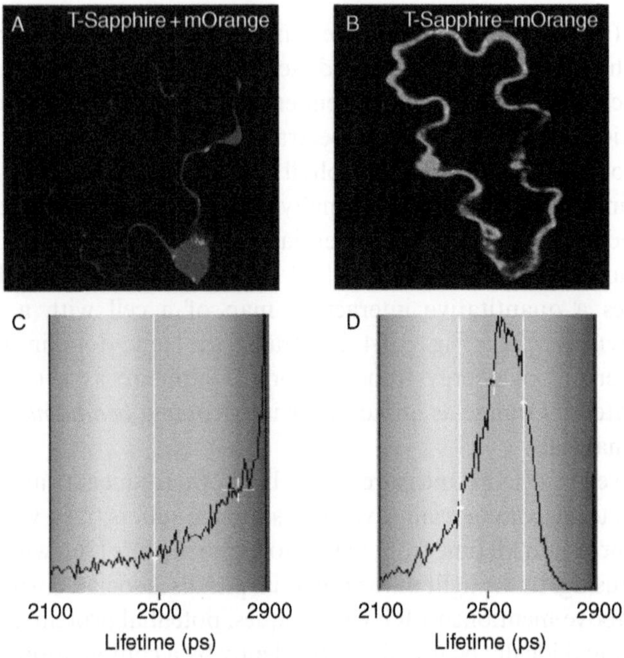

Fig. 10.4. Time-domain FLIM between T-Sapphire and mOrange. A chimeric construct harboring T-Sapphire and mOrange fluorophore cDNA's separated by a 12 amino acid linker, under the control of constitutive cauliflower mosaic virus (CaMV) 35S promoter, was delivered into *N. benthamiana* leaf epidermal cells by ballistic-mediated particle bombardment [47]. At the same time, T-Sapphire and mOrange cDNA's on separate plasmids, driven by CaMV 35S promoters, were co-bombarded into control *N. benthamiana* leaf epidermal cells. The lifetime images of the donor T-Sapphire were obtained using the 405 nm diode laser modulated at 80 MHz and the T-Sapphire fluorescence was selectively imaged using a 520DF40 bandpass emission filter. The microscope setup, lifetime image calculation, and the image processing were performed as per the time-domain FLIM system from Becker and Hickl [47, 69]. Only pixels with more than 50 photons were used for calculating the lifetime. The lifetime of T-Sapphire coexpressed with mOrange was measured as 2.9 ns on average (A and C; in agreement with the lifetime of wild-type GFP; 3.16 ns; [72, 73]) whereas the average lifetime of T-Sapphire in the chimeric construct was found to be around 2.45 ns, yielding an average of around 15% energy transfer to mOrange (B and D; Eq. 10.1; [47]). (C) and (D) represent the temporal histograms and the fluorescence lifetime pixel values in (A) and (B) respectively between 2.1 ns (strong interaction) and 2.9 ns (no interaction). (See Color Insert.)

GFP–RFP as a donor–acceptor FRET pair for sensing intermolecular interactions in living cells. Using an FSPIM approach, Shah and colleagues [51] provided evidence that *Arabidopsis thaliana* somatic embryogenesis receptor kinase 1 (AtSERK1) is targeted to the plasma membrane and forms homodimers in cowpea mesophyll protoplasts. In a later study the same group provided further evidence that AtSERK1 interacts with the kinase-associated protein phosphatase (KAPP) in intracellular vesicles in a phosphorylation-dependant manner [27]. Although in both studies only FSPIM was used to validate either the homodimerization of AtSERK1, or its interaction with KAPP1, immunoblots were used to provide convincing evidence that both the donor and the acceptor fluorophores were expressed to similar levels in the cells, thus underlining the detection of sensitized emission rather than direct acceptor excitation being the result of unequal expression levels. Immink et al. [24] used FSPIM and frequency-domain FLIM to study the homo- and heterodimerization of various plant MADS box transcription factors (Floral binding protein 2, 5, 9, and 11) in petunia mesophyll protoplasts. Using a combination of acceptor photobleaching and time-domain FLIM, Bhat and colleagues [53] reported the in planta interaction between putative transcriptional coactivators from maize viz., *Zm*GCN5 and *Zm*ADA2 in cowpea mesophyll protoplasts (Fig. 10.2).

All these early in planta FRET–FLIM studies used the protoplast system (heterologous or homologous) for the delivery and transient expression of the donor–acceptor tagged gene products [22]. However, it is best to study in planta interactions at an organismal level in a tissue of physiological and biological relevance. A few recent studies have addressed this important issue and the field is slowly moving from the protoplast system to intact plant tissues for studying and characterizing stimulus-dependant protein–protein interactions [23, 25, 28, 30, 47, 54, 55]. One of the first studies to report in planta FRET in intact barley leaf epidermal tissues was the interaction between a plant specific heptahelical protein MLO (involved in resistance against powdery mildew pathogen) and cytosolic calcium sensor protein Calmodulin [23]. The authors used acceptor photobleaching and time-domain FLIM to study the dynamics of the interaction between MLO and Calmodulin during the growth and invasion of the fungal pathogen into intact cells (Fig. 10.3). In another study Feys and colleagues [55] reported in planta interaction between EDS1 (enhanced disease susceptibility 1) and SAG 101 (senescence associated gene 101) – two components involved in plant defense signaling – in intact leaf epidermal cells of *Arabidopsis*. Since then several more papers have been published studying FRET in intact plant tissues [25, 28, 30, 47, 54]. Given the obvious need, it is anticipated that the use of intact tissues to understand biological questions in plant sciences will prevail over the use of protoplast systems in the near future.

10.6. Beyond protein–protein interactions—development and use of FRET-based nanosensors

Determining the dynamic fluctuations of different metabolites (e.g., sugars), ions (e.g., Ca^{2+}) and nutrients (e.g., phosphorus, nitrogen) in different cellular and subcellular compartments is of critical importance to understanding metabolic networks and fluxes [74]. The first genetically encoded sensor for measuring changes in cytosolic calcium levels was the FRET-based *cameleon* sensor consisting of CFP and YFP at the either end of the molecule along with the calcium-binding domain protein calmodulin and the calcium–calmodulin binding M13-peptide in the middle [75] (see also Chapter 5). Upon calcium binding to the calmodulin, the M13 peptide binds to the calcium–calmodulin unit, inducing a conformational change that brings the CFP and YFP close together, resulting in an increased energy transfer from CFP to YFP [76]. Since the acceptor:donor molecular ratio in *cameleon* is always 1:1 (both being fused into one protein), the YFP:CFP fluorescence intensity ratio directly reports the cytosolic calcium concentration [18]. The concept was successfully used to determine calcium concentrations in living plant cells in response to a range of stimuli [16, 18, 64, 77, 78]. Until recently no sensors to measure the concentrations of primary metabolites and nutrients were available in living cells [74]. The situation has changed rapidly in the last few years with the finding that members of the bacterial periplasmic-binding protein superfamily (PBPs) that recognize hundreds of substrates with high affinity and specificity [79], can undergo a significant conformational change upon ligand binding in FRET-based assays [80]. This ligand-specific conformational change in PBPs has been used to develop sensors for measuring maltose, ribose, and glucose uptake and homeostasis in living animal cells by fusion of an individual sugar-binding PBP with a suitable donor–acceptor FRET pair [74, 81–86]. Similarly, a nanosensor to monitor glutamate levels inside and at the surface of cells (viz., FLIPE for fluorescent indicator protein for glutamate) was developed using the bacterial glutamate-binding protein ybeJ as a recognition element fused to eCFP and Venus (improved YFP, see Section 10.3) as donor and acceptor [87]. FRET sensors for measuring cellular phosphate (Pi) levels were engineered by fusing a *Synechococcus* phosphate-binding protein (PiBP) to CFP and Venus as donor and acceptor [88]. These sensors, known as FLIPPi (fluorescent indicator protein for inorganic phosphate), are highly sensitive to different concentrations of intracellular phosphate and are thus suitable for studying Pi uptake and translocation, and regulatory mechanisms controlling compartmental Pi homeostasis [88]. There have been numerous reports about the use of these nanosensors in studying plant metabolic networks and fluxes. A study describing the use of FLIPglu – a glucose nanosensor—was used to study glucose uptake and metabolism in living plant cells [89]. Likewise, a sensor viz., FLIPsuc-4μ for detecting sucrose, the major sugar used by plants to supply energy, was generated by fusing putative sucrose-binding proteins to eCFP and eYFP [90]. More recently a FRET-based nano-sensor that responds to changes in amino acid concentrations was described paving way for studying in vivo dynamics of metabolism and compartmentalization in plants [91].

10.7. Protocols for protoplast transfection and particle bombardment of leaf epidermal cells

Protoplasts from Tobacco BY2 (*Nicotiana tabacum* cv *Bright Yellow* [26]), Cowpea (*Vigna unguiculata* [27, 51, 53]), Petunia (*Petunia hybridia* [24]), Maize (Zea mays [92]) as well as intact leaf tissues from *Arabidopsis* (*A. thaliana* [30, 55]), Barley (*Hordeum vulgare* [23, 25, 28, 54]), and *N. benthamiana* [47] have been successfully used to study in planta protein–protein interactions by FRET-based methods. Depending on the protoplasts or the intact cells, the gene delivery may be mediated either by PEG-based transfection (for the former) or agroinfiltration and/or particle bombardment-mediated gene transfer into intact cells. Here I give simple protocols for protoplast transfection and particle bombardment-mediated gene transfer. Unless indicated otherwise, all the reagents and chemicals are available from general vendors like Sigma, Fluka, or Fisher Scientific etc.

10.7.1. Mesophyll protoplast isolation and transfection

Most protocols for the isolation of protoplasts from Cowpea, Petunia, or Tobacco mesophyll cells are based on [93, 94] and others [95–97]. After the isolation of protoplasts from the cell type of choice, the transfection is carried out as follows (based on [51, 96]):

1. Gently add equimolar amounts of purified plasmid DNA from the donor and the acceptor FRET pairs (up to 10 μg in total) with approximately 2×10^6 freshly isolated protoplasts in 75–150 ml of ice cold Ca–Mannitol solution (0.6 M mannitol, 10 mM $CaCl_2$; pH 5.3–5.8).
2. Immediately add 0.6 ml of PEG solution (40% (w/v) PEG Mw 6000, 0.6 M mannitol, 0.1 M $Ca(NO_3)_2$).
3. After 5–25 s gentle shaking, add 4.5 ml of Mg–Mannitol solution (0.5 M mannitol, 15 mM $MgCl_2$, and 0.1% MES; pH 5.3–5.7).

4. Incubate at room temperature for 20 min.
5. Wash the protoplasts 3 times with Mg–Mannitol solution and incubate for 12–24 h.
6. Analyze the protoplasts under a fluorescence microscope suitable for FRET measurements (e.g., Leica SP2/SP5 or Zeiss LSM 510 Meta CLSM).

10.7.2. Particle bombardment-mediated gene transfer

Delivery of DNA encoding fluorescently-tagged fusion proteins (coated onto gold or tungsten particles) into intact plant cells using a particle inflow gun [98] has also been successfully used to study in planta interactions [23, 25, 28, 30, 54, 55]. This method of gene delivery has the advantage that intact cells (over) expressing the proteins in their natural environment are readily available for studying dynamic interactions under different kinds of biotic and antibiotic stimuli. However an associated limitation is that only proteins that are normally expressed in the epidermal cell layer can be studied by this method. If these criteria can be met, this method is much less laborious than the protoplast preparation and transfection. Fusion protein expression using this method can be easily achieved in any intact plant tissue (such as *N. benthamiana*, barley, or *Arabidopsis* leaves) by following the steps given below [29]:

1. Sterilize gold or tungsten particles (Bio-Rad, USA) of around 1 μm diameter with 70% ethanol followed by several washes with sterile water in an eppendorf tube. After the last wash, pellet the particles, discard the supernatant, and resuspend the pellet in 50% glycerol to have a final gold concentration of 60 mg/ml.
2. Mix 50 μl of fine gold particle suspension successively with donor/acceptor plasmid DNA (up to 10 μg in total), 50 μl of 2.5 M $CaCl_2$, and 20 μl of 0.1 M Spermidine while vortexing.
3. After 2–3 min vortexing, allow the particles to settle down for 1 min and then pellet for 2 s.

4. Discard the supernatant and resuspend the pellet in 140 μl of 70% ethanol.
5. Vortex at low speed for 2 s and then pellet for 2 s.
6. Repeat the step 4 and 5 two times but using 100% instead of 70% ethanol.
7. In the final step, resuspend the gold particles in around 15–20 μl 100% ethanol and keep the samples on ice till further use.
8. Place detached leaves from Tobacco (2–3 weeks old), *Arabidopsis* (4–5 weeks old), or Barley (1 week old) onto 1% agar plates.
9. Deliver the plasmid-coated gold or tungsten particles into the detached leaves using a particle delivery system of choice (e.g., Bio-Rad® Biolistic PDS-1000 HE particle delivery system) following the manufacture's protocol.
10. Place the bombarded leaves in a growth chamber for 12–24 h and analyze under a fluorescence microscope suitable for FRET measurements. Detached Arabidopsis leaves undergo senescence very quickly and thus analysis should be carried out no later than 24 h after bombardment.

Excellent protocols for transient expression of fluorescently tagged proteins in *N. benthamiana* leaves using *Agrobacterium*-mediated infiltration are available [99, 100].

10.8. Functional assays and stable plant transformation

Protoplast transfections using PEG-, agroinfiltration-, or particle-mediated gene transfer methods are frequently used for quick transient gene expression driven by strong constitutive promoters. In these systems, ectopic protein expression and/or overexpression may lead to false FRET signals in different subcellular compartments [22]. Thus, it is of utmost importance to maintain near native protein levels in the transfected cells by using relatively weak or endogenous gene promoters. It is also important to validate the biological functionality of the tagged protein, ideally by complementing mutant phenotypes [22]. Additionally, in order to study protein functions and interactions deep within the mesophyll cell layers, or in other specific tissues, it may be necessary to stably transform the respective mutants. For FRET experiments, ideally, this means generating stable transgenic lines for each potential partner protein in respective mutant backgrounds and then crossing the two fluorescent lines and isolating homozygous lines expressing both donor–acceptor FRET pairs in a double mutant background. Even though this procedure is very time consuming and laborious, it represents an ideal scenario for studying FRET-based protein associations in clean mutant backgrounds and deep within the plant tissues. Additionally, since the fusion protein expression takes place in respective (double) null mutant backgrounds, there is no problem of untagged endogenous gene copies out competing the fluorescently tagged interaction partners [22]. However, it may not always be possible to fulfill all of the conditions specified above. For example, mutants corresponding to one or the other interaction partner may not be available because of lethality. Nonetheless, an effort should be made to fulfill as many of the above criteria as possible [22]. Many protocols for the stable transformation of the plant model Arabidopsis using *Agrobacterium tumefaciens* are currently available [101].

10.9. Outlook

The successful use of intrinsically fluorescent proteins (despite their bulkiness) to follow protein dynamics and associations with subcellular resolution has greatly increased our understanding of the physiological processes occurring in living (plant) cells. Undoubtedly FRET-based methods have emerged as convenient tools, not only to study protein–protein interactions and/or protein conformational changes, but also to follow the strength and dynamics of these interactions in real time and in different subcellular compartments. This is a clear advantage over other currently popular fluorescence-based methods such as bi-fluorescence complementation assays (variously called split YFP or split GFP), which are not able to capture these dynamic events [102]. However, it is not easy to study more than one potential FRET-based interaction/s at a time within a single cell [103], although recently a single cell experiment with three distinct FRET couples was described [104]. This is a drawback since many biologically meaningful interactions occur in multiprotein complexes involving more than two interacting proteins [22].

FRET-based nanosensors have been successfully used to monitor steady state levels of metabolites, nutrients, and ions in mammalian cells [74, 87]. Recently FRET-based glucose, sucrose, and amino acid nanosensors have been developed to study the metabolism of glucose, sucrose, and amino acid uptake and metabolism in plant cells [80, 89, 91]. The enormous potential of these nanosensors will be crucial for understanding ion (e.g., calcium), metabolite (e.g., sugars), hormone (e.g., auxins, gibberellins etc.), and nutrient (e.g., nitrogen, potassium, phosphorus) requirements and homeostasis in living plant tissues.

To summarize, FRET remains one of the few available techniques to study in vivo protein dynamics and associations in real time. Although the plant sciences community has been slow to embrace the application of this technology, FRET-based plant cell biology is bound to grow increasingly in coming years. This is partly due to the recent commercial availability of relatively inexpensive, but robust FRET and FLIM capable microscope systems e.g., Zeiss LSM 510 META (Carl Zeiss, Germany) or Leica SP2/SP5 AOBS (Leica Microsystems, Germany) confocal laser scanning microscopes with optional time domain FLIM hardware from Becker and Hickl (Berlin, Germany).

Acknowledgments

I am grateful to Dr Oliver Griesbeck (Max Planck Institute of Neurobiology Am Klopferspitz, Martinsried, Germany) and Prof. Roger Tsien (Howard Hughes Medical Institute, University of California, San Diego, USA) for the kind donation of cDNA clones encoding T-Sapphire and mOrange, respectively. The GCN5-ADA2 FRET and FSPIM work was done at Microspectroscopy center, WAU, Netherlands in collaboration with Dr Jan-Willem Borst and the figures are published here with due permission from Dr Borst. I wish to thank Laurent Noël (LIPM-INRA/CNRS, Castanet, France), Roger Innes (Indiana University Bloomington, USA) and Sidney Shaw (Indiana University, Bloomington) for critically reading the manuscript and for helpful comments.

References

[1] Day, R. N. (1998). Visualization of Pit-1 transcription factor interactions in the living cell nucleus by fluorescence resonance energy transfer microscopy. Mol. Endocrinol. *12*, 1410–9.

[2] Gilroy, S. (1997). Fluorescence Microscopy of Living Plant Cells. Ann. Rev. Plant Physiol. Plant. Mol. Biol. *48*, 165–90.

[3] Yuste, R. (2005). Fluorescence microscopy today. Nat. Methods *2*, 902–4.

[4] Pfleger, K. D. and Eidne, K. A. (2006). Illuminating insights into protein-protein interactions using bioluminescence resonance energy transfer (BRET). Nat. Methods *3*, 165–74.

[5] Magliery, T. J. and Regan, L. (2006). Reassembled GFP: Detecting protein-protein interactions and protein expression patterns. Methods Biochem. Anal. *47*, 391–405.

[6] Chalfie, M., Tu, Y., Euskirchen, G., Ward, W. W. and Prasher, D. C. (1994). Green fluorescent protein as a marker for gene expression. Science *263*, 802–5.

[7] Heim, R. and Tsien, R. Y. (1996). Engineering green fluorescent protein for improved brightness, longer wavelengths and fluorescence resonance energy transfer. Curr. Biol. *6*, 178–182.

[8] Prasher, D. C., Eckenrode, V. K., Ward, W. W., Prendergast, F. G. and Cormier, M. J. (1992). Primary structure of the Aequorea victoria green-fluorescent protein. Gene *111*, 229–33.

[9] Tsien, R. Y. (1998). The green fluorescent protein. Ann. Rev. Biochem. *67*, 509–44.

[10] Pollok, B. A. and Heim, R. (1999). Using GFP in FRET-based applications. Trends Cell Biol. *9*, 57–60.

[11] Ellenberg, J., Lippincott-Schwartz, J. and Presley, J. F. (1999). Dual-colour imaging with GFP variants. Trends Cell Biol. *9*, 52–6.

[12] Matz, M. V., Fradkov, A. F., Labas, Y. A., Savitsky, A. P., Zaraisky, A. G., Markelov, M. L. and Lukyanov, S. A. (1999). Fluorescent proteins from nonbioluminescent Anthozoa species. Nat. Biotechnol. *17*, 969–73.
[13] Campbell, R. E., Tour, O., Palmer, A. E., Steinbach, P. A., Baird, G. S., Zacharias, D. A. and Tsien, R. Y. (2002). A monomeric red fluorescent protein. Proc. Natl. Acad. Sci. USA *99*, 7877–82.
[14] Shaner, N. C., Campbell, R. E., Steinbach, P. A., Giepmans, B. N., Palmer, A. E. and Tsien, R. Y. (2004). Improved monomeric red, orange and yellow fluorescent proteins derived from *Discosoma* sp. red fluorescent protein. Nat. Biotechnol. *22*, 1567–72.
[15] Shaner, N. C., Steinbach, P. A. and Tsien, R. Y. (2005). A guide to choosing fluorescent proteins. Nat. Methods *2*, 905–9.
[16] Fricker, M., Runions, J. and Moore, I. (2006). Quantitative fluorescence microscopy: from art to science. Ann. Rev. Plant Biol. *57*, 79–107.
[17] Chapman, S., Oparka, K. J. and Roberts, A. G. (2005). New tools for in vivo fluorescence tagging. Curr. Opin. Plant. Biol. *8*, 565–73.
[18] Gadella, T. W., Jr., van der Krogt, G. N. and Bisseling, T. (1999). GFP-based FRET microscopy in living plant cells. Trends Plant Sci. *4*, 287–91.
[19] Muller, E. G., Snydsman, B. E., Novik, I., Hailey, D. W., Gestaut, D. R., Niemann, C. A., O'Toole, E. T., Giddings, T. H., Jr., Sundin, B. A. and Davis, T. N. (2005). The organization of the core proteins of the yeast spindle pole body. Mol. Biol. Cell *16*, 3341–52.
[20] Miyawaki, A. (2003). Visualization of the spatial and temporal dynamics of intracellular signaling. Dev. Cell. *4*, 295–305.
[21] Silvius, J. R. and Nabi, I. R. (2006). Fluorescence-quenching and resonance energy transfer studies of lipid microdomains in model and biological membranes. Mol. Membr. Biol. *23*, 5–16.
[22] Bhat, R. A., Lahaye, T. and Panstruga, R. (2006). The visible touch: in planta visualization of protein-protein interactions by fluorophore-based methods. Plant Methods *2*, 12.
[23] Bhat, R. A., Miklis, M., Schmelzer, E., Schulze-Lefert, P. and Panstruga, R. (2005). Recruitment and interaction dynamics of plant penetration resistance components in a plasma membrane microdomain. Proc. Natl. Acad. Sci. USA *102*, 3135–40.
[24] Immink, R. G., Gadella, T. W., Jr., Ferrario, S., Busscher, M. and Angenent, G. C. (2002). Analysis of MADS box protein-protein interactions in living plant cells. Proc. Natl. Acad. Sci. USA *99*, 2416–21.
[25] Kwon, C., Neu, C., Pajonk, S., Yun, H. S., Lipka, U., Humphry, M., Bau, S., Straus, M., Kwaaitaal, M. Rampelt, H. et al. (2008). Co-option

of a default secretory pathway for plant immune responses. Nature *451*, 835–40.
[26] Mas, P., Devlin, P. F., Panda, S. and Kay, S. A. (2000). Functional interaction of phytochrome B and cryptochrome 2. Nature *408*, 207–11.
[27] Shah, K., Russinova, E., Gadella, T. W., Jr., Willemse, J. and De Vries, S. C. (2002). The *Arabidopsis* kinase-associated protein phosphatase controls internalization of the somatic embryogenesis receptor kinase 1. Genes Dev. *16*, 1707–20.
[28] Shen, Q. H., Saijo, Y., Mauch, S., Biskup, C., Bieri, S., Keller, B., Seki, H., Ulker, B., Somssich, I. E. and Schulze-Lefert, P. (2007). Nuclear activity of MLA immune receptors links isolate-specific and basal disease-resistance responses. Science *315*, 1098–103.
[29] Shirasu, K., Nielsen, K., Piffanelli, P., Oliver, R. and Schulze-Lefert, P. (1999). Cell-autonomous complementation of mloresistance using a biolistic transient expression system. Plant J. *17*, 293–6.
[30] Wenkel, S., Turck, F., Singer, K., Gissot, L., Le Gourrierec, J., Samach, A. and Coupland, G. (2006). CONSTANS and the CCAAT box binding complex share a functionally important domain and interact to regulate flowering of Arabidopsis. Plant Cell *18*, 2971–84.
[31] Chinnayelka, S. and McShane, M. J. (2005). Microcapsule biosensors using competitive binding resonance energy transfer assays based on apoenzymes. Anal. Chem. *77*, 5501–11.
[32] Lichlyter, D. J., Grant, S. A. and Soykan, O. (2003). Development of a novel FRET immunosensor technique. Biosens. Bioelectron *19*, 219–26.
[33] Maurel, D., Kniazeff, J., Mathis, G., Trinquet, E., Pin, J. P. and Ansanay, H. (2004). Cell surface detection of membrane protein interaction with homogeneous time-resolved fluorescence resonance energy transfer technology. Anal. Biochem. *329*, 253–62.
[34] Nagy, P., Vereb, G., Sebestyen, Z., Horvath, G., Lockett, S. J., Damjanovich, S., Park, J. W., Jovin, T. M. and Szollosi, J. (2002). Lipid rafts and the local density of ErbB proteins influence the biological role of homo- and heteroassociations of ErbB2. J. Cell. Sci. *115*, 4251–62.
[35] Poupot, M., Griffe, L., Marchand, P., Maraval, A., Rolland, O., Martinet, L., L'Faqihi-Olive, F. E., Turrin, C. O., Caminade, A. M. Fournie, J. J. et al. (2006). Design of phosphorylated dendritic architectures to promote human monocyte activation. Faseb J. *20*, 2339–51.
[36] Hallworth, R., Currall, B., Nichols, M. G., Wu, X. and Zuo, J. (2006). Studying inner ear protein-protein interactions using FRET and FLIM. Brain Res. *1091*, 122–31.

[37] Zacharias, D. A., Violin, J. D., Newton, A. C. and Tsien, R. Y. (2002). Partitioning of lipid-modified monomeric GFPs into membrane microdomains of live cells. Science *296*, 913–6.

[38] Kremers, G. J., Goedhart, J., van Munster, E. B. and Gadella, T. W., Jr. (2006). Cyan and yellow super fluorescent proteins with improved brightness, protein folding, and FRET Forster radius. Biochemistry *45*, 6570–80.

[39] Nguyen, A. W. and Daugherty, P. S. (2005). Evolutionary optimization of fluorescent proteins for intracellular FRET. Nat. Biotechnol. *23*, 355–60.

[40] Rizzo, M. A., Springer, G. H., Granada, B. and Piston, D. W. (2004). An improved cyan fluorescent protein variant useful for FRET. Nat. Biotechnol. *22*, 445–9.

[41] Merzlyak, E. M., Goedhart, J., Shcherbo, D., Bulina, M. E., Shcheglov, A. S., Fradkov, A. F., Gaintzeva, A., Lukyanov, K. A., Lukyanov, S. Gadella, T. W. et al. (2007). Bright monomeric red fluorescent protein with an extended fluorescence lifetime. Nat. Methods *4*, 555–7.

[42] Shaner, N. C., Lin, M. Z., McKeown, M. R., Steinbach, P. A., Hazelwood, K. L., Davidson, M. W. and Tsien, R. Y. (2008). Improving the photostability of bright monomeric orange and red fluorescent proteins. Nat. Methods *5*, 545–51.

[43] Shcherbo, D., Merzlyak, E. M., Chepurnykh, T. V., Fradkov, A. F., Ermakova, G. V., Solovieva, E. A., Lukyanov, K. A., Bogdanova, E. A., Zaraisky, A. G. Lukyanov, S. et al. (2007). Bright far-red fluorescent protein for whole-body imaging. Nat. Methods *4*, 741–6.

[44] Zapata-Hommer, O. and Griesbeck, O. (2003). Efficiently folding and circularly permuted variants of the Sapphire mutant of GFP. BMC Biotechnol. *3*, 5.

[45] Dixit, R., Cyr, R. and Gilroy, S. (2006). Using intrinsically fluorescent proteins for plant cell imaging. Plant J. *45*, 599–615.

[46] Adjobo-Hermans, M. J., Goedhart, J. and Gadella, T. W., Jr. (2006). Plant G protein heterotrimers require dual lipidation motifs of G{alpha} and G{gamma} and do not dissociate upon activation. J. Cell Sci. *119*, 5087–97.

[47] Bayle, V., Nussaume, L. and Bhat, R. A. (2008). Combination of novel GFP mutant TSapphire and DsRed variant mOrange to set up a versatile in planta FRET–FLIM assay. Plant Physiol *148*, 51–60.

[48] Fujiwara, M. T., Nakamura, A., Itoh, R., Shimada, Y., Yoshida, S. and Moller, S. G. (2004). Chloroplast division site placement requires

dimerization of the ARC11/AtMinD1 protein in Arabidopsis. J. Cell. Sci. *117*, 2399–2410.

[49] Kato, N., Pontier, D. and Lam, E. (2002). Spectral profiling for the simultaneous observation of four distinct fluorescent proteins and detection of protein-protein interaction via fluorescence resonance energy transfer in tobacco leaf nuclei. Plant Physiol. *129*, 931–42.

[50] Seidel, T., Golldack, D. and Dietz, K. J. (2005). Mapping of C-termini of V-ATPase subunits by in vivo-FRET measurements. FEBS Lett. *579*, 4374–82.

[51] Shah, K., Gadella, T. W., Jr., van Erp, H., Hecht, V. and de Vries, S. C. (2001). Subcellular localization and oligomerization of the Arabidopsis thaliana somatic embryogenesis receptor kinase 1 protein. J. Mol. Biol. *309*, 641–55.

[52] Vermeer, J. E., Van Munster, E. B., Vischer, N. O. and Gadella, T. W., Jr. (2004). Probing plasma membrane microdomains in cowpea protoplasts using lipidated GFP-fusion proteins and multimode FRET microscopy. J. Microsc. *214*, 190–200.

[53] Bhat, R. A., Borst, J. W., Riehl, M. and Thompson, R. D. (2004). Interaction of maize Opaque-2 and the transcriptional co-activators GCN5 and ADA2, in the modulation of transcriptional activity. Plant Mol. Biol. *55*, 239–52.

[54] Elliott, C., Muller, J., Miklis, M., Bhat, R. A., Schulze-Lefert, P. and Panstruga, R. (2005). Conserved extracellular cysteine residues and cytoplasmic loop-loop interplay are required for functionality of the heptahelical MLO protein. Biochem J. *385*, 243–54.

[55] Feys, B. J., Wiermer, M., Bhat, R. A., Moisan, L. J., Medina-Escobar, N., Neu, C., Cabral, A. and Parker, J. E. (2005). Arabidopsis senescence-associated gene101 stabilizes and signals within an enhanced disease susceptibility1 complex in plant innate immunity. Plant Cell *17*, 2601–13.

[56] Karpova, T. S., Baumann, C. T., He, L., Wu, X., Grammer, A., Lipsky, P., Hager, G. L. and McNally, J. G. (2003). Fluorescence resonance energy transfer from cyan to yellow fluorescent protein detected by acceptor photobleaching using confocal microscopy and a single laser. J. Microsc. *209*, 56–70.

[57] Immink, R. G., Ferrario, S., Busscher-Lange, J., Kooiker, M., Busscher, M. and Angenent, G. C. (2003). Analysis of the petunia MADS-box transcription factor family. Mol. Genet. Genomics *268*, 598–606.

[58] Tonaco, I. A., Borst, J. W., de Vries, S. C., Angenent, G. C. and Immink, R. G. (2006). In vivo imaging of MADS-box transcription factor interactions. J. Exp. Bot. *57*, 33–42.
[59] Berney, C. and Danuser, G. (2003). FRET or No FRET: A Quantitative Comparison. Biophys. J. *84*, 3992–4010.
[60] Gordon, G. W., Berry, G., Liang, X. H., Levine, B. and Herman, B. (1998). Quantitative fluorescence resonance energy transfer measurements using fluorescence microscopy. Biophys. J. *74*, 2702–13.
[61] Xia, Z. and Liu, Y. (2001). Reliable and global measurement of fluorescence resonance energy transfer using fluorescence microscopes. Biophys. J. *81*, 2395–2402.
[62] Patterson, G., Day, R. N. and Piston, D. (2001). Fluorescent protein spectra. J. Cell Sci. *114*, 837–8.
[63] Bhat, R. A., Riehl, M., Santandrea, G., Velasco, R., Slocombe, S., Donn, G., Steinbiss, H. H., Thompson, R. D. and Becker, H. A. (2003). Alteration of GCN5 levels in maize reveals dynamic responses to manipulating histone acetylation. Plant J. *33*, 455–69.
[64] Gadella, T. W., Jr., Vereb, G., Jr., Hadri, A. E., Rohrig, H., Schmidt, J., John, M., Schell, J. and Bisseling, T. (1997). Microspectroscopic imaging of nodulation factor-binding sites on living Vicia sativa roots using a novel bioactive fluorescent nodulation factor. Biophys. J. *72*, 1986–96.
[65] Goedhart, J., Hink, M. A., Visser, A. J., Bisseling, T. and Gadella, T. W., Jr. (2000). In vivo fluorescence correlation microscopy (FCM) reveals accumulation and immobilization of Nod factors in root hair cell walls. Plant J. *21*, 109–19.
[66] Bastiaens, P. I. and Jovin, T. M. (1996). Microspectroscopic imaging tracks the intracellular processing of a signal transduction protein: fluorescent-labeled protein kinase C beta I. Proc. Natl. Acad. Sci. USA *93*, 8407–12.
[67] Kim, M. C., Panstruga, R., Elliott, C., Muller, J., Devoto, A., Yoon, H. W., Park, H. C., Cho, M. J. and Schulze-Lefert, P. (2002). Calmodulin interacts with MLO protein to regulate defence against mildew in barley. Nature *416*, 447–51.
[68] Valentin, G., Verheggen, C., Piolot, T., Neel, H., Coppey-Moisan, M. and Bertrand, E. (2005). Photoconversion of YFP into a CFP-like species during acceptor photobleaching FRET experiments. Nat. Methods *2*, 801.
[69] Becker, W., Bergman, A., Biskup, C., Kelbaukas, L., Zimmer, T., Kloecker, N. and Bendorf, K. (2003). High resolution TCSPC lifetime imaging. Proc. SPIE *4963*, 1–10.

[70] Elangovan, M., Day, R. N. and Periasamy, A. (2002). Nanosecond fluorescence resonance energy transfer-fluorescence lifetime imaging microscopy to localize the protein interactions in a single living cell. J. Microsc. *205*, 3–14.
[71] Bastiaens, P. I. and Squire, A. (1999). Fluorescence lifetime imaging microscopy: Spatial resolution of biochemical processes in the cell. Trends Cell Biol. *9*, 48–52.
[72] Striker, G., Subramaniam, V., Seidel, C. A. M. and Volkmer, A. (1999). Photochromicity and fluorescence lifetimes of green fluorescent protein. J. Phys. Chem. B *103*, 8612–17.
[73] Volkmer, A., Subramaniam, V., Birch, D. J. and Jovin, T. M. (2000). One- and two-photon excited fluorescence lifetimes and anisotropy decays of green fluorescent proteins. Biophys. J. *78*, 1589–98.
[74] Fehr, M., Okumoto, S., Deuschle, K., Lager, I., Looger, L. L., Persson, J., Kozhukh, L., Lalonde, S. and Frommer, W. B. (2005a). Development and use of fluorescent nanosensors for metabolite imaging in living cells. Biochem. Soc. Trans. *33*, 287–90.
[75] Miyawaki, A., Llopis, J., Heim, R., McCaffery, J. M., Adams, J. A., Ikura, M. and Tsien, R. Y. (1997). Fluorescent indicators for Ca2+ based on green fluorescent proteins and calmodulin. Nature *388*, 882–7.
[76] Miyawaki, A., Griesbeck, O., Heim, R. and Tsien, R. Y. (1999). Dynamic and quantitative Ca2+ measurements using improved cameleons. Proc. Natl. Acad. Sci. USA *96*, 2135–40.
[77] Allen, G. J., Kwak, J. M., Chu, S. P., Llopis, J., Tsien, R. Y., Harper, J. F. and Schroeder, J. I. (1999). Cameleon calcium indicator reports cytoplasmic calcium dynamics in Arabidopsis guard cells. Plant J. *19*, 735–47.
[78] Wymer, C. L., Bibikova, T. N. and Gilroy, S. (1997). Cytoplasmic free calcium distributions during the development of root hairs of Arabidopsis thaliana. Plant J. *12*, 427–39.
[79] Tam, R. and Saier, M. H., Jr. (1993). Structural, functional, and evolutionary relationships among extracellular solute-binding receptors of bacteria. Microbiol. Rev. *57*, 320–46.
[80] Looger, L. L., Lalonde, S. and Frommer, W. B. (2005). Genetically encoded FRET sensors for visualizing metabolites with subcellular resolution in living cells. Plant Physiol. *138*, 555–7.
[81] Fehr, M., Ehrhardt, D. W., Lalonde, S. and Frommer, W. B. (2004a). Minimally invasive dynamic imaging of ions and metabolites in living cells. Curr. Opin. Plant Biol. *7*, 345–51.

[82] Fehr, M., Frommer, W. B. and Lalonde, S. (2002). Visualization of maltose uptake in living yeast cells by fluorescent nanosensors. Proc. Natl. Acad. Sci. USA *99*, 9846–51.

[83] Fehr, M., Lalonde, S., Ehrhardt, D. W. and Frommer, W. B. (2004b). Live imaging of glucose homeostasis in nuclei of COS-7 cells. J. Fluoresc. *14*, 603–9.

[84] Fehr, M., Lalonde, S., Lager, I., Wolff, M. W. and Frommer, W. B. (2003). In vivo imaging of the dynamics of glucose uptake in the cytosol of COS-7 cells by fluorescent nanosensors. J. Biol. Chem. *278*, 19127–33.

[85] Fehr, M., Takanaga, H., Ehrhardt, D. W. and Frommer, W. B. (2005b). Evidence for high-capacity bidirectional glucose transport across the endoplasmic reticulum membrane by genetically encoded fluorescence resonance energy transfer nanosensors. Mol. Cell. Biol. *25*, 11102–12.

[86] Lager, I., Fehr, M., Frommer, W. B. and Lalonde, S. (2003). Development of a fluorescent nanosensor for ribose. FEBS Lett. *553*, 85–9.

[87] Okumoto, S., Looger, L. L., Micheva, K. D., Reimer, R. J., Smith, S. J. and Frommer, W. B. (2005). Detection of glutamate release from neurons by genetically encoded surface-displayed FRET nanosensors. Proc. Natl. Acad. Sci. USA *102*, 8740–5.

[88] Gu, H., Lalonde, S., Okumoto, S., Looger, L. L., Scharff-Poulsen, A. M., Grossman, A. R., Kossmann, J., Jakobsen, I. and Frommer, W. B. (2006). A novel analytical method for in vivo phosphate tracking. FEBS Lett. *580*, 5885–93.

[89] Deuschle, K., Chaudhuri, B., Okumoto, S., Lager, I., Lalonde, S. and Frommer, W. B. (2006). Rapid metabolism of glucose detected with FRET glucose nanosensors in epidermal cells and intact roots of Arabidopsis RNA-silencing mutants. Plant Cell *18*, 2314–25.

[90] Lager, I., Looger, L. L., Hilpert, M., Lalonde, S. and Frommer, W. B. (2006). Conversion of a putative Agrobacterium sugar-binding protein into a FRET sensor with high selectivity for sucrose. J. Biol. Chem. *281*, 30875–83.

[91] Bogner, M. and Ludewig, U. (2007). Visualization of arginine influx into plant cells using a specific FRET-sensor. J. Fluoresc. *17*, 350–60.

[92] Zelazny, E., Borst, J. W., Muylaert, M., Batoko, H., Hemminga, M. A. and Chaumont, F. (2007). FRET imaging in living maize cells reveals that plasma membrane aquaporins interact to regulate their subcellular localization. Proc. Natl. Acad. Sci. USA *104*, 12359–64.

[93] Hibi, T., Rezelman, G. and Van Kammen, A. (1975). Infection of cowpea mesophyll protoplasts with cowpea mosaic virus. Virology *64*, 308–18.

[94] Rottier, P. J., Rezelman, G. and van Kammen, A. (1979). The inhibition of cowpea mosaic virus replication by actinomycin D. Virology *92*, 299–309.
[95] Nagata, T., Okabe, K., Takebe, I. and Matsui, C. (1981). Delivery of tobacco mosaic virus RNA into plant protoplasts mediated by reverse-phase evaporation vesicles (liposomes). Mol. Genet. Genomics *184*, 161–5.
[96] Negrutiu, I., Shillito, R., Potrykus, I., Biasini, G. and Sala, F. (1987). Hybrid genes in the analysis of transformation conditions. Plant Mol. Biol. *8*, 363–73.
[97] Rezelman, G., van Kammen, A. and Wellink, J. (1989). Expression of cowpea mosaic virus mRNA in cowpea protoplasts. J. Gen. Virol. *70*, 3043–50.
[98] Vain, P., Murillo, J., Rathus, C., Nemes, C. and Finer, J. J. (1993). Development of the particle inflow gun. Plant Cell Tissue. Org. Cult. *33*, 237–46.
[99] Romeis, T., Ludwig, A. A., Martin, R. and Jones, J. D. (2001). Calcium-dependent protein kinases play an essential role in a plant defence response. Embo J. *20*, 5556–67.
[100] Witte, C. P., Noel, L. D., Gielbert, J., Parker, J. E. and Romeis, T. (2004). Rapid one-step protein purification from plant material using the eight-amino acid StrepII epitope. Plant Mol. Biol. *55*, 135–47.
[101] Weigel, D. and Glazebrook, J. (2002). Arabidopsis-a laboratory manual. Cold Spring Harbor Laboratory Press, Cold Spring Harbor, New York.
[102] Jach, G., Pesch, M., Richter, K., Frings, S. and Uhrig, J. F. (2006). An improved mRFP1 adds red to bimolecular fluorescence complementation. Nat. Methods *3*, 597–600.
[103] Galperin, E., Verkhusha, V. V. and Sorkin, A. (2004). Three-chromophore FRET microscopy to analyze multiprotein interactions in living cells. Nat. Methods *1*, 209–17.
[104] Piljic, A. and Schultz, C. (2008). Simultaneous recording of multiple cellular events by FRET. ACS Chem. Biol. *3*, 156–60.

Laboratory Techniques in Biochemistry and Molecular Biology, Volume 33
FRET and FLIM Techniques
T. W. J. Gadella (Editor)

CHAPTER 11

Biomedical FRET–FLIM applications

Phill B. Jones, Brian J. Bacskai, and Bradley T. Hyman

Harvard Medical School, Massachusetts General Hospital,
Mass General Institute for Neurodegenerative Disorders,
Charlestown, Massachusetts 02129

Over the past decade or so, technological and biological advances have led to enormous leaps forward in the area of biomedical microscopy. The introduction of affordable lifetime microscopy systems has enabled in situ measurements of protein dynamics with exquisite resolution, well beyond the limitations of diffraction limited optics. The most developed use for the technology is as a nanometer scale ruler for the detection and quantification of interactions and conformational changes of proteins, lipids, DNA, RNA, and other biologically important molecules. In addition, remote sensing of the local environment, including pH, O_2, and Ca^{2+} sensing, using custom designed biosensors, offers new insight into the cellular environment, function, and metabolism. As longer lifetime fluorophores are developed and more robust methods to suppress autofluorescence and other potential artifacts become available, it seems likely that FRET–FLIM will become an even more commonly used approach in biological sciences. In addition, the use of intrinsic tissue autofluorescence and exogenous disease-specific stains may offer the ability to noninvasively diagnose conditions as diverse as cancer and Alzheimer's disease, using endoscopes or tomography, in a clinical setting.

DOI: 10.1016/S0075-7535(08)00011-9

11.1. Introduction

The introduction and diversification of genetically encoded fluorescent proteins (FPs) [1] and the expansion of available biological fluorophores have propelled biomedical fluorescent imaging forward into new era of development [2]. Particular excitement surrounds the advances in microscopy, for example, inexpensive time-correlated single photon counting (TCSPC) cards for desktop computers that do away with the need for expensive and complex racks of equipment and compact infrared femtosecond pulse length semiconductor lasers, like the *Mai Tai*, mode locked titanium sapphire laser from Spectra physics, or the similar *Chameleon* manufactured by Coherent, Inc., that enable multiphoton excitation.

When many types of molecules are exposed to electromagnetic radiation, they absorb one or more photons, which excite an electron from the ground, or minimum energy state to some higher energy state. The distinctive quality of fluorophores is that the electron decays to a slightly lower energy state before falling back to ground and emitting a photon. The fact that the molecule decays to ground state from a slightly lower level gives rise to the most immediately striking quality of fluorescence: the emitted light has lower energy per photon than the absorbed light. The resulting color shift is known as the Stokes shift. Since quantum transitions are never instantaneous, but decay randomly with an average dwell time associated with them, a populations of molecules glows with decreasing intensity for a predictable amount of time. The average length of time that molecules spend in the excited state is known as the lifetime (τ). It can be calculated from the sum of all radiative and nonradiative decay rates away from the excited state (see Chapter 1).

The lifetime, therefore, depends not only on the intrinsic properties of the fluorophore but also the characteristics of the environment. For example, any agent that removes energy from the excited state (i.e., dynamic quenching by oxygen) shortens the lifetime of the fluorophore. This general process of increasing the nonradiative decay rates is referred to as *quenching*.

Förster resonance energy transfer (FRET) is a form of quenching. For a fluorophore (donor) to be quenched by another molecule (acceptor), three criteria must be met:

1. The two molecules must be in close proximity (~10nm).
2. The emission spectrum of the donor must overlap with the excitation spectrum of the acceptor.
3. The emission dipole moment must not be orthogonal to the acceptor absorption dipole moment (see Chapter 1).

FRET is an extremely useful phenomenon when it comes to the analysis of molecular conformations and interactions. For the analysis of interactions, in which two separate molecules are labeled with an appropriate pair of fluorophores, an interaction can be shown by observing FRET. Further, FRET can be used as a type of spectroscopic ruler to measure the closeness of interactions. Proteins, lipids, enzymes, DNA, and RNA can all be labeled and interactions documented. This general method can be applied not only to questions of cellular function like kinase dynamics [3] but also to disease pathways, for example, the APP–PS1 interaction that is important in Alzheimer's disease (AD) [4]. Alternatively, two parts of a molecule of interest can be labeled with a donor and acceptor fluorophore. Using this technique, changes in protein conformation and differences between isoforms of proteins can be measured, as well as protein cleavage.

Stepping outside of the subject of biochemical protein dynamics; there is also a healthy literature on the use of sensing using fluorophores. Spectral and lifetime characteristics of fluorophores are dependent on their environment, for example, pH, O_2, and Ca^{2+}, these features are a useful tool, particularly in the study of the basic biology of the cell (see for instance Chapter 4).

11.2. Rationale

Before the widespread adoption of FRET imaging techniques, the only optical, in situ method of assessing whether two proteins were interacting, was a colocalization study. Here, two proteins would be stained with two separate dyes, either fluorescently or with simple chromophores. Using multichannel microscopy, it could be confirmed that the proteins were at least occupying the same physical space within the resolution of the instrument.

Measuring FRET by fluorescence lifetime imaging microscopy (FRET–FLIM) offers the ability to see beyond the resolution of the optical system (~10–100 times that of modern far field microscopes [5]). FRET efficiency can be used as a proxy for molecular distance, thereby allowing the easy detection and somewhat more challenging quantification of molecular interactions. Although many types of assay exist, FRET–FLIM is a highly suitable technique that is capable of in situ measurements of molecular interactions and conformation in living and fixed cells.

The impact of FLIM goes beyond FRET. Lifetime imaging can be used as a method of remote sensing of the intracellular and intercellular environment and even as a passive diagnostic instrument, using autofluorescence (AF) as an endogenous contrast agent, capable of categorizing tissue and possibly identifying diseased tissue.

11.3. Methods

11.3.1. Frequency domain

In frequency domain instruments, the sample is excited by a light source that is oscillating in intensity sinusoidally. The optimal (angular) frequency ($=2\pi f$) for this oscillation is close to the typical inverse of the lifetime of the sample. Since there is a delay between the fluorescence excitation and emission, the emitted light will be a sinusoid of the same form as the exciting light but phase shifted with a reduced mean intensity and reduced modulation (peak-to-trough amplitude). The modulation (τ_M) and phase (τ_φ) lifetimes can be calculated and for a single homogeneous fluorescent lifetime, should be equal (see Chapter 2). In cases where there is more than one lifetime, global analysis or the phasor approach can be applied to resolve two lifetime components at a single frequency, or more complicated experimental acquisition strategies can be employed using a range of modulation frequencies to obtain a set of dispersion relationships [6, 7] (see Chapter 2 for more information).

11.3.2. Time domain

In contrast, time domain instruments attempt to directly measure the decay characteristics of a fluorophore of interest by excitation with ultrashort light pulses and monitoring the decay using either TCSPC [5] or a time gated image intensifier [8].

There is significant debate about the relative merits of frequency and time domain. In principle, they are related via the Fourier transformation and have been experimentally verified to be equivalent [9]. For some applications, frequency domain instrumentation is easier to implement since ultrashort light pulses are not required, nor is deconvolution of the instrument response function, however, signal to noise ratio has recently been shown to be theoretically higher for time domain. The key advantage of time domain is that multiple decay components can, at least in principle, be extracted with ease from the decay profile by fitting with a multiexponential function, using relatively simple mathematical methods.

A popular FLIM analysis program is SPCImage from Becker and Hickl (GmbH). This software works by performing a nonlinear minimization of the goodness of fit parameter (χ^2) using a Levenberg–Marquardt method on a pixel-wise basis. The function it uses is not a simple multiexponential model but a convolution of the decay profile with a measured or estimated instrument response function and several correction factors for incomplete decays, background light, back scattering, imperfect rejection of excitation light, and second harmonic generation. The software returns matrices of the percentage contribution from each exponential component and the fitted lifetime; it can display histograms of those parameters, as well as pseudocolored maps. Becker and Hickl have made it possible to fix the lifetimes of one or more components to constrain the fit using known lifetimes which can be determined from control experiments.

An alternative method of analyzing the data from a FLIM experiment is global analysis. Here, certain decay parameters are forced to be globally invariant within a whole section, region of interest, or even sets of experiments. This can be done in two ways, either by summing up the pixels of interest, or the preferred method of simultaneously fitting both spatially variant fractional contributions and the spatially invariant lifetimes [10]. The advantages of global analysis are that it is faster than computationally intensive pixel-wise analysis and that, the signal to noise is drastically increased. There is, however, an inevitable loss of spatial lifetime information. Peter Verveer and Phillippe Bastiens have demonstrated global analysis in the frequency domain and shown that multiexponential decays can be recovered even when taking single frequency domain measurements [10–12]. In the time domain, researchers at King's college have analyzed samples using global analysis [13].

Recently, Raluca Niesner et al. [14] have published a report in which they developed a noniterative method of fitting multiexponential decays which does not discard information and is nonrestrictive, thereby eliminating the difficulties with excess time being taken by computationally expensive iterative fitting routines.

11.3.3. Nonlifetime-based FRET techniques

It is possible to make FRET measurements using much simpler and less mathematically intense measurement methods. If the excitation light were not to excite the acceptor fluorophore directly, the acceptor should only emit light when the two fluorophores are in close enough proximity for FRET. In addition, the brightness of the donor should also be reduced because the lifetime should be reduced. These steady-state techniques (filter FRET) and spectral imaging are highlighted in Chapters 7 and 8, respectively.

An alternative method of detecting protein–protein interaction or monitoring protein cleavage is two color fluorescence correlation spectroscopy (FCS) [15, 16]. Stated briefly, if two fluorophores of different color are present in the same protein or complex, fluorescence fluctuations caused by molecules randomly entering and exiting a focal volume as a result of Brownian motion will be highly correlated. If two fluorphores are diffusing independently, there should, in principle be no crosscorrelation. Moreover, the presence of FRET will affect both the auto- and crosscorrelation function requiring the use of FCS–FRET analysis [17, 18]. The advantage of using crosscorrelation is that it merely requires the concomitant movement of the fluorescent tags and is, therefore, not limited to a range of interepitope distances. The major disadvantage is that a very low concentration of protein is required whether in vitro or in cells, and that they must be subject to Brownian motion, thereby limiting the technique to conditions which may or may not be physiologically or pathologically relevant, depending on the protein and disease being studied, and completely ruling out the use of fixed histological samples.

11.4. Materials

11.4.1. Labeling strategies for biomedical applications

Since about the mid-1990s, the green fluorescent protein (GFP), and its spectral variants [1] have become some of the most exciting molecules in microscopy, biochemistry, and cell biology (see Chapter 5).

Alternatively, proteins can be labeled selectively using amine-reactive dyes. Particularly, cysteine and lysines can be modified covalently with a variety of commercially available fluorophores including Texas Red, Oregon Green, and Cy3 [19] (see also Chapter 6 for small molecule FRET probes, and Chapter 12 depicting a variety of alternative molecular labeling approaches). If direct labeling (genetic encoded or chemical) is not possible or available, immunolabeling can also be considered. Immunolabeling has been in use in biology for the purposes of histology for over 40 years [20, 21] and is now ubiquitous. Using this technique, a protein can be fluorescently immunolabeled with a diverse range of commercially available biological fluorophores. However, considering the size of antibodies in relation to the Förster radii of common fluorophore pairs, labeled FAB fragments are the method of choice for immunolabeling.

Although the dependence of fluorescent brightness and lifetime of many fluorophores on their immediate environment, most particularly pH, O_2, Ca^{2+} concentration, and temperature, is often considered a drawback for studies of molecular conformation and interactions, some researchers are using these properties as novel remote sensing tools. Perhaps the most developed of these technologies is the intracellular calcium (Ca^{2+}) indicators that have been pioneered by Tsien [22, 23]. There are essentially two basic approaches. Synthetic dyes like Quin-2 [24, 25], calcium crimson [26], Calcium Green, Fluo-3 [27], and Oregon Bapta Green [28] change fluorescent decay profiles dependent on their calcium binding state. FRET-based sensors [29–31] make use of two genetically encoded variants of GFP, with a connector that changes conformation under specific conditions thereby altering FRET efficiency and lifetime. The use of lifetime rather than intensity to monitor events is more robust and reliable due to the concentration independence of lifetime measurements and lack of susceptibility to photobleaching effects.

Oxygen concentration can be mapped with the long lived ruthenium-based oxygen sensors [32]. In a related field, significant literature exists on the sensing of reactive oxygen species, which can become harmful in the absence of antioxidants [33]. Performing such measurements using intensity based techniques would be impossible as it would require accurate calibration for unquenched intensity and probe concentration. Finally, FLIM has been applied to pH sensing in cells [34] and in the skin stratum corneum [35].

Biological fluorescent dyes and FPs, although indispensable tools of modern biomedical research suffer from several drawbacks. Particularly, they are prone to photobleaching, which is a problem for quantitative imaging and FRET–FLIM experiments [36], and they tend to have broad overlapping spectra and generally do not have a wide range of lifetimes.

These shortcomings have spurred the search for novel fluorescent labels based on nanoparticle technologies [37]. Among the most exciting developments are quantum dots (QDs). These tiny semiconductor crystals (1–10nm) luminesce due to quantum confinement. QDs are much more photostable than conventional biological dyes and can be tuned to a wide range of wavelengths [38, 39]. They also have a broad excitation and narrow emission spectra, which in principle would aid in imaging multiple fluorophores, using a single wavelength of excitation and multispectral detection. They also are highly photostable and are resistant to metabolic degradation [40] due to the fact that they are not organic. Some concern has been raised concerning the use of highly toxic semiconductors which may well impede their acceptance as medical diagnostic tool in the future.

Dye doped silica nanoparticles are conventional biological dyes encapsulated in a ceramic matrix to protect them from oxygen, enhance chemical stability, and allows the surface of the nanoparticle to be modified to enhance the hydrophilic qualities and improve cell uptake [37].

11.5. Biomedical applications

FRET–FLIM has been applied to numerous biological problems, centering on protein–protein interactions, protein conformation, posttranslational modifications, and activation state of enzymes, with lipid microdomains. Each of these applications takes advantage of the combination of exquisite localization of two fluorophores in close proximity and the spatial resolution of microscopy approaches. Fluorophores have been introduced into living or fixed samples so that imaging of living cells, fixed cells, and even pathological specimens can be informative.

11.5.1. Approaches

Posttranslational modifications can be monitored in several ways. One popular approach is to utilize antibodies labeled with a fluorophore and directed against a specific posttranslational modification, like a phosphotyrosine epitope, and a second antibody directed against the protein of interest. For example, this has been applied to protein kinase C alpha, and in our lab to transmembrane proteins like glutamate receptors or the low density lipoprotein receptor-related protein (LRP) [41–43]. FRET approaches allow exquisite subcellular resolution such as imaging individual dendrites and even dendritic spines in neurons. Ligand–receptor interactions or their subsequent phosphorylation event can be directly imaged in this fashion [44]. This approach uses labeled antibodies or sometimes a combination of antibodies and labeled living color FPs like the GFP family, which has become increasingly popular as a source of the fluorescent moiety in living cell populations [45].

The use of pairs of matched spectral variants of GFP, like cyan and yellow or GFP and mRed, to differentially tag proteins and look for interactions, is now in routine use. The approach can be readily applied to homodimerization of molecules that differ only by their living color or epitope tag. For example, homomultimerization of a viral coat protein can be observed by imaging a mixture of cyan and yellow tagged homomeric molecules [46], whereas α-synuclein stacking can be detected by utilizing α-synuclein transcripts that encoded different epitope tags for detection by immunostaining [47].

Outside of FRET, the basic technology of FLIM can be applied to other problems, like the remote sensing of the local environment [33, 48].

11.5.2. Protein dynamics

The advantage of FRET techniques and how they can be used in synergy with multispectral colocalization was shown by [49]. In this study into the relationship between auxilin and dynamin during endocytosis, cells expressing the two proteins were immunolabeled with Alexafluor 488 and Alexafluor 568, respectively. To observe whether the interaction is dependent on clathrin-mediated endocytosis, cells were treated with β-methyl-cyclodextrin, which inhibits endocytosis. To detect if the interaction is GTP dependent, cells were infected with adenovirus expressing a mutant form of dynamin that fails to bind to GTP. Despite the fact that there was no discernable difference between the three cases in terms of the colocalization of proteins, both manipulations eliminated the FRET signal as measured by FLIM. This elegant experiment showed that the dynamin–auxilin interaction is dependent on clathrin-mediated endocytosis and that the formation of a complex with GTP is a vital step in the interaction of the two proteins (Fig. 11.1).

In AD, proteolytic cleavage of amyloid precursor protein (APP) to form the amyloid-β (Aβ) peptide involves β- and γ-secretases. Aβ then forms extracellular plaques, which are a major pathological hallmark. To study the formation of the peptide, H4 cells were transfected with two different isoforms of APP and LRP. APP770 contains a Kunitz protease inhibitor (KPI) and was previously known to interact with LRP, this was verified using donor dequenching (Fig. 11.2) [50]. Using the same technique, a non-KPI containing isoform (APP695) was also found, against expectations, to interact with LRP. Even more intriguing, the authors found that APP and LRP interacted both in their extracellular and intracellular domains.

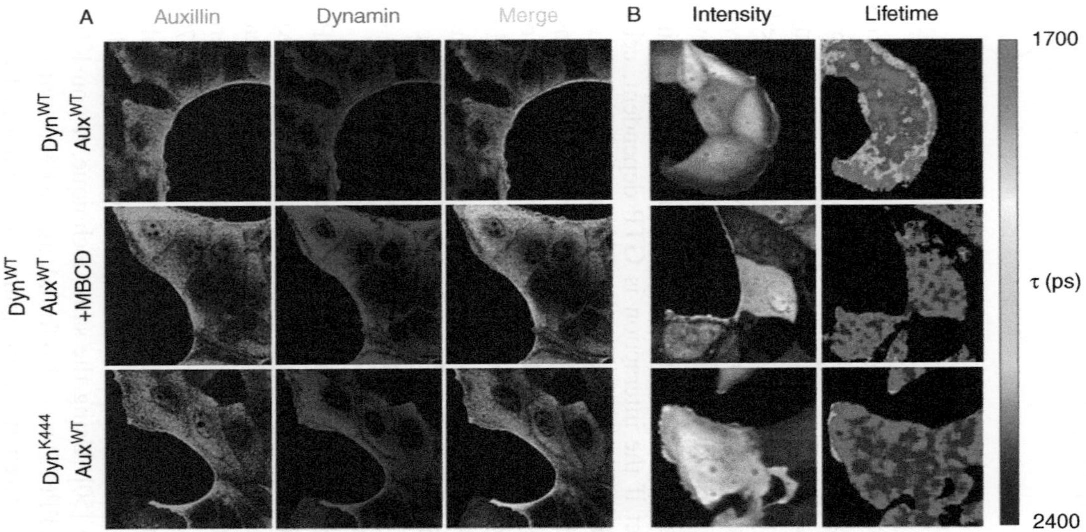

Fig.11.1. Identification of GTP- and endocytosis-dependent dynamin–auxilin interactions as detected with a confocal colocalization study and a two-photon FLIM experiment. Auxilin and dynamin colocalize at the edge of a monolayer of inner medullary collecting duct (IMCD) cells to within the resolution of the instrument (A). No difference is detectable in the presence of β-methyl-cyclodextrin (MBCD), which depletes cholesterol from the plasma membrane and inhibits endocytosis. Equally, the mutant version of dynamin (dyn^{K44A}) that cannot bind with GTP, also appears to be colocalized with auxilin. FLIM data (B) showed that endocytosis inhibition and prevention of GTP binding eliminated the interaction between the proteins even though the spatial localization remains unchanged [49]. (See Color Insert.)

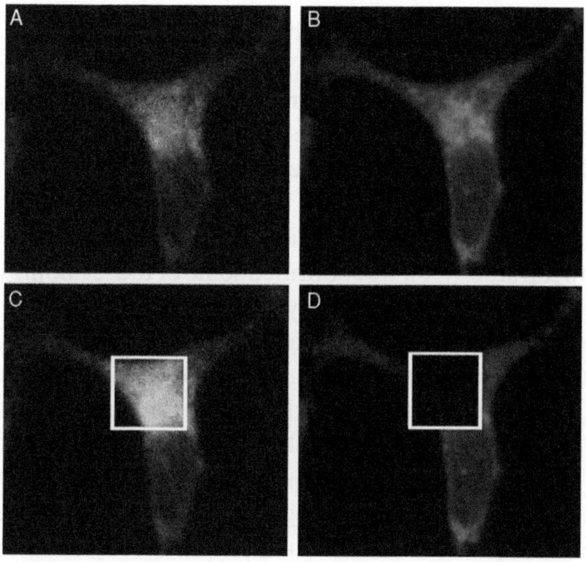

Fig.11.2. A donor dequenching experiment that demonstrates that extracellular domains of APP and LRP closely associated with one another. H4 cells were transfected with APP770 and LRP and immunolabeled with Fluorescein (A) and Cy3 (B), respectively. When the Cy3 is photobleached in the area marked by the white rectangle, the signal from fluorescein increases (C), while the signal from Cy3 disappears (D). The increase in donor fluorescence is calculated to be 51%, which shows that the ectodomains of APP770 and LRP, are in close proximity and therefore likely to be interacting. This result has implications for the field of Alzheimer's disease research as it helps elucidate the nature of APP processing into amyloid-β [50].

Protein–protein interactions between heterodimeric protein pairs that form only transient interactions can be detected, γ-secretase is presenilin-1 (PS1) dependent [51–53]. PS1 is a 467-amino acid, 9-transmembrane domain protein. Over 100 documented single point mutations are known to cause autosomal-dominant familial AD (FAD) [54], in which the ratio of the more fibrilogenic variety of Aβ (Aβ42) to the less fibrilogenic variety (Aβ40) is increased. Chinese hamster ovary (CHO) cells were stably transfected with human APP and either wild type or mutant PS1 [4, 55].

The APP was immunolabeled with fluorescein and PS1 was stained with a FRET partner, Cy3. The C- and N-termini of PS1 were also labeled, to detect differences in protein conformation between the mutant and wild type proteins. We found differences between immature PS1 holoprotein and mature PS1 heterodimer conformation and in the proximity of PS1–APP interactions that were dependent on FAD mutations. Figure 11.3 shows example data from this study. In a follow-up study [56], we confirmed some of those findings during the course of developing an analysis method for high throughput time domain plate readers. Similar interactions between APP and other metabolic partners, including β-site of APP cleaving enzyme (BACE), can be demonstrated and studied using similar techniques [57]. It has been subsequently shown that a class of molecules believed to allosterically impact γ-secretase also alters the FRET–FLIM signal. These data, taken together are consistent with the idea that changes in protein conformation can be directly monitored as a readout of both direct and indirect drug effects on targets [58].

FRET–FLIM has also been applied to visualize protein–DNA interactions [59], protein–ligand interactions, and protein–membrane interactions [60, 61].

11.5.3. Calcium sensing

The technique described in Section 11.5.2 above can be employed to create biological signaling units with protein biosensors [62, 63], changing shape upon binding of a ligand—for example, calcium or chloride [64]—and thereby changing relative conformation and FRET efficiency [65]. Alternatively, synthetic dyes, which change decay profile in a predictable manner dependent on environment, can be employed.

Ionized calcium (Ca^{2+}) is the most common signal transduction element in cells [66]. Excitable cells, like neurons, contain voltage-dependent Ca^{2+} channels, which enable these cells to drastically increase cytosolic calcium levels. Rapid fluctuations in presynaptic

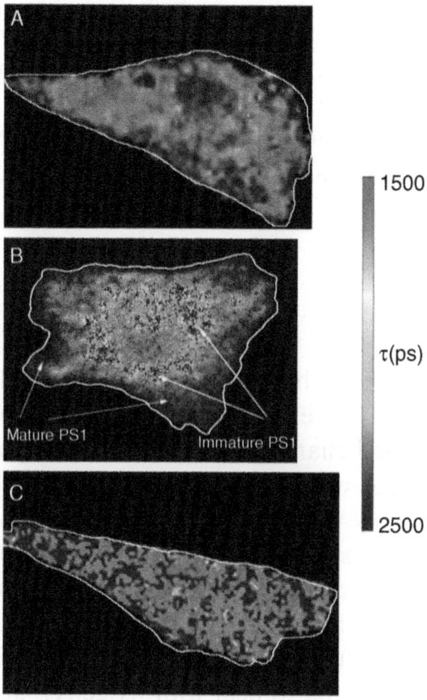

Fig.11.3. FLIM images that show a protein conformational difference between PS1 holoprotein and mature heterodimer. Chinese hamster ovary (CHO) cells were double immunostained with antibodies to the PS1 NT (stained with FITC) and to the ER resident protein BiP (stained with cy3), which does not interact with PS1. The lifetime of FITC was found to be (≈2300ps) (A). The same FRET pair was targeted to PS1 CT and NT (B) and a reduction in lifetime is clearly shown, indicating a change in protein conformation, close to the plasma membrane whereas immature PS1, located near the nucleus, exhibits no such strong FRET interaction. An average lifetime of ≈1700ps across the cell was observed. The equivalent image for D257A mutant PS1 expressing cells, a mutation known to inhibit PS1 endoproteolysis, and diminish amyloid-β production (C) shows no spatial heterogeneity. In addition, the mutation apparently abolishes the FRET signal [4]. (See Color Insert.)

ion concentrations associated with action potentials [67] as well as astrocyte communication via Ca^{2+} waves [68] can be observed. There is a significant list of cellular proteins that are triggered by

Ca^{2+} [66], making calcium monitoring a vital tool for cellular biologists. For example, calmodulin-dependent protein phosphatase, calcineurin can be monitored using a probe based on the regulatory domain of the nuclear factor of activated T cells [69]. This approach can be taken further to monitor multiple calcium-dependent events in real time [70].

11.5.4. Autofluorescence

Although tissue AF is ubiquitous, and undoubtedly of potential clinical interest [71–73], relatively little work has been dedicated to the identification and characterization of AF components.

One of the issues which have impeded the study of tissue AF is the complex decay profile that it exhibits. It is often assumed that fluorescence decay is necessarily a simple exponential. This assumption is not necessarily valid, as is explained in Chapter 4.

In many cases, AF is regarded as a nuisance in FLIM studies, to be avoided by careful cell preparation or manipulation [13]. A common way of trying to deal with tissue AF in industry standard FLIM analysis software like SPCImage (Becker and Hickl, GmbH) is to treat the AF as a monoexponential. This is not an ideal solution; as previously mentioned, AF does not fit well to a monoexponential decay; and multiexponential fitting is by nature ill-posed [74], resulting in large uncertainties with multiple solutions being indistinguishable from the true answer. This is compounded by the large variability within most samples of the AF decay function, making interpretation and separation from exogenous fluorophore components difficult. Attempting to improve the fit by adding the degree of stretch parameter (β) (see Section 11.5.5) further exacerbates the ill-posedness of the problem; stretched exponentials are almost indistinguishable from multiexponentials when working with even mildly noisy data [56].

Using quenching agents like Nile Blue and Sudan Black are effective in suppressing AF for confocal or widefield microscopy

[75–78] and may be of help for filter FRET. For FLIM, however, the technique does not work as it affects the decays of the exogenous fluorophores in an unpredictable fashion and causes excessive spot heating.

More successful approaches to this issue, including the use of FRET pairs with longer lifetimes, have been developed [79], which go some way to minimizing the effect of AF on the fit, and preprocessing of the data before analysis. We have been developing a suite of tools for FLIM data analysis which consistently help reduce the confounding affects of AF. Our first technique involves postprocessing the matrix that is output from the FLIM analysis software and treating it as if it is the sum of multiple Guassian distributions [80], which can be used to separate multiple lifetimes, even when the lifetimes are similar. In addition, populations with different FRET efficiencies can be automatically segmented based on the probability distributions of the overlapping Gaussians. An example of this technique is shown in Fig. 11.4 applied to intracellular aggregates of the tau protein, known as neurofibrillary tangles, which are a major pathological hallmark of AD [81] and are strongly correlated with cell death. In a second technique, we have used the nonexponential feature of AF to identify it and remove it from FLIM images [80]. Figure 11.4D shows the application of this technique to the experimental data and shows good agreement with the multiple Gaussian method.

11.5.5. Clinical applications

The stretched exponential function, $A = A_0 \exp(-t/\tau)^{\beta}$, has been applied to the fluorescence of unstained tissue [82–84]. In particular, researchers at Paul French's group at Imperial college [82], show that the use of the stretched exponential, the parameters of mean, and the heterogeneity parameter (the inverse of the degree of stretch, β) gives better tissue contrast and better fit than the mono- or multiexponential models.

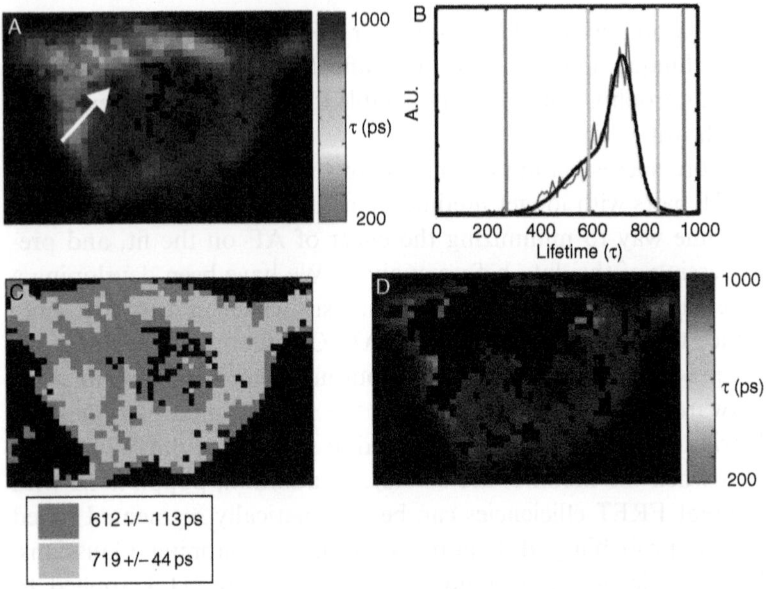

Fig. 11.4. Distinguishing FRET from autofluorescence. This neurofibrillary tangle (NFT) in human brain tissue was stained with an Alexa-488 (A488) labeled anti-tau antibody. A Cy3 acceptor labeled secondary was used to guarantee FRET between A488 and Cy3. In this particular neuron, a large aggregate of lipofuscin autofluorescence in the center of the tangle was observed and confirmed by visual inspection via wide field optical imaging (not shown). The FLIM data was fit with a biexponential function with the lifetime of the noninteracting A488 fixed at 1890ps, as determined from sections stained with donor alone. The pseudocolor image of the shorter component (A) shows the lipofuscin aggregate in the center of the neuron (white arrow). As described in the discussion, the autofluorescence resulted in a broad distribution of lifetimes that overlapped with the FRETing lifetime of interest. By applying multi-Gaussian analysis, the two contributions could be distinguished (B). The red vertical lines bound the broader, slightly shorter distribution, and the green vertical lines bound the narrow peak that the A488 molecules that are FRETing. The image was segmented (C); the spatial distribution of the broader distribution (RED), conforms well to the location of the observed autofluoresce. A goodness of fit filter was applied, where all pixels that have a relatively poor fit to the monoexponential model are discarded, with a cutoff of $\chi^2<2$ (D). We see excellent agreement between the visual inspection, segmentation via multiple Gaussian analysis and goodness of fit filtering. (See Color Insert.)

This type of modeling, as well as wide field FLIM [84–86] and endoscopy [72] have tremendous potential for clinical diagnosis. Although still at the preclinical stage, various groups have looked at the differences in fluorescence spectrum and decay profiles in breast cancer [87, 88], the gastrointestinal tract [89], and skin cancer [90, 91] (see also Chapter 4).

An endoscope, coupled with video rate wide field imager, operating in either the time or frequency domain [92] could be an invaluable tool for early detection of cancer or possibly other diseases whose AF signatures are yet to be studied.

For areas of the body that are not easily accessible by endoscope, the development of infrared fluorescent markers with disease-specific affinities, for example, near-infrared amyloid-specific fluorophores [93] or antibodies to cancerous cells [94], may eventually lead to other optics-based bedside clinical instruments. Imaging in three dimensions, through highly scattering skin and bone will inevitably require tomographic reconstructions. Traditional continuous wave approaches to tomography are incapable of distinguishing fluorescence lifetime from yield [95]. The combination of tomography, and FLIM [96] or time domain [97, 98] offers the possibility of both whole animal imaging in the research laboratory setting and diagnosis of a range of conditions based on either exogenous dyes for aggregated proteins, or as with the endoscopes, the autofluorescent properties of pathological tissue.

11.5.6. Outlook

Taken together, the field is now well placed to design new biosensors, examine protein–protein and protein–lipid interactions, and sensitively determine protein conformation in living tissues at submicron resolution. These interactions are either impossible or extraordinarily difficult to examine in other ways, and the subcellular resolution of FRET–FLIM that allows detection of interactions in specific subcellular compartments may provide insight that would be difficult to determine using standard biochemical subcellular fractionation techniques. As longer lifetime fluorophores are developed, and more robust methods to suppress AF and other potential artifacts become available, it seems likely that FRET–FLIM will become an even more commonly used approach in biological sciences. In addition, the use of intrinsic tissue AF will in the future offer an ability to noninvasively characterize and even diagnose tissue. This technique is likely to be initially applied to endoscopic or skin cancer diagnosis but potential exists to expand the technique to other applications.

11.6. Conclusions

Fluorescence microscopy continues to make a massive impact in the field of biomedical research as the necessary technology continues to advance in both sophistication and cost effectiveness. The most developed use for the new technology is as a kind of nanometer scale ruler for the detection and quantification of interactions and conformational changes of proteins, lipids, DNA, RNA, and other biologically important molecules. The technology has matured to the point that it is now making solid contributions to the understanding of normal and pathological cellular function.

The two most established methods of labeling a molecule or pair of molecules to make a FRET measurement are immunolabeling and fusion to genetically encoded FPs like GFP. Although these are well established techniques, they have certain drawbacks and so novel sensors like QDs and labeling techniques, like cysteine-reactive fluorophores continue to be developed and offer great promise for the future.

There remain challenges to overcome in the use of FLIM and FRET in the general laboratory environment. A clear understanding of the nature of fluorescence decay is necessary for the correct interpretation of FLIM data and care must be taken to use negative controls appropriately. The contamination of lifetime data by autofluorescent proteins remains a serious issue that can lead to the misinterpretation of data unless negative control experiments are conducted with great care. The solution to this issue will require close collaboration between physicists that develop FLIM analysis tools and the biologists who apply them in real biological systems.

In addition, a new generation of lifetime-based techniques is expanding the usefulness of FLIM. Remote sensing of the local environment, including pH, O_2, and Ca^{2+} sensing offers new insight into the cellular environment, function, and metabolism. Although in its infancy, the study of AF as a means of tissue characterization and diagnosis of disease continues to proceed and early results suggest that noninvasive identification of cancer cells by their fluorescent signatures may be possible.

References

[1] Tsien, R. Y. (1998). The green fluorescent protein. Annu. Rev. Biochem. *67*, 509–44.

[2] Bastiaens, P. I. H. and Squire, A. (1999). Fluorescence lifetime imaging microscopy: Spatial resolution of biochemical processes in the cell. Science *9*, 48–52.

[3] Ni, Q., Titov, D. V. and Zhang, J. (2006). Analyzing protein kinase dynamics in living cells with FRET reporters. Methods *40*, 279–86.

[4] Berezovska, O., Lleo, A., Herl, L. D., Frosch, M. P., Stern, E. A., Bacskai, B. J. and Hyman, B. T. (2005). Familial Alzheimer's disease presenilin 1 mutations cause alterations in the conformation of presenilin and interactions with amyloid precursor protein. J. Neurosci. *25*, 3009–17.

[5] Suhling, K., French, P. M. and Phillips, D. (2005). Time-resolved fluorescence microscopy. Photochem. Photobiol. Sci. *4*, 13–22.

[6] Gratton, E. and Limkeman, M. (1983). A continuously variable frequency cross-correlation phase fluorometer with picosecond resolution. Biophys. J. *44*, 315–24.

[7] Lakowicz, J. R. and Maliwal, B. P. (1985). Construction and performance of a variable-frequency phase-modulation fluorometer. Biophys. Chem. *21*, 61–78.

[8] Dowling, K., Hyde, S. C. W., Dayel, M. J. and Barry, N. P.Dainty, J. C., Hughes, A. J., Lever, M. J., Dymoke-Bradshaw, A. K. L., Hares, J. D. and Kellett, P. A. (1997). Fluorescence lifetime imaging for medicine and biology using a gated image intensifier. IEE Semin. Dig. *1997*, 15/1–15/4.
[9] Hedstrom, J., Sedarous, S. and Prendergast, F. G. (1988). Measurements of fluorescence lifetimes by use of a hybrid time-correlated and multifrequency phase fluorometer. Biochemistry *27*, 6203–8.
[10] Verveer, P. J., Squire, A. and Bastiaens, P. I. H. (2000). Global analysis of fluorescence lifetime imaging microscopy data. Biophys. J. *78*, 2127–37.
[11] Verveer, P. J. and Bastiaens, P. I. (2003). Evaluation of global analysis algorithms for single frequency fluorescence lifetime imaging microscopy data. J. Microsc. *209*, 1–7.
[12] Verveer, P. J., Squire, A. and Bastiaens, P. I. (2001). Improved spatial discrimination of protein reaction states in cells by global analysis and deconvolution of fluorescence lifetime imaging microscopy data. J. Microsc. *202*, 451–6.
[13] Peter, M. and Ameer-Beg, S. M. (2004). Imaging molecular interactions by multiphoton FLIM. Biol. Cell *96*, 231–6.
[14] Raluca Niesner, B. P., Schlüsche, P. and Gericke, K. H. (2004). Noniterative Biexponential fluorescence lifetime imaging in the investigation of cellular metabolism by means of NAD(P)H autofluorescence. Chem. Phys. Chem. *5*, 1141–9.
[15] Webb, W. (2001). Fluorescence correlation spectroscopy: Inception, biophysical experimentations, and prospectus. Appl. Opt. *40*, 3969–83.
[16] Webb, W. W. (1976). Applications of fluorescence correlation spectroscopy. Biophysics *9*, 49–68.
[17] Kohl, T., Heinze, K. G., Kuhlemann, R., Koltermann, A. and Schwille, P. (2002). A protease assay for two-photon crosscorrelation and FRET analysis based solely on fluorescent proteins. Proc. Natl Acad. Sci. USA *99*, 12161–6.
[18] Remaut, K., Lucas, B., Braeckmans, K., Sanders, N. N., De Smedt, S. C. and Demeester, J. (2005). FRET–FCS as a tool to evaluate the stability of oligonucleotide drugs after intracellular delivery. J. Control Release *103*, 259–71.
[19] Kobilka, B. K. and Gether, U. (2002). Use of fluorescence spectroscopy to study conformational changes in the beta 2-adrenoceptor. Methods Enzymol. *343*, 170–82.
[20] Kanitakis, J. and Thivolet, J. (1987). Immunolabeling methods in cutaneous histopathology. Principles and practical applications. Ann. Pathol. *7*, 79–97.
[21] Mikhailov, I. F. and Stanislavskii, E. S. (1963). Staining isolated bacterial structures with fluorescent antibodies. Science *40*, 74–9.

[22] Tsien, R. Y. (1980). New calcium indicators and buffers with high selectivity against magnesium and protons: Design, synthesis, and properties of prototype structures. Biochemistry *19*, 2396–404.
[23] Tsien, R. Y., Pozzan, T. and Rink, T. J. (1982). Calcium homeostasis in intact lymphocytes: Cytoplasmic free calcium monitored with a new, intracellularly trapped fluorescent indicator. J. Cell Biol. *94*, 325–34.
[24] Lakowicz, J. R., Szmacinski, H., Nowaczyk, K. and Johnson, M. L. (1992). Fluorescence lifetime imaging of calcium using Quin-2. Cell Calcium *13*, 131–47.
[25] Lakowicz, J. R., Szmacinski, H., Nowaczyk, K., Lederer, W. J., Kirby, M. S. and Johnson, M. L. (1994). Fluorescence lifetime imaging of intracellular calcium in COS cells using Quin-2. Cell Calcium *15*, 7–27.
[26] Herman, B., Wodnicki, S., Kwon, S., Perisamy, A., Gordon, G. W., Mahanajan, N. and Xue Feng, W. (1997). Recent developments in monitoring calcium and protein interactions in cells using fluorescence lifetime microscopy. J. Fluoresc. *7*, 85–92.
[27] Sanders, R., Gerritsen, H. C., Draaijer, A., Houpt, P. M. and Levine, Y. K. (1994). Fluorescence lifetime imaging of free calcium in single cells. Bioimaging *2*, 131–8.
[28] Wilms, C. D., Schmidt, H. and Eilers, J. (2006). Quantitative two-photon Ca^{2+} imaging via fluorescence lifetime analysis. Cell Calcium *40*, 73–9.
[29] Fan, X., Majumder, A., Reagin, S. S., Porter, E. L., Sornborger, A. T., Keith, C. H. and Lauderdale, J. D. (2007). New statistical methods enhance imaging of cameleon fluorescence resonance energy transfer in cultured zebrafish spinal neurons. J. Biomed. Opt. *12*, 034017.
[30] Liu, X., Gong, H., Li, X. and Zhou, W. (2008). Monitoring calcium concentration in neurons with cameleon. J. Biosci. Bioeng. *105*, 106–9.
[31] Truong, K., Sawano, A., Mizuno, H., Hama, H., Tong, K. I., Mal, T. K., Miyawaki, A. and Ikura, M. (2001). FRET-based in vivo Ca^{2+} imaging by a new calmodulin-GFP fusion molecule. Nat. Struct. Biol. *8*, 1069–73.
[32] Gerritsen, H. C., Sanders, R., Draaijer, A., Ince, C. and Levine, Y. K. (1997). Fluorescence lifetime imaging of oxygen in living cells. J. Fluoresc. *7*, 11–6.
[33] Gomes, A., Fernandes, E. and Lima, J. L. F. C. (2005). Fluorescence probes used for detection of reactive oxygen species. J. Biochem. Biophys. Methods *65*, 45–80.
[34] Sanders, R., Draaijer, A., Gerritsen, H. C., Houpt, P. M. and Levine, Y. K. (1995). Quantitative pH imaging in cells using confocal fluorescence lifetime imaging microscopy. Anal. Biochem. *227*, 302–8.
[35] Behne, M. J., Meyer, J. W., Hanson, K. M., Barry, N. P., Murata, S., Crumrine, D., Clegg, R. W., Gratton, E., Holleran, W. M. Elias, P. M. et al. (2002). NHE1 regulates the stratum corneum permeability barrier

homeostasis. Microenvironment acidification assessed with fluorescence lifetime imaging. J. Biol. Chem. *277*, 47399–406.

[36] Tramier, M., Zahid, M., Mevel, J. C., Masse, M. J. and Coppey-Moisan, M. (2006). Sensitivity of CFP/YFP and GFP/mCherry pairs to donor photobleaching on FRET determination by fluorescence lifetime imaging microscopy in living cells. Microsc. Res. Tech. *69*, 933–9.

[37] Sharma, P., Brown, S., Walter, G., Santra, S. and Moudgil, B. (2006). Nanoparticles for bioimaging. Adv. Colloid Interface Sci. *123–6*, 471–85.

[38] Guzelian, A. A., Katari, J. E. B., Kadavanich, A. V., Banin, U., Hamad, K., Juban, E., Alivisatos, A. P., Wolters, R. H., Arnold, C. C. and Heath, J. R. (1996). Synthesis of size-selected, surface-passivated InP nanocrystals. J. Phys. Chem. *100*, 7212–9.

[39] Weller, H., Koch, U., Gutierrez, M. and Henglein, A. (1984). Photochemistry of colloidal metal sulfides. 7. Absorption and fluorescence of extremely small ZnS particles (the world of the neglected dimensions). Ber. Bunsenges. Phys. Chem. *88*, 649.

[40] Jaiswal, J. K., Mattoussi, H., Mauro, J. M. and Simon, S. M. (2003). Long-term multiple color imaging of live cells using quantum dot bioconjugates. Nat. Biotechnol. *21*, 47–51.

[41] Hallett, P. J., Spoelgen, R., Hyman, B. T., Standaert, D. G. and Dunah, A. W. (2006). Dopamine D1 activation potentiates striatal NMDA receptors by tyrosine phosphorylation-dependent subunit trafficking. J. Neurosci. *26*, 4690–700.

[42] Ng, T., Squire, A., Hansra, G., Bornancin, F., Prevostel, C., Hanby, A., Harris, W., Barnes, D., Schmidt, S. Mellor, H. et al. (1999). Imaging protein kinase Calpha activation in cells. Science *283*, 2085–9.

[43] Peltan, I. D., Thomas, A. V., Mikhailenko, I., Strickland, D. K., Hyman, B. T. and von Arnim, C. A. (2006). Fluorescence lifetime imaging microscopy (FLIM) detects stimulus-dependent phosphorylation of the low density lipoprotein receptor-related protein (LRP) in primary neurons. Biochem. Biophys. Res. Commun. *349*, 24–30.

[44] Verveer, P. J., Wouters, F. S., Reynolds, A. R. and Bastiaens, P. I. (2000). Quantitative imaging of lateral ErbB1 receptor signal propagation in the plasma membrane. Science *290*, 1567–70.

[45] Pepperkok, R., Squire, A., Geley, S. and Bastiaens, P. I. (1999). Simultaneous detection of multiple green fluorescent proteins in live cells by fluorescence lifetime imaging microscopy. Curr. Biol. *9*, 269–72.

[46] van Kuppeveld, F. J., Melchers, W. J., Willems, P. H. and Gadella, T. W., Jr. (2002). Homomultimerization of the coxsackievirus 2B protein in living cells visualized by fluorescence resonance energy transfer microscopy. J. Virol. *76*, 9446–56.

[47] Klucken, J., Outeiro, T. F., Nguyen, P., McLean, P. J. and Hyman, B. T. (2006). Detection of novel intracellular alpha-synuclein oligomeric species by fluorescence lifetime imaging. FASEB J. *20*, 2050–7.
[48] Demaurex, N. and Frieden, M. (2003). Measurements of the free luminal ER Ca(2+) concentration with targeted "cameleon" fluorescent proteins. Cell Calcium *34*, 109–19.
[49] Sever, S., Skoch, J., Newmyer, S., Ramachandran, R., Ko, D., McKee, M., Bouley, R., Ausiello, D., Hyman, B. T. and Bacskai, B. J. (2006). Physical and functional connection between auxilin and dynamin during endocytosis. EMBO J. *25*, 4163–74.
[50] Kinoshita, A., Whelan, C. M., Smith, C. J., Mikhailenko, I., Rebeck, G. W., Strickland, D. K. and Hyman, B. T. (2001). Demonstration by fluorescence resonance energy transfer of two sites of interaction between the low-density lipoprotein receptor-related protein and the amyloid precursor protein: Role of the intracellular adapter protein Fe65. J. Neurosci. *21*, 8354–61.
[51] Fortini, M. E. (2002). Gamma-secretase-mediated proteolysis in cell-surface-receptor signalling. Nat. Rev. Mol. Cell Biol. *3*, 673–84.
[52] Fraser, P. E., Yang, D. S., Yu, G., Levesque, L., Nishimura, M., Arawaka, S., Serpell, L. C., Rogaeva, E. and St George-Hyslop, P. (2000). Presenilin structure, function and role in Alzheimer disease. Biochim. Biophys. Acta *1502*, 1–15.
[53] Selkoe, D. J. (2001). Presenilin, Notch, and the genesis and treatment of Alzheimer's disease. Proc. Natl Acad. Sci. USA *98*, 11039–41.
[54] Selkoe, D. J. (1996). Amyloid beta-protein and the genetics of Alzheimer's disease. J. Biol. Chem. *271*, 18295–8.
[55] Berezovska, O., Bacskai, B. J. and Hyman, B. T. (2003). Monitoring proteins in intact cells. Sci. Aging Knowl. Environ. *2003*, PE14.
[56] Jones, P. B., Herl, L., Berezovska, O., Kumar, A. T. N., Bacskai, B. J. and Hyman, B. T. (2006). Time-domain fluorescent plate reader for cell based protein–protein interaction and protein conformation assays. J. Biomed. Opt. *11*, 054024–10.
[57] von Arnim, C. A., Kinoshita, A., Peltan, I. D., Tangredi, M. M., Herl, L., Lee, B. M., Spoelgen, R., Hshieh, T. T., Ranganathan, S. Battey, F. D. et al. (2005). The low density lipoprotein receptor-related protein (LRP) is a novel beta-secretase (BACE1) substrate. J. Biol. Chem. *280*, 17777–85.
[58] Lleo, A., Berezovska, O., Growdon, J. H. and Hyman, B. T. (2004). Clinical, pathological, and biochemical spectrum of Alzheimer disease associated with PS-1 mutations. Am. J. Geriatr. Psychiatry *12*, 146–56.
[59] Cremazy, F. G., Manders, E. M., Bastiaens, P. I., Kramer, G., Hager, G. L., van Munster, E. B., Verschure, P. J., Gadella, T. J., Jr. and van Driel, R. (2005). Imaging in situ protein–DNA interactions in the cell nucleus using FRET–FLIM. Exp. Cell Res. *309*, 390–6.

[60] Berezovska, O., Jack, C., McLean, P., Aster, J. C., Hicks, C., Xia, W., Wolfe, M. S., Weinmaster, G., Selkoe, D. J. and Hyman, B. T. (2000). Rapid Notch1 nuclear translocation after ligand binding depends on presenilin-associated gamma-secretase activity. Ann. N.Y. Acad. Sci. *920*, 223–6.

[61] McLean, P. J., Kawamata, H., Ribich, S. and Hyman, B. T. (2000). Membrane association and protein conformation of alpha-synuclein in intact neurons. Effect of Parkinson's disease-linked mutations. J. Biol. Chem. *275*, 8812–6.

[62] Giepmans, B. N., Adams, S. R., Ellisman, M. H. and Tsien, R. Y. (2006). The fluorescent toolbox for assessing protein location and function. Science *312*, 217–24.

[63] Tsien, R. Y. (2006). Breeding and building molecules to spy on cells and tumors. Keio J. Med. *55*, 127–40.

[64] Jose, M., Nair, D. K., Reissner, C., Hartig, R. and Zuschratter, W. (2007). Photophysics of Clomeleon by FLIM: Discriminating excited state reactions along neuronal development. Biophys. J. *92*, 2237–54.

[65] Calleja, V., Ameer-Beg, S. M., Vojnovic, B., Woscholski, R., Downward, J. and Larijani, B. (2003). Monitoring conformational changes of proteins in cells by fluorescence lifetime imaging microscopy. Biochem. J. *372*, 33–40.

[66] Clapham, D. E. (1995). Calcium signaling. Cell *80*, 259–268.

[67] Rusakov, D. A. (2006). Ca^{2+}-dependent mechanisms of presynaptic control at central synapses. Neuroscientist *12*, 317–26.

[68] Scemes, E. and Giaume, C. (2006). Astrocyte calcium waves: What they are and what they do. Glia *54*, 716–25.

[69] Newman, R. H. and Zhang, J. (2008). Visualization of phosphatase activity in living cells with a FRET-based calcineurin activity sensor. Mol. Biosyst. *4*, 496–501.

[70] Piljic, A. and Schultz, C. (2008). Simultaneous recording of multiple cellular events by FRET. ACS Chem. Biol. *3*, 156–60.

[71] Glanzmann, T., Ballini, J. P., van den Bergh, H. and Wagnieres, G. (1999). Time-resolved spectrofluorometer for clinical tissue characterization during endoscopy. Rev. Sci. Instrum. *70*, 4067–77.

[72] Munro, I., McGinty, J., Galletly, N., Requejo-Isidro, J., Lanigan, P. M., Elson, D. S., Dunsby, C., Neil, M. A., Lever, M. J. Stamp, G. W. et al. (2005). Toward the clinical application of time-domain fluorescence lifetime imaging. J. Biomed. Opt. *10*, 051403.

[73] Wagnieres, G. A., Star, W. M. and Wilson, B. C. (1998). In vivo fluorescence spectroscopy and imaging for oncological applications. Photochem. Photobiol. *68*, 603–32.

[74] Parker, R. L. and Song, Y. Q. (2005). Assigning uncertainties in the inversion of NMR relaxation data. J. Magn. Reson. *174*, 314–24.

[75] Kikugawa, K., Beppu, M. and Sato, A. (1995). Extraction and purification of yellow-fluorescent lipofuscin in rat kidney. Gerontology *41*(Suppl. 2), 1–14.

[76] Kikugawa, K., Beppu, M., Sato, A. and Kasai, H. (1997). Separation of multiple yellow fluorescent lipofuscin components in rat kidney and their characterization. Mech. Ageing Dev. *97*, 93–107.

[77] Pearce, A. (1972). *In* Histochemistry, Theoretical and Applied, Appendix 26. Churchill Livingstone, Edinburgh, pp. 1383–1385.

[78] Schnell, S. A., Staines, W. A. and Wessendorf, M. W. (1999). Reduction of lipofuscin-like autofluorescence in fluorescently labeled tissue. J. Appendix Histochem. Cytochem. *47*, 719–30.

[79] Schobel, U., Egelhaaf, H. J., Brecht, A., Oelkrug, D. and Gauglitz, G. (1999). New donor–acceptor pair for fluorescent immunoassays by energy transfer. Bioconjug. Chem. *10*, 1107–14.

[80] Jones, P. B., Rozkalne, A., Meyer-Luehmann, M., Spires-Jones, T. L., Makarova, A., Kumar, A. T. N., Berezovska, O., Bacskai, B. B. and Hyman, B. T. (2008). Two postprocessing techniques for the elimination of background autofluorescence for fluorescence lifetime imaging microscopy. J. Biomed. Opt. *13*, 014008.

[81] Hyman, B. T., Augustinack, J. C. and Ingelsson, M. (2005). Transcriptional and conformational changes of the tau molecule in Alzheimer's disease. Biochim. Biophys. Acta *1739*, 150–7.

[82] Elson, D., Requejo-Isidro, J., Munro, I., Reavell, F., Siegel, J., Suhling, K., Tadrous, P., Benninger, R., Lanigan, P. McGinty, J. et al. (2004). Time-domain fluorescence lifetime imaging applied to biological tissue. Photochem. Photobiol. Sci. *3*, 795–801.

[83] Lee, K. C. B., Siegel, J., Webb, S. E. D., Lévéque-Fort, S., Cole, M. J., Jones, R., Dowling, K., Lever, M. J. and French, P. M. W. (2001). Application of the stretched exponential function to fluorescence lifetime imaging. Biophys. J. *81*, 1265–74.

[84] Siegel, J., Elson, D. S., Webb, S. E., Lee, K. C., Vlandas, A., Gambaruto, G. L., Leveque-Fort, S., Lever, M. J., Tadrous, P. J. Stamp, G. W. et al. (2003). Studying biological tissue with fluorescence lifetime imaging: Microscopy, endoscopy, and complex decay profiles. Appl. Opt. *42*, 2995–3004.

[85] Cole, M. J., Siegel, J., Webb, S. E., Jones, R., Dowling, K., Dayel, M. J., Parsons-Karavassilis, D., French, P. M., Lever, M. J. Sucharov, L. O. et al. (2001). Time-domain whole-field fluorescence lifetime imaging with optical sectioning. J. Microsc. *203*, 246–57.

[86] Requejo-Isidro, J., McGinty, J., Munro, I., Elson, D. S., Galletly, N. P., Lever, M. J., Neil, M. A., Stamp, G. W., French, P. M. Kellett, P. A. et al. (2004). High-speed wide-field time-gated endoscopic fluorescence-lifetime imaging. Opt. Lett. *29*, 2249–51.
[87] Provenzano, P. P., Rueden, C. T., Trier, S. M., Yan, L., Ponik, S. M., Inman, D. R., Keely, P. J. and Eliceiri, K. W. (2008). Nonlinear optical imaging and spectral-lifetime computational analysis of endogenous and exogenous fluorophores in breast cancer. J. Biomed. Opt. *13*, 031220.
[88] Tadrous, P. J., Siegel, J., French, P. M., Shousha, S., Lalani el, N. and Stamp, G. W. (2003). Fluorescence lifetime imaging of unstained tissues: Early results in human breast cancer. J. Pathol. *199*, 309–17.
[89] Mayinger, B. (2004). Endoscopic fluorescence spectroscopic imaging in the gastrointestinal tract. Gastrointest. Endosc. Clin. N. Am. *14*, 487–505, viii–ix.
[90] Brancaleon, L., Durkin, A. J., Tu, J. H., Menaker, G., Fallon, J. D. and Kollias, N. (2001). In vivo fluorescence spectroscopy of nonmelanoma skin cancer. Photochem. Photobiol. *73*, 178–83.
[91] Smith, P. W. (2002). Fluorescence emission-based detection and diagnosis of malignancy. J. Cell. Biochem. Suppl. *39*, 54–9.
[92] Esposito, A., Gerritsen, H. C., Oggier, T., Lustenberger, F. and Wouters, F. S. (2006). Innovating lifetime microscopy: A compact and simple tool for life sciences, screening, and diagnostics. J. Biomed. Opt. *11*, 34016.
[93] Hintersteiner, M., Enz, A., Frey, P., Jaton, A. L., Kinzy, W., Kneuer, R., Neumann, U., Rudin, M., Staufenbiel, M. Stoeckli, M. et al. (2005). In vivo detection of amyloid-[beta] deposits by near-infrared imaging using an oxazine-derivative probe. Nat. Biotechnol. *23*, 577–83.
[94] Qu, X., Wang, J., Zhang, Z., Koop, N., Rahmanzadeh, R. and Huttmann, G. (2008). Imaging of cancer cells by multiphoton microscopy using gold nanoparticles and fluorescent dyes. J. Biomed. Opt. *13*, 031217.
[95] Kumar, A. T., Raymond, S. B., Bacskai, B. J. and Boas, D. A. (2008). Comparison of frequency-domain and time-domain fluorescence lifetime tomography. Opt. Lett. *33*, 470–2.
[96] Kumar, A. T., Skoch, J., Bacskai, B. J., Boas, D. A. and Dunn, A. K. (2005). Fluorescence-lifetime-based tomography for turbid media. Opt. Lett. *30*, 3347–9.
[97] Bloch, S., Lesage, F., McIntosh, L., Gandjbakhche, A., Liang, K. and Achilefu, S. (2005). Whole-body fluorescence lifetime imaging of a tumor-targeted near-infrared molecular probe in mice. J. Biomed. Opt. *10*, 054003.
[98] Godavarty, A., Sevick-Muraca, E. M. and Eppstein, M. J. (2005). Three-dimensional fluorescence lifetime tomography. Med. Phys. *32*, 992–1000.

Laboratory Techniques in Biochemistry and Molecular Biology, Volume 33
FRET and FLIM Techniques
T. W. J. Gadella (Editor)

CHAPTER 12

Reflections on FRET imaging: Formalism, probes, and implementation

Elizabeth A. Jares-Erijman[1] and Thomas M. Jovin[2]

[1]*Departamento de Química Orgánica, Facultad de Ciencias Exactas y Naturales (FCEyN), Universidad de Buenos Aires (UBA), Ciudad Universitaria, Pabellón II/Piso 3, 1428 Buenos Aires, Argentina*
[2]*Laboratory of Cellular Dynamics, Max Planck Institute for Biophysical Chemistry, am Fassberg 11, 37077 Göttingen, Germany*

Imaging based on FRET can exploit all the modalities of fluorescence phenomena to provide information about the spatial distribution and states of association and function of biomolecules in the cellular context. In this chapter, we deal with fundamental aspects of the FRET formalism in the search for more direct and robust determinations under certain circumstances. A "FRET calculator" is introduced that permits a thorough exploration of these principles even under conditions of extreme nonlinearity, that is, saturation of donor and/or acceptor excited states. The implementation of numerous new strategies for FRET determinations requires new "hardware," which we briefly review, as well as "software," both in the literal sense and in reference to new organic and nanoparticle probes. We supply a comprehensive tabulation of new and potential expression systems, most of which need to be explored systematically in and for FRET. A final attempt at "crystal-balling," the future prospects for FRET technology and its applications emphasizes the obvious: it is difficult to make predictions except for stating that images will get finer, faster, and more multidimensional. There are 149 citations from the literature.

DOI: 10.1016/S0075-7535(08)00012-0

Nothing exists except atoms and empty space; everything else is opinion
– Democritus

12.1. Introduction

The preceding 11 chapters in this volume cover every conceivable aspect of FRET, its theory, implementation, and application. As the authors of the final chapter, we are faced with the daunting, unenviable challenge of coming up with additional and useful material. The advantage is that we can be arbitrary in our selection of topics. These we have divided into six sections, starting with issues regarding FRET formalism and its simulation (a "FRET calculator" is introduced), then passing to a brief discussion of novel FRET techniques and, more extensively, probes. We conclude with some conjectures about future developments. In this connection, we take note of an issue of *Nature Chemical Biology* (Oct. 2007, Focus on molecular metrics) dedicated to the *spatiotemporal mechanisms of life.* The editorial states: "… in biological systems the fourth dimension, time, can be born from the three dimensions of space." This rather deterministic view implies that knowledge of the identities and positions of the (macro) molecules and small molecules constituting the cellular milieu will facilitate the prediction of ensuing molecular reactions and redistributions. The scientific tools currently available for such endeavors, as in the case of weather prediction, are fairly primitive and preclude a very successful outcome. To the degree that progress is being made, however, we can regard Förster resonance energy transfer as an indispensable member of the armamentarium available to the cellular biologist because it provides information about molecular properties and distributions on the spatial scale corresponding to the size of the molecules themselves. Thus, it constitutes an inherent "nanotool" with the ultimate sensitivity and selectivity afforded by the underlying fluorescence phenomena. In the fourth to fifth century B.C., Democritus devised the atomic theory by postulating that "atoms" ("molecules" in modern translation) of different shapes and other characteristics collide and interlock, thereby forming more complex structures. FRET addresses these issues by virtue of its dependence on distance, conformation, orientation, and spectral characteristics. One caveat: our presentation is not intended as a comprehensive guide to any of these subjects. However, we have selected citations, the majority from 2006 to 2008, with the intention of orienting the reader in a manner complementary to that of the preceding chapters.

Less is more

– Ludwig Mies van der Rohe

12.2. The FRET formalism revisited

In his thorough review of FRET fundamentals, R. Clegg (Chapter 1) discusses the "monkey business" that characterizes FRET. We take up this issue by first "reconsidering" the definition of the Förster radius, R_o, along lines explored in our two previous reviews of FRET and imaging [1, 2]. We do so because of the inherent limitations of the numerous FRET methods based on relative donor characteristics, as embodied in expressions for the FRET "efficiency" E (see other chapters and the classification scheme introduced in [1, 2]). This quantity is given by the following expressions applicable for a given DA pair and used universally by the FRET community.

$$
\begin{aligned}
k_t &= \frac{(R_o/r_{DA})^6}{\tau_D}; E \equiv k_t\tau_{DA} = \left[1 + (r_{DA}/R_o)^6\right]^{-1}; \\
R_o^6 &= c_o J \kappa^2 n^{-4} Q_D \\
c_o &= 8.8 \times 10^{-28} \text{ for } R_o \text{ in nm and} \qquad (12.1) \\
J &= 10^{17} \int q_D(\lambda)\varepsilon_A(\lambda)\lambda^4 d\lambda \text{ in nm}^6 \times \text{mol}^{-1} \\
Q_D &= k_f\tau_D; \tau_D^{-1} = k_f + k_{nr} + k_{isc} + k_{bl}; \tau_{DA}^{-1} = \tau_D^{-1} + k_t
\end{aligned}
$$

Thus, E is defined as the product of the energy transfer rate constant, k_t, and the fluorescence lifetime, τ_{DA}, of the donor experiencing quenching by the acceptor. The other quantities in Eq. (12.1) are the DA separation, r_{DA}; the DA overlap integral, J; the refractive index of the transfer medium, n; the orientation factor, κ^2; the normalized (to unit area) donor emission spectrum, $q_D(\lambda)$; the acceptor extinction coefficient, $\varepsilon_A(\lambda)$; and the unperturbed donor quantum yield, Q_D.

The reciprocal of the FRET-unperturbed donor lifetime, τ_D, is given by the sum of all rate constants for deactivation. These parameters have been extensively discussed in earlier chapters. We note in passing that the constants with extreme values in Eq. (12.1) disappear if one expresses the absorption (excitation) spectrum of the acceptor in terms of the molecular absorption cross-section, $\sigma_A(\lambda) = 10^{17}\ln[10]\, N_{Av}^{-1} \times \varepsilon_A(\lambda)(\mathrm{nm}^2/\mathrm{molecule})$.

$$R_o^6 = J_\sigma \kappa^2 n^{-4} Q_D;\ J_\sigma = 10^{-4}\ \ln[10] \int q_D(\lambda)\sigma_A(\lambda)\lambda^4 \mathrm{d}\lambda (R_o \text{ and } \lambda \text{ in nm}) \qquad (12.2)$$

We now focus our attention on the presence of the unperturbed donor quantum yield, Q_D, in the definition of R_o^6 [Eq. (12.1)]. We have pointed out previously [1, 2] that τ_D appears both in the numerator and denominator of k_t and, therefore, cancels out. In fact, τ_D is absent from the more fundamental expression representing the essence of the Förster relationship, namely the ratio of the rate of energy transfer, k_t, to the radiative rate constant, k_f [Eq. (12.3)]. Thus, this quantity can be expressed in the form of a "simplified" Förster constant we denote as Γ_o. We propose that Γ_o is better suited to FRET measurements based on *acceptor* (±donor) properties in that it avoids the arbitrary introduction into the definition of R_o of a quantity (τ_D) that can vary from one position to another in an unknown and indeterminate manner (for example due to changes in refractive index, [3]), and thereby bypasses the requirement for an estimation of E [Eq. (12.1)].

$$\frac{k_t}{k_f} = \left(\frac{\Gamma_o}{r_{DA}}\right)^6; \Gamma_o^6 = c_o J \kappa^2 n^{-4} = J_\sigma \kappa^2 n^{-4}; \left(\frac{\Gamma_o}{r_{DA}}\right)^6 \rightarrow \rho_i \Omega_i \qquad (12.3)$$

The ratio k_t/k_f has no inherent upper bound, in contrast to the 0–1 range of E; the parameters are related by the identity $E = Q_D \times (k_t/k_f)/[1 + Q_D \times (k_t/k_f)]$. In addition, while R_o can be "engineered" via Q_D (e.g., by altering k_{nr} with a quencher; see Chapter 1), Γ_o and k_t/k_f are invariant under such manipulations.

The focus of many if not most, FRET determinations is on the DA separation, r_{DA}, as an index of molecular proximity and association. However, FRET is sensitive to *all* of the indicated parameters [Eq. (12.3)] and can be employed in a given situation for systematically exploiting or evaluating any of them, alone or in combination. In addition, in images of structures with components exhibiting FRET, every pixel (2D) or voxel (3D) may exhibit arbitrary degrees of heterogeneity with respect to composition and/or molecular environment. It follows that the common assumption in FRET imaging of an invariant Q_D, and/or J, κ^2, and n^{-4}, may often be improper and thus misleading. Rather, in devising the strategy for any given FRET-based experiment, one should consider the suitability or even necessity of factoring $(\Gamma_o/r_{DA})^6$ into a particular targeted parameter of interest (ρ_i) and an associated "partial FRET constant" Ω_i [Eq. (12.3)]. Concrete examples are (1) changes in conformation or orientation $(\rho = \kappa^2)$, (2) spectral perturbations of the microenvironment and/or binding $(\rho = J)$, and (3) changes in temperature $\left(\rho = r_{DA}^{-6}, \kappa^2, J, n^{-4}\right)$. The maximal dynamic range of the FRET process occurs for $\rho_i\Omega_i \approx 1$ (see Fig. 12.1 in [1, 2]).

We stated earlier that Eq. (12.3) might be more suitable than Eq. (12.1) in FRET determinations based on acceptor (±donor) properties, for example, sensitized emission (see also Chapters 7 and 8). In such cases, k_t can be expressed directly in terms of measured experimental parameters, without the need for estimating E. One example is given by Eq. (12.4), valid in the regime of low excitation intensity (negligible donor or acceptor saturation). For FilterFRET (Chapter 7), one performs excitations at/near the donor and acceptor absorption peaks (intensities $I_{D,\text{ex}}$ and $I_{A,\text{ex}}$, respectively), and acquires emission signals $(f_{\text{exc,em}})$ in the two corresponding emission bands (D,A). Under such conditions, the ratio $\left(k_t/k_f\right)$ at every image position is given by

$$\frac{k_t}{k_f} = Q_A^{-1}\left[\frac{d_{D,D}}{d_{A,A}}\right]F; F \equiv \left(\frac{f_{D,A}}{f_{D,D}}\right) - \left[\frac{\varepsilon_{A,D}}{\varepsilon_{A,A}}\right]\left(\frac{f_{A,A}}{f_{D,D}}\right)\frac{I_{D,\text{ex}}}{I_{A,\text{ex}}} - \left[\frac{d_{D,A}}{d_{D,D}}\right], \tag{12.4}$$

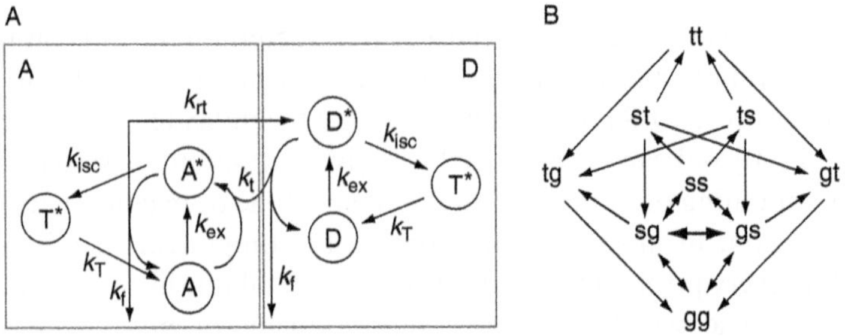

Fig. 12.1. Photophysical dynamics for donor–acceptor FRET pair interconverting between ground, and excited singlet (asterisk) and triplet (T) states. (A) Donor (D) and acceptor (A) manifolds. The rate constants for resonance energy transfer between D and A are k_t and k_{rt} (forward, reverse, respectively). The radiative rate constants for D and A are k_f. See text and [1] for further definitions and discussion. Nonradiative decay is not shown as a separate pathway, neither is photobleaching represented. (B) Virtual paired states of paired D and A. **g**, ground state; **s**, excited singlet state; **t**, excited triplet state. Paired states: left, donor; right, acceptor. The arrows represent the pathways depicted in (A). Bold arrows, transitions involved in FRET.

in which the fixed (constant) ratios in square brackets represent (1) extinction coefficients $\varepsilon_A(\lambda)$ (or, equivalently, absorption cross-sections $\sigma_A(\lambda)$) of the acceptor at the two excitation wavelengths, and (2) relative detection efficiencies $(d_{A \text{ or } D, A \text{ or } D \text{ emission band}})$ reflecting both $q_D(\lambda)$ and $q_A(\lambda)$ and wavelength-dependent instrument response functions. The second and third terms in the expression for F [Eq. (12.4)] represent direct acceptor excitation and spillover of donor emission in the acceptor emission band, respectively; inverse overlap, $f_{A,D}$, can also be incorporated into Eq (12.4). Note that the presumably (generally) invariant acceptor quantum yield, Q_A, appears as a factor in Eq. (12.4) instead of the often more variable reference donor quantum yield, Q_D. (In any event, Q_A, as opposed to Q_D, can be evaluated in double labeled samples from measurements of the acceptor lifetime using direct excitation.) Most importantly, the "k_t/k_f" image provides a direct measure of $(\Gamma_o/r_{DA})^6$ or of its factored form $\rho_i\Omega_i$ [Eq. (12.3)]. Images of reliable relative values can also be based on the function F alone [Eq. (12.4)], assuming constancy of the proportionality constants. By way of comparison, Eq. (12.4) is also written using the formalisms described in Chapter 7 [Eq. (12.4a)] and 8 [Eq. (12.4b)], clearly demonstrating that the ratio k_t/k_f is directly related to other representations of experimental image and spectral data (bold letters). The reader is referred to these chapters for the definition of the other symbols used in Eqs. (12.4a and b)

$$\frac{k_t}{k_f} = \frac{1}{\mathrm{Q}_A}\frac{\ell^{\mathrm{d}}\mathrm{F}_D^{\mathrm{d}}\mathrm{g}^{\mathrm{d}}}{\ell^{\mathrm{s}}\mathrm{F}_A^{\mathrm{s}}\mathrm{g}^{\mathrm{s}}} \times F; F = \frac{\mathbf{S}}{\mathbf{D}} - \gamma\frac{\mathbf{A}}{\mathbf{D}} - \beta, \tag{12.4a}$$

$$\frac{k_t}{k_f} = \frac{\Delta \mathrm{a}}{\mathrm{d}_{\mathrm{apparent}}}\frac{1}{Q_A}\left(k^2 - k^1\right)^{-1}, \tag{12.4b}$$

As in the case of most FRET imaging formalisms, Eq. (12.4) applies to the donor at a particular x, y, x coordinate j as a virtual species with an apparent $k_t(j)$; it applies for arbitrary absolute and relative local concentrations of donor and acceptor but only if population distributions are properly considered (see other chapters). For example, assuming a population of donors, a fraction α of which are in a unique DA environment (e.g., as a bound complex) and the rest free, one can generate an expression for α in terms of known parameters and experimental signals (varying from point to point), k_t/k_f, Q_D, and Q_A. Related expressions can be formulated in terms of donor and acceptor emission anisotropies and lifetimes, and for conditions of nonlinearity such as ground state depletion of the donor and/or acceptor [1] (see Section 12.3).

12.3. The "FRET Calculator"

The simplest scheme incorporating coupled FRET donor and acceptor manifolds in which both species can adopt the ground, singlet excited, and triplet excited states is depicted in Fig. 12.1a. Resonance energy transfer occurs between the S1 states, and reversibility ($S1_A \rightarrow S1_D$) is allowed for the case of homoFRET. The scheme in panel b depicts the interconnections between all possible combinations of DA states, a mode of representation that greatly facilitates the treatment of the concerted transfer/deactivation underlying the FRET process (see also [1]). With a "FRET calculator" we seek to compute the population dynamics of this closed system under conditions of pulsed, periodic, or steady-state excitation of the donor and/or acceptor. The treatment should allow for arbitrary degrees of saturation of all ground and excited states so as to accommodate FRET strategies based on finite depletion of the ground state(s). Numerical solutions to the differential equations corresponding to the systems depicted in Fig. 12.1 are easy to obtain. Yet we impose an additional requirement on our calculator, namely the generation of *analytical* expressions for the time course of the individual concentration functions, as well as of their integrals, derivatives, and combinations.

A program such as *Mathematica* (Wolfram Research) easily and efficiently produces numerical solutions to the stated problem. However, these provide little insight into the details of the dynamics, and differentiation or integration of such data tends to be awkward and slow. Analytical expressions permit a much better perception of the mechanisms operating under given conditions, with identification of the major species and pathways. They also permit the representation of arbitrary combinations (sums, ratios) of signals, including the weighting factors corresponding to absorption and/or fluorescent parameters required to simulate experimental signals. Such a tool would greatly facilitate the exploration of new FRET approaches and strategies, be it for studies in solution or in an imaging environment.

Unaware of the existence of relevant literature on this subject, we pursued the means for implementing the proposed FRET calculator using the tools provided by *Mathematica.* Success was achieved by resorting to solutions of the differential equations according to well-known techniques based on Laplace transforms. A major challenge in this approach arose from the complexity of the coefficients of the polynomial terms in the Laplace transforms of the concentration variables; the largest of these had >9000 elements involving all of the rate constants depicted in Fig. 12.1. This difficulty was overcome, albeit laboriously, leading to compiled functions and parametrized equations capable of providing very rapid solutions. The FRET calculator accepts as input the numerical array of rate constants and the specification of the desired time-dependent functions. It then quickly outputs the corresponding analytical expressions in terms of (real) exponential and trigonometric functions. Two additional input parameters dictate the degree of "fine structure" that is reported, that is, to what degree similar terms are collapsed or small amplitude components are eliminated. The results are plotted immediately and can be manipulated further. An example of an input set of parameters and some of the corresponding output is shown in Fig. 12.2.

Let the punishment fit the crime
– The Mikado, W.S. Gilbert and A. Sullivan

12.4. Novel FRET methods

We have previously proposed a classification scheme for FRET imaging techniques based on the measurement parameter and strategy for data acquisition [1, 2]. New entries are constantly appearing as the technology (probes, instruments) advances. In the imaging context, the implementation of FRET requires particular attention to issues such as speed and sensitivity, optical sectioning, and choice of photophysical parameter(s). For example, does one confine the measurements to the conventional low light-level "linear" regime or deliberately exploit saturation and other excited state dynamics [1, 2, 4, 5]?

The most sophisticated techniques require time-resolved measurements (lifetime, anisotropy, spectra) either in the time or frequency domain ([6–10]; for a focused journal issue on the subject see [11]). Thus, the significance of new, versatile, commercially available light

(A)

Parameters (rates: ns^{-1})	Donor		Acceptor	
Direct exc	ked	0.1	kea	0.00002
Singlet decay	ked	0.5	kda	0.3
Singlet → triplet	kiscd	0.001	kisca	0.01
Energy transfer (FRET)	kt	0.5	krt	0.001
Triplet decay	kTd	0.001	kTa	0.001
•	•	•	•	•
Tolerances	evtol	10^{-5}	vlim	10^{-7}
Plot time scales	t1	10	t2	150

(B)

x1 = "sg"; s1 = "val";
x2 = "sg"; s2 = "int";
x3 = "sg"; s3 = "der";
x4 = "(gs+ss+ts)/(sg+ss+st)"; s4 = "val";
x5 = "(gs+ss+ts)/(sg+ss+st)"; s5 = "der";

(C)

$$\text{val}\,[sg(t)] = (0.0327343 + 1.80861\times10^{-6}\,e^{-0.601839\,t} + 0.0109148\,e^{-0.339713\,t} + 2.33468\times10^{-6}\,e^{-0.311014t} + 0.0321732\,e^{-0.00269219t} + 0.0210977\,e^{-0.00173522\,t} - 0.00267864\,e^{-0.00117997\,t} - 0.0942455\,e^{-0.990944\,t}\,\text{Cos}\,[0.140784\,t] + 0.0741405\,e^{-0.990944\,t}\,\text{Sin}\,[0.140784t]$$

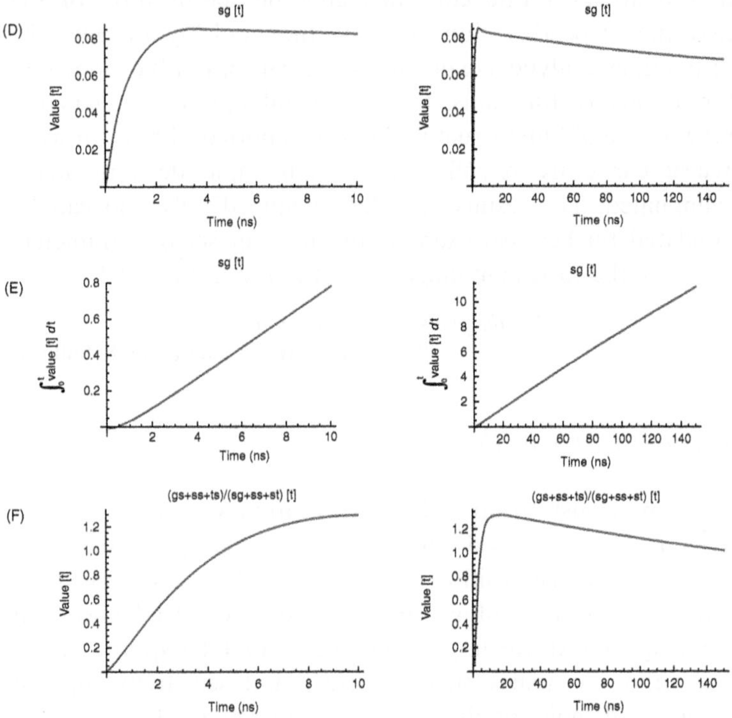

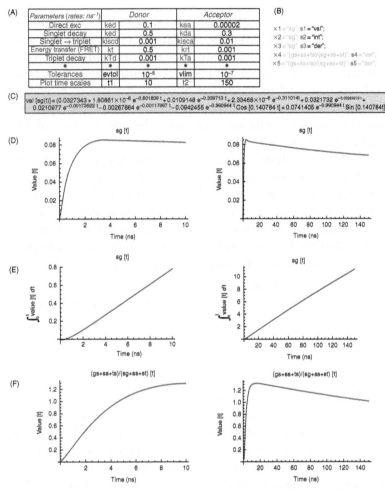

Fig. 12.2. FRET calculator. (A) Table of input parameters; arbitrary concentration and time units. The rate constants selected for singlet decay (kdd = k_D, excluding FRET; see Eq. (12.1)) and for FRET (kt = k_t) are the same, corresponding to a FRET efficiency $E = 0.5$. The acceptor has been assigned quite different parameters, except for triplet decay (kT = k_T). The data are plotted for the indicated time scales (10, 150). (B) specification of desired output signals (see Fig. 12.1 b): x1, sg = excited donor&ground state acceptor; x2: integral of sg; x3: derivative of sg; x4: ratio of the sum of all states

sources cannot be overemphasized. These include very high power LEDs (single or arrays), photonic lattice LEDs ([12], commercial PhlatLight (Luminus)), a vast array of CW and pulsed solid-state lasers, photonic crystal white-light lasers [13], spectrally programmable light engines [14], and multiband (2–6) light engines based on luminescent materials (Lumencor). Multiple lines are also available in current solid-state lasers (Omicron, Cobolt, etc.). The Omicron model line includes the DeepStar (10^6 on/off contrast ratio, high speed analog/digital modulation) and LDM_DC180/PS350/A350 (ns light pulse shaping, optimal for depletion and saturation strategies [1]). On the detector end, the advent of electron multiplying CCDs and the emerging on-chip phase-sensitive CCDs [7] provide almost ideal characteristics with respect to sensitivity, temporal resolution, and system simplification. Random access low-noise CMOS and hybrid cameras are being developed, as are multianode, hybrid, and solid-state detector arrays with minimal transit time spread, optimal for TCSPC determinations.

A significant development in the FRET imaging field has been the systematic implementation of spectral resolution [15–20], including D–A population analysis [8, 19] (see also Chapter 8), often in the context of single-molecule determinations [21–26]; see focus issue on this subject, *Nature Methods*, June 2008. Invariably, photobleaching phenomena [21, 27, 28] intervene either as a hindrance (that can be minimized, CLEM [29]) or a facilitation of the FRET determination [1, 30]. The equally important issue of background suppression or compensation can be achieved by novel means: "photon-free" (bio)chemical instead of photonic excitation

with excited acceptor to the sum of all states with excited donor ("acceptor/donor" ratio, see Eq. (12.4) in text; x5: derivative of x4. (C) analytical equation for the time course of sg. (D) sg over two time ranges. (E) integral of sg; (F) acceptor/donor ratio. The reader is encouraged to rationalize the time course of these signals in terms of the pathways shown in Fig. 12.1. A new FRET determination can be derived from the properties of x5.

(BRET; [31, 32]), lock-in detection techniques exploiting optical switches [33], and schemes for alternating D/A excitation (ALEX [34]). The increased attention to quantitative FRET imaging encompasses the use of polarization [35–39], the perennial issue of calibration and standards [40–44], and practical guides to operational principles and protocols ([45, 46] and other references above). The fundamental distinctions between the requirements for live and fixed cell imaging cannot be overemphasized, as is exemplified in a report of erroneous FRET determinations with visible fluorescent proteins (VFPs) in fixed cells [47].

In the quest for optimal FRET resolution in three dimensions [48, 49], one requires sophisticated algorithms for analyzing complex distributions [48, 50, 51]. An issue of particular relevance when multiple probes [52–54] and multiparametric detection [26, 55] are employed. Strategies for spatial superresolution [56] have spawned a family of acronyms: RESOLFT, STED, STORM, astigmatic STORM, PALM, fPALM, sptPALM, PALMIRA ([57–59] and other references), some of which involve systematic photoconversion or destruction. Undoubtedly, these techniques will be applied systematically to FRET imaging, and, conversely, one can anticipate that FRET mechanisms will be exploited for *achieving* superresolution.

12.5. FRET probes

It is instructive to consider that a photon has an "adjustable" energy of 200 attojoules/λ or 1000 eV/λ (λ is in nm). Accordingly, a blue (400 nm) photon constitutes an energy packet of 0.5 attojoules, a value 10× the energy released by the hydrolysis of one ATP molecule to ADP. Thus, photons are powerful "reagents" and their conversion of one form (excitation "color") to another (emission "color") via the action of "enzymatic fluorophores" [1] releases quantities of energy that can exert strong influences upon cellular processes. By the same token, spectral differences can be exploited systematically for implementing FRET, implying that the selection of donor (D) and acceptor (A) probes will depend on the method being employed. Table 12.1 summarizes some of the factors and considerations that would apply in most situations (see also Chapters 5 and 6).

It is apparent from Table 12.1 that the selection of the most appropriate FRET probes for a certain application requires careful consideration of the photophysical properties of the dyes as well as the availability of methods for the specific labeling of the

TABLE 12.1
Ground and excited state properties and processes affecting FRET parameters

Property or process	J	n	κ^2	r_{DA}	E
Donor/acceptor absorption (cross-section)	+				
Donor/acceptor anisotropies			+		
Donor/acceptor quantum yields					+
Donor–acceptor separation				+	+
Donor/acceptor Stokes shifts	+				
Donor-to-acceptor refractive index		+			
Excited-state reactions					
Binding (ligand, association)	+	+	+	+	
Blinking (dark states)	+				+
Conformational transition	+	+	+	+	+
Diffusion (rotational)			+	+	
Diffusion (translational)			+	+	
Electron transfer	+				+
Emission S1 → S0	+	+	+	+	+
Excimer formation	+				
Intersystem crossing S1 → T1, T1 → S1	+				
J-aggregation	+				
Nonradiative deactivation					+
Phosphorescence T1 → S0					+
Photochromism	+				+
Photodegradation/photobleaching	+				+
Photoionization	+				+
Photoreaction, photodecaging	+		+		+
Proton transfer	+				+
Resonance energy transfer D → A, A → D					+
Solvatochromism (solvent relaxation)	+	+			
Spectrochromism	+	+			

biomolecule of interest. We have considered some of these issues previously [2] and confine ourselves here to some general comments. Generally, one seeks fluorophores with large extinction coefficients (for single and/or multiphoton absorption) and high emission quantum yields (which may or may not correspond to large radiative rate constants, k_f, depending on the desired range of lifetimes). Large Stokes shifts, that is, separation between absorption and emission bands [for a selection of such probes see the websites of Seta Biochemicals (SetaTau dyes), Active Motif (Chromeo dyes), Dyomics (Megastokes dyes), Kodak (X-Sight dyes)] of both the donor and acceptor facilitate the establishment of the measurement windows required for heteroFRET, for example, according to Eq. (12.4). In contrast, homoFRET [60–62] requires a large overlap between the D ($\equiv$ A) excitation and emission spectra in order to achieve a finite value of the overlap integral J [Eqs. (12.1) and (12.2)]. High photostability is another desirable property for the donor and acceptor, yet is not essential if one can resort to donor and/or acceptor photobleaching FRET techniques or if the acceptor can be regenerated from a large pool [63]. With the exception of CALI protocols, one seeks to avoid toxic photobleaching products yet the problem can be acute in point scanning systems using high irradiances. Radical scavengers and triplet "depletors" are available and under constant development.

The time window available for FRET is dictated by the lifetime of the donor. Is there an "optimal" lifetime? If "very short", it is more difficult to measure in FLIM. If it is "very long", the levels of fluorescence are low ("k_f-limited", [1]). In addition, the lifetime is a relevant parameter when one is interested in dynamics, either of binding, conformational change, or diffusion (translational, rotational). These processes influence FRET via the parameters κ^2 and r_{DA} (Table 12.1). Long lifetimes are useful in luminescence RET (LRET) and can help to reduce background and increase signal-to-noise ratios.

High or low anisotropy values, which are more desirable? A low steady-state anisotropy of the donor is generally attributed to rapid rotational motion and thus as justification for the assumption of $\kappa^2 = 2/3$. However, this value strictly speaking only applies for rapid and isotropic motion of both donor and acceptor during the excited state lifetime of the *donor*. Fluorescent proteins never satisfy this condition, because their lifetimes are on the order of 3–4 ns compared to rotational correlation times of >20 ns for the monomeric protein, and of course, greater for a dimer, tetramer, or fusion construct. Such a circumstance is disadvantageous for quantitative distance estimations by heterotransfer, less so in the assessment of homo- and heterodimerization by homoFRET [62, 64]. Other important properties to consider are the dependency on the parameters defining the molecular microenvironment such as polarity, H-bonding, pH, ionic strength, and microviscosity.

With these principles in mind, we refer the reader to valuable recent reviews, original reports, and discussions [65–68] of probes consisting of proteins, organic dyes, and nanoparticles such as quantum dots (QDs, [2, 68–77]) and other intriguing particles: plastic microspheres and nanoparticles [78], multifunctional encoded particles [79], nanodisk codes [80], nano-flares [81], E-PEBBLES [82], C-dots [83], nanocrystal (NC)-encoded microbeads [84], and fluorescent superparamagnetic nanoparticles (FMNP) [85]. Some manifestation of FRET is almost always involved [65, 67, 68, 72, 78, 86–89]. Newer, brighter VFPs span the entire visible spectrum from the blue to the red and offer much improved photostability and a variety of DA FRET pairs [90–97]. A nonfluorescent variant provides an acceptor with a long "reach" (large R_o, [98]). A recent, extraordinary development of the Miyawaki group, a consistent contributor (along with other labs) of highly innovative VFP-based biosensors, is a series of constructs reporting on cell cycle dynamics during differentiation, morphogenesis, and cell death [99]. While not FRET-based, these probes can conceivably be extended for sensing cell cycle dependent protein-interactions by FRET. There exist intriguing possibilities for exploiting other genetically encoded fluorophores, such as the pyoverdin siderophores [100], in expression systems and FRET.

The general topic of FRET-based biosensors is dealt with in detail in other chapters. The websites of numerous firms, including Amersham, AnaSpec, Atto-Tec, Active Motif, Bangs Laboratories, Denovo Biolabels, Dyomics, Few, Kodak, Marker Gene Technologies, Invitrogen, SSS Optical Technologies, and other Internet sites (e.g., www.fluorescence-resource.com/index.cfm?fig=Reagents) feature organic dye and nanoparticle reagents.

A central consideration in the application of FRET in studies of live cells is the means for incorporating the FRET probe into the desired target without perturbing structure and impairing function while selecting loci (ends, internal) suitable for evaluating the conformational transitions and interactions of interest. Strategies for achieving such specificity have been based on peptide segments or proteins that are genetically fused to the target and are recognized by small-molecule probes. One relies either on a direct interaction and spontaneous chemical reaction of the given structure or the intervention of an enzymatic activity mediating the conjugation of the small molecule carrying the fluorophore. Intracellular labeling requires that externally applied dyes be membrane permeable, sometimes constituting a kinetic block limiting the temporal window in which cellular processes are accessible for study. Noncovalent complexes must have adequate photo- and thermodynamic stability, although this requirement that may be relaxed if the label is fluorogenic and can be regenerated by exchange after photobleaching (FAP probes; [63]).

Table 12.2 provides a classification of expression systems in actual or potential (our considered opinion) use, ordered according to decreasing size (kDa, or number of amino acids) of the protein/peptide tag introduced into the target protein. For the related large class of "activity-based probes" (ABPs) see [112] and references therein. The tags and the mechanisms required for introduction of their conjugate probe(s) are indicated. Information is supplied for each system: the target amino acid(s), and whether (1) the conjugate is covalent, (2) the probe is fluorogenic, and (3) the fluorophore is restricted or arbitrary in nature. Schemes of the reactions

Table 12.2

Expression systems with direct or indirect fluorescence readout

Expression system VFPs: intrinsic probes; others: primary or secondary extrinsic probes	Target			Probe		
	Target # aa	Target kDa	Target aa	Covalent	Fluorogenic	Arbitrary
1. Visible fluorescent proteins from jellyfish and corals (VFPs)	238	27				
1a. Spectral variants						
1b. Dual-VFP or VFP-luciferase constructs for sensing ions, pH, covalent modification,. . .						
1c. Photoactivatable, photoconvertible, photochromic VFPs					+	
1d. Bimolecular complementation (BiFC, 1/2 + 1/2 or 2/3 + 1/3 VFP molecules)					+	
2. Mutant haloalkane dehalogenase (HaloTag protein) [probe-substituted haloalkanes]		33	asp	+		+
3. Cutinase [probe-labeled suicide inhibitor, *p*-nitrophenyl phosphonate (pNPP)][1]		22	ser	+		+
4. Mutant O^6-alkylguanine-DNA-alkyltransferase (SNAP/CLIP-tags) [*p*-subt benzyl-guanine/cytosine]		20	cys	+		+
5. Dihydrofolate reductase (DHFR) [methotrexate-linked probe—LigandLink]		18				+
6. Zn-finger domain (ZifQNK) [maleimide-probe + metal/redox dependent reversible thiol protection][2]		18; 32	cys	+	+	+
7 Fluorogen activating protein (FAP)[3]; infinite afinity antibodies[4] [fluorogenic ligands & reagents]		11–15; 24			+	

(continued)

TABLE 12.2 (*continued*)

Expression system VFPs: intrinsic probes; others: primary or secondary extrinsic probes	Target			Probe		
	Target # aa	Target kDa	Target aa	Covalent	Fluorogenic	Arbitrary
8 Acyl and peptidyl carrier protein (ACP, PCP) [CoA-linked probes + phosphopantetheinyl transferase]	11–80	1.2–9	ser	+		+
9 Biotin ligase acceptor protein (AP) [ketone analog of biotin + biotin ligase]	15	1.7	lys			+
10 Biotin-mimetic peptide (Nano-tags) [avidin, streptavidin, anti-biotin linked probes]	9–15	1–1.7				+
11 Tetracysteine (-CC$x_{2,4}$CC-) motifs [bisarsenical derivatives of fluorescent probes]	6–12	0.7–1.3	4 cys		+	
12 Transglutaminase recognition sequence (Q-tag) [probe-linked primary amine + transglutaminase]	6–7	0.7	gln			+
13 LCTPSR... [aminooxy- or hydrazide-functionalized probes + formylglycine-generating enzyme][5]	6–13	0.7–1.4	cys	+		+
14 Tetraaspartate tag (–CA_6D_4-) [probe linked-chloroacetyl-Zn(II)-DpaTyr][6]	11	1.2	cys	+		+
15 Oligohistidine [NTA, trifunctional NTA-linked photoreactive probes]	4–6	0.7	his4–6	–, +		+
16 Sortase recognition sequence [oligoglycine-linked probes + sortase-transpeptidase][7]	6	0.6	thr	+		+
17 N-terminal cysteine (cleavage by protease, intein; ubiquitin fusion) [thioester-coupled probes][8]	1	-	cys			+
18 C-terminal aa (RGAA) [puromycin-coupled probes + ribosomal peptidyl transferase][9]	1–4	-		+		

19 Binding site [aptamers + beacons: peptides and nucleic acids with bound or binding fluorophore]		−, +	+
20 Binding site [trifunctional reagent: probe-photo-affinity reagent-ligand + protein trp as FRET donor]	+	−, +	+
21 Binding site [nanoparticle-linked ligands (e.g., Quantum Dots; factors, peptides, modifiers]		−, +	+
22 Reaction above → azide [Cu-free click chemistry: probe-linked difluorinated cyclooctenes (DI-FOs)][10]	+	−, +	+

First entry on each line is the tag. Square brackets enclose the probe "donor" + transfer enzyme (where applicable) or reaction. Probe: organic fluorophores, nanoparticles (can be QDs as in methods 2, 3, 8, 9) or a bridging/recognition moiety. If desired, the latter can serve in a second orthogonal reaction ("piggyback" strategy). See text. Biotin readout: probes linked to avidin, streptavidin, anti-biotin

[1] [101];
[2] [102];
[3] [63];
[4] [103];
[5] [104];
[6] [105];
[7] [106, 107];
[8] [108, 109];
[9] [110];
[10] [111]; one of the reaction listed in the table introduces an azide at a specific site of the protein target.

For further references see text and [2].

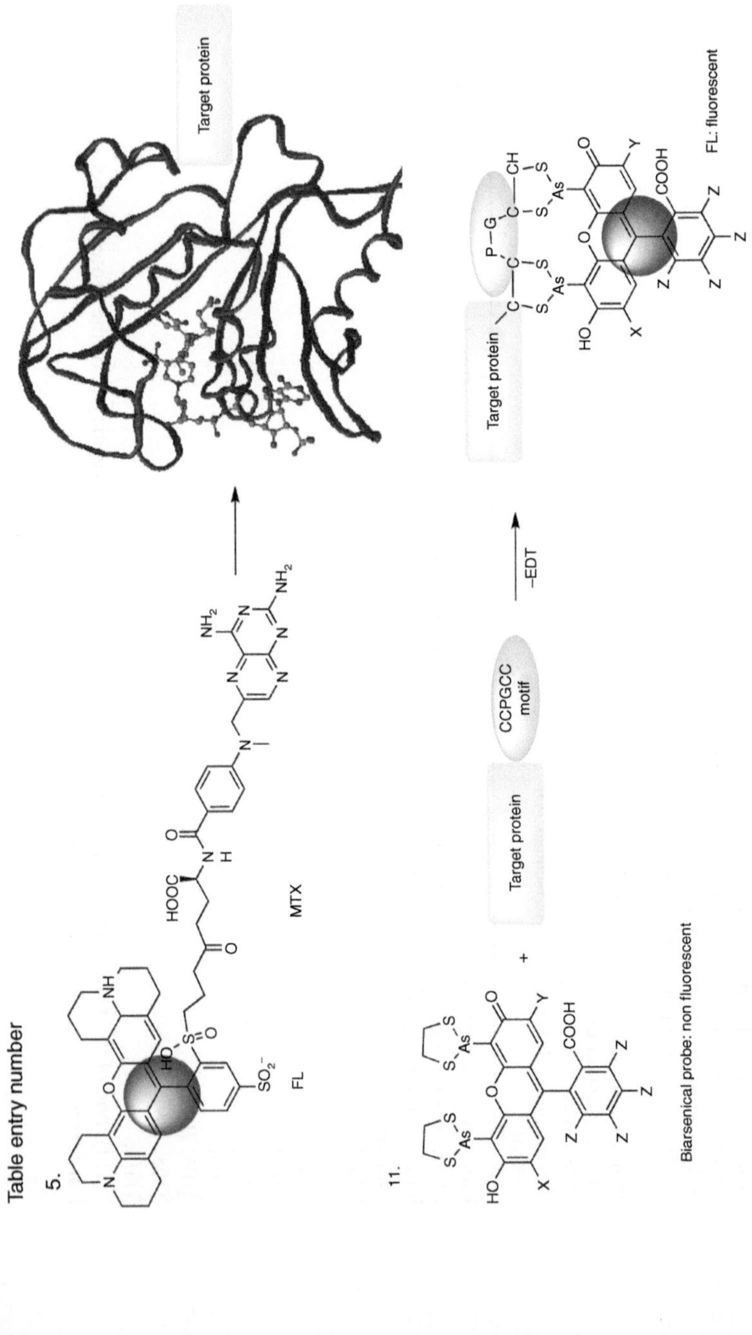
Table entry number
5.
FL
MTX
Target protein
11.
Target protein
CCPGCC motif
−EDT
Target protein
Biarsenical probe: non fluorescent
FL: fluorescent

involved in selected methods, some of which are highly chemoselective, are depicted in Figs. 12.3 and 12.4. In the case of noncovalent complexes, the protein of interest is fused to a protein tag capable of binding a small-molecule ligand. A very recent and intriguing development (method 7) is the "fluorescent activation" of fluorogenic probes by single-chain antibody tags [63], and "infinite affinity" antibodies [103] are closely related in principle. Other approaches include enzymes as target domains, for example, dihydrofolate reductase (DHFR; Table 12.2, Fig. 12.3, method 5), which binds the small-molecule ligand methotrexate with picomolar affinity. Methotrexate can be conjugated to a variety of dyes in a position of the molecule that does not influence its interaction with the DHFR. The probes attached to a protein tag by a covalent bond require the intervention of enzymatic or spontaneous highly chemoselective chemical reactions. A disadvantage of such systems is the need for introducing (expressing) enzymes or other components intracellularly or applying them externally in cases restricted to cell surface modification, as in the case of ACP (Table 12.2, method 8). In addition to the individual reactions featured in Table 12.2, one can conceive of a dual reaction "piggyback" (our designation) strategy for greatly extending the range of possibilities. The procedures would involve a site-specific protein modification with a reactive/recognition moiety (which could even be a short pepide sequence) followed by a second, different probe-transfer reaction.

Of increasing utility have been the tetracysteine tags specific for fluorogenic, biarsenical ligands (e.g., FlAsH, Table 12.2, Fig. 12.3, method 11; [72, 113, 114]). The latter bind with high affinity and selectivity to very short peptide sequences incorporating two dicysteine elements separated by 2 or more residues. The variety of dyes

---

Fig. 12.3. Some expression systems involving noncovalent (binding) linkages. Numbers correspond to entries (methods) in Table 12.2: (5) labeling with methotrexate derivatives that recognize a dihydrofolatereductase segment, and (11) labeling with biarsenical derivatives binding to a 6–12 tetracysteine motif.

is restricted by the requirement for accommodating the biarsenical substructure without loss of the fluorescence scaffold, although in some applications a nonfluorescent ligand may be preferable [113, 114]. Two recent innovations include an improved carboxyFlAsH (CrAsH, [115]) and a photostable Cy3-based biarsenical (AsCy3, [116]) with orthogonal binding specificity; see also [114]. A useful "bipartite tetracysteine display" strategy for detecting (mis)folded proteins in cells has been reported [117], as well as a strategy for label transfer to detect protein–protein interactions [118]. The sensitivity and selectivity of the FlAsH reagents have been deemed insufficient in at least some studies of living cells [119], yet they offer unique capabilities such as in correlative optical (fluorescence) and ultrastructural electron microscopy [72]. Our own work with this expression system has included the development [120] of fluorinated fluorescein-based biarsenicals (F2-FlAsH, F4-FlAsH) with improved photostability, reduced pH sensitivity, new emission wavelengths, and larger R_o values (Table 12.3); and studies of amyloid aggregation in living cells with tetracysteine tag fusions to α-synuclein [121]. Imaging based on both heteroFRET and homoFRET signals revealed the intimate protein association presumed to underlie the cytotoxic processes responsible for neuropathology in Parkinson's and other neurodegenerative diseases.

We now provide a brief description of the (bio)chemical mechanisms corresponding to selected schemes in Table 12.2 (see Fig. 12.4 and also [2]). In method 2, a mutant of wildtype dehalogenase (293 amino acids) reacts with chloroalkyl-derivatives undergoing a nucleophilic substitution reaction in which an ester is formed between the aliphatic substrate and aspartate. Virtually, any dye can be introduced with this method. However, the HaloTag is the largest of those featured in Table 12.2 and more data are required with respect to the labeling specificity. Method 3 is based on spontaneous chemical reaction of a "suicide" ligand to the cutinase tag, while method 4 involves the alkylation of Cys145 of human DNA O^6-alkylguanine transferase (hAGT) constituting the SNAP-tag with O^6-benzylguanine derivatives of a fluorophore. The

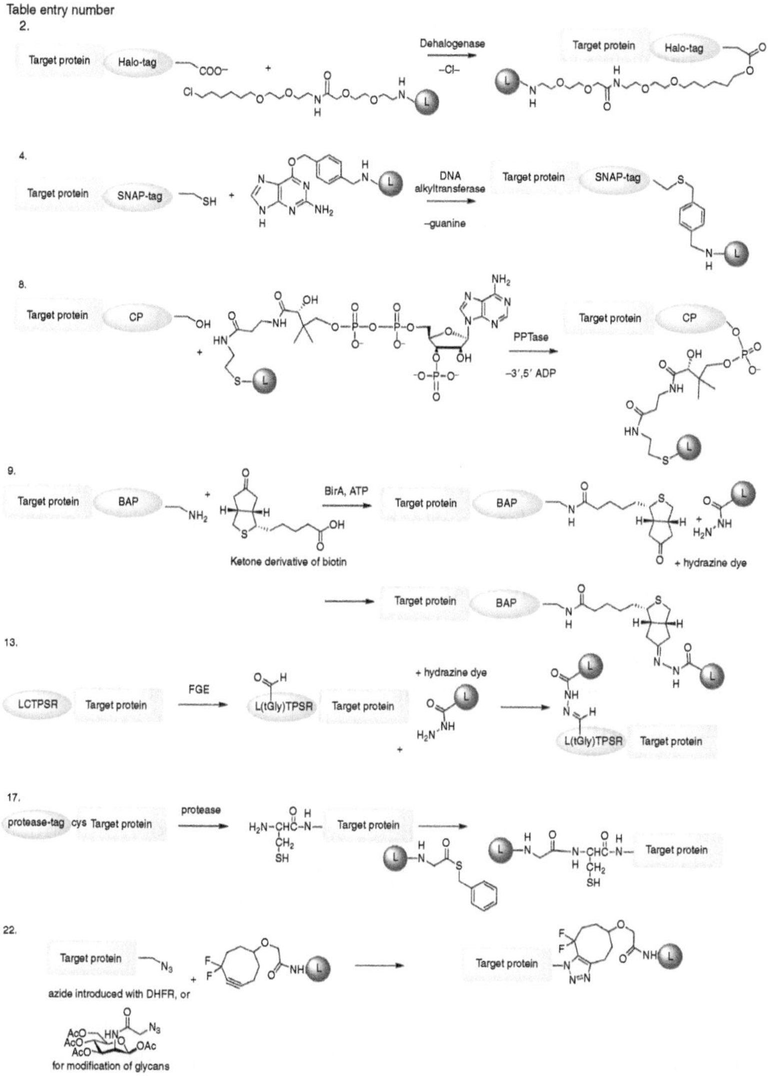

Fig. 12.4. Expression systems involving covalent linkages. Numbers correspond to entries (methods) in Table 12.2: (2) labeling of dehalogenase fusion proteins (i.e., HaloTag) with aliphatic chlorides, (4) labeling of AGT (O_6-alkylguanine-DNA alkyltransferase) fusion proteins with benzylguanine (BG)

TABLE 12.3
FRET parameters (J, Ro) for various biarsenical reagents. Taken from [120]

	Acceptor			
	ReAsH		F4FlAsH	
Donor	$10^{14}\cdot J$	R_o nm (Q_D)	$10^{13}\cdot J$	R_o nm (Q_D)
FlAsH	5.0	3.9 (0.4)	1.5	4.7 (0.4)
F2FlAsH	3.4	4.1 (0.8)	1.6	5.4 (0.8)
F4FlAsH	6.5	4.1 (0.4)	1.2	4.5 (0.4)
	F2FlAsH		FlAsH	
	$10^{13}\cdot J$	R_o nm (Q_D)	$10^{14}\cdot J$	R_o nm (Q_D)
F2FlAsH	1.4	5.2 (0.8)	6.4	4.1 (0.8)

guanine moiety is lost in the reaction, leaving the O^6 substituent bearing the dye covalently attached to the peptide tag. The newer CLIP-tags achieve the same end with cytosine derivatives and thus offer orthogonal labeling. An important advantage of the method is that the derivatization sites on the benzylguanine are permissive such that nearly any probe can be introduced without affecting the specificity or rate of the binding reaction. In addition, labeling is rapid. Unfortunately, the size of the tag is relatively large (Table 12.2). The SNAP tag been utilized in time-resolved FRET determinations [33, 122]. In method 8, the 80 aa (recently abbreviated to 11 [123]) protein-tagging sequence derived from either peptide carrier protein (PCP) or acyl carrier protein (ACP), is labeled using a phosphopantetheine transferase (PPtase). The latter transfers phosphopantetheine-linked probes from coenzyme

derivatives, (8) labeling with CoA derivatives via a PPTase, (9) labeling with a ketone derivative of biotin, in a first step, followed by a second step of reaction of the carbonyl group with a hydrazide dye, (13) labeling in two steps, oxidation of one amino acid generates an aldehyde group that is subsequently labeled with a hydrazide dye, (17) labeling with thioester derivatives of amino terminal cysteine groups, (22) a copper-free version of click chemistry for the labeling of glycans, peptides and proteins.

A (CoA) to a serine sidechain of ACP or PCP, forming a specific covalent link. The method shares advantages cited for the AGT, such as short labeling time, biorthogonality (multiple tag/probe combinations), and probe permissivity. In our experience, the ACP system has proven to be extremely effective and versatile, providing excellent labeling suitable for FRET (D. Arndt-Jovin, unpublished data). Method 9 is based on the biotinylation of peptide-tags (only 15 aa) by biotin ligase, for example, *Escherichia coli* BirA that utilizes a ketone as a substrate for conjugation to the tag. The keto group is subsequently reacted with a hydrazide linked to a fluorophore. In method 13, a formylglycine-generating enzyme (FGE) oxidizes a cysteine to formylglycine. The newly formed aldehyde reacts with hydrazides as in the previous approach. These two-step labeling methods are primarily restricted to cell-surface modification. The second step is also very slow; the rate of hydrazone formation at pH 7 is $< 1\ M^{-1}\ s^{-1}$, that is, more than three orders of magnitude slower than the labeling of AGT or ACP fusion proteins. Thus, millimolar concentrations of the dye are required in order to achieve a reasonable extent of labeling. Method 17 is based on the labeling of an N-terminal cysteine generated in a protein of interest by protein splicing, for example by inteins, or most recently, with a short peptide tag recognized by a specific protease that can be coexpressed with the fusion protein. The N-terminal cysteine reacts with a thioester-functionalized probe in a two-step mechanism that is initiated by reaction with the thiol group followed by intramolecular reordering to form a stable peptide bond. This method is versatile and should work with any cell type, but can require large fusion segments, even if only transiently. The rates of splicing and thioester ligation are also slow, restricting the methodology to biological processes taking place over hours. The use of click chemistry [124] is very attractive since it offers the possibility of biorthogonal labeling at specific sites of a given biomolecule or of different biomolecules due to the unique characteristics of the reaction. Its general application in cells was has been precluded by the toxicity of copper, a required catalyzer. The advent of a copper free click chemistry [111] now permits the facile introduction of an azido derivative of tetra-acetylmanosamine into glycans or of an azide group into a peptide or protein. The authors employed DHFR to achieve the transfer but we would propose (method 22, Fig. 12.4) the use of other methods listed in Table 12.2 as well. In the second stage of method 22, a difluorinated cyclooctyne (DIFO) reacts with the azide via a [3 + 2] dipolar cycloadition. The kinetics are comparable to that of reactions catalyzed by copper. That is, the reaction occurs in minutes; toxicity would not be anticipated [111].

The exploitation of the above expression systems in FRET requires the coherent selection of donor and acceptor moieties, from both the spectroscopic and biological perspective: relative expression levels, compartmentalization, and temporal evolution of the system under study. Very advantageous are combinations of small fluorophores with VFPs, as well as with fluorescent nanoparticles, particularly QDs.

Given their unique properties (brightness, photostability, narrowband emission, and broadband absorption), and surprising bio (chemical)compatibility in spite of their relatively large size, QDs are ideally suited for FRET imaging. Commercially available nanoparticles are coated with a polymeric layer that isolates the core-shell structure from the aqueous environment, thereby attaining a diameter of 15–20 nm, or double that value if covered with PEG. Is this considerable size important such that QDots operate more as a platform rather than as a molecular entity? In fact, QDs have been shown to be useful as FRET donors (even individually; [125, 126]) and acceptors in numerous applications [67], although there have been relatively few reports of imaging applications. In our own work on cellular signaling mediated by growth factors (EGF, NGF) and their receptors, QDs conjugated with ligands have revealed the existence of novel transport and trafficking mechanisms. In one instance, a biotinylated organic dye FRET acceptor served to demonstrate that retrograde transport but not endocytosis occurs on filopodia, cellular extensions with a core of actin filaments [127].

As stated earlier, QD donors have been applied extensively in FRET-based assays of enzymatic activities, with the systematic introduction of multifunctionality being a major issue ([128] and references therein). In combination with selected VFP acceptors, up to 90% transfer can be achieved [129, 130]. QD donors function in cells [131] and can transfer to QD acceptors [132]. The distance dependence of FRET operating via surface-bound or nearby small-molecule acceptors reflects the complex interplay between factors such as stoichiometry, spatial distribution and orientation, composition (passivation shell, capping moieties), and shape. [QDs emitting at >600 nm tend to be nonspherical and demonstrate finite emission polarization.] QDs utilized as acceptors pose a problem due their very large absorption cross-sections extending (and increasing) into the UV, making it difficult to selectively excite potential donors. However, we have confirmed that QDs can operate as acceptors by fluorescence lifetime analysis of dimers and multimers (Miskoski et al., manuscript in preparation). In an intriguing recent application [133], we encountered a novel phenomenon of "funnel" FRET in which numerous donors are presumed to relay their excitation energy by a two-stage hetero-homotransfer FRET mechanism to QD "antennas," thereby providing enhanced sensitivity and selectivity coupled with positional superresolution in imaging experiments.

QDs excited by single or multiple photons exhibit multixponential decays, in our hands generally with two components of <10 and 10–25 ns. This property renders the nanoparticles useful reagents for the systematic modulation of emission via a technique we have denoted as photochromic FRET (pcFRET, Fig. 12.5; see [1, 2, 134]). The method was developed in order to circumvent inherent limitations in the quantitative determination of FRET in cells, particularly in imaging applications. We have classified pcFRET as a member of the family of so-called "acceptor depletion" methods [1, 2] due to the reversible change in the structure, and thus spectroscopic properties, of the photochromic compound (e.g., a diheteroarylethene) serving as the FRET acceptor. The key

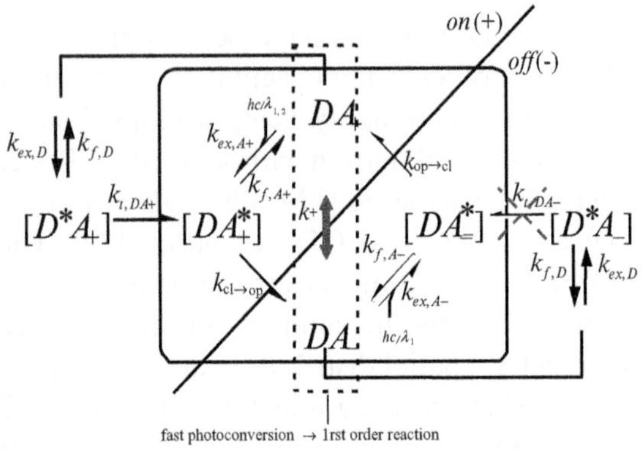

Fig. 12.5. pcFRET. Photophysical–photochromic scheme for a donor + photochromic acceptor pair (*D*,*A*). The central region depicts the ring closure and opening reactions of the photochromic acceptor. On the *on* (left side) of the slanted line, *A* is in the FRET-competent (absorbing, high *J*) closed form. On the *off* (right side) of the slanted line, *A* is in the FRET-incompetent (nonabsorbing, low *J*) open form. The FRET reactions feeding into the photochromic cycle are shown on the outside. Only D^*A_+ is active in this process (D^*A_- is crossed out for this reason). Because the excited state processes are rapid, the scheme reduces to an apparent first-order reaction (grey vertical arrow) driven by cyclical UV and visible irradiation. For more explanation, see [134] from which this diagram was adapted.

feature of pcFRET is the modulation of donor emission by cyclical transformations of the pc acceptor leading to a corresponding variation in the donor-acceptor overlap integral *J* (and thus Förster transfer distance R_0) between the two states (Fig. 12.5). Near-ultraviolet illumination leads to the conversion of the colorless open form to a closed structure featuring an absorption band overlapping the emission band of the fluorescent donor, thereby quenching the latter (FRET *on* state). Cycloreversion by exposure to visible light reverses the process, that is, restores the FRET *off* state (Fig. 12.5). The pcFRET principles underlie probe developments aimed at very sensitive FRET-based and superresolution microscopy [135–137].

The implementation of pcFRET in microscopy has been hindered by the paucity of suitable compounds capable of undergoing multiple cycles without fatigue in an aqueous environment (although protein conjugation helps; [135–137]). Photoreversible VFPs such as Dronpa [138] have not yet been employed in pcFRET. However, the combination of the photostable diheteroarylethenes and QDs offer a system applicable to many imaging situations. Such an experiment [139] is featured in Fig. 12.6, in which a reversible modulation of the emission of CdSe/ZnS QDs was achieved by binding a photochromic biotinylated diheteroarylethene derivative to QDs bearing conjugated streptavidin, causing an intensity decrease as a consequence of energy transfer to the closed form of the acceptor. Interconversion between the open and closed forms by cyclical irradiation at 365 and 546 nm led to deactivation and activation, respectively, of the FRET process,

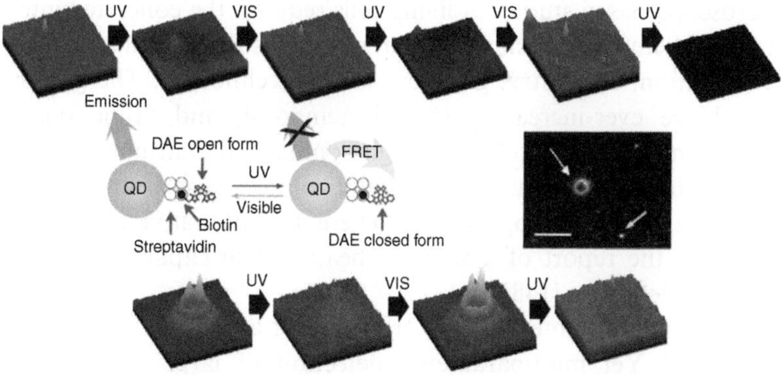

Fig. 12.6. Imaging QD emission modulated by pcFRET. Surface plots of a selected region containing aggregates and single QDs. The surface exhibited fluorescence shells (QDs coated microspheres), and single and aggregated QDs. After a pre-irradiation period with UV and visible light the QDs displayed a reversible decrease of emission upon irradiation with UV light (as a result of pcFRET) and an increase of emission when irradiated with visible light. Surface plots of an area about a bright spot corresponding to an aggregate (yellow arrow) are shown at the top. Bottom, surface plots of a 2.4 μm microsphere (white arrow). White bar, 5 μm. Data from [139]. (See Color Insert.)

with a corresponding modulation of the QD emission observed both in solution and by sequential wide-field imaging.

We close this section on actual and potential FRET probes by making reference in passing to certain specialized but very interesting topics: through-bond FRET [87, 140], "amplification FRET" based on ingenious coupling of molecular-beacon/aptamer constructs and enzymatic readout [141, 142], decaging/photoactivation FRET [143, 144], targeting of QDs and other nano-microstructures to cells [70, 145], and "smart polymers" as potential modulatable mediators of FRET probes [146, 147].

12.6. *Quo vadis*

This volume demonstrates unequivocally that the exploitation of FRET and the other associated fluorescence parameters for microscope-based studies of living cells requires the concerted integration of emerging technologies in numerous fields such as optics, computation, chemistry, genetics, and biotechnology. The goal is to achieve ever-increasing spatial, temporal, and "functional" resolution. Unfortunately, although commercial instruments offer outstanding performance, the cost and complexity is at times overwhelming for the biological user. There are alternatives, as exemplified by the report of a simple, cheap system capable of single molecule detection [148].

There is undoubtedly some truth in the axiom that "seeing is believing." Yet multiparametric detection of large fields (cells, tissues, organisms) exhibiting complex spatiotemporal patterns of behavior is often incompatible with visual observation. By not insisting on ergonomic designs, manufacturers could devote attention to highly desirable innovations such as large-field high NA "objectives" and modular, multiported "microscopes" devoid of oculars (an intriguing example is the iMIC of Till Photonics).

Schemes for achieving 3D spatial superresolution (see Section 12.4) will continue to proliferate, and at the same time microscopes will become more "intelligent." For example, a capability for real-time sample-driven adaptive sectioning is being incorporated in our own developmental microscope based on spatially structured concerted illumination/detection using spatial light modulators (PAM; [149]). Such innovations incorporating adaptive photophysical and optical sectioning strategies will facilitate the acquisition of the statistically significant data required for applications in systems biology and other diverse fields, including the nonbiological. In this context, the prospects for FRET are "bright" and we can anticipate such adjectives as "binary," "integrating," "superresolution" in the next tabulations of methods for implementing this fascinating technique. *Stay tuned...*

Acknowledgments

This work was supported by the Max Planck Society (Toxic Protein Conformation Project; Partner Group Grant to EJ-E), Volkswagen Foundation (grant I/79986), DFG Research Center for Molecular Physiology of the Brain (Excellence Cluster "Microscopy on the nanometer scale," grant EXC117), ANPCyT, and the University of Buenos Aires. We are indebted to numerous colleagues, including other authors of this volume, for productive discussions and research activities related to the topics of the chapter.

References

[1] Jares-Erijman, E. A. and Jovin, T. M. (2003). FRET imaging. Nat. Biotechnol. *21*, 1387–95.

[2] Jares-Erijman, E. A. and Jovin, T. M. (2006). Imaging molecular interactions in living cells by FRET microscopy. Curr. Opin. Chem. Biol. *10*, 409–16.

[3] van Manen, H. J., Verkuijlen, P., Wittendorp, P., Subramaniam, V., van den Berg, T. K., Roos, D. and Otto, C. (2008). Refractive index sensing of green fluorescent proteins in living cells using fluorescence lifetime imaging microscopy. Biophys. J. *94*, L67–9.

[4] Beutler, M., Makrogianneli, K., Vermeij, R. J., Keppler, M., Ng, T., Jovin, T. M. and Heintzmann, R. (2008). satFRET: Estimation of Förster resonance energy transfer by acceptor saturation. Eur. J. Biophys. DOI: 10.1007/s00249-008-0361-5.

[5] Heintzmann, R., Jovin, T. M. and Cremer, C. (2002). Saturated patterned excitation microscopy—a novel concept for optical resolution improvement. J. Opt. Soc. Am. A *19*, 1599–609.

[6] Colyer, R. A., Lee, C. and Gratton, E. (2008). A novel fluorescence lifetime imaging system that optimizes photon efficiency. Microsc. Res. Tech. *71*, 201–13.

[7] Esposito, A., Gerritsen, H. C., Oggier, T., Lustenberger, F. and Wouters, F. S. (2006). Innovating lifetime microscopy: A compact and simple tool for life sciences, screening, and diagnostics. J. Biomed. Opt. *11*, 34016.

[8] Padilla-Parra, S, Auduge, N, Coppey-Moisan, M and Tramier, M. (2008). Quantitative FRET analysis by fast acquisition time domain FLIM at high spatial resolution in living cells. Biophys. J DOI: 10.1529/biophysj.108.131276.

[9] Sanden, T., Persson, G., Thyberg, P., Blom, H. and Widengren, J. (2007). Monitoring kinetics of highly environment sensitive states of fluorescent molecules by modulated excitation and time-averaged fluorescence intensity recording. Anal. Chem. *79*, 3330–41.

[10] Spriet, C., Trinel, D., Riquet, F., Vandenbunder, B., Usson, Y. and Heliot, L. (2008). Enhanced FRET contrast in lifetime imaging. Cytometry A *73A*, 745–53.

[11] Diaspro, A. E. (2007). Special Issue: Advanced multiphoton and fluorescence lifetime imaging techniques. Microsc. Res. Tech. *70*, 397–492.

[12] Rakich, P., Dahlem, M., Tandon, S. N., Ibanescu, M., Soljacic, M., Petrich, G. S., Joannopoulos, J. D., Kolodziejski, L. A. and Ippen, E. P. (2006). Achieving centimetre-scale super-collimation in a large area 2D photonic crystal. Nat. Mater. *5*, 93–6.

[13] Tada, J., Kono, T., Suda, A., Mizuno, H., Miyawaki, A., Midorikawa, K. and Kannari, F. (2007). Adaptively controlled supercontinuum pulse from a microstructure fiber for two-photon excited fluorescence microscopy. Appl. Opt. *46*, 3023–30.

[14] MacKinnon, N. (2005). Spectrally programmable light engine for in vitro or in vivo molecular imaging and spectroscopy. Appl. Opt. *44*, 2033–44.

[15] Forde, T. S. and Hanley, Q. S. (2006). Spectrally resolved frequency domain analysis of multi-fluorophore systems undergoing energy transfer. Appl. Spectrosc. *60*, 1442–52.

[16] Hanley, Q. S., Murray, P. I. and Forde, T. S. (2006). Microspectroscopic fluorescence analysis with prism-based imaging spectrometers: Review and current studies. Cytometry A *69*, 759–66.
[17] Ramanujan, V. K., Biener-Ramanujan, E., Armmer, K., Centonze, V. E. and Herman, B. A. (2006). Spectral kinetics ratiometry: A simple approach for real-time monitoring of fluorophore distributions in living cells. Cytometry A *69*, 912–9.
[18] Thaler, C. and Vogel, S. S. (2006). Quantitative linear unmixing of CFP and YFP from spectral images acquired with two-photon excitation. Cytometry A *69*, 904–11.
[19] Wlodarczyk, J., Woehler, A., Kobe, F., Ponimaskin, E., Zeug, A. and Neher, E. (2008). Analysis of FRET-signals in the presence of free donors and acceptors. Biophys. J. *94*, 986–1000.
[20] Zheng, J. (2006). Spectroscopy-based quantitative fluorescence resonance energy transfer analysis. Methods Mol. Biol. *337*, 65–77.
[21] Eggeling, C., Widengren, J., Brand, L., Schaffer, J., Felekyan, S. and Seidel, C. A. (2006). Analysis of photobleaching in single-molecule multicolor excitation and Forster resonance energy transfer measurements. J. Phys. Chem. A *110*, 2979–95.
[22] Kalinin, S., Felekyan, S., Antonik, M. and Seidel, C. A. (2007). Probability distribution analysis of single-molecule fluorescence anisotropy and resonance energy transfer. J. Phys. Chem. B *111*, 10253–62.
[23] Lee, N. K., Kapanidis, A. N., Koh, H. R., Korlann, Y., Ho, S. O., Kim, Y., Gassman, N., Kim, S. K. and Weiss, S. (2007). Three-color alternating-laser excitation of single molecules: Monitoring multiple interactions and distances. Biophys. J. *92*, 303–12.
[24] Nir, E., Michalet, X., Hamadani, K. M., Laurence, T. A., Neuhauser, D., Kovchegov, Y. and Weiss, S. (2006). Shot-noise limited single-molecule FRET histograms: Comparison between theory and experiments. J. Phys. Chem. B *110*, 22103–24.
[25] Rasnik, I., McKinney, S. A. and Ha, T. (2005). Surfaces and orientations: Much to FRET about? Acc. Chem. Res. *38*, 542–8.
[26] Widengren, J., Kudryavtsev, V., Antonik, M., Berger, S., Gerken, M. and Seidel, C. A. (2006). Single-molecule detection and identification of multiple species by multiparameter fluorescence detection. Anal. Chem. *78*, 2039–50.
[27] Kong, X., Nir, E., Hamadani, K. and Weiss, S. (2007). Photobleaching pathways in single-molecule FRET experiments. J. Am. Chem. Soc. *129*, 4643–54.
[28] Widengren, J., Chmyrov, A., Eggeling, C., Lofdahl, P. A. and Seidel, C. A. (2007). Strategies to improve photostabilities in ultrasensitive fluorescence spectroscopy. J. Phys. Chem. A *111*, 429–40.

[29] Hoebe, R. A., Van Oven, C. H., Gadella, T. W., Jr, Dhonukshe, P. B., Van Noorden, C. J. and Manders, E. M. (2007). Controlled light-exposure microscopy reduces photobleaching and phototoxicity in fluorescence live-cell imaging. Nat. Biotechnol. *25*, 249–53.

[30] Van Munster, E. B., Kremers, G. J., Adjobo-Hermans, M. J. and Gadella, T. W., Jr. (2005). Fluorescence resonance energy transfer (FRET) measurement by gradual acceptor photobleaching. J. Microsc. *218*, 253–62.

[31] Coulon, V., Audet, M., Homburger, V., Bockaert, J., Fagni, L., Bouvier, M. and Perroy, J. (2008). Subcellular imaging of dynamic protein interactions by bioluminescence resonance energy transfer. Biophys. J. *94*, 1001–9.

[32] Xu, X., Soutto, M., Xie, Q., Servick, S., Subramanian, C., von Arnim, A. G. and Johnson, C. H. (2007). Imaging protein interactions with bioluminescence resonance energy transfer (BRET) in plant and mammalian cells and tissues. Proc. Natl. Acad. Sci. USA *104*, 10264–9.

[33] Mao, S., Benninger, R. K., Yan, Y., Petchprayoon, C., Jackson, D., Easley, C. J., Piston, D. W. and Marriott, G. (2008). Optical lock-in detection of FRET using synthetic and genetically encoded optical switches. Biophys. J. *94*, 4515–24.

[34] Santoso, Y., Hwang, L. C., Le Reste, L. and Kapanidis, A. N. (2008). Red light, green light: Probing single molecules using alternating-laser excitation. Biochem. Soc. Trans. *36*, 738–44.

[35] Cohen-Kashi, M., Namer, Y. and Deutsch, M. (2006). Fluorescence resonance energy transfer imaging via fluorescence polarization measurement. J. Biomed. Opt. *11*, 34015.

[36] Fisz, J. J. (2007). Another look at magic-angle-detected fluorescence and emission anisotropy decays in fluorescence microscopy. J. Phys. Chem. A *111*, 12867–70.

[37] Fisz, J. J. (2007). Fluorescence polarization spectroscopy at combined high-aperture excitation and detection: Application to one-photon-excitation fluorescence microscopy. J. Phys. Chem. A *111*, 8606–21.

[38] Fixler, D., Namer, Y., Yishay, Y. and Deutsch, M. (2006). Influence of fluorescence anisotropy on fluorescence intensity and lifetime measurement: theory, simulations and experiments. IEEE Trans. Biomed. Eng. *53*, 1141–52.

[39] Rizzo, M. A. and Piston, D. W. (2005). High-contrast imaging of fluorescent protein FRET by fluorescence polarization microscopy. Biophys. J. *88*, L14–6.

[40] Chen, H., Puhl, H. L., III, Koushik, S. V., Vogel, S. S. and Ikeda, S. R. (2006). Measurement of FRET efficiency and ratio of donor to acceptor concentration in living cells. Biophys. J. *91*, L39–41.

[41] Domingo, B., Sabariegos, R., Picazo, F. and Llopis, J. (2007). Imaging FRET standards by steady-state fluorescence and lifetime methods. Microsc. Res. Tech. *70*, 1010–21.
[42] Schwartz, J. W., Piston, D. and DeFelice, L. J. (2006). Molecular microfluorometry: Converting arbitrary fluorescence units into absolute molecular concentrations to study binding kinetics and stoichiometry in transporters. Handb. Exp. Pharmacol. 23–57.
[43] Vogel, S. S., Thaler, C. and Koushik, S. V. (2006). Fanciful FRET. Sci STKE 2006, re2.
[44] Wallrabe, H., Chen, Y., Periasamy, A. and Barroso, M. (2006). Issues in confocal microscopy for quantitative FRET analysis. Microsc. Res. Tech. *69*, 196–206.
[45] Nagy, P., Vereb, G. and Damjanovich, S. (2006). Measuring FRET in flow cytometry and microscopy. Curr. Prot. Microsc. 12.18.11–12.18.13.
[46] Szöllösi, J., Damjanovich, S., Nagy, P., Vereb, G. and Mátyus, L. (2006). Principles of resonance energy transfer. Curr. Prot. Cell Biol. 1.12.11–11.12.16.
[47] Rodighiero, S., Bazzini, C., Ritter, M., Furst, J., Botta, G., Meyer, G. and Paulmichl, M. (2008). Fixation, mounting and sealing with nail polish of cell specimens lead to incorrect FRET measurements using acceptor photobleaching. Cell Physiol. Biochem. *21*, 489–98.
[48] Hoppe, A. D., Shorte, S. L., Swanson, J. A. and Heintzmann, R. (2008). Three-dimensional FRET reconstruction microscopy for analysis of dynamic molecular interactions in live cells. Biophys. J. *95*, 400–18.
[49] van Rheenen, J., Langeslag, M. and Jalink, K. (2004). Correcting confocal acquisition to optimize imaging of fluorescence resonance energy transfer by sensitized emission. Biophys. J. *86*, 2517–29.
[50] Li, E., Placone, J., Merzlyakov, M. and Hristova, K. (2008). Quantitative measurements of protein interactions in a crowded cellular environment. Anal. Chem. *80*, 5976–85.
[51] Posokhov, Y. O., Merzlyakov, M., Hristova, K. and Ladokhin, A. S. (2008). A simple "proximity" correction for Forster resonance energy transfer efficiency determination in membranes using lifetime measurements. Anal. Biochem. *380*, 134–6.
[52] Allen, M. D. and Zhang, J. (2008). A tunable FRET circuit for engineering fluorescent biosensors. Angew. Chem. Int. Ed. Engl. *47*, 500–2.
[53] Fazekas, Z., Petras, M., Fabian, A., Palyi-Krekk, Z., Nagy, P., Damjanovich, S., Vereb, G. and Szollosi, J. (2008). Two-sided fluorescence resonance energy transfer for assessing molecular interactions of up to three distinct species in confocal microscopy. Cytometry A *73*, 209–19.

[54] Piljic, A. and Schultz, C. (2008). Simultaneous recording of multiple cellular events by FRET. ACS Chem. Biol. *3*, 156–60.
[55] Kudryavtsev, V., Felekyan, S., Wozniak, A. K., Konig, M., Sandhagen, C., Kuhnemuth, R., Seidel, C. A. and Oesterhelt, F. (2007). Monitoring dynamic systems with multiparameter fluorescence imaging. Anal. Bioanal. Chem. *387*, 71–82.
[56] Heintzmann, R. and Ficz, G. (2007). Breaking the resolution limit in light microscopy. Briefings Funct. Genom. Proteom. *5*, 289–301.
[57] Bates, M., Huang, B., Dempsey, G. T. and Zhuang, X. (2007). Multicolor super-resolution imaging with photo-switchable fluorescent probes. Science *317*, 1749–53.
[58] Hell, S. W. (2007). Far-field optical nanoscopy. Science *316*, 1153–8.
[59] Hess, S. T., Girirajan, T. P. K. and Mason, M. S. (2006). Ultra-high resolution imaging by fluorescence photoactivation localization microscopy. Biophys. J. *91*, 4258–72.
[60] Marushchak, D. and Johansson, L. B. (2005). On the quantitative treatment of donor–donor energy migration in regularly aggregated proteins. J. Fluoresc. *15*, 797–803.
[61] Tramier, M. and Coppey-Moisan, M. (2008). Fluorescence anisotropy imaging microscopy for homo-FRET in living cells. Methods Cell Biol. *85*, 395–414.
[62] Yeow, E. K. L. and Clayton, A. H. A. (2007). Enumeration of oligomerization states of membrane proteins in living cells by homo-FRET spectroscopy and microscopy: Theory and application. Biophys. J. *92*, 3098–104.
[63] Szent-Gyorgyi, C., Schmidt, B. F., Creeger, Y., Fisher, G. W., Zakel, K. L., Adler, S., Fitzpatrick, J. A., Woolford, C. A., Yan, Q., Vasilev, K. V., Berget, P., Bruchez, M. et al. (2007). Fluorogen activating proteins: Technology for imaging and assaying cell surface proteins. Nat. Biotechnol. *26*, 235–40.
[64] Szabó, A., Horváth, G., Szöllösi, J. and Nagy, P. (2008). Quantitative characterization of the large-scale association of ErbB1 and ErbB2 by flow cytometric homo-FRET measurements. Biophys. J. *95*, 2086–96.
[65] Demchenko, A. P. (2005). Optimization of fluorescence response in the design of molecular biosensors. Anal. Biochem. *343*, 1–22.
[66] O'Hare, H. M., Johnsson, K. and Gautier, A. (2007). Chemical probes shed light on protein function. Curr. Opin. Struct. Biol. *17*, 488–94.
[67] Sapsford, K. E., Berti, L. and Medintz, I. L. (2006). Materials for fluorescence resonance energy transfer analysis: Beyond traditional donor–acceptor combinations. Angew. Chem. Int. Ed. Engl. *45*, 4562–89.

[68] Waggoner, A. (2006). Fluorescent labels for proteomics and genomics. Curr. Opin. Chem. Biol. *10*, 62–6.

[69] Bakalova, R., Zhelev, Z., Aoki, I. and Kanno, I. (2007). Designing quantum dot probes. Nat. Photonics *1*, 487–9.

[70] Chen, I., Choi, Y. A. and Ting, A. Y. (2007). Phage display evolution of a peptide substrate for yeast biotin ligase and application to two-color quantum dot labeling of cell surface proteins. J. Am. Chem. Soc. *129*, 6619–25.

[71] Clarke, S. J., Hollmann, C. A., Aldaye, F. A. and Nadeau, J. L. (2008). Effect of ligand density on the spectral, physical, and biological characteristics of CdSe/ZnS quantum dots. Bioconjug. Chem. *19*, 562–8.

[72] Giepmans, B. N., Adams, S. R., Ellisman, M. H. and Tsien, R. Y. (2006). The fluorescent toolbox for assessing protein location and function. Science *312*, 217–24.

[73] Howarth, M., Takao, K., Hayashi, Y. and Ting, A. Y. (2005). Targeting quantum dots to surface proteins in living cells with biotin ligase. Proc. Natl. Acad. Sci. USA *102*, 7583–8.

[74] Landegren, U., Schallmeiner, E., Nilsson, M., Fredriksson, S., Baner, J., Gullberg, M., Jarvius, J., Gustafsdottir, S., Dahl, F. Soderberg, O. et al. (2004). Molecular tools for a molecular medicine: Analyzing genes, transcripts and proteins using padlock and proximity probes. J. Mol. Recog. *17*, 194–7.

[75] Pinaud, F., Michalet, X., Bentolila, L. A., Tsay, J. M., Doose, S., Li, J. J., Iyer, G. and Weiss, S. (2006). Advances in fluorescence imaging with quantum dot bio-probes. Biomaterials *27*, 1679–87.

[76] Smith, A. M., Ruan, G., Rhyner, M. N. and Nie, S. (2006). Engineering luminescent quantum dots for in vivo molecular and cellular Imaging. Ann. Biomed. Eng. *34*, 3–14.

[77] Zhang, Y., So, M.-K., Loening, A. M., Yao, H., Gambhir, S. S. and Rao, J. (2006). HaloTag protein-mediated site-specific conjugation of bioluminescent proteins to quantum dots. Angew. Chem. Int. Ed. Engl. *45*, 4936–40.

[78] Borisov, S. M., Mayr, T., Karasyov, A. A., Klimant, I., Chojnacki, P., Moser, C., Nagl, S., Schaeferling, M., Stich, M. I. Vasylevska, G. S. et al. (2007). New plastic microparticles and nanoparticles for fluorescence sensing and encoding. In: Fluorescence of Supermolecules, Polymers, and Nanosystems (Berberan, M. N., ed.). Springer, Berlin, pp. 431–63.

[79] Pregibon, D. C., Toner, M. and Doyle, P. S. (2007). Multifunctional encoded particles for high-throughput biomolecule analysis. Science *315*, 1393–6.

[80] Qin, L., Banholzer, M. J., Millstone, J. E. and Mirkin, C. A. (2007). Nanodisk codes. Nano Lett. *7*, 3849–52.
[81] Seferos, D. S., Giljohann, D. A., Hill, H. D., Prigodich, A. E. and Mirkin, C. A. (2007). Nano-flares: Probes for transfection and mRNA detection in living cells. J. Am. Chem. Soc. *129*, 15477–9.
[82] Tyner, K. M., Kopelman, R. and Philbert, M. A. (2007). "Nanosized voltmeter" enables cellular-wide electric field mapping. Biophys. J. *93*, 1163–74.
[83] Burns, A., Ow, H. and Wiesner, U. (2006). Fluorescent core-shell silica nanoparticles: Towards "Lab on a Particle" architectures for nanobiotechnology. Chem. Soc. Rev. *35*, 1028–42.
[84] Sukhanova, A and Nabiev, I. (2008). Fluorescent nanocrystal-encoded microbeads for multiplexed cancer imaging and diagnosis. Crit. Rev. Oncol. Hematol DOI: 10.1016/j.critrevonc.2008.05.006.
[85] Zhang, B., Cheng, J., Gong, X., Dong, X., Liu, X., Ma, G. and Chang, J. (2008). Facile fabrication of multi-colors high fluorescent/superparamagnetic nanoparticles. J. Colloid Interface Sci. *322*, 485–90.
[86] Johnsson, N. and Johnsson, K. (2007). Chemical tools for biomolecular imaging. ACS Chem. Biol. *2*, 31–8.
[87] Loudet, A. and Burgues, S. K. (2007). BODIPY dyes and their derivatives: Syntheses and spectroscopic properties. Chem. Rev. *107*, 4891–932.
[88] Miyawaki, A. and Karasawa, S. (2007). Memorizing spatiotemporal patterns. Nat. Chem. Biol. *3*, 598–601.
[89] Piston, D. W. and Kremers, G.-J. (2007). Fluorescent protein FRET: The good, the bad and the ugly. Trends Biochem. Sci. *32*, 407–14.
[90] Ai, H. W., Shaner, N. C., Cheng, Z., Tsien, R. Y. and Campbell, R. E. (2007). Exploration of new chromophore structures leads to the identification of improved blue fluorescent proteins. Biochemistry *46*, 5904–10.
[91] Kremers, G. J., Goedhart, J., van den Heuvel, D. J., Gerritsen, H. C. and Gadella, T. W., Jr. (2007). Improved green and blue fluorescent proteins for expression in bacteria and mammalian cells. Biochemistry *46*, 3775–83.
[92] Merzlyak, E. M., Goedhart, J., Shcherbo, D., Bulina, M. E., Shcheglov, A. S., Fradkov, A. F., Gaintzeva, A., Lukyanov, K. A., Lukyanov, S. Gadella, T. W. et al. (2007). Bright monomeric red fluorescent protein with an extended fluorescence lifetime. Nat. Methods *4*, 555–7.
[93] Olenych, S. G., Claxton, N. S., Ottenberg, G. K. and Davidson, M. W. (2007). The fluorescent protein color palette. Curr. Protoc Cell Biol. Chapter 21: Unit 21.5.

[94] Shaner, N. C., Patterson, G. H. and Davidson, M. W. (2007). Advances in fluorescent protein technology. J. Cell Sci. *120*, 4247–60.
[95] Shimozono, S. and Miyawaki, A. (2008). Engineering FRET constructs using CFP and YFP. Methods Cell Biol. *85*, 381–93.
[96] Tsutsui, H., Karasawa, S., Okamura, Y. and Miyawaki, A. (2008). Improving membrane voltage measurements using FRET with new fluorescent proteins. Nat. Methods *5*, 683–85.
[97] van der Krogt, G. N., Ogink, J., Ponsioen, B. and Jalink, K. (2008). A comparison of donor–acceptor pairs for genetically encoded FRET sensors: Application to the Epac cAMP sensor as an example. PLoS ONE *3*, e1916.
[98] Ganesan, S., Ameer-Beg, S. M., Ng, T. T., Vojnovic, B. and Wouters, F. S. (2006). A dark yellow fluorescent protein (YFP)-based resonance energy-accepting chromoprotein (REACh) for Forster resonance energy transfer with GFP. Proc. Natl. Acad. Sci. USA *103*, 4089–94.
[99] Sakaue-Sawano, A., Kurokawa, H., Morimura, T., Hanyu, A., Hama, H., Osawa, H., Kashiwagi, S., Fukami, K., Miyata, T. Miyoshi, H. et al. (2008). Visualizing spatiotemporal dynamics of multicellular cell-cycle progression. Cell *132*, 487–98.
[100] Budzikiewicz, H., Schafer, M., Fernandez, D. U., Matthijs, S. and Cornelis, P. (2007). Characterization of the chromophores of pyoverdins and related siderophores by electrospray tandem mass spectrometry. Biometals *20*, 135–44.
[101] Bonasio, R., Carman, C. V., Kim, E., Sage, P. T., Love, K. R., Mempel, T. R., Springer, T. A. and von Andrian, U. H. (2007). Specific and covalent labeling of a membrane protein with organic fluorochromes and quantum dots. Proc. Natl. Acad. Sci. USA *104*, 14753–8.
[102] Smith, J. J., Conrad, D. W., Cuneo, M. J. and Hellinga, H. W. (2005). Orthogonal site-specific protein modification by engineering reversible thiol protection mechanisms. Protein Sci. *14*, 64–73.
[103] Butlin, N. G. and Meares, C. F. (2006). Antibodies with infinite affinity: Origins and applications. Acc. Chem. Res. *39*, 780–7.
[104] Carrico, I. S., Carlson, B. I. and Bertozzi, C. R. (2007). Introducing genetically encoded aldehydes into proteins. Nat. Chem. Biol. *3*, 321–2.
[105] Nonaka, H., Tsukiji, S., Ojida, A. and Hamachi, I. (2007). Non-enzymatic covalent protein labeling using a reactive tag. J. Am. Chem. Soc. *129*, 15777–9.
[106] Popp, M. W., Antos, J. M., Grotenbreg, G. M., Spooner, E. and Ploegh, H. L. (2007). Sortagging: A versatile method for protein labeling. Nat. Chem. Biol. *3*, 707–8.

[107] Tanaka, T., Yamamoto, T., Tsukiji, S. and Nagamune, T. (2008). Site-specific protein modification on living cells catalyzed by sortase. Chembiochem *9*, 802–7.

[108] Busch, G. K., Tate, E. W., Gaffney, P. R., Rosivatz, E., Woscholski, R. and Leatherbarrow, R. J. (2008). Specific N-terminal protein labelling: Use of FMDV 3C protease and native chemical ligation. Chem. Commun. 3369–71.

[109] Chattopadhaya, S., Abu Bakar, F. B., Srinivasan, R. and Yao, S. Q. (2008). In vivo imaging of a bacterial cell division protein using a protease-assisted small-molecule labeling approach. Chembiochem *9*, 677–80.

[110] Kobayashi, T., Shiratori, M., Nakano, H., Eguchi, C., Shirai, M., Naka, D. and Shibui, T. (2007). Short peptide tags increase the yield of C-terminally labeled protein. Biotechnol. Lett. *29*, 1065–73.

[111] Baskin, J. M., Prescher, J. A., Laughlin, S. T., Agard, N. J., Chang, P. V., Miller, I. A., Lo, A., Codelli, J. A. and Bertozzi, C. R. (2007). Copper-free click chemistry for dynamic in vivo imaging. Proc. Natl. Acad. Sci. U.S.A. *104*, 16793–97.

[112] Blum, G., von Degenfeld, G., Merchant, M. J., Blau, H. M. and Bogyo, M. (2007). Noninvasive optical imaging of cysteine protease activity using fluorescently quenched activity-based probes. Nat. Chem. Biol. *3*, 668–77.

[113] Bhunia, A. K. and Miller, S. C. (2007). Labeling tetracysteine-tagged proteins with a SplAsH of color: A modular approach to bis-arsenical fluorophores. Chembiochem *8*, 1642–5.

[114] Soh, N. (2008). Selective chemical labeling of proteins with small fluorescent molecules based on metal-chelation methodology. Sensors *8*, 1004–24.

[115] Cao, H., Chen, B., Squier, T. C. and Uljana Mayer, M. U. (2006). CrAsH: A biarsenical multi-use affinity probe with low non-specific fluorescence. Chem. Comm. 2601–3.

[116] Cao, H., Xiong, Y., Wang, T., Chen, B., Squier, T. C. and Mayer, M. U. (2007). A red Cy3-based biarsenical fluorescent probe targeted to a complementary binding peptide. J. Am. Chem. Soc. *129*, 8672–3.

[117] Luedtke, N. W., Dexter, R. J., Fried, D. B. and Schepartz, A. (2007). Surveying polypeptide and protein domain conformation and association with FlAsH and ReAsH. Nat. Chem. Biol. *3*, 779–84.

[118] Liu, B., Archer, C. T., Burdine, L., Gillette, T. G. and Kodadek, T. (2007). Label transfer chemistry for the characterization of protein–protein interactions. J. Am. Chem. Soc. *129*, 12348–9.

[119] Hearps, A. C., Pryor, M. J., Kuusisto, H. V., Rawlinson, S. M., Piller, S. C. and Jan, D. A. (2007). The biarsenical dye lumiotrade exhibits a reduced ability to specifically detect tetracysteine-containing proteins within live cells. J. Fluor. *17*, 593–7.

[120] Spagnuolo, C. C., Vermeij, R. J. and Jares-Erijman, E. A. (2006). Improved photostable FRET-competent biarsenical-tetracysteine probes based on fluorinated flluoresceins. J. Am. Chem. Soc. *128*, 12040–3041.

[121] Roberti, M. J., Bertoncini, C. W., Klement, R., Jares-Erijman, E. A. and Jovin, T. M. (2007). Fluorescence imaging of amyloid formation in living cells by a functional, tetracysteine-tagged α-synuclein. Nat. Methods *4*, 345–51.

[122] Maurel, D., Comps-Agrar, L., Brock, C., Rives, M. L., Bourrier, E., Ayoub, M. A., Bazin, H., Tinel, N., Durroux, T. Prezeau, L. et al. (2008). Cell-surface protein–protein interaction analysis with time-resolved FRET and snap-tag technologies: Application to GPCR oligomerization. Nat. Methods *5*, 561–7.

[123] Zhou, Z., Cironi, P., Lin, A. J., Xu, Y., Hrvatin, S., Golan, D. E., Silver, P. A., Walsh, C. T. and Yin, J. (2007). Genetically encoded short peptide tags for orthogonal protein labeling by Sfp and AcpS phosphopantetheinyl transferases. ACS Chem. Biol. *2*, 337–46.

[124] Wolfbeis, O. S. (2007). The click reaction in the luminescent probing of metal ions, and its implications on biolabeling techniques. Angew. Chem. Int. Ed. Engl. *46*, 2980–2.

[125] Hohng, S. and Ha, T. (2005). Single-molecule quantum-dot fluorescence resonance energy transfer. Chemphyschem *6*, 956–60.

[126] Roy, R., Hohng, S. and Ha, T. (2008). A practical guide to single-molecule FRET. Nat. Methods *5*, 507–16.

[127] Lidke, D. S., Lidke, K. A., Rieger, B., Jovin, T. M. and Arndt-Jovin, D. J. (2005). Reaching out for signals: Filopodia sense EGF and respond by directed retrograde transport of activated receptors. J. Cell Biol. *170*, 619–26.

[128] Susumu, K., Uyeda, H. T., Medintz, I. L., Pons, T., Delehanty, J. B. and Mattoussi, H. (2007). Enhancing the stability and biological functionalities of quantum dots via compact multifunctional ligands. J. Am. Chem. Soc. *129*, 13988–96.

[129] Dennis, A. M. and Bao, G. (2008). Quantum dot-fluorescent protein pairs as novel fluorescence resonance energy transfer probes. Nano Lett. *8*, 1439–45.

[130] Hering, V. R., Gibson, G., Schumacher, R. I., Faljoni-Alario, A. and Politi, M. J. (2007). Energy transfer between CdSe/ZnS core/shell quantum dots and fluorescent proteins. Bioconjug. Chem. *18*, 1705–8.
[131] McGrath, N. and Barroso, M. (2008). Quantum dots as fluorescence resonance energy transfer donors in cells. J. Biomed. Opt. *13*, 031210/1–031210/9.
[132] Baer, R. and Rabani, E. (2008). Theory of resonance energy transfer involving nanocrystals: The role of high multipoles. J. Chem. Phys. *128*, 184710.
[133] Roberti, M. J., Morgan, M., Pietrasanta, L., Jovin, T. M. and Jares-Erijman, E. A. (2008). Quantum dots as efficient triggers-sensors of α-synuclein amyloid aggregation in living cells. submitted.
[134] Giordano, L., Jovin, T. M., Irie, M. and Jares-Erijman, E. A. (2002). Diheteroarylethenes as thermally stable photoswitchable acceptors in photochromic fluorescence resonance energy transfer (pcFRET). J. Am. Chem. Soc. *124*, 7481–9.
[135] Fölling, J., Polyakova, S., Belov, V., van Blaaderen, A., Bossi, M. L. and Hell, S. W. (2008). Synthesis and characterization of photoswitchable fluorescent silica nanoparticles. Small *4*, 134–42.
[136] Sakata, T., Jackson, D. K., Mao, S. and Marriott, G. (2008). Optically switchable chelates: Optical control and sensing of metal ions. J. Org. Chem. *73*, 227–33.
[137] Soh, N., Yoshida, K., Nakajima, H., Nakano, K., Imato, T., Fukaminato, T. and Irie, M. (2007). A fluorescent photochromic compound for labeling biomolecules. Chem. Commun. 5206–8.
[138] Dedecker, P., Hotta, J.-I., Flors, C., Sliwa, M., Uji, I. H., Roeffaers, M. B. J., Ando, R., Mizuno, H., Miyawaki, A. and Hofkens, J. (2007). Subdiffraction imaging through the selective donut-mode depletion of thermally stable photoswitchable fluorophores: Numerical analysis and application to the fluorescent protein Dronpa. J. Am. Chem. Soc. *129*, 16132–41.
[139] Miskoski, S, Giordano, L, Etchehon, M. H., Menendez, G., Lidke, K. A., Hagen, G. M., Jovin, T. M. and Jares-Erijman, E. A. (2006). Spectroscopic modulation of multifunctionalized quantum dots for use as biological probes and effectors. Proc SPIE 6096, 60960X60961–60967.
[140] Jiao, G., Thoresen, L., Kim, T., Haaland, W., Gao, F., Topp, M., Hochstrasser, R., Metzker, M. and Burgess, K. (2006). Syntheses, photophysical properties, and application of through-bond energy-transfer cassettes for biotechnology. Chemistry *12*, 7816–26.

[141] Li, Y., Zhou, X. and Ye, D. (2008). Molecular beacons: An optimal multifunctional biological probe. Biochem. Biophys. Res. Commun. *373*, 457–61.

[142] Shlyahovsky, B., Weizmann, Y., Nowarski, R., Kotler, M. and Willner, I. (2006). Spotlighting of cocaine by autonomous aptamer-based machine. J. Am. Chem. Soc. *129*, 3814–5.

[143] Ellis-Davies, G. C. (2007). Caged compounds: Photorelease technology for control of cellular chemistry and physiology. Nat. Methods *4*, 619–28.

[144] Zheng, G., Guo, Y. M. and Li, W. H. (2007). Photoactivatable and water soluble FRET dyes with high uncaging cross section. J. Am. Chem. Soc. *129*, 10616–7.

[145] Skirtach, A. G. and Kreft, O. (2007). Stimuli-sensitive nanotechnology for drug delivery. In: Nanotechnology in Drug Delivery (de Villiers, M. M., Aramwit, P. and Kwon, G. S., eds.). pp. 117–130. Springer, Berlin.

[146] Hoffman, A. S. and Stayton, P. S. (2007). Conjugates of stimuli-responsive polymers and proteins. Prog. Polymer Sci. *32*, 922–32.

[147] Pelah, A. and Jovin, T. M. (2007). Polymeric actuators for biological aapplications. Chemphyschem *8*, 1757–60.

[148] Protasenko, V., Hull, K. L. and Kuno, M. (2005). Demonstration of a low-cost, single-molecule capable, multimode optical microscope. Chem. Educator *10*, 269–82.

[149] Hagen, G., Lidke, K., Rieger, B., Caarls, W., Arndt-Jovin, D. and Jovin, T. (2007). Dynamics of membrane receptors: Single molecule tracking of quantum dot liganded epidermal growth factor. In: Single Molecule Dynamics (Ishii, Y. and Yanagida, T., eds.). Wiley, Orlando.

Laboratory Techniques in Biochemistry and Molecular Biology, Volume 33
FRET and FLIM Techniques
T. W. J. Gadella (Editor)

Glossary

Explanation of symbols used in the text

D	donor
A	acceptor
I_D	fluorescence intensity of the donor (in absence of the acceptor)
I_{DA}	fluorescence intensity of the donor (in presence of the acceptor), also called I_{D-S}
I_A	fluorescence intensity of the acceptor (in absence of the donor)
I_{AD}	fluorescence intensity of the acceptor (in presence of the donor)
I_S	net sensitized emission of the acceptor ($=I_{AD} - I_A$).
k_t	energy transfer rate
k_f	fluorescence decay rate
k_{ic}	internal conversion (or nonradiative decay) rate
k_{isc}	inter system crossing (or triplet state conversion) rate
k_{bl}	photobleaching rate
Q_D	fluorescence quantum yield of the donor
Q_A	fluorescence quantum yield of the acceptor
τ_D	fluorescence lifetime of the donor (in absence of the acceptor)
τ_{DA}	fluorescence lifetime of the donor (in presence of the acceptor)
$\tau_{bl,D}$	bleach time constant of the donor (in absence of the acceptor)
$\tau_{bl,DA}$	bleach time constant of the donor (in presence of the acceptor)
λ_{ex}	excitation wavelength
$F(\lambda)$	corrected fluorescence emission spectrum
$q(\lambda)$	normalized (to unit area) fluorescence emission spectrum

ε_D absorbance of the donor
ε_A absorbance of the acceptor
$\varepsilon(\lambda)$ absorbance spectrum
J the overlap integral ($= \int q_D(\lambda)\varepsilon_A(\lambda)\lambda^4 d\lambda$)
κ^2 orientation factor
R_0 Förster radius
r_{DA} distance between donor and acceptor
E energy transfer efficiency
f_D fraction of donor molecules displaying FRET
E_D the apparent FRET efficiency based on the donor signal ($=f_D E$)
f_A fraction of acceptor molecules accepting FRET
E_A the apparent FRET efficiency based on the acceptor signal ($=f_A E$)

Index

A

B

C

D

E

F

G

H

I

J

K

L

M

N

O

P

Q

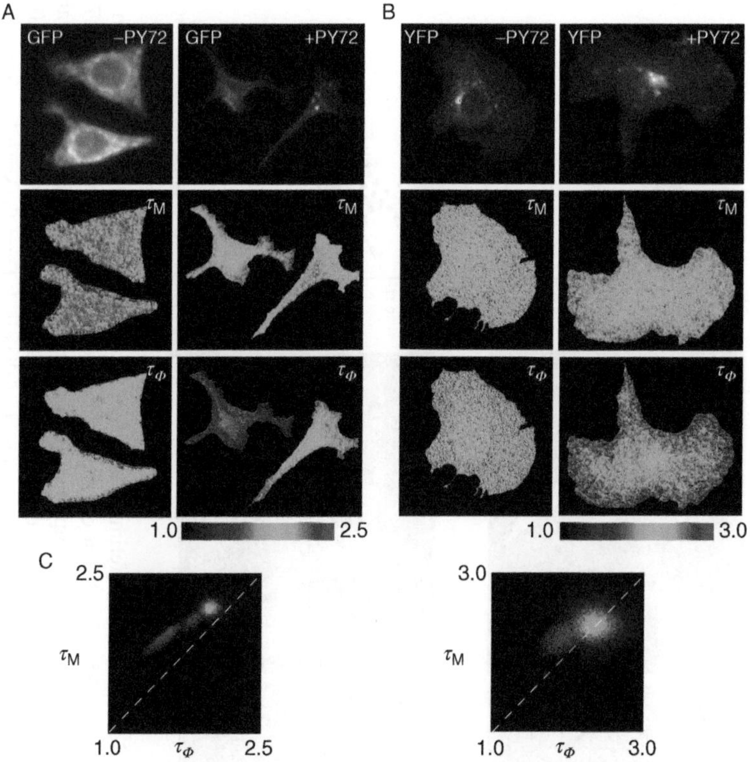

Fig. 2.4. (A) FLIM measurements of EGFR-GFP in the absence (left panels) and presence (right panels) of PY72-Cy3. (B) FLIM measurements of EGFR-YFP in the absence (left panels) and presence (right panels) of PY72-Cy3.5. The scaling of the pseudocolored lifetime images in panel (A) and (B) are indicated with the color bar in ns. Top panels: GFP intensity; Middle: modulation lifetimes; Bottom: phase lifetimes. (C) 2D histograms of τ_ϕ versus τ_M for EGFR-GFP (left) and EGFR-YFP (right). Red: samples incubated with PY72-Cy3 or PY72-Cy3.5, respectively; Green: control samples without PY72.

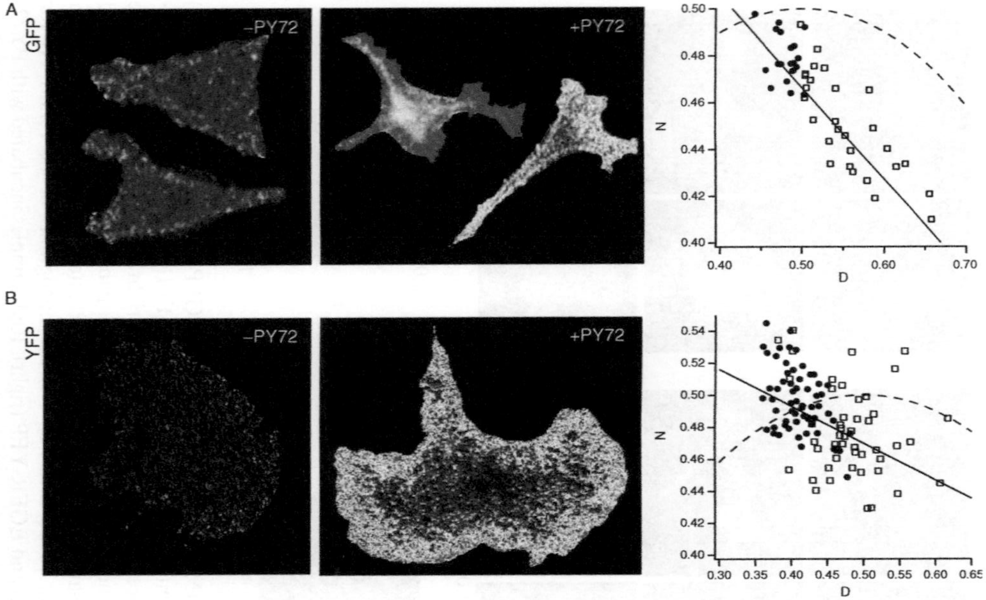

Fig. 2.5. (A) The relative concentrations of the short lifetime component, in samples expressing EGFR-GFP in the absence (left panel) and presence (middle panel) of PY72-Cy3. The calculated lifetimes values of the two species were 0.7 and 2.2 ns. (B) The relative concentrations of the short lifetime component, in samples expressing EGFR-YFP in the absence (left panel) and presence (middle panel) of PY72-Cy3.5. The calculated lifetimes values of the two species were 1.0 and 2.4 ns. The right panels show a plot of N_i versus D_i for a subset of the pixels from the samples shown.

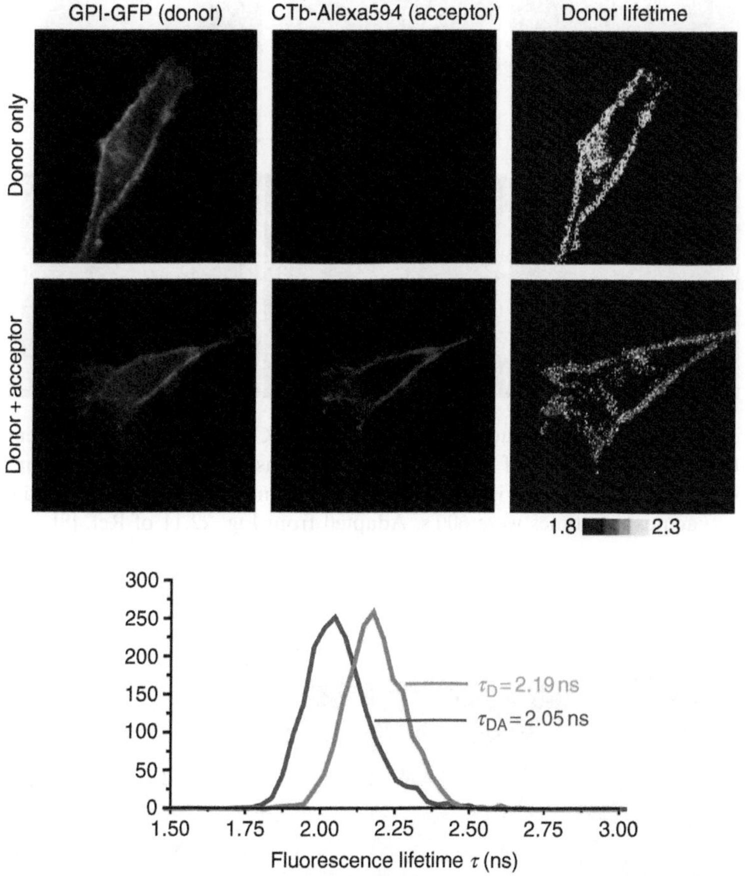

Fig. 3.11. FRET–FLIM experiment to study colocalization of two lipid raft markers, GPI-GFP and CTB-Alexa594. The rows of images show intensity and lifetime images of donor-labeled and donor + acceptor-labeled cells. The histogram shows the lifetime distribution of the whole cells. The FRET efficiency is ~6%.

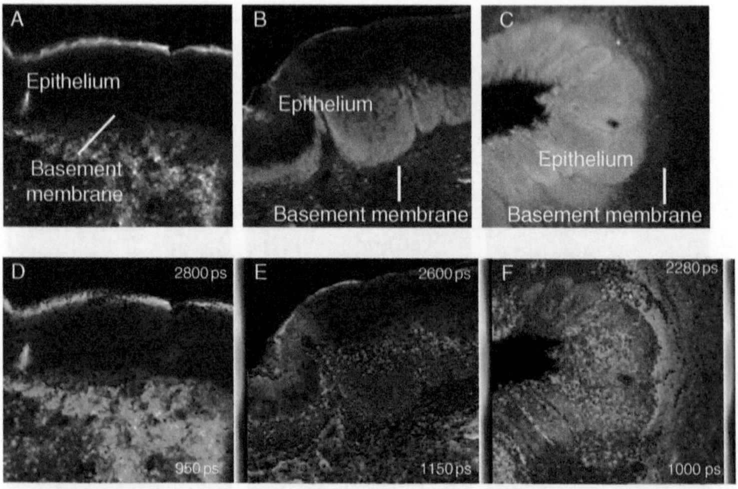

Fig. 4.1. Multiphoton fluorescence intensity (A–C) and TCSPC fluorescence lifetime images (D–F) of fresh unstained sections of human cervical biopsy excited at 740 nm and imaged between 385 and 600 nm. The individual acquisition times were 600 s. Adapted from Fig. 22.11 of Ref. [8].

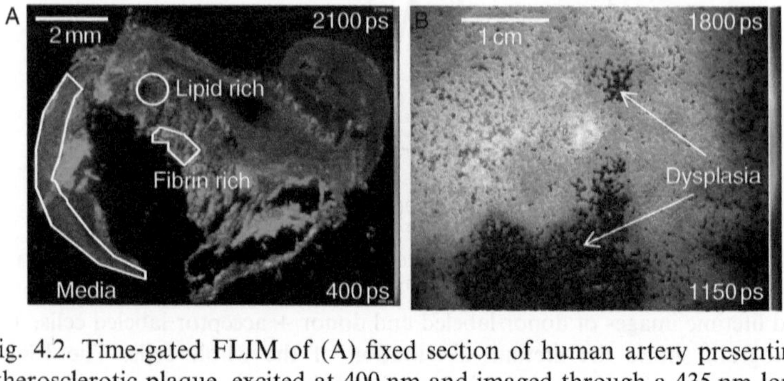

Fig. 4.2. Time-gated FLIM of (A) fixed section of human artery presenting atherosclerotic plaque, excited at 400 nm and imaged through a 435 nm long pass filter, and (B) freshly resected human colon tissue presenting dysplasia (precancerous state), excited at 355 nm and imaged through a 375 nm long pass filter.

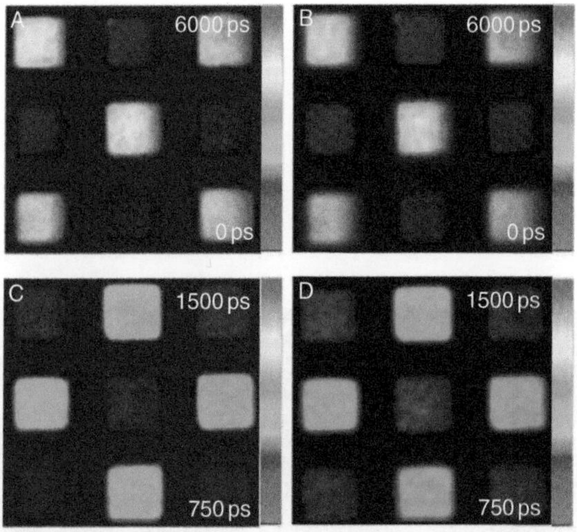

Fig. 4.3. FLIM of multiwell plate with alternate wells containing (A, B) Stilbene_1 and Coumarin_314 and (C, D) Stilbene_1 and Stilbene_3. Figures (A, C) were acquired using 8 time gates (400 ps gate width, ~10 s total acquisition time) followed by WNLLS fitting; figures (B, D) were calculated using RLD from two time-gated images (gate width 2.4 ns, acquired at 16 frames per second). All acquisition excited at 355 nm and imaged through (A, B) 435 nm and (C, D) 375 nm long pass filters respectively. Adapted from Fig. 2 of Ref. [18].

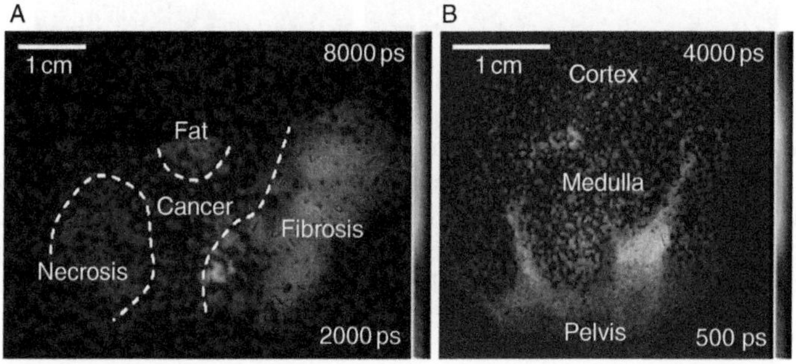

Fig. 4.4. RLD FLIM of (A) unstained freshly resected human pancreas imaged though a macroscope at 7.7 frames per second and (B) unstained sheep's kidney imaged through a rigid endoscope at 7 frames per second. Both samples illuminated at excitation wavelength of 355 nm and fluorescence imaged through a 375 nm long pass filter. Adapted from Fig. 3 of Ref. [18].

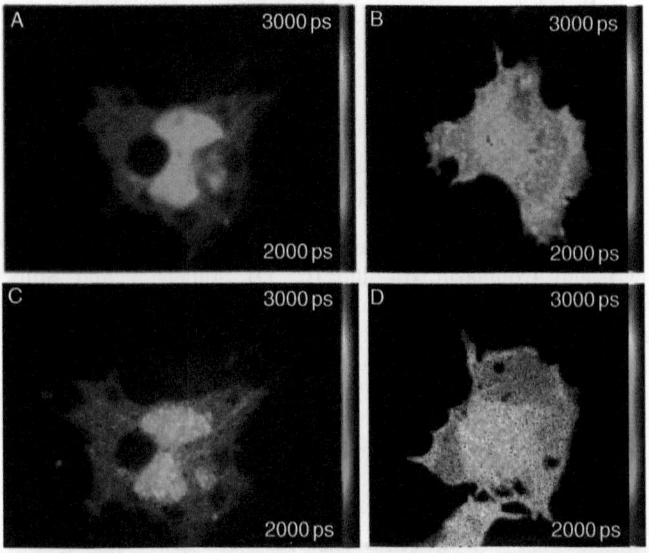

Fig. 4.5. FLIM images of live COS cells transfected with EGFP acquired in (A, B) 10 s and (C, D) 1 s using (A, C) wide-field time-gated Nipkow microscope (10 time-gates of 1 ns gate width, calculated using WNLLS fitting) and (B, D) TCSPC-enabled confocal microscope using 64 time bins with a maximum pixel count rate of 2 million photons per second. Excitation wavelength 488 nm, provided by a frequency doubled Ti:Sapphire laser, imaged through 500–550 nm emission band filter. Adapted from Fig. 2 of Ref. [33].

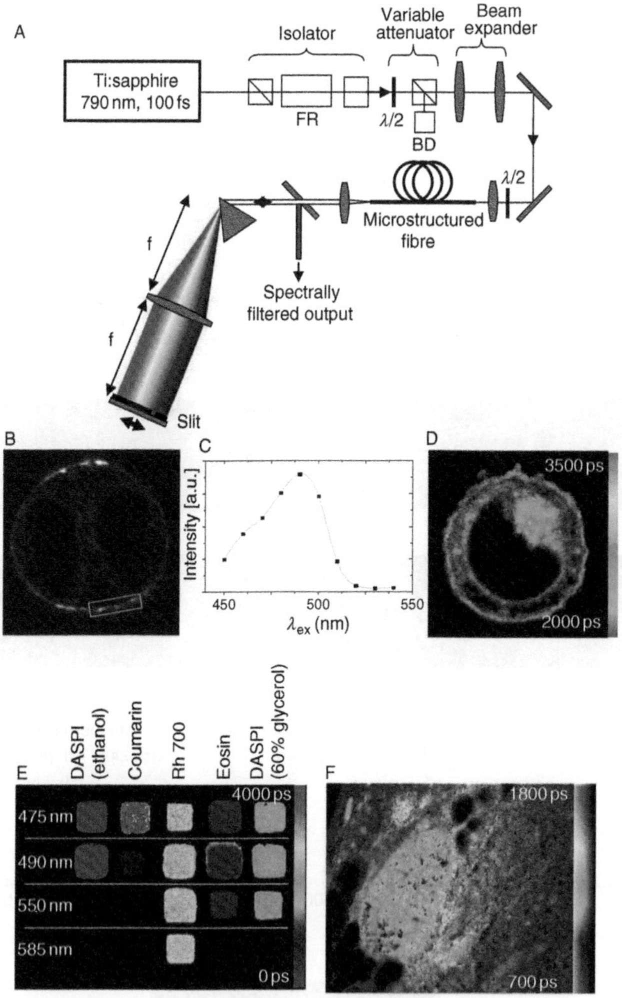

Fig. 4.6. (A) Schematic of femtosecond Ti:Sapphire laser-pumped tunable continuum source (TCS); (B) fluorescence intensity image; and (C) in situ excitation spectrum of a GPI–GFP transfected human B-cell imaged using a confocal microscope (from region in box) using long pass filter with cut-off 20 nm beyond excitation wavelength. (D) FLIM image of a 721.221 cell transfected with HLA-Cw6 linked to CFP, acquired on a TCSPC-enabled confocal microscope excited using the TCS; (E) wide-field time-gated fluorescence lifetime images of a multiwell plate sample array acquired at different TCS excitation wavelengths through long pass filters (shown on left), and (F) a wide-field fluorescence lifetime image of a fixed unstained section of human pancreas excited at 440–450 nm using the TCS and imaged beyond 470 nm. Adapted from Figs. 1 and 3–6 of Ref. [35].

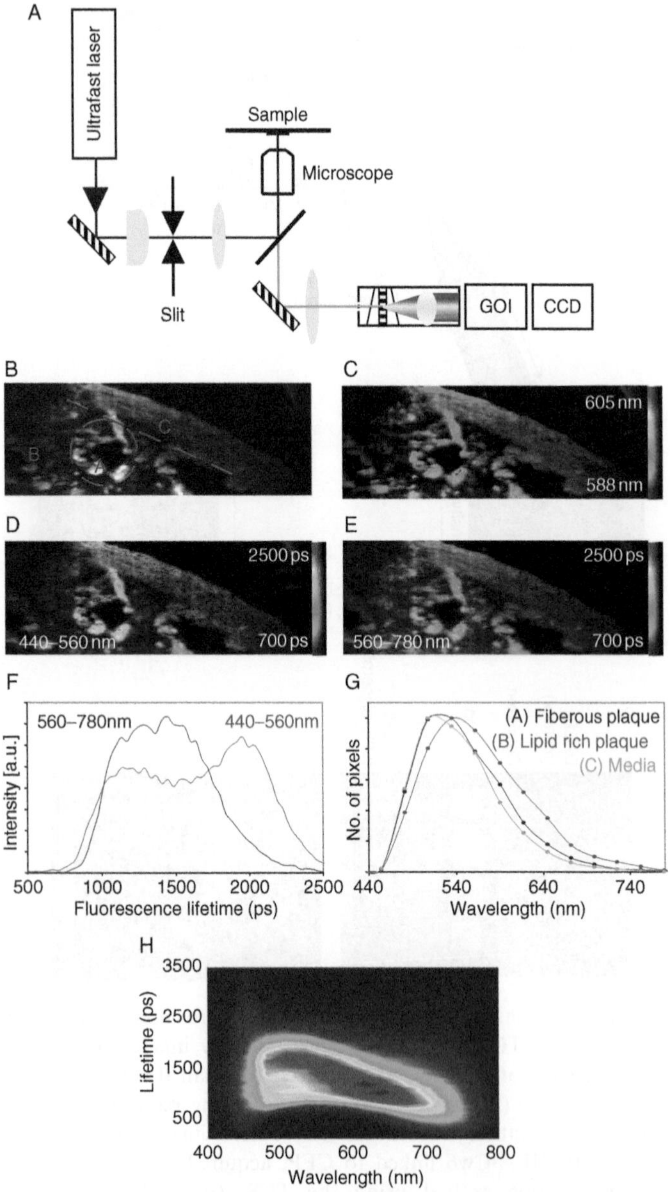

Fig. 4.7. Application of hyperspectral FLIM to an unstained fixed section of human artery excited at 400 nm: (A) schematic of a line-scanning hyperspectral FLIM microscope; (B) integrated intensity image marked with regions of A-fiberous plaque, B-lipid rich plaque and C-media; (C) central wavelength image; (D, E) fluorescence lifetime images calculated by integrating over

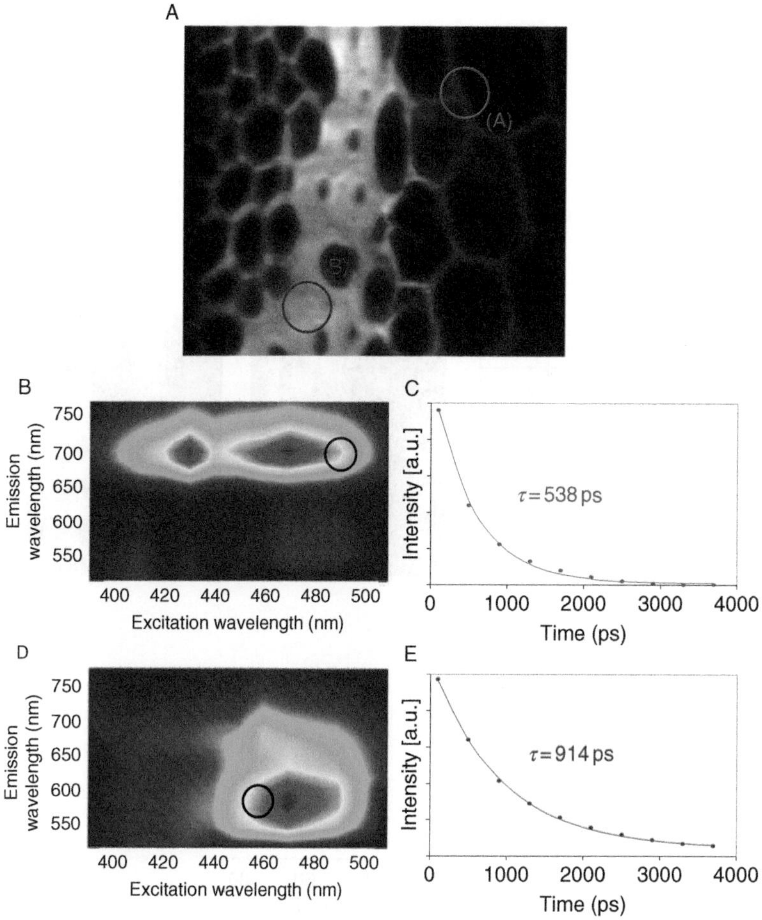

Fig. 4.8. Fluorescence lifetime of a stained section of Convallaria resolved with respect to lifetime, excitation and emission wavelength: (A) intensity image integrated over the time-resolved excitation-emission matrix (EEM); (B, D) time-integrated EEM from areas A and B respectively in (A); (C) fluorescence decay profile for λ_{ex}~490 nm and λ_{em}~700 nm corresponding to area A; (E) fluorescence decay profile for λ_{ex}~460 nm and λ_{em}~570 nm corresponding to area B.

---

emission spectral windows 440-560 and 560-780 nm respectively; (F) fluorescence lifetime histograms comparing these two spectrally integrated FLIM images, (G) time-integrated emission spectra for areas A and B in intensity image and (H) 2-D histogram of fluorescence lifetime and emission wavelength for the whole data set. Adapted from Fig. 3 of Ref. [56].

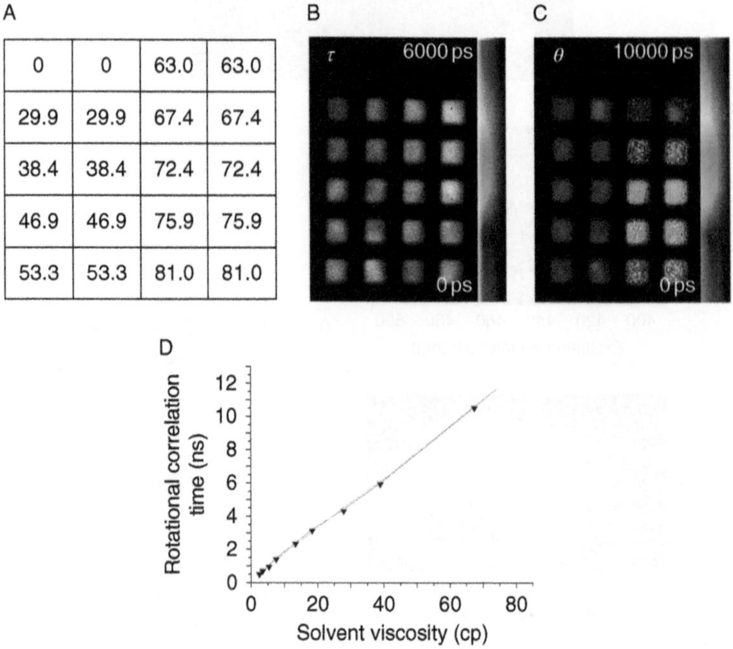

Fig. 4.10. (A) Schematic of percentage weights of glycerol in composite solvents corresponding to array of fluorescein solutions of varying viscosity; (B) fluorescence lifetime; (C) rotational correlation time images of this array; and (D) plot of the rotational correlation time as a function of viscosity for this sample array exited at 470 nm: the straight line fit yields a fluorophore radius of 0.54 nm for fluorescein. Adapted from Fig. 2 of Ref. [64].

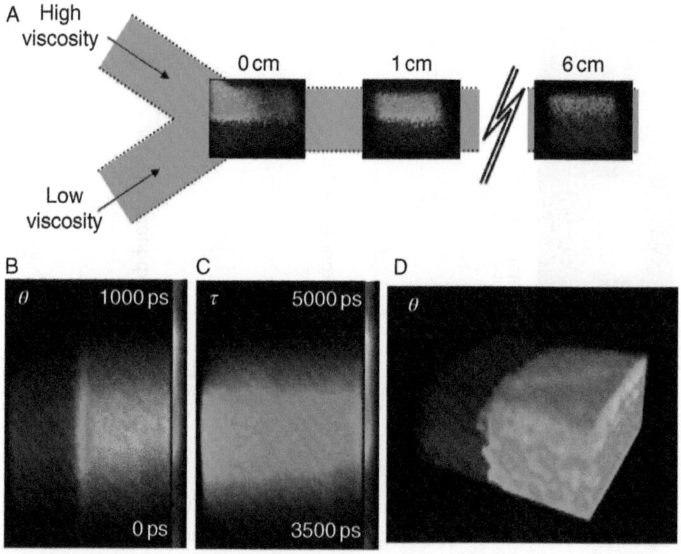

Fig. 4.11. (A) Schematic of microfluidic Y-junction mixer of 50 μm channel width with inset rotational correlation time images (40× magnification) recorded at 0, 1, and 6 cm after the junction, together with 60× magnification images of (B) rotational correlation time, (C) fluorescence lifetime, and (D) 3-D rendered image of rotational correlation time recorded immediately after the junction, excited 780 nm. Adapted from Figs. 5 and 6 [20].

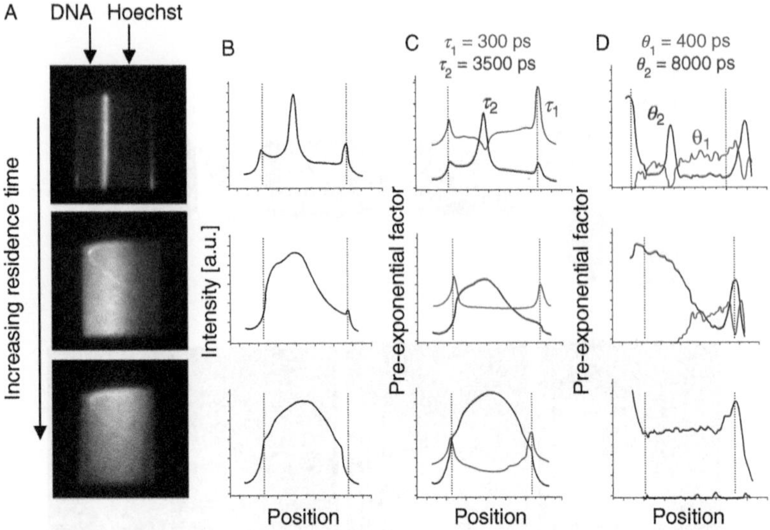

Fig. 4.12. Polarization resolved imaging, excited at 780 nm, applied to ligand binding in a microfluidic device using the fluorescent molecule Hoechst 332258 and a nonfluorescing DNA plasmid (5.8 kbp): (A) integrated intensity images (residence times of 300 ms, 7.2 s, and $t \to \infty$ from top to bottom) and cross-sectional plots (vertically averaged over image) of (B) intensity, (C) preexponential factors resulting from a double exponential fit (lifetimes fixed to 300 and 3500 ps after global analysis) to the decay of fluorescence intensity, and (D) preexponential factors resulting from a double exponential fit (rotational correlation times fixed to 500 and 10,000 ps after global analysis) to decay of the fluorescence anisotropy. The dashed lines show the location of the channel walls. Adapted from Figs. 2–4 of Refs. [67].

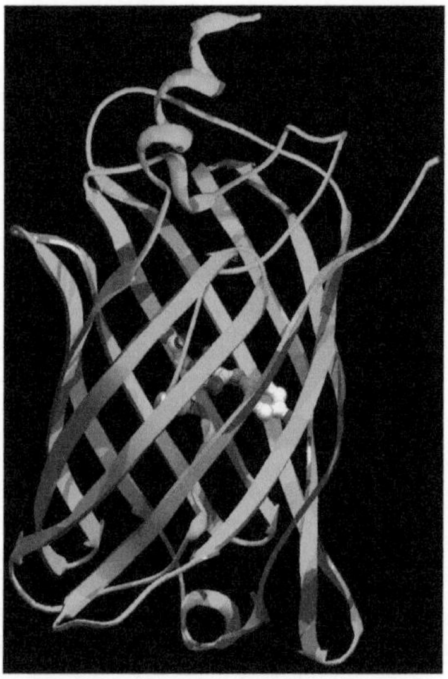

Fig. 5.1. Ribbon diagram of a fluorescent protein (citrine, PDB entry 1HUI) crystal structure. The chromophore is buried in the protein's interior and shown in balls and sticks representation.

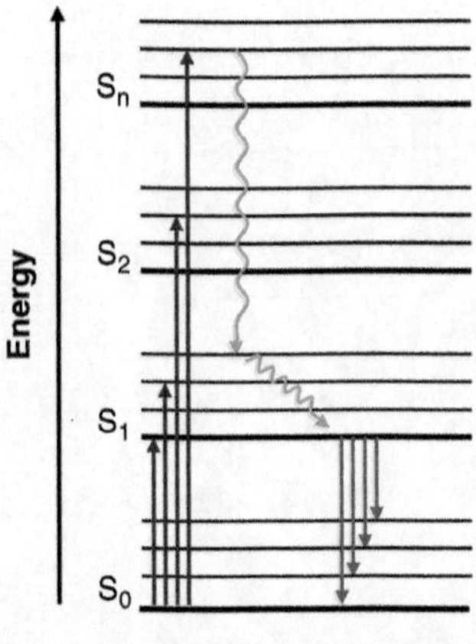

Fig. 6.1. Jablonski diagram, representing electron energy levels of fluorophores and transitions after photon excitation. S = electronic state, different lines within each state represent different vibrational levels. Blue arrows represent absorption events, green arrows depict internal conversion or heat dissipation, and orange arrows indicate fluorescence emission. Intersystem crossing into triplet states has been omitted for simplicity (see also Chaps. 1 and 12).

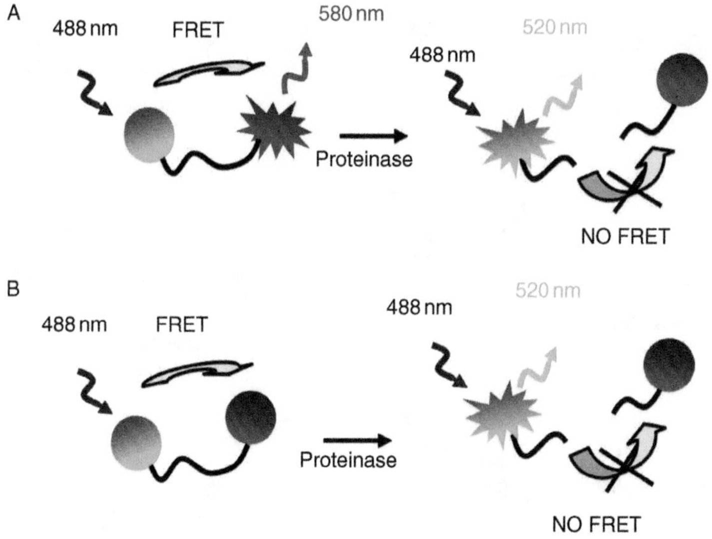

Fig. 6.11. Two types of FRET probes. (A) Ratiometric probes are formed by two fluorescent molecules that allow determination of emission ratio. (B) Quenched probes feature a donor fluorophore and a quencher. The emission increase of the donor after release of the acceptor is detected. Both types are frequently used to build proteinases probes.

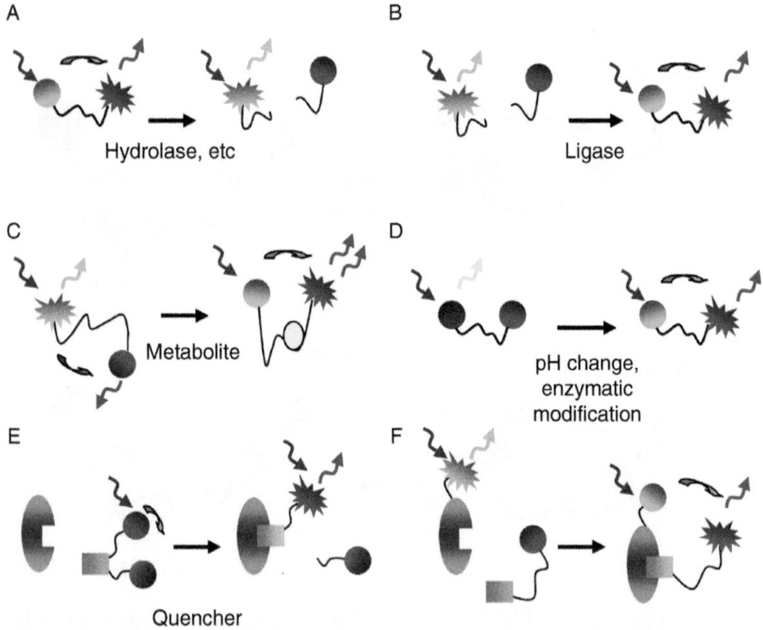

Fig. 6.13. Different designs of FRET sensors. (A) Substrates for hydrolytic enzymes. (B) Sensors for bond formation. (C) Sensors based on conformational or structural change. (D) Environmentally sensitive probes. (E) Quenched activity-based probe to monitor small molecule–enzyme interaction. (F) Small molecule–enzyme interaction using a labeled protein.

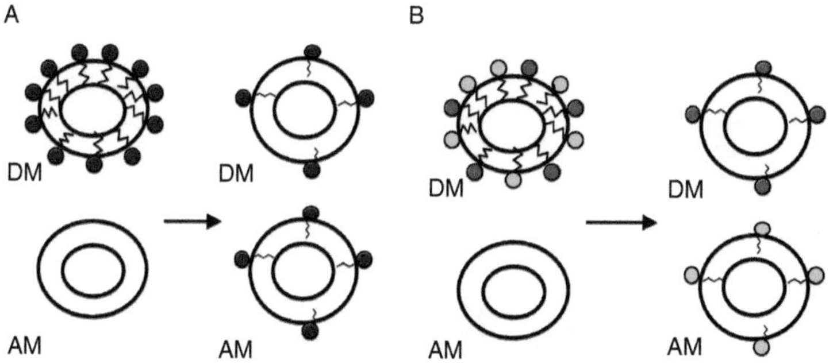

Fig. 6.20. Types of assays to visualize lipid trafficking in membranes. (A) Self-quenching method. Here the self-quenching is released upon transfer to unlabeled acceptor membranes that are usually in large excess. (B) FRET assays. Here the donor membrane contains a transferable lipid (green) that is quenched by FRET to a non transferable acceptor lipid (red). Upon transfer to an unlabeled acceptor membrane the green-labeled lipid becomes unquenched. DM = donor membrane, AM = acceptor membrane.

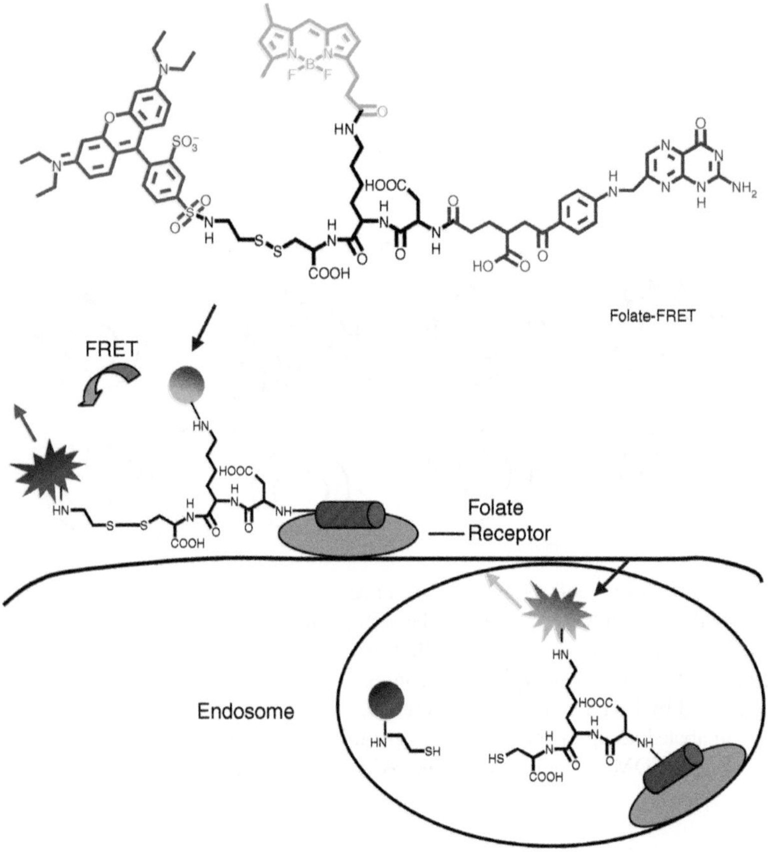

Fig. 6.22. Folate-FRET sensor structure and its application to measure disulfide bond reduction in endosomes. The molecule contains the folate moiety which is recognized by the folate receptor situated at the plasma membrane. This recognition leads to endocytosis and after some time to cleavage of the probe. The latter is monitored by a loss in FRET.

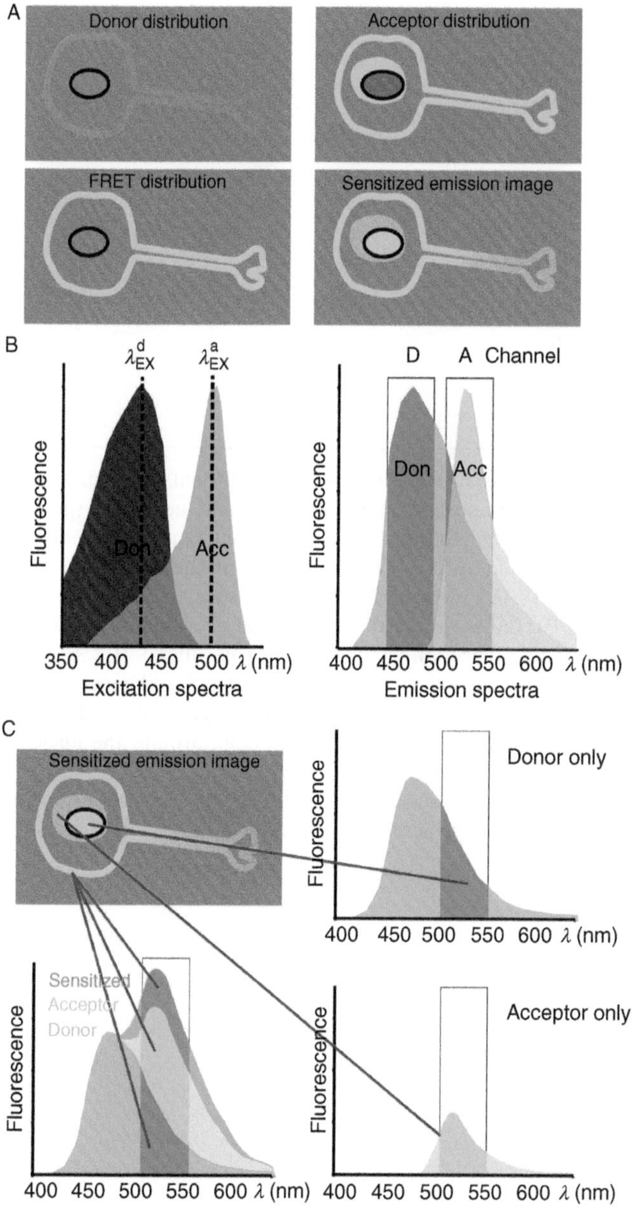
A
Donor distribution
Acceptor distribution
FRET distribution
Sensitized emission image
B
λ_{EX}^{d}
λ_{EX}^{a}
Fluorescence
Don
Acc
350 400 450 500 λ (nm)
Excitation spectra
D A Channel
Don
Acc
400 450 500 550 600 λ (nm)
Emission spectra
C
Sensitized emission image
Donor only
Sensitized
Acceptor
Donor
Acceptor only
400 450 500 550 600 λ (nm)

Fig. 7.1. *Detecting sensitized emission.* (A) Neuronal cell expressing donors (blue) at the membrane and in the nucleus, and acceptors (green) at the membrane and around the nucleus. Lower abundance of the donor in the axon is also depicted. FRET can only occur at the membrane of this cell (lower left panel). (B) Normalized excitation spectra of donors (CFP) and acceptors (YFP). Indicated are the donor- and acceptor excitation lines (left panel) and the two detection channels (bandpass filters). (C) Appearance of signals in the s.e. image. Whereas FRET is restricted to the membrane, due to leak-through of donor signal in the s.e. channel (e.g., in the nucleus; rightmost spectrum) and false excitation of acceptors (e.g., around the nucleus; lower right spectrum and figure B) additional signals are apparent. Note that leak-through and cross-excitation are not restricted to areas stained with either donors or acceptors alone (lower left panel).

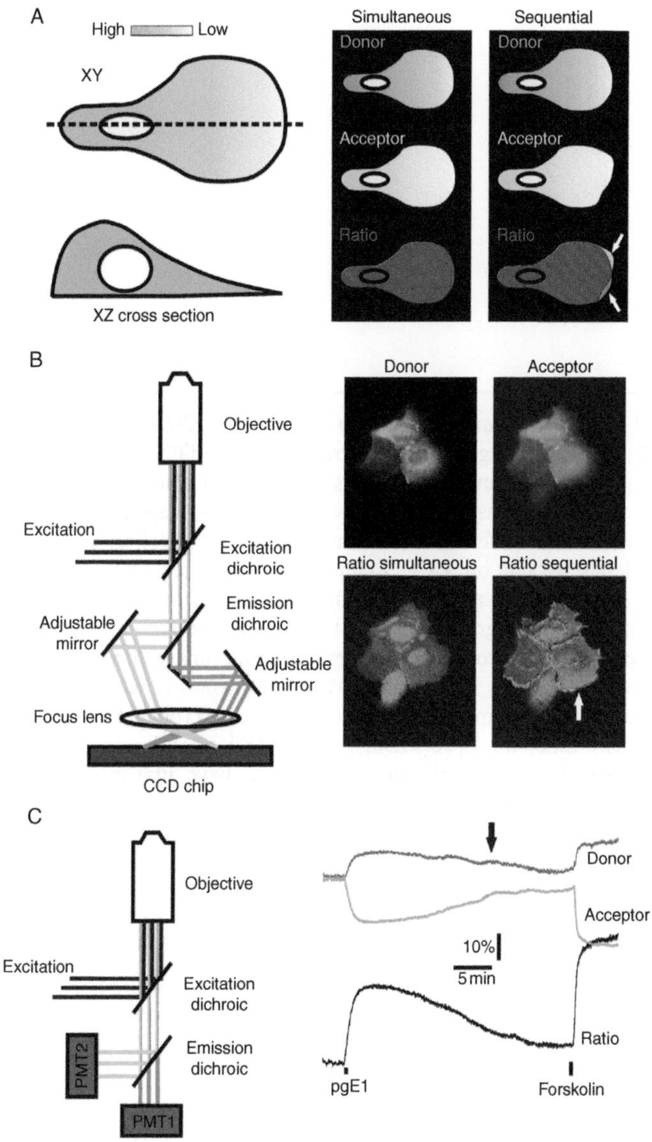

A
High
Low
XY
XZ cross section
Simultaneous
Sequential
Donor
Acceptor
Ratio
B
Objective
Excitation
Excitation dichroic
Emission dichroic
Adjustable mirror
Adjustable mirror
Focus lens
CCD chip
Donor
Acceptor
Ratio simultaneous
Ratio sequential
C
Objective
Excitation
Excitation dichroic
Emission dichroic
PMT2
PMT1
Donor
Acceptor
10%
5 min
Ratio
pgE1
Forskolin

Fig. 7.T1. *Emission ratio imaging of the cAMP FRET sensor CFP-Epac-YFP* [8]. (A) Ratio imaging largely corrects intensity differences that are due to cell morphology. However, if channels are collected consecutively, any shape changes of the cells cause errors (arrows) in the ratio image that may easily be mistaken for FRET differences. (B). Simultaneous image collection. **D** and **S** are projected side by side on a CCD chip using a commercially available image splitting device fitted to a widefield epifluorescence microscope. The ratio image (lower left photomicrograph) is calculated using Image J; no attempt was made to correct for bleed-through. The lower right ratio image shows errors due to cell movement (images taken 5 s apart) (C) Emission ratioing using a beam splitter and dual photometer setup.Traces represent **D** (blue), **S** (green), and the ratio (black). Since the Epac sensor looses FRET upon binding to cAMP, in this case the ratio was calculated as **D/S** to have upward ratio changes correspond with increased [cAMP]. Note that excitation fluctuations (arrow) disappear in the ratio. All data courtesy of B. Ponsioen.

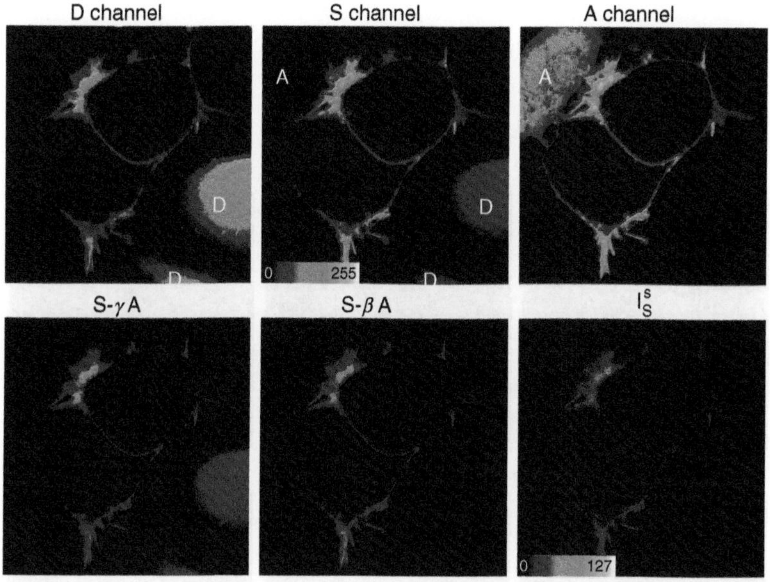

Fig. 7.2. *Sensitized emission calculated from confocal images.* Cells expressing CFP- and YFP-tagged Pleckstrin homology (PH) domains were seeded together with control cells expressing either CFP (marked "D") or YFP (marked "B"). Top row shows raw input files and illustrates donor leak-through (middle panel, "D") and cross-excitation (middle panel, "A"). In the bottom row, **S** images are corrected for cross-excited YFP (left), for CFP leak-through (middle) or according to Eq. (7.8) (right panel). The contrast of the I_S^s panel is stretched twofold as indicated. All images in this chapter are collected with Leica TCS SP2 or SP5 confocal microscopes except for Fig. 7.T1, which was acquired with a Leica ASMDW wide-field epifluorescence microscope equipped with dual-view adapter (Optical Insights). Image acquisition and specimen refocusing were automated from within a custom-made Visual Basic (v6.0) program by calling commands from the Leica macro tool package. ROIs were manually assigned to cells expressing only CFP or YFP and from these, correction factors were measured and calculated. Using these factors, sensitized emission was calculated as outlined in the text.

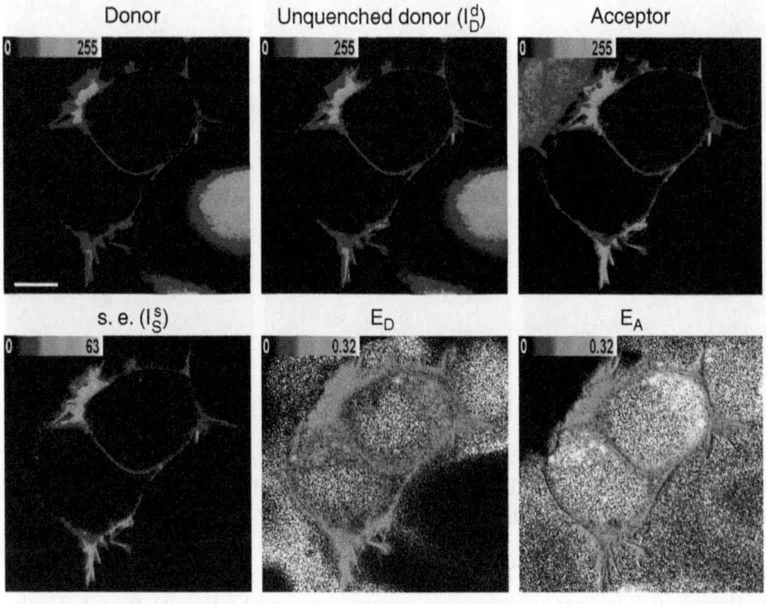

Fig. 7.3. Fret efficiency. The unquenched donor image (top row, middle panel), as calculated according to Eq. (7.13), and the acceptor image (top right panel) are used to normalize the s.e. image. The resulting images E_D and E_A (Eqs. (7.10) and (7.11), respectively) are quantitative, as detailed in the text. Unfiltered, raw data are shown. Scale bar is 12 μm.

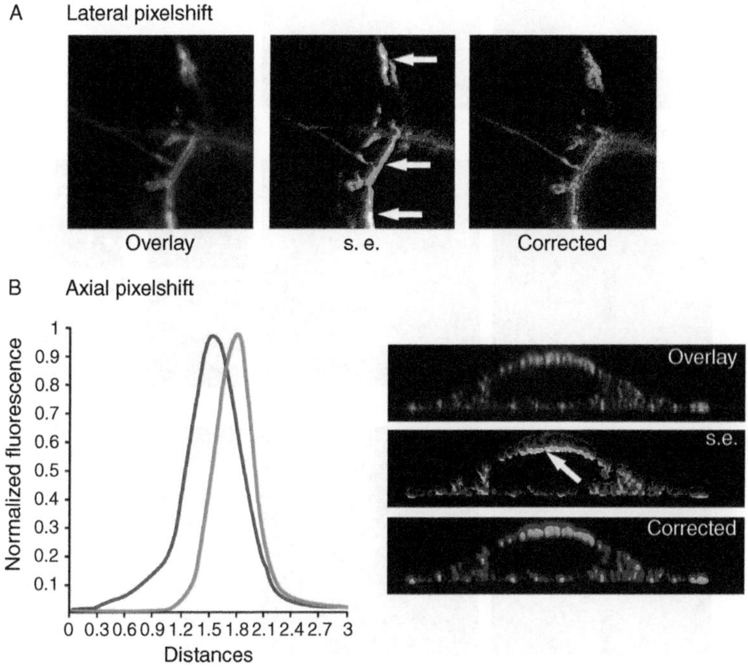

Fig. 7.5. *Effects of poor co-registration on calculated FRET images.* (A) Typical artifacts due to improper alignment (left) of raw input images caused by switching between unmatched filter cubes. The consistent appearance of high FRET values at the right side of bright structures (middle) is a sure indication to check image alignment. Right panel, proper alignment of the images corrects FRET artifacts. (B) Left panel, profiles of fluorescence intensities in a confocal X/Z image of the green emission (525 nm) of a 0.17-microm bead was registered using a HCX PL APO CS 63× objective upon 430-nm (blue line) and 514-nm (red line) excitation. Both scans use detection at 525 nm, demonstrating the extent of axial offset. Right panels, confocal images were acquired from a cell expressing CFP- and YFP-tagged membrane anchors. Top image, green–red overlay illustrates axial offset. Erroneous values (middle image) in the calculated I_S^s (s.e.) image are effectively corrected by using the refocusing macro routine (lower image). Shown are extreme examples.

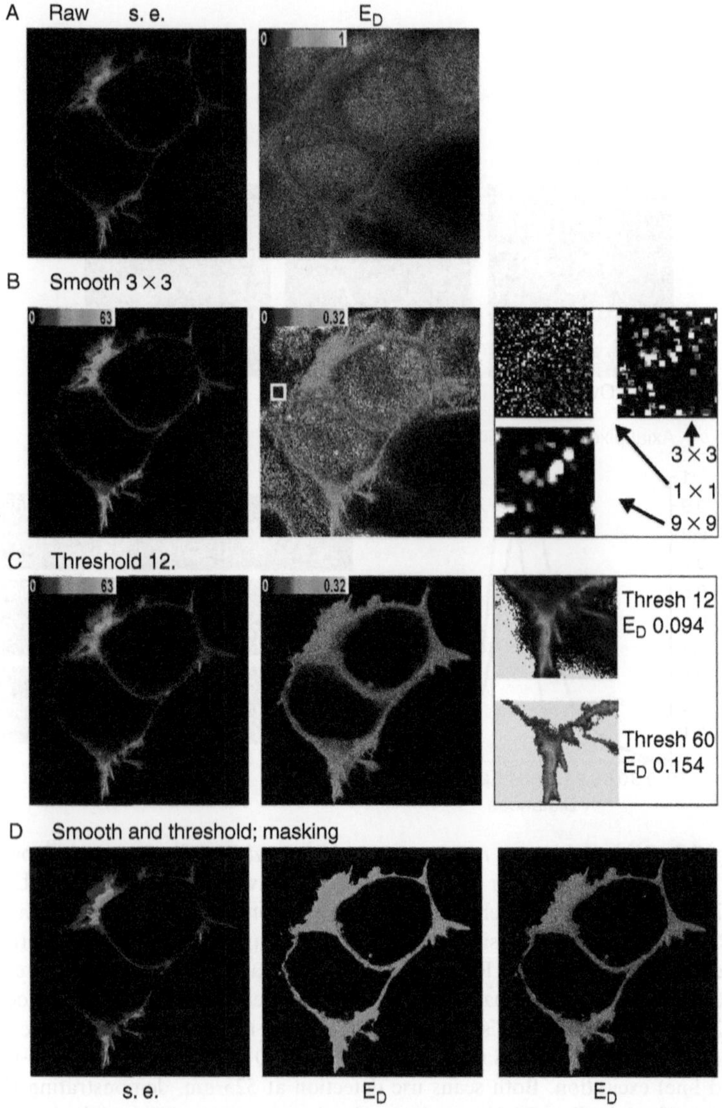

Fig. 7.8. *Postacquisition improvements.* (A) Unaltered "raw" 1024 × 1024 confocal s.e. and E_D images. Note the appearance of excessive noise in E_D at low-signal locations. (B) Lateral averaging (smoothing) using a 3 × 3 kernel cleans up the s.e. image but is less effective on the E_D image. Note the difference in Look-Up Table. Right panel: detail of the middle panel (yellow box) showing how smoothing with the indicated kernel sizes influences E_D. (C) Application of lower threshold on the input images rejects any pixels for which either donor or acceptor intensity is below 12 gray levels.

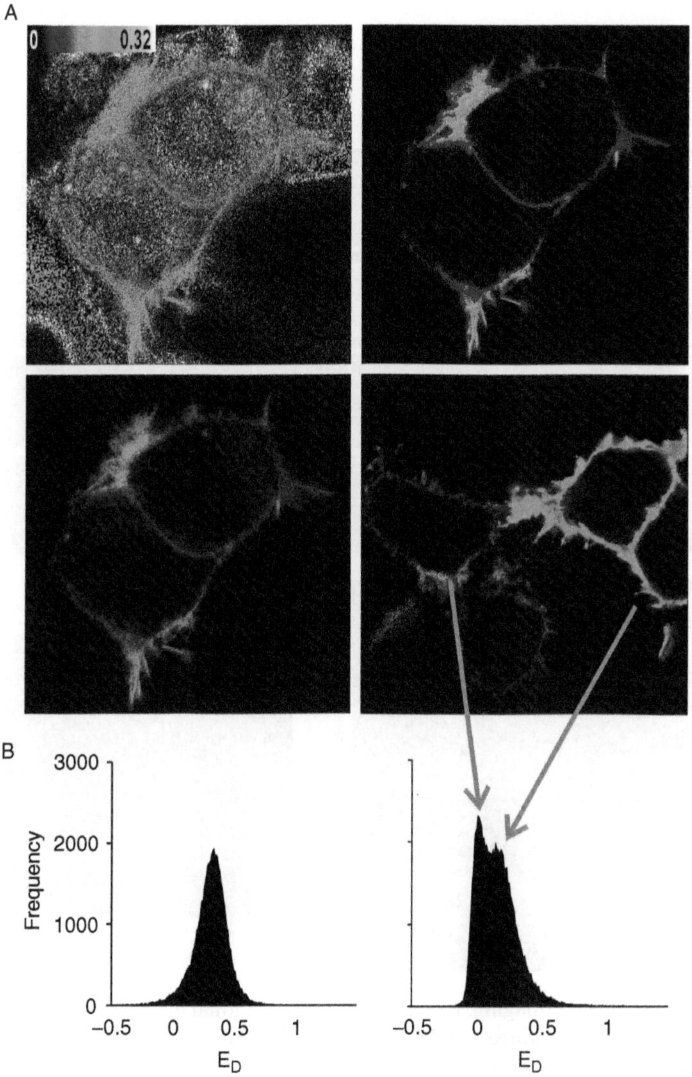

Fig. 7.9. *Further FRET efficiency analysis.* (A) Unbiased display of FRET efficiency. The E_D image (upper left panel) is modulated with an intensity

---

Right panel: the detail taken from the lower portion of the middle panel demonstrates that thresholding significantly influences calculated E_D values. Yellow pixels are excluded from the calculations. (D) A combination of smoothing and thresholding with settings as in B and C cleans up the s.e. and E_D images. Right panel, a mask, derived from the smoothed and thresholded E_D image, is applied to the unfiltered E_D data to preserve fine details in the FRET image.

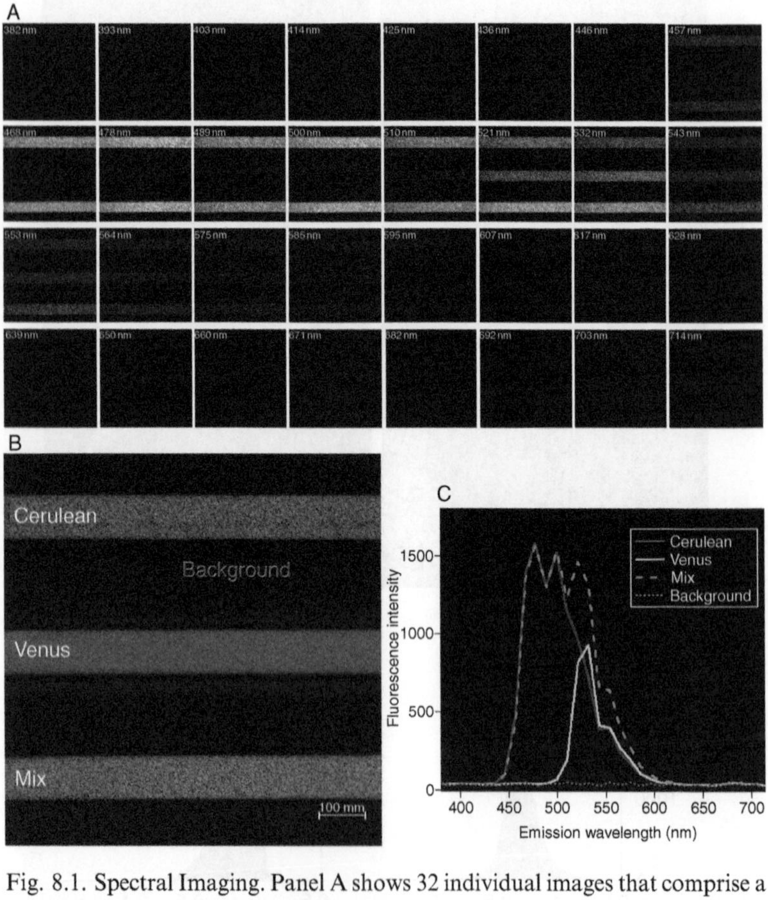

Fig. 8.1. Spectral Imaging. Panel A shows 32 individual images that comprise a spectral image of three capillaries containing from top to bottom 10 μM Cerulean, 10 μM Venus, and a mixture of Cerulean and Venus of unknown

---

picture (in this case, s.e., upper right panel) to yield the lower left image. See text for further details. Lower right panel, example with several cells displaying different E_D. (B) Frequency histograms of E_D in the pictures in the lower panels in A. Note that whereas the distribution on the left hand displays merely stochastic noise (compare Fig. 7.7C), the rightmost histogram reveals the heterogeneous FRET efficiency in the cells of the corresponding image.

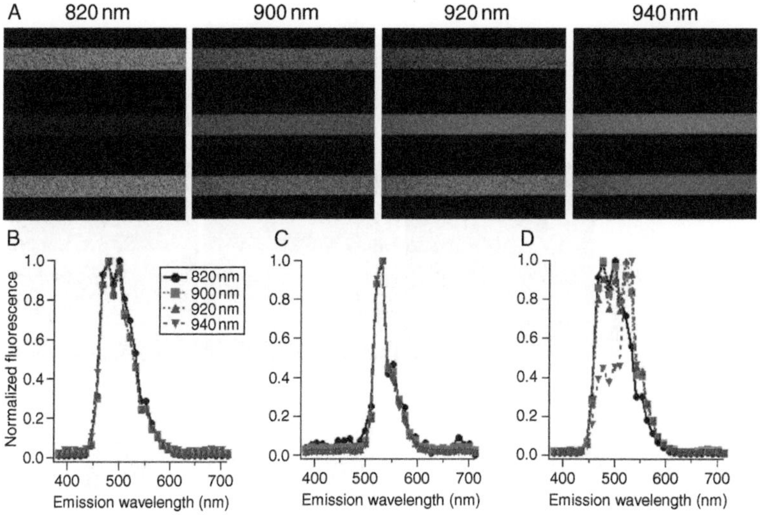

Fig. 8.2. Spectral Images can change with excitation wavelength. Spectral images of the same three capillaries as depicted in Fig. 8.1 were imaged at four different excitation wavelengths: 820, 900, 920, and 940 nm (A). Normalized emission spectra (at all four excitation wavelengths) from ROI's centered over the Cerulean capillary (B), the Venus capillary (C), and the mixture capillary (D) are co-plotted. Note that the color and normalized emission spectra do not change for a sample containing a single fluorophore. In contrast the color and normalized spectrum changed as a function of excitation wavelength in the capillary containing both fluorophores.

---

stoichiometry acquired on a Zeiss 510 META/NLO laser scanning confocal microscope with 900 nm two-photon excitation. Each image was acquired from a single node of a 32 node photomultiplier array, and represents sequential ~10 nm portion of the emission spectrum that are detected on each node after being dispersed by a diffraction grating. The emission wavelength measured by each node is indicated in yellow. Panel B shows a color-coded representation of the images shown in panel A. The color (wavelength) is the intensity weighted average of the 32 individual emission wavelengths, and the brightness is proportional to the total number of photons detected at each pixel. Each capillary is labeled as is a background region. Size bar is 100 μm. Panel C shows emission spectra calculated measured from ROI centered over the Cerulean capillary, Venus capillary, mixture capillary or from a background region.

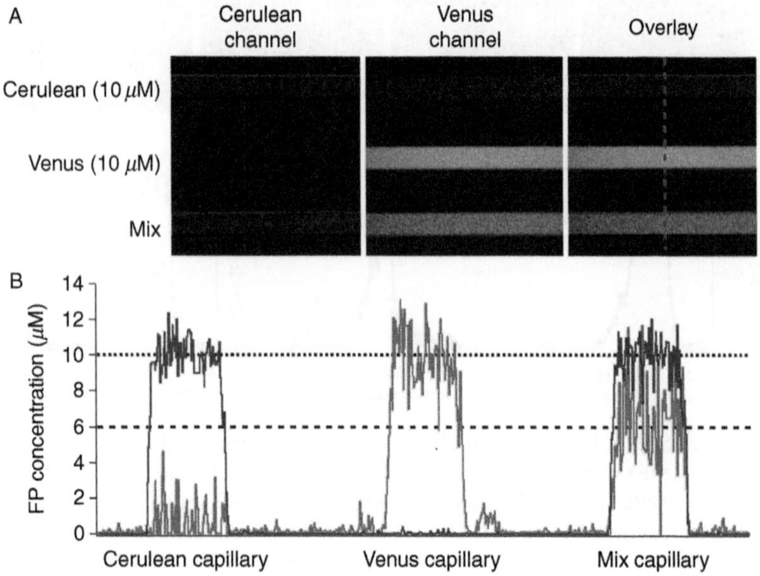

Fig. 8.4. Linear unmixing with curve fitting algorithms. Linear unmixing image processing software utilize least-square curve fitting routines to fit a spectra from each pixel of a spectral image to the linear unmixing equation and predict values for the abundance factors for each fluorophore. These abundance factors for Cerulean and Venus are then multiplied by the concentration of the individual reference spectrum samples (10 μM) to produce a Cerulean Channel image (blue), a Venus Channel image (yellow), and an overlay image (panel A). Now it can be seen that the top and bottom capillaries have the same concentration of Cerulean. The middle and bottom capillaries both have Venus, but at different concentrations. The dashed red line indicates the location of a line scan across the two image channels. Line scan plots for each channel are useful for measuring the actual concentration of fluorophores observed in a sample, and are plotted in panel B.

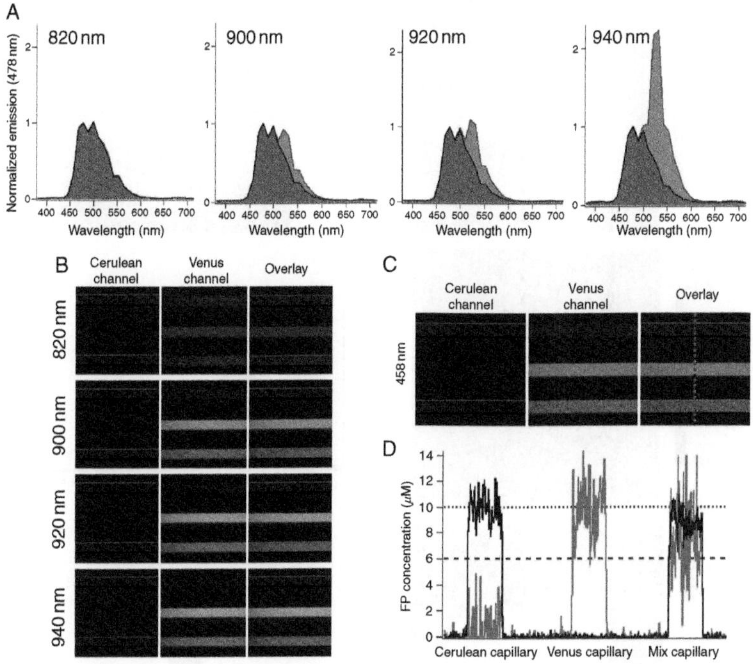

Fig. 8.5. The effects of excitation wavelength on linear unmixing. In panel A we see the Cerulean normalized emission spectra of the mixture capillary (red) overlaid with the normalized emission spectrum of Cerulean alone (blue) at the four different two-photon excitation wavelengths used in Fig. 8.2. Note that with 820 nm excitation the mix spectrum was not significantly different than the Cerulean spectrum, while with 900, 920, and 940 nm excitation they were different. When the four spectral images depicted in Fig. 8.2 were processed by linear unmixing they produced the four unmixing channel-sets depicted in panel B. The unmixed images with 900, 920, and 940 nm excitation were all identical, while with 820 nm excitation erroneous results were obtained. In panels C the results of linear unmixing of a spectral image of the same samples, obtained with one-photon excitation at 458 nm is shown. These images looked identical to the images obtained with 900, 920, and 940 nm two-photon excitation, and a line scan across the one-photon image confirmed this conclusion (compare panel D with Fig. 8.4B).

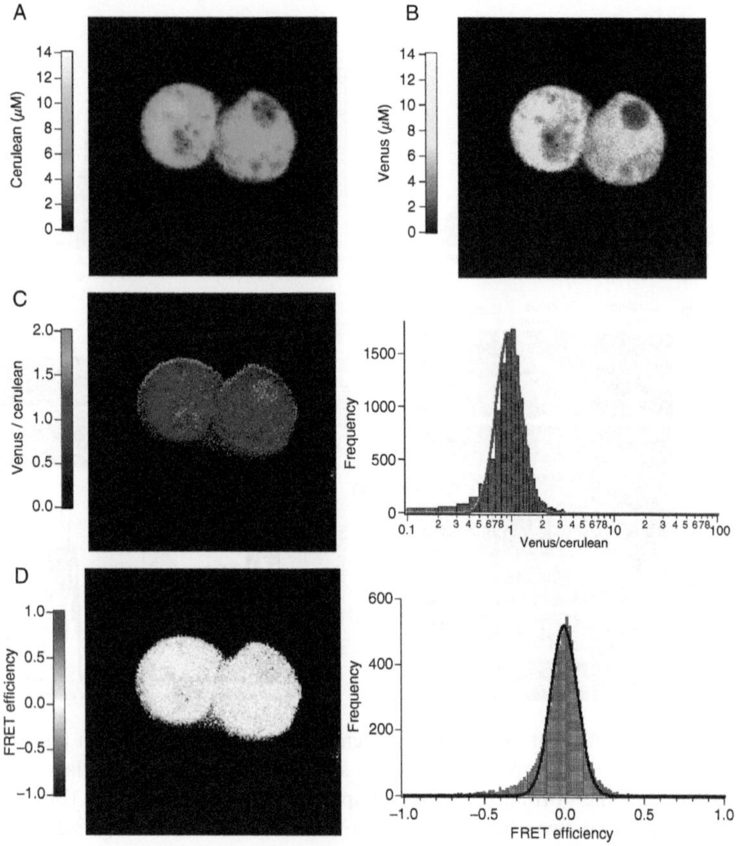

Fig. 8.6. sRET analysis of CTV, a Cerulean-Venus construct with a low FRET efficiency. sRET analysis is based on linear unmixing of two spectral images obtained at two different excitation. Spectral images of cells expressing the CTV construct were acquired with 890 and 940 nm excitation. These spectral images, and their matching reference spectra were processed using the sRET algorithm to produce (A) a Cerulean concentration image, (B) a Venus concentration Image, (C) a Venus/Cerulean ratio image, and (D) a FRET-efficiency image. The graphs in panels C and D show frequency histograms of the pixel values for the corresponding images. The red trace in panel C is a log-normal fit to the Venus/Cerulean histogram, while the black trace in panel D is a Gaussian fit to the FRET-efficiency histogram.

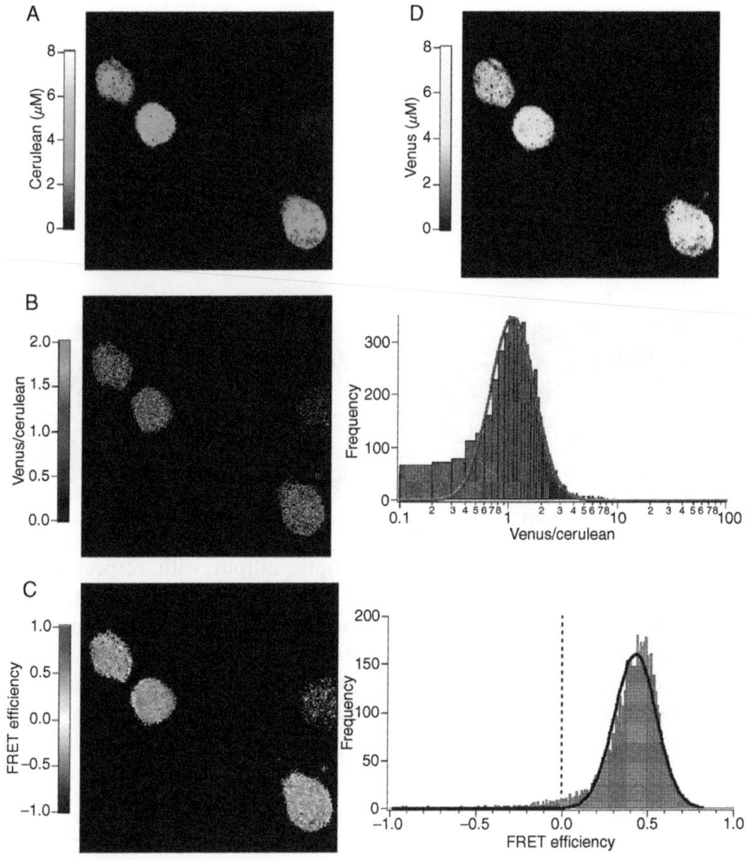

Fig. 8.7. sRET analysis of C5V, a Cerulean-Venus construct with a high FRET efficiency. Spectral images of cells expressing C5V, and their matching reference spectra were processed using the sRET algorithm to produce (A) a Cerulean concentration image, (B) a Venus concentration Image, (C) a Venus/Cerulean ratio image, and (D) a FRET-efficiency image. The graphs in panels C and D show frequency histograms of the pixel values for the corresponding images.

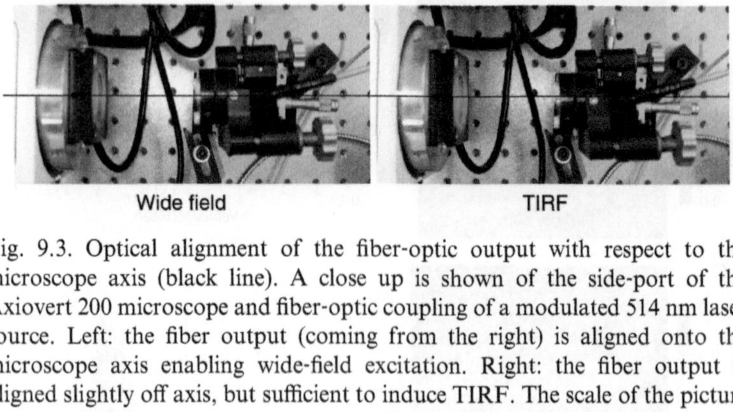

Fig. 9.3. Optical alignment of the fiber-optic output with respect to the microscope axis (black line). A close up is shown of the side-port of the Axiovert 200 microscope and fiber-optic coupling of a modulated 514 nm laser source. Left: the fiber output (coming from the right) is aligned onto the microscope axis enabling wide-field excitation. Right: the fiber output is aligned slightly off axis, but sufficient to induce TIRF. The scale of the picture can be inferred from the optical table M6 screw mounts separated by 1 inch.

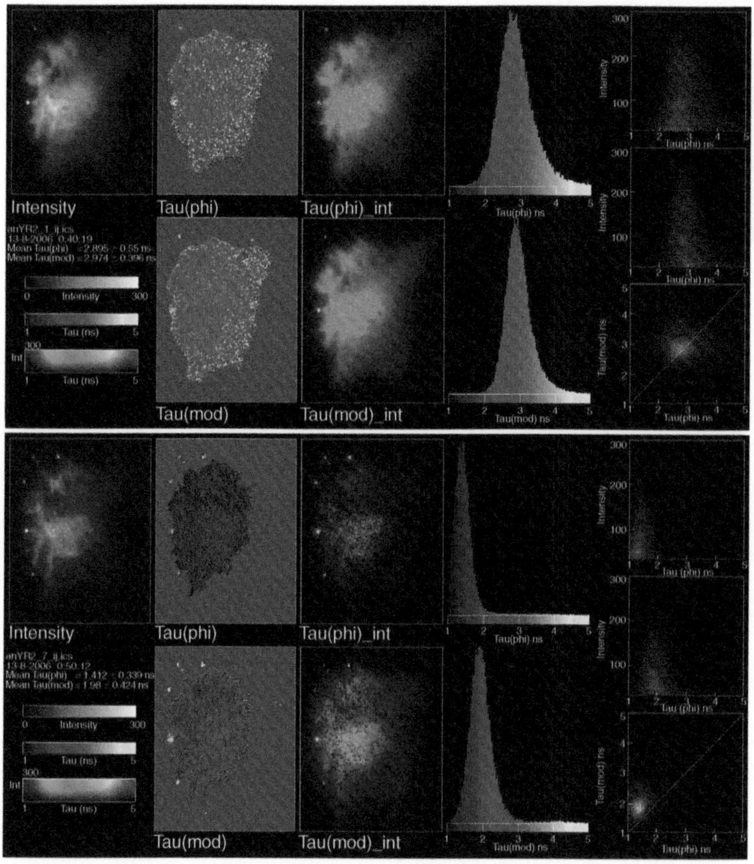

Fig. 9.4. Annexin-A4 relocation and aggregation detected by FRET TIR-FLIM. The top image represents the situation before relocation with no FRET, the lower image represents the situation 5 min after adding ionomycin inducing relocation and aggregation of annexin A4-EYFP and annexin A4-mCherry reflected by a substantial drop of the annexin A4-EYFP fluorescence lifetime. The bright fluorescent dots are fluorescent latex beads (0.1 μm diameter) that were incorporated into the medium for efficient focusing the plasma membrane area before translocation. Only beads adhered to the glass are excited in TIRF. The images represent from left to right: the DC fluorescence intensity image, the phase lifetime image, the pseudocolored intensity-phase lifetime image and the phase lifetime distribution histogram. Below are from left to right the modulation lifetime image, the pseudocolored intensity-modulation lifetime image, and the modulation lifetime distribution histogram. The color bars depicting fluorescence intensity (arbitrary units), fluorescence lifetime (between 1 and 5 ns), and pseudocolored intensity-lifetimes are shown in the lower left corner of the image. At the right side of

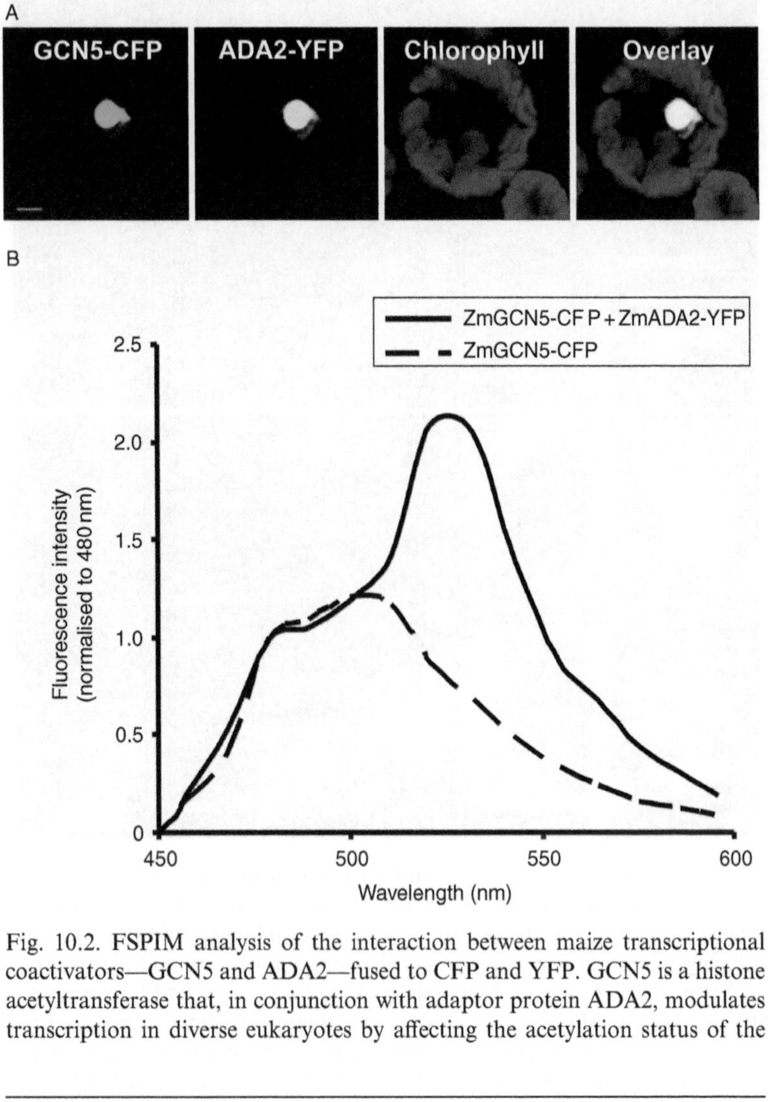

Fig. 10.2. FSPIM analysis of the interaction between maize transcriptional coactivators—GCN5 and ADA2—fused to CFP and YFP. GCN5 is a histone acetyltransferase that, in conjunction with adaptor protein ADA2, modulates transcription in diverse eukaryotes by affecting the acetylation status of the

---

the images three dotplots or two-dimensional histograms are shown depicting from top to bottom the correlation between the fluorescence intensity with the phase lifetime; the fluorescence intensity with the modulation lifetime; and the correlation between phase and modulation lifetime. The composite layout was generated using a standard macro (Lifetime6) in image J using an input image stack of three images: the calculated DC intensity image, the phase, and the modulation fluorescence lifetime images.

core histones in nucleosomes [63]. CFP- and YFP-tagged proteins, expressed in protoplasts, were excited by the 458 nm and the 514 nm laser lines sequentially. CFP fluorescence was selectively detected by an HFT 458 dichroic mirror and BP 470–500 band pass emission filter while YFP fluorescence was selectively detected by using an HFT 514 dichroic mirror and BP 535–590 band pass emission filter. In both cases the chlorophyll autofluorescence was filtered out and detected in another channel using a LP650 long pass filter. A 25 × Plan–Neofluar water immersion objective lens (N.A 0.8) was used for scanning protoplasts. In order to avoid crosstalk between CFP and YFP, images were acquired in the multichannel tracking mode and analyzed with Zeiss LSM510 Meta software. (A) Colocalization of *Zm*GCN5-CFP and *Zm*ADA2-YFP in cowpea protoplasts imaged by a confocal laser-scanning microscope. Nuclear localized *Zm*GCN5-CFP and *Zm*ADA2-YFP along with chlorophyll autofluorescence and an overlay image of all the three is shown. (B) Nuclei of cowpea protoplasts expressing either *Zm*GCN5-CFP alone, or in combination with *Zm*ADA2-YFP, were subjected to FSPIM analysis as described [27, 64, 65]. For FSPIM, a 435 nm line from a 100 W Hg Arc lamp light source was selected by inserting a 435DF10 bandpass filter from Omega (Brattleboro, VT) in the excitation light path. Subsequently the excitation light was reflected onto the sample by an Omega 430DCLP dichroic mirror. Residual excitation light was rejected using a Schott (Mainz, Germany) GG455 longpass filter. Images were acquired using a 20 × Plan Neofluar objective. Spectra were obtained using a spectrograph (Chromex 250; Chromex, Albuquerque) with 150 grooves/mm grating set at a central wavelength of 500 nm. Protoplasts expressing CFP and YFP fusion proteins were aligned across the entrance slit of the spectrograph (set at 200 μm width corresponding to 10-μm width in the object plane). Acquisition time was 2–3 s. Data collection and background subtraction was performed as described [65]. Representative spectra normalized to the detected intensity at 480 nm from the nuclei of protoplasts expressing *Zm*GCN5-CFP + *Zm*ADA2-YFP (solid line) or *Zm*GCN5-CFP alone (dashed line) are shown.

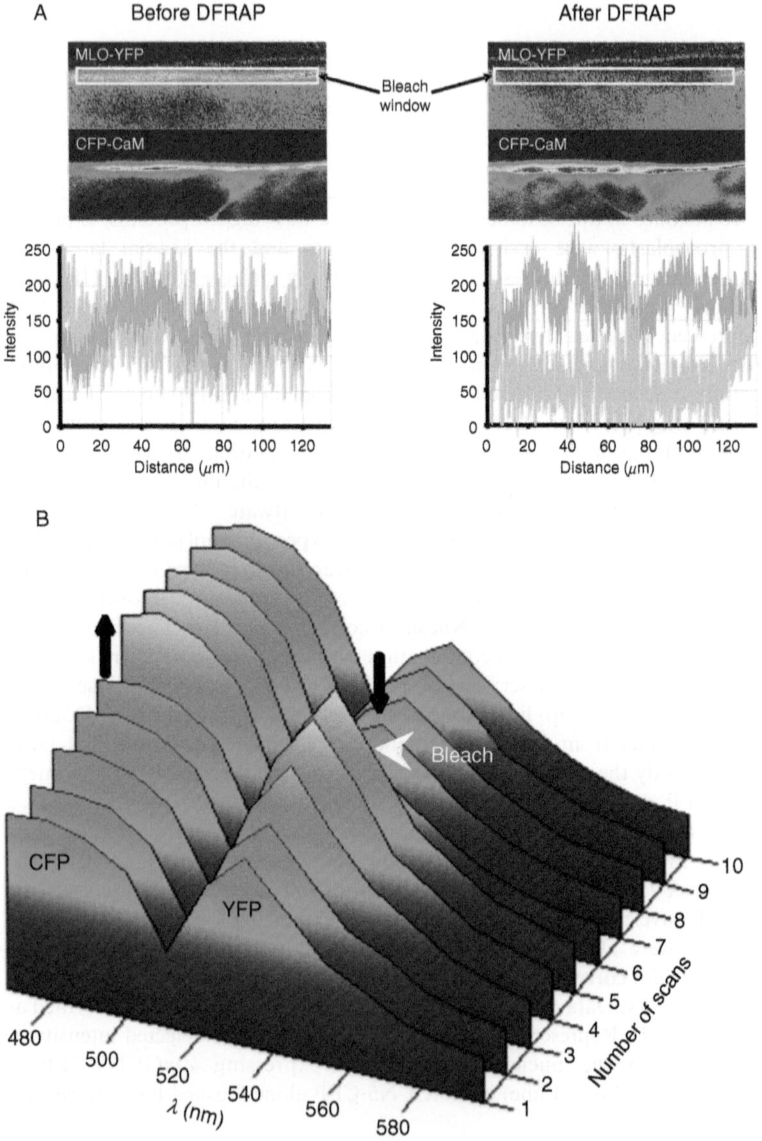

Fig. 10.3. Acceptor photobleaching analysis of interaction between barley MLO and calmodulin. Barley MLO is a plant-specific integral membrane protein that associates with the cytosolic calcium sensor protein Calmodulin (CaM) and functions as a modulator of defense against the common powdery

mildew pathogen [67]. Wild-type MLO fused to YFP was coexpressed with CaM fused to CFP in barley leaf epidermal cells [23]. 458 and 514 nm argon laser lines were used to excite CFP and YFP respectively. The experimental conditions and filter settings were the same as in Figure 10.2 except that a 40 × Plan–Neofluar water immersible objective lens (N.A 0.75) was used for scanning the leaf epidermal cells. Acceptor photobleaching was performed as described [22, 53, 56]. (A) Intensity-based acceptor photobleaching on MLO–CaM interaction. Following a rapid photobleaching of the acceptor YFP (white rectangular bleach window in the top panels), the donor CFP intensity before and after acceptor photobleaching was compared. The photochemical destruction of the acceptor YFP abolishes any energy transfer from CFP–CaM to MLO–YFP resulting in a rapid increase in the CFP–CaM emission intensity (compare line plot intensity profile panels in bottom panels; light gray–MLO–YEP; dark grey–CFP–CaM) indicating a FRET and hence an association between MLO and CaM (see also [23]) (B) Spectral acceptor photobleaching analysis on MLO–CaM interaction. A 458 nm laser argon line was used to co-excite both CFP–CaM and MLO–YFP and fluorescence spectra were recorded in a lambda mode from 470 to 600 nm using a Zeiss LSM510 Meta confocal laser-scanning microscope. The rapid photobleaching pulse was given at the 5th scan (white arrow head). With the bleaching of the YFP (530 nm, black downward arrow), the CFP emission spectrum at 480 nm sharply increases (upwards black arrow), indicative of FRET and hence the association between MLO and CaM. The dichroic mirror used to reflect 514 nm laser line causes the dip in the emission spectra.

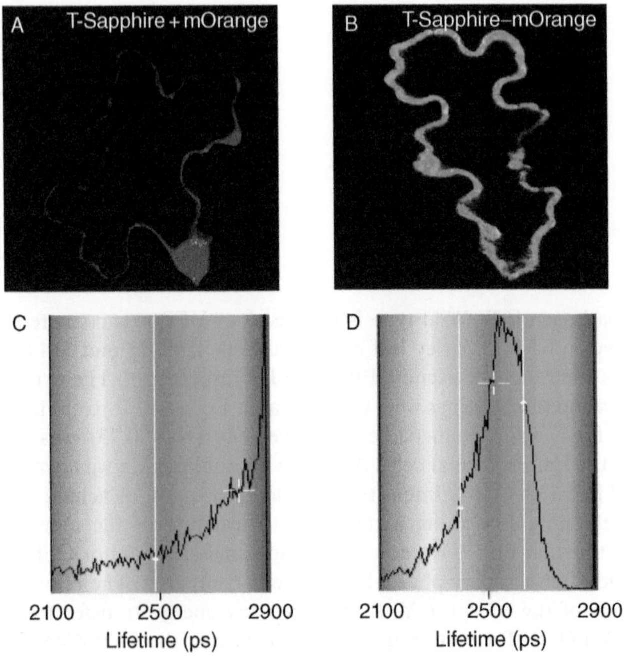

Fig. 10.4. Time-domain FLIM between T-Sapphire and mOrange. A chimeric construct harboring T-Sapphire and mOrange fluorophore cDNA's separated by a 12 amino acid linker, under the control of constitutive cauliflower mosaic virus (CaMV) 35S promoter, was delivered into *N. benthamiana* leaf epidermal cells by ballistic-mediated particle bombardment [47]. At the same time, T-Sapphire and mOrange cDNA's on separate plasmids, driven by CaMV 35S promoters, were co-bombarded into control *N. benthamiana* leaf epidermal cells. The lifetime images of the donor T-Sapphire were obtained using the 405 nm diode laser modulated at 80 MHz and the T-Sapphire fluorescence was selectively imaged using a 520DF40 bandpass emission filter. The microscope setup, lifetime image calculation, and the image processing were performed as per the time-domain FLIM system from Becker and Hickl [47, 69]. Only pixels with more than 50 photons were used for calculating the lifetime. The lifetime of T-Sapphire coexpressed with mOrange was measured as 2.9 ns on average (A and C; in agreement with the lifetime of wild-type GFP; 3.16 ns; [72, 73]) whereas the average lifetime of T-Sapphire in the chimeric construct was found to be around 2.45 ns, yielding an average of around 15% energy transfer to mOrange (B and D; Eq. 10.1; [47]). (C) and (D) represent the temporal histograms and the fluorescence lifetime pixel values in (A) and (B) respectively between 2.1 ns (strong interaction) and 2.9 ns (no interaction).

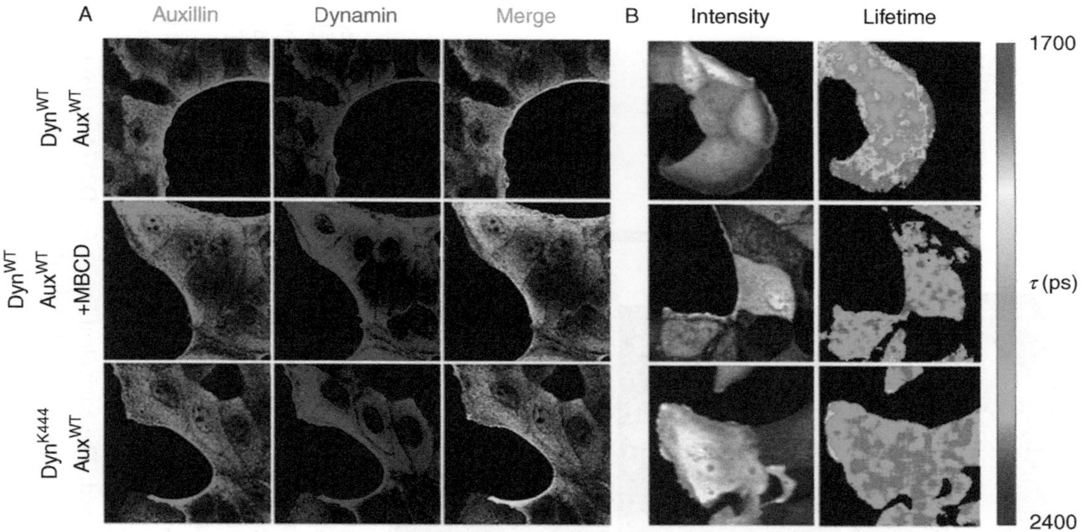

Fig.11.1. Identification of GTP- and endocytosis-dependent dynamin–auxilin interactions as detected with a confocal colocalization study and a two-photon FLIM experiment. Auxilin and dynamin colocalize at the edge of a monolayer of inner medullary collecting duct (IMCD) cells to within the resolution of the instrument (A). No difference is detectable in the presence of β-methyl-cyclodextrin (MBCD), which depletes cholesterol from the plasma membrane and inhibits endocytosis. Equally, the mutant version of dynamin (dyn^{K44A}) that cannot bind with GTP, also appears to be colocalized with auxilin. FLIM data (B) showed that endocytosis inhibition and prevention of GTP binding eliminated the interaction between the proteins even though the spatial localization remains unchanged [49].

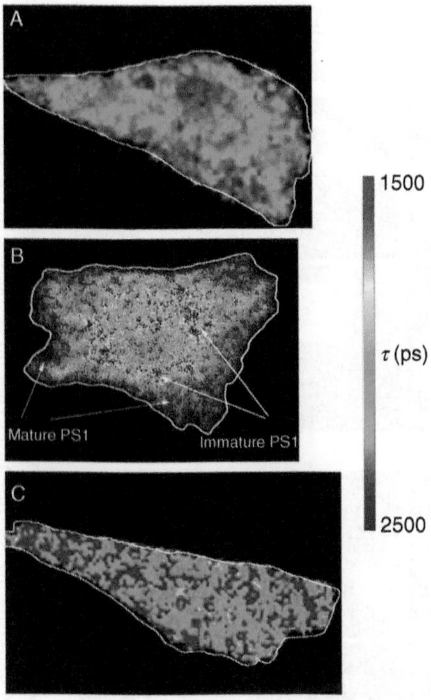

Fig.11.3. FLIM images that show a protein conformational difference between PS1 holoprotein and mature heterodimer. Chinese hamster ovary (CHO) cells were double immunostained with antibodies to the PS1 NT (stained with FITC) and to the ER resident protein BiP (stained with cy3), which does not interact with PS1. The lifetime of FITC was found to be (≈2300ps) (A). The same FRET pair was targeted to PS1 CT and NT (B) and a reduction in lifetime is clearly shown, indicating a change in protein conformation, close to the plasma membrane whereas immature PS1, located near the nucleus, exhibits no such strong FRET interaction. An average lifetime of ≈1700ps across the cell was observed. The equivalent image for D257A mutant PS1 expressing cells, a mutation known to inhibit PS1 endoproteolysis, and diminish amyloid-β production (C) shows no spatial heterogeneity. In addition, the mutation apparently abolishes the FRET signal [4].

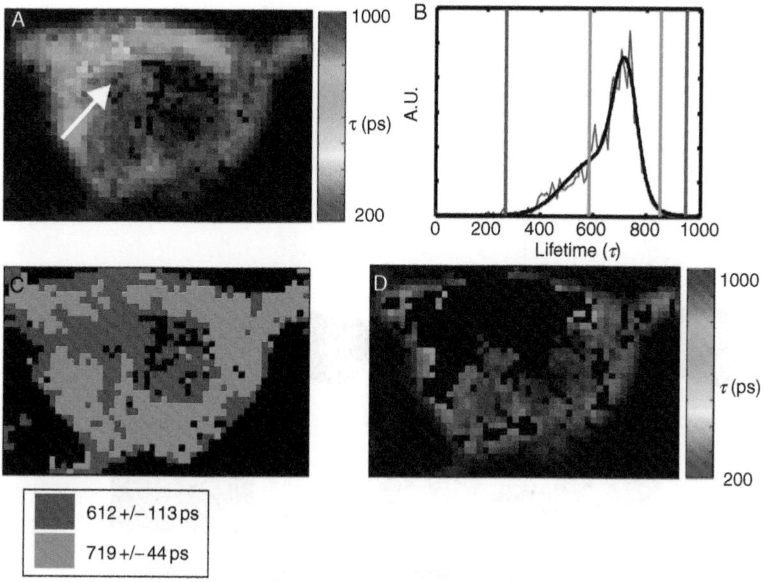

Fig. 11.4. Distinguishing FRET from autofluorescence. This neurofibrillary tangle (NFT) in human brain tissue was stained with an Alexa-488 (A488) labeled anti-tau antibody. A Cy3 acceptor labeled secondary was used to guarantee FRET between A488 and Cy3. In this particular neuron, a large aggregate of lipofuscin autofluorescence in the center of the tangle was observed and confirmed by visual inspection via wide field optical imaging (not shown). The FLIM data was fit with a biexponential function with the lifetime of the noninteracting A488 fixed at 1890ps, as determined from sections stained with donor alone. The pseudocolor image of the shorter component (A) shows the lipofuscin aggregate in the center of the neuron (white arrow). As described in the discussion, the autofluorescence resulted in a broad distribution of lifetimes that overlapped with the FRETing lifetime of interest. By applying multi-Gaussian analysis, the two contributions could be distinguished (B). The red vertical lines bound the broader, slightly shorter distribution, and the green vertical lines bound the narrow peak that the A488 molecules that are FRETing. The image was segmented (C); the spatial distribution of the broader distribution (RED), conforms well to the location of the observed autofluoresce. A goodness of fit filter was applied, where all pixels that have a relatively poor fit to the monoexponential model are discarded, with a cutoff of $\chi^2<2$ (D). We see excellent agreement between the visual inspection, segmentation via multiple Gaussian analysis and goodness of fit filtering.

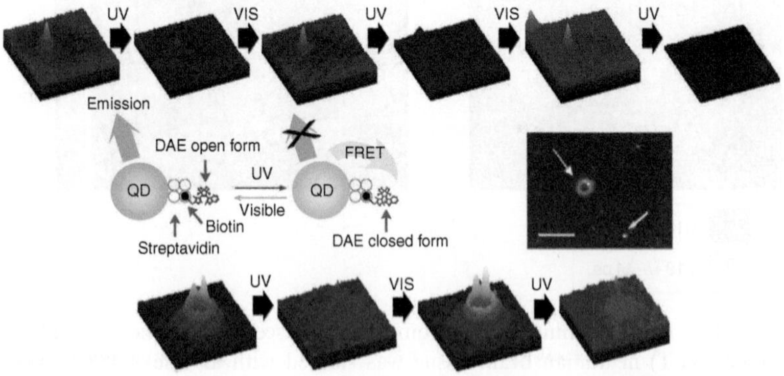

Fig. 12.6. Imaging QD emission modulated by pcFRET. Surface plots of a selected region containing aggregates and single QDs. The surface exhibited fluorescence shells (QDs coated microspheres), and single and aggregated QDs. After a pre-irradiation period with UV and visible light the QDs displayed a reversible decrease of emission upon irradiation with UV light (as a result of pcFRET) and an increase of emission when irradiated with visible light. Surface plots of an area about a bright spot corresponding to an aggregate (yellow arrow) are shown at the top. Bottom, surface plots of a 2.4 μm microsphere (white arrow). White bar, 5 μm. Data from [139].

Printed and bound by CPI Group (UK) Ltd, Croydon, CR0 4YY
16/10/2024
01774872-0001